T0331044

Stationary
Stochastic Models
An Introduction

World Scientific Series on Probability Theory and Its Applications

Print ISSN: 2737-4467
Online ISSN: 2737-4475

Series Editors: Zenghu Li *(Beijing Normal University, China)*
Yimin Xiao *(Michigan State University, USA)*

Published:

Vol. 4   *Stationary Stochastic Models: An Introduction*
by Riccardo Gatto (University of Bern, Switzerland)

Vol. 3   *Introduction to Probability Theory: A First Course on the
Measure-Theoretic Approach*
by Nima Moshayedi (University of California, Berkeley, USA)

Vol. 2   *Introduction to Stochastic Processes*
by Mu-Fa Chen (Beijing Normal University, China) and
Yong-Hua Mao (Beijing Normal University, China)

Vol. 1   *Random Matrices and Random Partitions: Normal Convergence*
by Zhonggen Su (Zhejiang University, China)

World Scientific Series on
**Probability Theory and
Its Applications**

Volume 4

Stationary
Stochastic Models

An Introduction

Riccardo Gatto

University of Bern, Switzerland

World Scientific

NEW JERSEY · LONDON · SINGAPORE · BEIJING · SHANGHAI · HONG KONG · TAIPEI · CHENNAI · TOKYO

Published by

World Scientific Publishing Co. Pte. Ltd.

5 Toh Tuck Link, Singapore 596224

USA office: 27 Warren Street, Suite 401-402, Hackensack, NJ 07601

UK office: 57 Shelton Street, Covent Garden, London WC2H 9HE

Library of Congress Cataloging-in-Publication Data
Names: Gatto, Riccardo, author.
Title: Stationary stochastic models : an introduction /
   Riccardo Gatto, University of Bern, Switzerland.
Description: New Jersey : World Scientific, [2022] | Series: World Scientific series on
   probability theory and its applications, 2737-4467 ; volume 4 |
   Includes bibliographical references and index.
Identifiers: LCCN 2022006675 | ISBN 9789811251832 (hardcover) |
   ISBN 9789811251849 (ebook for institutions) | ISBN 9789811251856 (ebook for individuals)
Subjects: LCSH: Stationary processes. | Stochastic processes.
Classification: LCC QA274.3 .G38 2022 | DDC 519.2/32--dc23/eng20220422
LC record available at https://lccn.loc.gov/2022006675

British Library Cataloguing-in-Publication Data
A catalogue record for this book is available from the British Library.

For any available supplementary material, please visit
https://www.worldscientific.com/worldscibooks/10.1142/12710#t=suppl

Desk Editors: Jayanthi Muthuswamy/Lai Fun Kwong

Typeset by Stallion Press
Email: enquiries@stallionpress.com

Printed in Singapore

Preface

Temporal measurements or observations appear in various scientific fields such as engineering, meteorology, oceanography, biology, finance, insurance, epidemiology, etc. The statistical analysis of this type of data is more reliable and substantially simplified after transformation to stationary data. Indeed, stationary data have no trend, i.e. constant mean, they have constant variance and covariance, between any two points separated by a fixed time lag. The mathematical models for these data are called stationary stochastic models. We follow the terminology of statistics and call these models "stationary time series", when time is discrete. When time is assumed continuous, we call them "continuous time stationary processes" or simply "stationary processes". Stationary models play a central role in predictions or forecasts.

This book introduces the mathematical foundations of stationary models. An important aspect is the unified presentation of stationary time series and continuous time stationary processes. Although these two topics share common concepts and similar theoretical results, they are mostly given in separate books. For instance, the spectral theory in these two situations shares the same main ideas. The joint presentation allows introduction of these theoretical concepts progressively, by starting with the simpler discrete time models and then passing to more sophisticated continuous time processes. The consequential theory of continuous time stochastic processes can thus be approached with concepts of stationarity already in mind. Mathematical rigor is important in this text and a detailed mathematical appendix assists the reader.

This book originates from personal notes written for lectures on time series analysis and on the theory of continuous time stochastic processes. These lectures have been held by the author during several years, at the

Institute of Mathematical Statistics and Actuarial Science of the University of Bern. Initial lectures were mainly based on distinguished references on stationary models: Cramér and Leadbetter (1967), Brockwell and Davis (1991) and Lindgren (2012). Consequently, this book inherits ideas and concepts from these references, that have however been recast, newly illustrated and completed for the scope of this text. The two major parts of this book are Chapter 2, which has been used for a one semester lecture on time series analysis, and Chapter 3, which has been employed for a one semester lecture on continuous time stationary processes. These two lectures have been given to students of mathematics or statistics, at upper level Bachelor or Master level. A list of misprints and remarks is available online at: "`http://www.stat.unibe.ch`".

The author is thankful to his teaching assistants, in particular to Katrin Gysel, for her assistance in typing parts of this text and for the elaboration of various numerical examples, and to Federico Pianoforte, for his assistance in typing other parts and for various technical suggestions. The author is thankful to his students, for their numerous questions, to Sreenivasa Rao Jammalamadaka, for many discussions in the related field of directional statistics, to Georg Lindgren, for a technical discussion, to institute colleagues Johanna Ziegel and Ilya Molchanov, for supporting the preparation of this book. The author is grateful to World Scientific, in particular to Lai Fun Kwong, for her assistance during submission and publication processes, and to the Editor of the *World Scientific Series on Probability Theory and Its Applications*. An acknowledgment is addressed to the Reviewers, for their remarks and suggestions that lead to various improvements and extensions. Personal thanks are given to my son Raffaele and daughter Valérie Regina, for making most of my spare time interesting and entertaining.

Riccardo Gatto
Bern, February 2022

About the Author

 **Riccardo Gatto** holds a doctoral degree from University of Geneva and a habilitation degree in mathematical statistics from University of Bern, Switzerland. He has held academic positions in statistics departments at University of Neuchâtel, University of California at Santa Barbara, ESSEC business school and currently at University of Bern. He has been teaching various lectures in mathematical statistics and probability, including in time series and stochastic processes. His research topics include stationary processes, first passage of stochastic processes, actuarial risk processes, asymptotic and large deviations methods as well as statistics for directional data.

Professor Gatto has authored research articles in various journals of statistics and applied probability, in particular in *Biometrika, Journal of Statistical Computation and Simulation, Journal of the American Statistical Association, Mathematical Methods of Statistics, Methodology and Computing in Applied Probability* and *Stochastic Models.* He has also authored a book on stochastic models for actuarial risk and has been serving as associate editor for *Journal of Statistical Computation and Simulation.*

Contents

List of Figures

List of Tables

Chapter 1

Introduction

1.1 Stationary stochastic models and outline

Many scientific studies rely on data obtained upon repeatedly measuring or observing the evolution of a phenomenon during time. This happens mostly in meteorology, oceanography, earthquake studies, naval architecture, financial studies, actuarial risk analysis, epidemiology, etc. These temporal data often possess mean and variance that vary substantially over time. In this situation they are referred to as nonstationary data. Yet, obtaining reliable predictions or forecasts directly from nonstationary data is a difficult task. So nonstationary data are transformed to stationary data, that possess: constant mean, constant variance and covariance, when taken between any two points separated by a fixed time lag. Mathematical models for data with the above properties are simpler to obtain and they allow for easier statistical inference. They are called stationary stochastic models and we distinguish two main categories. When time is assumed discrete (and usually represented by the integers $\mathbb{Z}$), these models are called stationary time series. When time is continuous (and usually represented by the real numbers $\mathbb{R}$), these models are called (continuous time) stationary processes.

This book introduces the mathematical foundations of stationary time series and of continuous time stationary processes. The particularity of these stationary models is that their analysis can be carried out in two different ways: in time domain and in frequency domain. The analysis of a stationary model in time domain studies the fluctuations with respect to (w.r.t.) time. Time domain analysis alone may be unfavorable or insufficient in some situations. The analysis in frequency domain, also called spectral analysis, may be more suitable in these situations. It consists in identifying the frequencies and the amplitudes of the various harmonics (or sinusoids) of the stationary model.

The central object for the analysis in time domain is the autocovariance function (a.c.v.f.), namely the covariance between any two values of the signal, re-expressed as a function of the time lag between these two values. The central function for the analysis in frequency domain is the spectral distribution, which indicates the mean magnitude of the frequency of each one of the harmonics that are present in the stationary signal.

With more detail, the essence of spectral analysis is that any stationary signal can be decomposed as sum of harmonics, with different random phases, namely angles of retardation, and amplitudes. The signal is the stationary model that has been estimated by the data and it can be discrete or continuous. These harmonics are uncorrelated and their mean magnitudes give rise to the spectral distribution. The spectral distribution summarizes the most relevant information carried by the stationary signal. It is the analogue in frequency domain of the a.c.v.f. in the time domain: a.c.v.f. and spectral distribution form a Fourier pair. Transitions from time to frequency domain and back to time domain are obtained by the Fourier transform and its inverse transform.

The first developments of the theory of stationary models appeared at the end of the 19-th century, with the analysis of data in frequency domain. Interestingly, the analysis in time domain began only later. Some initial statistical theory for analysis of periodic phenomena was given by Fisher (1929). The theory of stationary time series began with Yule (1921) and Slutsky (1927), who proposed moving average (MA) time series in order to model cyclic variations. The autoregressive (AR) time series model was introduced by Yule (1927) with a study on the periodicity of sunspots. The AR time series was further analyzed by Walker (1931). Then Wold (1938) introduced the more general autoregressive and moving average (ARMA) time series, model, which has nowadays become the basic model of time series and which is an important part of this book. Some pioneer theoretical developments of continuous time stationary processes can be found in Cramér (1940, 1942). Rice (1944, 1945, 1954) introduced the spectral decomposition of stationary Gaussian processes, in the context of noise in radio transmission. A more detailed historical overview with focus on spectral analysis in statistics can be found in Brillinger (1993).

As already mentioned, the main feature of this book is the integration of two topics that are usually presented in separate volumes: stationary time series and continuous time stationary processes. There are several references that mainly present time series and some important ones are: Priestley (1981), Brockwell and Davis (1991, 2002), Anderson (1994),

Montgomery et al. (2015), Broemeling (2019), which includes Bayesian aspects, Shumway and Stoffer (2019), which includes computational aspects, and the original monograph of Breckling (1989), which considers time series of directions. Regarding continuous time stationary processes, some important references are: Cramér and Leadbetter (1967), Yaglom (1987a,b), Lindgren (2012), Koopmans (1995), Lindgren et al. (2014) and Brémaud (2014). This book starts with a thorough study of ARMA time series. It introduces the concept of stationarity with discrete time only, thus by avoiding theoretical concepts of continuous time stochastic processes. In particular, the spectral analysis is introduced in a simple way.

In the light of these considerations, the structure of this book is the following. Given the importance of Fourier analysis for the spectral analysis of stationary models, the remaining part of this introductory Chapter, precisely Section 1.2, presents a short introduction to Fourier series and Fourier transforms.

Then, Chapter 2 presents the analysis of stationary time series. The precise topics are the following. Section 2.1 provides various methods for obtaining stationary data starting from nonstationary data, for example through discrete differentiation. It also introduces the a.c.v.f., which is central quantity for the analysis of stationary time series in time domain. Section 2.2 presents the central models for stationary time series: the AR, the MA and the combination of these two models, namely the ARMA. Various properties of stationary ARMA time series such as causality and invertibility are analyzed. Section 2.3 studies the a.c.v.f. and introduces another practical function, the partial autocorrelation function (p.a.c.r.f.). Section 2.4 introduces the analysis in frequency domain of complex-valued time series. Now the central quantity is the spectral distribution. Herglotz's theorem is presented and the spectral distribution of ARMA time series is obtained. A short introduction to the theory of linear filters is given. Other common topics on time series are brought together Section 2.5: prediction, integrated ARMA (ARIMA) model, determination of degree of AR model, state space model, Kalman filter, integer-valued time series and nonlinear time series, that include generalized autoregressive conditionally heteroscedastic (GARCH) time series.

Chapter 3 presents the analysis of stationary stochastic processes with continuous time. The precise structure of the chapter is as follows. Some general notions on continuous time stochastic processes are required and introduced in Section 3.1. Section 3.2 reviews some important continuous time stochastic processes: Gaussian processes, Wiener process, selfsimilar processes, fractional Brownian motion, counting processes, compound processes, shot noise processes, point processes and Lévy processes. Mean square continuity and differentiability of stationary processes are analyzed in Section 3.3. Section 3.4 introduces stochastic integration, precisely the mean square integral. This type of integral allows for the definition of continuous time Gaussian white noise (WN), as generalized stochastic process. The concept of ergodicity is shortly introduced. Section 3.5 introduces the spectral distribution. Bochner's theorem and inversion formulae for the spectral distribution are presented. The sampling of a continuous time process is studied. The spectral decomposition of a stationary processes, in terms of a stochastic integral w.r.t. the spectral process, as increment, is presented in Section 3.6. The spectral theorem, which ensures the existence of the spectral process, is shown. Section 3.7 is devoted to the spectral analysis of Gaussian processes. The Gaussian WN and the Ornstein-Uhlenbeck process are obtained through the spectral decomposition. Section 3.8 provides the spectral analysis of counting processes based on Bartlett spectrum. One of the most interesting topics of this book is the theory of time invariant linear filters, which is mainly presented in Section 3.9. This theory appears first in Section 2.4, however in the context of time series only. The connection with differential equations is presented. Then shot noise processes are reformulated in terms of filtered point processes.

Some particular or more advanced topics on stationary time series and stationary processes are introduced in Chapter 4. Some of these topics are only briefly described, in order to encourage further reading. Section 4.1 introduces stationary random fields. They generalize stationary processes in the sense that the time domain is replaced by a multidimensional domain. Section 4.2 surveys time series for planar directions, called circular time series. Then, Section 4.3 presents the long range dependence or long memory of a process. This concept is described by the importance of low frequencies and the scarcity of high frequencies. Section 4.4 introduces spectral densities that are unbounded at the origin and in this way nonintegrable, giving rise to the concept of intrinsic stationarity. Section 4.5 briefly presents unstable systems, namely processes that are defined through an unstable differential equation. After that, the Hilbert transform and the envelope are presented in Section 4.6. The Hilbert transform is a stationary process obtained by the application of a particular filter. An important and practical topic is the stochastic simulation of stationary Gaussian processes.

This theme is presented in detail in Section 4.7, which provides four simulation algorithms that use: Choleski factorization, circulant embedding, spectral distribution and ARMA approximation. Basic notions of large deviations theory together with applications to AR time series can be found in Section 4.8. The last topic is given in Section 4.9, which provides results from information theory for time series and their spectral distributions.

This text is conceived to be mathematically complete and therefore accessible to a relatively large readership. So Appendix A provides an overview of several mathematical and probabilistic notions. For instance, some theory of Hilbert spaces is reviewed in Section A.1.1.

1.2 Fourier analysis

This book starts with a short presentation of Fourier analysis because it is the chief constituent in the study of stationary models. Fourier analysis provides a representation or an approximation of a function in terms of a sum or an integral of trigonometric functions. J. Fourier showed in 1822 that periodic functions can be re-expressed as infinite sums of harmonics, i.e. as Fourier series. This representation leads to the idea of a signal and its spectrum. The simplest example may be the following. While a musical instrument generates a steady note, a microphone together with an oscilloscope produce the graph of the air pressure $f_0(t)$ w.r.t. time t. Thus f_0 is a periodic function. If $\tau > 0$ is the period length, then $1/\tau$ is the frequency of the note. This elementary situation can be generalized to some more pleasant steady sound f with various overtones, namely τ-periodic harmonics whose frequencies are multiples of the fundamental frequency $1/\tau$, with different amplitudes and with different phases. The fundamental frequency is thus the lowest frequency of the periodic signal f. This decomposition yields the spectrum of the steady sound f. But not all signals are periodic. In a nonperiodic signal, all real frequencies can be present in the spectrum.

There are many references on Fourier analysis: a concise introduction is given in Chapters 1 and 2 of Blatter (1998), a basic book is Pinkus and Zafrany (1997) and an introduction with applications in physics is James (1995). This section summarizes basic and important results on Fourier analysis that are essential for the understanding of this book. The Fourier series is presented in Section 1.2.1 and the Fourier transform in Section 1.2.2. The presentation of the discrete Fourier transform is postponed to Section 2.4.5. An efficient algorithm for computing it, called Fast Fourier transform (FFT), is presented in Section A.10.

1.2.1 *Fourier series*

Let $f : \mathbb{R} \to \mathbb{C}$ be a 2π-periodic function that is square-integrable over $[0, 2\pi)$, in the sense that $\int_0^{2\pi} |f(t)|^2 \mathrm{d}t < \infty$. Consider another function g satisfying these properties. We define the scalar or inner product of these functions f and g by

$$\langle f, g \rangle = \frac{1}{2\pi} \int_0^{2\pi} f(t)\overline{g(t)}\mathrm{d}t,$$

and the $\mathcal{L}_2^\circ$-norm of f by

$$\|f\| = \langle f, f \rangle^{\frac{1}{2}}.$$

We denote the space of all such functions with this scalar product by $\mathcal{L}_2([0, 2\pi))$ or simply by $\mathcal{L}_2^\circ$. The space $\mathcal{L}_2^\circ$ is a Hilbert space consisting of equivalence classes, each one of them defined by all functions of norm equal to some given value. We refer to Section A.1 for details on Hilbert spaces.

A periodic function can be represented over the unit circle of the complex plane. Over this circle, the monomial $z \mapsto z^k$ can be conveniently reparametrized in terms of the variable t through the harmonic functions

$$e_k : \mathbb{R} \to \mathbb{C},$$
$$t \mapsto e^{\mathrm{i}kt}, \quad \forall k \in \mathbb{Z}. \tag{1.2.1}$$

We obtain directly the orthonormality relations

$$\langle e_j, e_k \rangle = \begin{cases} 1, & \text{if } j = k, \\ 0, & \text{if } j \neq k, \end{cases} \quad \forall j, k \in \mathbb{Z}.$$

We have thus obtained an orthonormal basis of $\mathcal{L}_2^\circ$. Then, $\forall k \in \mathbb{Z}$, we denote by

$$c_k = \hat{f}(k) = \langle f, e_k \rangle = \frac{1}{2\pi} \int_0^{2\pi} e^{-\mathrm{i}kt} f(t)\mathrm{d}t, \tag{1.2.2}$$

the k-th Fourier coefficient of f, which is precisely the k-th coordinate of f w.r.t. the orthonormal basis $\{e_k\}_{k\in\mathbb{Z}}$.

Let $n \in \mathbb{N}$. The finite sum

$$f_n = \sum_{k=-n}^{n} c_k e_k, \tag{1.2.3}$$

is the n-th Fourier approximation to f. It corresponds to the orthogonal projection of f onto the subspace of $\mathcal{L}_2^\circ$ of dimension $2n+1$ that is spanned by $e_{-n}, \ldots, e_n$. The Fourier series representation of f is given by

$$\sum_{k=-\infty}^{\infty} c_k e_k.$$

The next topic is the convergence of the above Fourier series: with few additional conditions, the series does converge to f, in mean square norm or pointwise.

Mean square convergence of Fourier series

We present some important results relating to mean square convergence of Fourier series. The Riemann-Lebesgue lemma tells that

$$\lim_{k \to \pm\infty} c_k = 0. \tag{1.2.4}$$

Parseval formula states that, for any other function $g \in \mathcal{L}_2^\circ$,

$$\sum_{k=-\infty}^{\infty} \hat{f}(k)\overline{\hat{g}(k)} = \langle f, g \rangle.$$

Consequently we have

$$\sum_{k=-\infty}^{\infty} |c_k|^2 = \|f\|^2. \tag{1.2.5}$$

In a Hilbert space, Pythagora's theorem holds and by combining it with (1.2.5), we have

$$\|f - f_n\|^2 = \|f\|^2 - \|f_n\|^2 \xrightarrow{n \to \infty} \|f\|^2 - \sum_{k=-\infty}^{\infty} |c_k|^2 = 0.$$

This means that

$$f_n \to f, \quad \text{in } \mathcal{L}_2^\circ,$$

which can be re-expressed as

$$f = \sum_{k=-\infty}^{\infty} c_k e_k, \quad \text{in } \mathcal{L}_2^\circ. \tag{1.2.6}$$

Pointwise convergence of Fourier series

Generally, (1.2.6) does not hold pointwise, but Carleson's theorem tells that convergence holds pointwise a.e. Also, Dirichlet's theorem tells that if f is piecewise continuous and possesses left and right derivatives, then

$$f_n(t) \stackrel{n\to\infty}{\longrightarrow} \frac{1}{2}\{f(t-) + f(t+)\}, \ \forall t \in \mathbb{R}. \tag{1.2.7}$$

So if f is also continuous, then pointwise convergence holds everywhere. Another result on pointwise convergence is that, if $f \in \mathcal{L}_2^\circ$ is continuous with bounded variation,[1] then

$$f_n(t) \stackrel{n\to\infty}{\longrightarrow} f(t), \quad \text{uniformly w.r.t. } t \in \mathbb{R}.$$

Consider now the Césaro sum[2] of the Fourier series, precisely

$$S_n(t) = \frac{1}{n} \sum_{j=0}^{n-1} f_n(t), \quad \forall t \in \mathbb{R}.$$

If f is continuous, then

$$S_n(t) \stackrel{n\to\infty}{\longrightarrow} f(t), \quad \text{uniformly w.r.t. } t \in \mathbb{R}. \tag{1.2.8}$$

1.2.2 　Fourier transform

In this section we consider the function $f : \mathbb{R} \to \mathbb{C}$, under two restrictions: first under $\|f\|_1 = \int_{-\infty}^\infty |f(t)|\mathrm{d}t < \infty$, that is in the space $\mathcal{L}_1 = \mathcal{L}_1(\mathbb{R})$, and then under $\|f\|_2 = \int_{-\infty}^\infty |f(t)|^2\mathrm{d}t < \infty$, that is in the Hilbert space

---

[1]This means that the total variation of f is bounded, i.e.

$$\sup_{\tau_n} \sum_{j=1}^n |f(t_j) - f(t_{j-1})| < \infty,$$

where τ_n denotes the partition $0 = t_0 < t_1 < \cdots < t_n < 2\pi$ and where the supremum is taken over all these partitions with vanishing mesh; the mesh being the largest distance between two consecutive point of the partition. The total variation of f is thus the length of the curve that is drawn by f over the complex plane.

[2]Let $\{x_j\}_{j\geq 1}$ be a real sequence and define the sequence of consecutive arithmetic means

$$s_n = \frac{1}{n} \sum_{j=1}^n x_j, \quad \forall n \in \mathbb{N}^*.$$

Roughly stated, $\{s_n\}_{n\geq 1}$ has better convergence properties than $\{x_n\}_{n\geq 1}$. In particular,

$$\lim_{n\to\infty} x_n = \lambda, \text{ for some } \lambda \in \mathbb{R} \implies \lim_{n\to\infty} s_n = \lambda.$$

The convergence of $\{s_n\}_{n\geq 1}$ is referred to Cesàro summability of the sequence $\{x_n\}_{n\geq 1}$; cf. e.g. Carothers (2000). A standard illustration is for $\{x_j\}_{j\geq 1} = \{0, 1, 0, 1, \ldots\}$.

$\mathcal{L}_2 = \mathcal{L}_2(\mathbb{R})$. We note that no one of $\mathcal{L}_1$ or $\mathcal{L}_2$ is subset of the other one. For instance, the function $f(t) = 1/\{\sqrt{|t|}(1+t^2)\}$ belongs to $\mathcal{L}_1$ but not to $\mathcal{L}_2$. Also, the function $f(t) = \operatorname{sinc} t$ of Definition A.8.1 is not in $\mathcal{L}_1$, as shown in (A.8.1), and it is nevertheless in $\mathcal{L}_2$, as explained in Section A.8.1. If the function f belongs to any one of the two spaces $\mathcal{L}_1$ and $\mathcal{L}_2$, then its Fourier transform is given by

$$\hat{f}(\alpha) = \int_{-\infty}^{\infty} e^{-i\alpha t} f(t) dt, \quad \forall \alpha \in \mathbb{R}. \tag{1.2.9}$$

Nonetheless, (1.2.9) needs some analysis regarding its existence and its pointwise validity, and it differs depending on whether $f \in \mathcal{L}_1$ or $f \in \mathcal{L}_2$.

Fourier transform in $\mathcal{L}_1$

The Fourier transform $\hat{f}(\alpha)$ represents the complex amplitude of the harmonic e_α, which is defined as in (1.2.1) with $k \in \mathbb{Z}$ replaced by $\alpha \in \mathbb{R}$, in the spectral decomposition of the function $f \in \mathcal{L}_1$. Thus, the Fourier transform inherits the nice interpretation of the Fourier coefficient (1.2.2) of a periodic function of $\mathcal{L}_2^\circ$. Indeed, if $f \in \mathcal{L}_1$ is approximately proportional to e_α, the harmonic with frequency α, then the integrand of (1.2.9) becomes $e^{-i\alpha t} f(t) \simeq e^{-i\alpha t} e^{i\alpha t} |f(t)| = |f(t)|$. Thus $\hat{f}(\alpha) \simeq \|f\|_1$, which becomes large given that $|f(t)|$ varies little.

One verifies directly that $\hat{f}$ is always continuous. Another important property is

$$\lim_{\alpha \to \pm\infty} \hat{f}(\alpha) = 0. \tag{1.2.10}$$

which is the analogue of the Riemann-Lebesgue lemma (1.2.4). Define by

$$f * g(t) = \int_{-\infty}^{\infty} f(t-s) g(s) ds, \quad \forall t \in \mathbb{R},$$

the convolution of $f, g \in \mathcal{L}_1$. One can easily show with Tonelli's theorem A.5.7 that

$$f * g \in \mathcal{L}_1.$$

One can also show that

$$\widehat{f * g}(\alpha) = \hat{f}(\alpha) \hat{g}(\alpha), \quad \forall \alpha \in \mathbb{R}.$$

For given $f, \hat{f} \in \mathcal{L}_1$, one defines the inverse Fourier transform of $\hat{f}$ by

$$f(t) = \frac{1}{2\pi} \int_{-\infty}^{\infty} e^{i\alpha t} \hat{f}(\alpha) d\alpha, \tag{1.2.11}$$

for t a.e. in $\mathbb{R}$ and, in particular, at any continuity point $t \in \mathbb{R}$ of f. Thus (1.2.11) gives the spectral decomposition of f in terms of the Riemann sum for $\alpha \in \mathbb{R}$ of harmonics $e_\alpha = e^{i\alpha \cdot}$ with complex amplitude $\hat{f}(\alpha)/(2\pi)$.

Fourier transform in $\mathcal{L}_2$

Let $f, g : \mathbb{C} \to \mathbb{R}$ in $\mathcal{L}_2$ and define the scalar product of f and g by

$$\langle f, g \rangle = \int_{-\infty}^{\infty} f(t)\overline{g(t)}\mathrm{d}t,$$

as well as the $\mathcal{L}_2$-norm of f by

$$\|f\|_2 = \langle f, f \rangle^{\frac{1}{2}}.$$

Cauchy-Schwarz inequality is given by

$$|\langle f, g \rangle| \le \|f\|_2 \|g\|_2.$$

Just like $\mathcal{L}_2^\circ$, $\mathcal{L}_2$ is an Hilbert space. But the nice geometrical interpretations obtained in $\mathcal{L}_2^\circ$ are no longer available in $\mathcal{L}_2$. The important difference w.r.t. Fourier series in $\mathcal{L}_2^\circ$ is that the functions e_α, $\forall \alpha \in \mathbb{R}$, that are obtained by extension of (1.2.1) from $\mathbb{Z}$ to $\mathbb{R}$, do not belong to $\mathcal{L}_2$. Consequently, the Fourier transform (1.2.9) no longer represents a scalar product, as it is with (1.2.2) in the context of Fourier series. In other terms, for given $\alpha \in \mathbb{R}$, we cannot write $\hat{f}(\alpha) = \langle f, e_\alpha \rangle$ and justify its existence by Cauchy-Schwarz inequality. This problem is overcome as follows. The Fourier transform (1.2.9) is initially defined as an operator $F : \mathcal{L}_1 \cap \mathcal{L}_2 \to \mathcal{L}_2$. From here, two important facts are exploited: $\mathcal{L}_1 \cap \mathcal{L}_2$ is dense in $\mathcal{L}_2$ and F is an isometry (cf. Definition A.1.3). These two facts allow for the application of an important result, according to which an isometric operator F admits an unique isometric extension to the closure of its domain of definition. Precisely, F admits the unique extension from $\mathcal{L}_1 \cap \mathcal{L}_2$ to its closure $\mathcal{L}_2$. This extension defines the Fourier transform of a function of $\mathcal{L}_2 \backslash (\mathcal{L}_1 \cap \mathcal{L}_2) = \mathcal{L}_2 \backslash \mathcal{L}_1$. In short, the Fourier transform in $\mathcal{L}_2$ is obtained in terms of an isometric operator $F : \mathcal{L}_2 \to \mathcal{L}_2$, which is thus a bijection. The isometry is obtained with Parseval-Plancherel formula, which is

$$\langle \hat{f}, \hat{g} \rangle = 2\pi \langle f, g \rangle, \quad \forall f, g \in \mathcal{L}_2. \tag{1.2.12}$$

This formula can be easily shown[3] and its straightforward corollary is

$$\|\hat{f}\|_2^2 = 2\pi \|f\|_2^2. \tag{1.2.13}$$

---

[3]Indeed,

$$\langle \hat{f}, \hat{g} \rangle = \int_{-\infty}^{\infty} \hat{f}(\alpha)\overline{\int_{-\infty}^{\infty} \mathrm{e}^{-\mathrm{i}\alpha t}g(t)\mathrm{d}t}\,\mathrm{d}\alpha = \int_{-\infty}^{\infty} \int_{-\infty}^{\infty} \mathrm{e}^{\mathrm{i}\alpha t}\hat{f}(\alpha)\mathrm{d}\alpha\overline{g(t)}\mathrm{d}t$$

$$= \int_{-\infty}^{\infty} 2\pi f(t)\overline{g(t)}\mathrm{d}t = 2\pi \langle f, g \rangle.$$

Thereupon, $\forall f \in \mathcal{L}_2$, one has $\hat{f} = Ff \in \mathcal{L}_2$ and bijectivity of F justifies the existence of the inverse Fourier transform $F^{-1}\hat{f}$, which takes the form of (1.2.11).

Note that there exists alternative versions or conventions for the Fourier transform (1.2.9). One finds for example, $\hat{f}(\alpha) = 1/\sqrt{2\pi} \int_{-\infty}^{\infty} e^{-i\alpha t} f(t)dt$ or $\hat{f}(\alpha) = 1/(2\pi) \int_{-\infty}^{\infty} e^{i\alpha t} f(t)dt$, $\forall \alpha \in \mathbb{R}$. Each convention leads to its corresponding version of the inverse Fourier transform (1.2.11).[4]

Regularity and decay of Fourier pair

This section analyzes the smoothness or the regularity and the decay along the tails of the Fourier pair f and $\hat{f}$. There is a qualitative relationship within the Fourier pair: the smoother f, the faster the decay of $\hat{f}$ along the tails.

Assume $f : \mathbb{C} \to \mathbb{R}$ continuously differentiable, integrable (i.e. in $\mathcal{L}_1$) and with derivative f', which is also assumed integrable. Let $\alpha \in \mathbb{R}$. By partial integration one obtains

$$\widehat{f'}(\alpha) = i\alpha \hat{f}(\alpha).$$

This formula can be iterated so to obtain

$$\widehat{f^{(k)}}(\alpha) = (i\alpha)^k \hat{f}(\alpha),$$

whenever f is integrable, $k \in \mathbb{N}^*$ times continuously differentiable and with these derivatives integrable. By applying Riemann-Lebesgue lemma (1.2.10) to the current situation, we obtain $\lim_{\alpha \to \pm\infty} |\alpha|^k \hat{f}(\alpha) = 0$, which is re-expressed as

$$\hat{f}(\alpha) = o\left(|\alpha|^{-k}\right), \quad \text{as } \alpha \to \pm\infty. \tag{1.2.14}$$

We can also obtain (1.2.14) through Parseval-Plancherel's corollary (1.2.13). Indeed, under the given assumptions and provided $f^{(k)} \in \mathcal{L}_2$, one finds the identity

$$\int_{-\infty}^{\infty} |\alpha|^{2k} \left|\hat{f}(\alpha)\right|^2 d\alpha = 2\pi \int_{-\infty}^{\infty} \left|f^{(k)}(t)\right|^2 dt < \infty.$$

This implies (1.2.14).

---

[4]These conventions are purely technical. The only rule that matters is that the product of the two constants before the two integrals is $1/(2\pi)$ and that $e^{i\alpha t}$ appears in any one of the two integrals and $e^{-i\alpha t}$ in the other one.

Now, we only assume f integrable and such that $\int_{-\infty}^{\infty} |t||f(t)|\mathrm{d}t < \infty$, which enforces a fast decay on the tails f. Denote by $\mathrm{id}_{\mathbb{R}}$ the identity function over $\mathbb{R}$. We can compute

$$\hat{f}'(\alpha) = \lim_{h \to 0} \frac{\hat{f}(\alpha+h) - \hat{f}(\alpha)}{h} = \int_{-\infty}^{\infty} e^{-i\alpha t} \lim_{h \to 0} \frac{e^{-iht} - 1}{h} f(t)\mathrm{d}t$$

$$= -i \int_{-\infty}^{\infty} e^{-i\alpha t} t f(t)\mathrm{d}t = -i\, \widehat{\mathrm{id}_{\mathbb{R}} f}(\alpha), \quad \forall \alpha \in \mathbb{R},$$

from Dominated convergence A.5.3, because $\forall h \in \mathbb{R}$ and $t \in \mathbb{R}$ such that $f(t) \neq 0$, we have

$$\left| e^{-i\alpha t} \frac{e^{-iht} - 1}{h} f(t) \right| \leq |tf(t)| \iff \left| e^{iht} - 1 \right| \leq |ht|.$$

As mentioned just before (1.2.10), $\widehat{\mathrm{id}_{\mathbb{R}} f}$ is continuous and therefore $\hat{f}'$ is continuous as well. This result can be generalized as follows. If $\int_{-\infty}^{\infty} |t|^k |f(t)|\mathrm{d}t < \infty$, for some $k \in \mathbb{N}^*$, then

$$\hat{f}^{(k)}(\alpha) = (-i)^k\, \widehat{\mathrm{id}_{\mathbb{R}}^k f}(\alpha), \quad \forall \alpha \in \mathbb{R}.$$

Thus $\hat{f}^{(k)}$ is continuous.

A related study, regarding the relations between the mean square regularity or smoothness of a stationary stochastic process and the tail behavior of the spectral distribution, is presented in Section 3.3.

Chapter 2

Stationary time series

This first of two central chapters of this book presents the stationary time series. It contains the following sections. Section 2.1 introduces methods for obtaining a stationary time series. It presents and analyzes the a.c.v.f. Section 2.2 presents the ARMA time series. Two important properties of stationary time series, causality and invertibility, are analyzed. Section 2.3 studies the a.c.v.f. and introduces the partial autocorrelation function (p.a.c.r.f.). Section 2.4 considers the analysis in frequency domain of complex-valued time series. Herglotz's theorem is presented and the spectral distribution of ARMA time series is obtained. The theory of linear filters is introduced Further special topics are presented in Section 2.5: prediction, ARIMA model, determination of AR order, Kalman filter and other types of time series models. The first parts of this chapter are based on Brockwell and Davis (1991).

2.1   Introduction

The term time series often designates a series of measurements obtained at equally spaced epochs. It is thus an ordered sample. Alternatively, a stochastic process with discrete time domain like the integers $\mathbb{Z}$ is often called time series in statistics. We consider this second case in which a time series is an ordered sequence of random variables that we denote by $\{X_k\}_{k \in \mathbb{Z}}$. This book always considers time series $\{X_k\}_{k \in \mathbb{Z}}$ with finite second moment, precisely such that

$$\mathsf{E}\left[X_k^2\right] < \infty, \quad \forall k \in \mathbb{Z}.$$

This is a simple but essential condition for most theorems and methods that are presented in this chapter. Indeed, the covariance between two random variables of the time series, taken at different epochs, does always exist whenever the second moment exists. These covariances are important quantities for analysis of a stationary time series. From the theoretical perspective, the existence of the second moment allows to use simple geometrical properties of the Hilbert space of square-integrable random variables $\mathcal{L}_2$ (see Section A.1), in order to develop important techniques and to demonstrate theoretical properties.

With these considerations, the structure of Section 2.1 is the following. Section 2.1.1 presents the concept of stationarity and the a.c.v.f. Section 2.1.2 explains some practical techniques for obtaining a stationary time series, from original nonstationary data. Section 2.1.3 provides the mathematical properties of the a.c.v.f. and introduces the empirical a.c.v.f.

2.1.1 *Stationarity and autocovariance function*

Let $\{X_k\}_{k \in \mathbb{Z}}$ be a time series for which $\mathsf{E}[X_k^2] < \infty$, $\forall k \in \mathbb{Z}$. The function $\gamma : \mathbb{Z} \times \mathbb{Z} \to \mathbb{R}$ defined by

$$\gamma(k,l) = \mathsf{cov}(X_k, X_l), \quad \forall k, l \in \mathbb{Z},$$

is called autocovariance function (a.c.v.f.) of the time series $\{X_k\}_{k \in \mathbb{Z}}$. However a simplified version of this function can be given for stationary time series.

The a.c.v.f. allows to define a central property for the study of a time series and for making predictions: the weak stationarity, where the adjective "weak" is omitted in an unambiguous situations.[1]

Definition 2.1.1 (Weak and strict stationarity). *The time series $\{X_k\}_{k \in \mathbb{Z}}$ is called (weakly) stationary if, $\forall k, l \in \mathbb{Z}$, $\mathsf{E}\left[X_k^2\right] < \infty$ and the two following properties hold:*

(1) $\mathsf{E}[X_k]$ does not depend on $k \in \mathbb{Z}$ (i.e. it is constant);
(2) $\gamma(k,l) = \gamma(0, k - l)$, $\forall k, l \in \mathbb{Z}$.

The time series $\{X_k\}_{k \in \mathbb{Z}}$ is called strictly stationary if

$$(X_{k_1}, \ldots, X_{k_n}) \sim (X_{k_1+h}, \ldots, X_{k_n+h}),$$

$\forall k_1 < \cdots < k_n \in \mathbb{Z}$, $n, h \in \mathbb{N}$.

The notation $U \sim V$ means that the random elements (variables, vectors, processes, etc.) U and V possess the same distribution.

---

[1] Weak stationarity is seldom called second order stationarity or wide sense stationarity.

Note that when $\{X_k\}_{k\in\mathbb{Z}}$ is stationary, the second moment $\mathsf{E}\left[X_k^2\right]$ does not depend on the time index k. Also, the a.c.v.f. at point $(k, k + h)$ does not depend on k but on $|h|$ only. Consequently, the following simplified definition of the a.c.v.f. and the following definition of the autocorrelation function (a.c.r.f.) of a stationary time series can be given.

Definition 2.1.2 (A.c.v.f. and a.c.r.f.). *Consider the stationary time series* $\{X_k\}_{k\in\mathbb{Z}}$. *The a.c.v.f. at time lag* $h \in \mathbb{Z}$ *is given by*

$$\gamma(h) = \gamma(k + h, k) = \mathsf{cov}(X_{k+h}, X_k), \quad \forall k \in \mathbb{Z}.$$

The autocorrelation function (a.c.r.f.) at time lag $h \in \mathbb{Z}$ *is given by*

$$\rho(h) = \frac{\mathsf{cov}(X_0, X_h)}{\sqrt{\mathsf{var}(X_0)\mathsf{var}(X_h)}} = \frac{\gamma(h)}{\gamma(0)},$$

provided that the time series has positive variance.

Thus, the a.c.r.f. measures the amount of linear dependence between X_0 and X_h, $\forall h \in \mathbb{Z}$. It does not depend on the unit in which the time series is measured.

Remark 2.1.3 (Weak and strict stationarity). *Consider strict stationarity of Definition 2.1.1 with* $n = 1$. *It implies that the distribution of* X_k *does not depend on* $k \in \mathbb{Z}$. *Although this assumption may seem too restrictive, various types of time series admit an equilibrium distribution, namely an asymptotic distribution as* $k \to \infty$. *Assuming strict stationarity is reasonable whenever the time series has become close to its equilibrium state.*

Consider strict stationarity of Definition 2.1.1 with $n = 2$. *The joint distribution of* (X_{k_1}, X_{k_2}) *depends on* k_1, k_2 *only through the lag* $k_2 - k_1$. *This implies weak stationarity. In particular,*

$$\mathsf{cov}(X_{k_2}, X_{k_1}) = \gamma(k_2 - k_1).$$

Weak stationarity is more appealing than strong stationarity for two main reasons. First, strong stationarity seems too restrictive for practical applications. Second, many important and useful properties are available under weak stationarity. There are of course time series that are weakly but not strictly stationary. One instance is given in Example 2.1.4 below. Conversely, there are time series without second moment that are strictly stationary.

Example 2.1.4 (Sum of two harmonics). *Let* $C, S \in \mathcal{L}_2$ *be uncorrelated, with mean 0 and variance 1. Define the time series*

$$X_k = C \cos k\theta + S \sin k\theta, \quad \forall k \in \mathbb{Z},$$

where $\theta \in [0, \pi]$ is the frequency. This time series has mean zero and a.c.v.f. given by

$$\gamma(k+h, k) = \text{cov}(X_{k+h}, X_k)$$
$$= \text{E}\left[\{C\cos(k+h)\theta + S\sin(k+h)\theta\}\{C\cos k\theta + S\sin k\theta\}\right]$$
$$= \text{E}\left[C^2\cos k\theta\cos(k+h)\theta + S^2\sin k\theta\sin(k+h)\theta\right]$$
$$= \cos h\theta, \ \forall h, k \in \mathbb{Z}.$$

Because the last expression is independent of k, the time series is weakly stationary.

Consider $\theta = \pi/2$. Then $\{X_0, X_1, X_2, X_3, X_4, X_5, X_6, X_7, \ldots\} = \{C, S, -C, -S, C, S, -C, -S, \ldots\}$. A necessary condition for strictly stationary is that (C, S) and $-(C, S)$ are identically distributed: without this symmetry, strict stationarity does not hold. Strict stationarity would hold for $C \sim S \sim \mathcal{N}(0, 1)$ and the time series would be called Gaussian.

Another interesting fact about this time series is that it can be periodic. We saw in the above paragraph that it can be periodic with period length 4. Alternatively, if the time unit is the month and $\theta = \pi/6$, then we obtain the practical periodic time series with period length is 12 and thus with yearly periodicity. A periodicity is often called seasonality. Thus, a stationary time series may have a seasonality. Further remarks on this aspect are given below.

The a.c.v.f. and a.c.r.f. provide a quantification and a standardized quantification of the dependence among the random variables at different time lags of the time series. These two functions have an important role for the prediction of future values of the time series, in terms of past and present values. These functions are central for the analysis in time domain. Alternatively, in frequency domain analysis of the time series, it is the spectral distribution (or spectrum) that inherits the central role. The frequency domain analysis is introduced in Section 2.4. The amplitudes and the frequencies of the different harmonics that constitute the stationary time series are analyzed. Whether to study a time series in the time domain in the frequency domain is in part a subjective choice. The analysis in frequency domain is often preferred in engineering applications, whereas time domain is often preferred for financial data.

The basic stationary time series is defined as follows.

Definition 2.1.5 (WN). *The time series $\{Z_k\}_{k\in\mathbb{Z}}$ is called white noise (WN) with variance σ^2, for some $\sigma > 0$, if it is real-valued, it has expectation null and if, $\forall k \in \mathbb{Z}$,*

$$\mathrm{cov}(Z_{k+h}, Z_k) = \begin{cases} \sigma^2, & \text{if } h = 0, \\ 0, & \text{otherwise.} \end{cases}$$

We denote any time series with these properties by WN(σ^2).

For example, $\ldots, Z_{-1}, Z_0, Z_1, \ldots$ can be independent $\mathcal{N}(0, \sigma^2)$. This important WN is called Gaussian WN. This definition will be extended to continuous time in Sections 3.4.2 and 3.4.4, where continuous time Gaussian WN is defined as generalized stochastic process. In statistics and signal processing, WN corresponds to a random signal for which all frequencies are equally represented. The term WN originates from the fact that white color reflects all visible wave frequencies of light.

We note that WN does not necessarily have independent terms. The following example illustrates this fact.

Example 2.1.6 (WN with dependent terms). *Let $\{Y_k\}_{k\in\mathbb{Z}}$ be* WN$(\sigma^2/2)$. *Define*

$$X_k = \begin{cases} Y_k - Y_{k-1}, & \text{if } k \in \{0, \pm 2, \ldots\}, \\ Y_{k+1} + Y_k, & \text{if } k \in \{\pm 1, \pm 3, \ldots\}. \end{cases}$$

The elements of $\{X_k\}_{k\in\mathbb{Z}}$ are generally dependent. Choose $k \in \{0, \pm 2, \ldots\}$, then $X_k = Y_k - Y_{k-1}$ and $X_{k-1} = Y_k + Y_{k-1}$ are clearly dependent. Let us show that $\{X_k\}_{k\in\mathbb{Z}}$ is nevertheless WN(σ^2). *We see directly that $\mathsf{E}[X_k] = 0$ and $\mathrm{var}(X_k) = \sigma^2$, $\forall k \in \mathbb{Z}$. It remains to show that $\mathsf{E}[X_k X_{k-h}] = 0$, $\forall k \in \mathbb{Z}$ and $h \in \mathbb{Z}^*$. We distinguish the three cases following cases.*

In order to have two even indices, let $k \in \{0, \pm 2, \ldots\}$ and $h \in \{2, 4, \ldots\}$, then

$$\begin{aligned} \mathsf{E}[X_k X_{k-h}] &= \mathsf{E}[(Y_k - Y_{k-1})(Y_{k-h} - Y_{k-h-1})] \\ &= \mathsf{E}[Y_k Y_{k-h}] - \mathsf{E}[Y_k Y_{k-h-1}] - \mathsf{E}[Y_{k-1} Y_{k-h}] + \mathsf{E}[Y_{k-1} Y_{k-h-1}] \\ &= 0. \end{aligned}$$

In order to have one even index and one odd index, let $k \in \{0, \pm 2, \ldots\}$ and $h \in \{1, 3, \ldots\}$, then

$$\begin{aligned} \mathsf{E}[X_k X_{k-h}] &= \mathsf{E}[(Y_k - Y_{k-1})(Y_{k-h+1} + Y_{k-h})] \\ &= \mathsf{E}[Y_k Y_{k-h+1}] - \mathsf{E}[Y_k Y_{k-h}] - \mathsf{E}[Y_{k-1} Y_{k-h+1}] - \mathsf{E}[Y_{k-1} Y_{k-h}] \\ &= 0. \end{aligned}$$

In order to have two odd indices, let $k \in \{\pm 1, \pm 3, \ldots\}$ and $h \in \{2, 4, \ldots\}$, then

$$\begin{aligned}
\mathsf{E}[X_k X_{k-h}] &= \mathsf{E}[(Y_{k+1} + Y_k)(Y_{k-h+1} + Y_{k-h})] \\
&= \mathsf{E}[Y_{k+1} Y_{k-h+1}] + \mathsf{E}[Y_{k+1} Y_{k-h}] + \mathsf{E}[Y_k Y_{k-h+1}] + \mathsf{E}[Y_k Y_{k-h}] \\
&= 0.
\end{aligned}$$

A time series is stationary if there are no systematic changes in the mean and in the covariance. Consequently, the variance is constant. Any systematic change in mean is called trend. Another important feature of a time series is the seasonality. A time series has a seasonality if it possesses a periodic additive component. Seasonality and stationarity may simultaneously exist in a time series. An illustration of this situation appears already in Example 2.1.4. Another illustration is given in the following Example 2.1.7.2. Nevertheless, it is preferred to analyze time series without seasonalities. In Section 2.1.2 we present methods for constructing stationary time series without seasonalities.

Two other interesting examples of stationary time series that relate to WN and to seasonality are the following.

Example 2.1.7 (Single harmonic: WN and seasonality). *Let*

$$X_k = a \cos(\eta + k\theta), \quad \forall k \in \mathbb{Z},$$

for some $a \in (0, \infty)$, let η be uniformly distributed over $[0, 2\pi)$ and let θ be a random variable with d.f. F over $[0, \pi)$ and independent of η. In order to control the stationarity of the time series $\{X_k\}_{k \in \mathbb{Z}}$, we verify the following three properties: $\mathsf{E}[X_k^2] < \infty$, $\forall k \in \mathbb{Z}$, $\mathsf{E}[X_k]$ does not depend on $k \in \mathbb{Z}$ and $\mathrm{cov}(X_k, X_{k+h})$ does not depend on $k \in \mathbb{Z}$, $\forall h \in \mathbb{Z}$.

The first property holds because $|X_k| \le a$ a.s., $\forall k \in \mathbb{Z}$. For the second property, we can use that

$$\mathsf{E}[X_k \,|\, \theta] = \frac{1}{2\pi} \int_0^{2\pi} a \cos(\eta + k\theta) \, \mathrm{d}\eta = \frac{a}{2\pi} \{\sin(2\pi + k\theta) - \sin k\theta\} = 0,$$

so that $\mathsf{E}[X_k] = \mathsf{E}[\mathsf{E}[X_k \,|\, \theta]] = 0$, $\forall k \in \mathbb{Z}$. For the last property, we have

$$\begin{aligned}
\mathrm{cov}(X_{k+h}, X_k) &= \mathsf{E}[X_{k+h} X_k] \\
&= \frac{1}{2\pi} \int_0^\pi \int_0^{2\pi} a \cos\{\eta + (k+h)\theta\} a \cos(\eta + k\theta) \, \mathrm{d}\eta \, \mathrm{d}F(\theta)
\end{aligned}$$

$$= \frac{a^2}{2\pi} \int_0^\pi \int_0^{2\pi} \cos(\eta + k\theta + h\theta) \cos(\eta + k\theta) \, \mathrm{d}\eta \, \mathrm{d}F(\theta)$$

$$= \frac{a^2}{2\pi} \int_0^\pi \int_0^{2\pi} \{\cos(\eta + k\theta) \cos h\theta - \sin(\eta + k\theta) \sin h\theta\}$$

$$\cos(\eta + k\theta) \, \mathrm{d}\eta \, \mathrm{d}F(\theta)$$

$$= \frac{a^2}{2\pi} \int_0^\pi \cos h\theta \underbrace{\int_0^{2\pi} \cos^2(\eta + k\theta) \, \mathrm{d}\eta}_{=\pi} \, \mathrm{d}F(\theta)$$

$$= \frac{a^2}{2} \int_0^\pi \cos h\theta \, \mathrm{d}F(\theta),$$

which does not depend on k. So the time series is stationary. Let us analyze two interesting special cases.

(1) First, if θ is uniformly distributed over $[0, \pi)$, then

$$\mathsf{cov}(X_{k+h}, X_k) = \begin{cases} \frac{a^2}{2}, & \text{if } h = 0, \\ 0, & \text{if } h = \pm 1, \pm 2, \dots. \end{cases}$$

In this case the time series $\{X_k\}_{k \in \mathbb{Z}}$ is in fact WN, in the sense of Definition 2.1.5.

(2) Another special case arises for θ uniformly distributed over $\{0, \pi/2\}$. By replacing θ by its possible values 0 and $\pi/2$, we see that the time series is 4-periodic. In other terms, the stationary time series $\{X_k\}_{k \in \mathbb{Z}}$ possesses a seasonality of length 4.

An important stationary time series is the following.

Example 2.1.8 (First order moving average). Let $\{Z_k\}_{k \in \mathbb{Z}}$ be an sequence of i.i.d. random variables with $\mathsf{E}[Z_1] = 0$ and $\sigma^2 = \mathsf{var}(Z_1) < \infty$. Let $\theta \in \mathbb{R}$ and let $\{X_k\}_{k \in \mathbb{Z}}$ be the time series defined by

$$X_k = Z_k + \theta Z_{k-1}, \quad \forall k \in \mathbb{Z}.$$

Then $\{X_k\}_{k \in \mathbb{Z}}$ is a strictly stationary time series with $\mathsf{E}[X_k] = 0$, $\forall k \in \mathbb{Z}$, and with a.c.v.f.

$$\gamma(h) = \begin{cases} \sigma^2(1 + \theta^2), & \text{if } h = 0, \\ \theta \sigma^2, & \text{if } |h| = 1, \\ 0, & \text{if } |h| \geq 2. \end{cases}$$

Together with weak and strict stationarity, another worth mentioning concept is that of dependence or correlation up to a given time lag.

Definition 2.1.9 (*h*-correlation and *h*-dependence). *Let $\{X_k\}_{k \in \mathbb{Z}}$ be a time series.*

(1) *Under weak stationarity (and thus necessarily in $\mathcal{L}_2$), $\{X_k\}_{k \in \mathbb{Z}}$ possesses correlation up to order h or, simply stated, is h-correlated, for some $h \in \mathbb{N}$, if*

$$\text{corr}(X_1, X_{1+l}) \begin{cases} \neq 0, & \text{for } l = h, \\ = 0, & \text{for } l = h+1, h+2, \ldots. \end{cases}$$

(2) *Under strict stationarity, $\{X_k\}_{k \in \mathbb{Z}}$ possesses dependence up to order h or, simply stated, is h-dependent, for some $h \in \mathbb{N}$, if*

$$X_1 \text{ and } X_{1+l} \begin{cases} \text{are dependent,} & \text{for } l = h, \\ \text{are independent,} & \text{for } l = h+1, h+2, \ldots. \end{cases}$$

Remarks 2.1.10. *(1) Weak stationarity allows to re-express Definition 2.1.9.1 in the following form, which may be more intuitive: $\forall k \in \mathbb{Z}$, the two sets*

$$\{\ldots, X_{k-1}, X_k\} \text{ and } \{X_{k+h+1}, X_{k+h+2}, \ldots\},$$

are termwise uncorrelated. The analogue statement, however with termwise independence, holds for Definition 2.1.9.2.

(2) When the time series is Gaussian, Definition 2.1.9.1 and 2.1.9.2 are confounded, according to Section A.8.3.

A trivial example is given by WN, which is 0-correlated. Another trivial case is any sequence of i.i.d. random variables, which 0-dependent. An important example is provided by Proposition 2.2.21 that follows, which gives the characterization of a central time series in terms of h-correlation.

We end this section with the following remark which mentions that, in simple situations, regression models can be useful in the context of time series.

Remark 2.1.11 (Regression and time series). *We show how a simple non-stationary time series can be modeled with a simple regression model. The method of Cochrane and Orcutt (1949) consists in obtaining a regression model with uncorrelated errors from one with errors subject to the first order autoregressive (AR) structure. We know from Gauss-Markov's theorem that the least-squares estimators of the regression parameters are inefficient in this case and thus we consider a transformation leading to white* *noise errors. Assume that* $\{X_k\}_{k\in\mathbb{Z}}$ *follows the regression model with correlated errors*

$$X_k = \alpha + \beta_0 k + \beta_1 z_k + Y_k, \quad \forall k \in \mathbb{Z}, \tag{2.1.1}$$

where

$$Y_k = \varphi Y_{k-1} + Z_k, \quad \forall k \in \mathbb{Z}, \tag{2.1.2}$$

is the first order AR time series, in which $\{Z_k\}_{k\in\mathbb{Z}}$ *represents white noise. Define the transformed time series* $\{X_k^*\}_{k\in\mathbb{Z}}$ *as follows,*

$$\underbrace{X_k - \varphi X_{k-1}}_{=X_k^*}$$
$$= \alpha + \beta_0 k + \beta_1 z_k + Y_k - \varphi\alpha - \varphi\beta_0(k-1) - \varphi\beta_1 z_{k-1} - \varphi Y_{k-1}$$
$$= \underbrace{\alpha(1-\varphi)}_{=\alpha^*} + \underbrace{\beta_0\varphi + \beta_0(1-\varphi)}_{=\beta_0^*}k + \beta_1\underbrace{(z_k - \varphi z_{k-1})}_{=z_k^*} + \underbrace{Y_k - \varphi Y_{k-1}}_{=Z_k},$$

$\forall k \in \mathbb{Z}$. *This gives the regression model with uncorrelated errors*

$$X_k^* = \alpha^* + \beta_0^* k + \beta_1 z_k^* + Z_k, \quad \forall k \in \mathbb{Z},$$

which is estimated by least-squares. The original parameters and estimators are obtained by $\alpha = (\alpha^* - \beta_0\varphi)/(1-\varphi)$ *and* $\beta_0 = \beta^*/(1-\varphi)$.

Assume that the time unit is the month and that $\{X_k\}_{k\in\mathbb{Z}}$ *has a seasonality of 12 months. In this setting one can consider the regression model*

$$X_k = \alpha + \beta_0 k + \beta_1 z_{k1} + \cdots + \beta_{11} z_{k11} + Y_k, \quad \forall k \in \mathbb{Z}, \tag{2.1.3}$$

where $z_{kj} = \mathsf{I}\{j = (k \bmod 12) - 1\}$, $\forall k \in \mathbb{Z}$, *for* $j = 1,\ldots,11$. *These indicators are sometimes called dummy variables. The matrix of explanatory variables, with a sample starting at time* $k = 1$ *and ending a some finite time, is thus given by*

$$\begin{pmatrix} 1 & 1 & 0 & 0 & \ldots & 0 \\ 1 & 2 & 1 & 0 & \ldots & 0 \\ 1 & 3 & 0 & 1 & 0 & 0 \\ \vdots & \vdots & \vdots & \vdots & \ddots & \vdots \\ 1 & 12 & 0 & 0 & \cdots & 1 \\ 1 & 13 & 0 & 0 & \ldots & 0 \\ \vdots & \vdots & \vdots & \vdots & \vdots & \vdots \end{pmatrix}.$$

The correlation of the errors can be eliminated by Cochrane-Orcutt's transform and the corresponding matrix becomes

$$
\begin{pmatrix}
1 & 2 & 1 & 0 & 0 & \ldots & 0 & 0 \\
1 & 3 & -\varphi & 1 & 0 & \ldots & 0 & 0 \\
1 & 4 & 0 & -\varphi & 1 & \ldots & 0 & 0 \\
\vdots & \vdots & \vdots & \vdots & \vdots & \ddots & \vdots & \vdots \\
1 & 11 & 0 & 0 & 0 & \cdots & 1 & 0 \\
1 & 12 & 0 & 0 & 0 & \cdots & -\varphi & 1 \\
1 & 13 & 0 & 0 & 0 & \cdots & 0 & -\varphi \\
1 & 14 & 1 & 0 & 0 & \ldots & 0 & 0 \\
\vdots & \vdots & \vdots & \vdots & \vdots & \vdots & \vdots & \vdots
\end{pmatrix}.
$$

The first order autocorrelation (2.1.2) can be detected by the test of Durbin and Watson (1950, 1951). The null hypothesis is $\mathrm{H}_0 : \varphi = 0$ *and the alternative is* $\mathrm{H}_1 : \varphi > 0$. *The test statistic of Durbin-Watson is given by*

$$
D_n = \frac{\sum_{k=2}^{n}(R_k - R_{k-1})^2}{\sum_{k=1}^{n} R_k^2},
$$

where R_k, *for* $k = 1, \ldots, n$, *are the least-squares residuals of the regression models (2.1.1) or (2.1.3) with a sample of size n. Critical values for this test are available. With the alternative* $\mathrm{H}_1 : \varphi < 0$, D_n *would be replaced by* $4 - D_n$.

2.1.2   *Construction of stationary time series*

As mentioned in the introduction, a stationary time series allows for simple and reliable statistical prediction. But in the practice, the first sample obtained is unsually not stationary and it may also have a seasonal variation. It is therefore preferable to transform such a sample to a stationary one that does show any seasonal effect. This section introduces commonly used techniques for obtaining a stationary time series without seasonality.

According to the classical decomposition model, the original nonstationary time series $\{X_k\}_{k \in \mathbb{Z}}$ can be decomposed as

$$
X_k = m_k + s_k + Y_k, \quad \forall k \in \mathbb{Z}, \tag{2.1.4}
$$

where $\{m_k\}_{k \in \mathbb{Z}}$ is the trend, which is a nonperiodic sequence with an unambiguous and simple pattern (such as linear or quadratic), $\{s_k\}_{k \in \mathbb{Z}}$ is the seasonality, which is a periodic sequence, and where the remaining part $\{Y_k\}_{k \in \mathbb{Z}}$ represents a stationary time series with mean zero.

Let us consider the sample $X_1, \ldots, X_n$ of the time series $\{X_k\}_{k \in \mathbb{Z}}$ following the model (2.1.4). Let $d \in \mathbb{N}$ such that $s_{k+d} = s_k$, $\forall k \in \mathbb{Z}$, i.e. the period or season length. Because the sum $\sum_{j=1}^{d} s_{k+j}$ does not depend on $k \in \mathbb{Z}$, by taking

$$s_k - \frac{1}{d} \sum_{j=1}^{d} s_j \quad \text{and} \quad m_k + \frac{1}{d} \sum_{j=1}^{d} s_j,$$

instead of the initial s_k and m_k in (2.1.4), respectively, we are lead to assume, without loss of generality, that $\sum_{j=1}^{d} s_j = 0$.

Denote by γ be the a.c.v.f. of the stationary part $\{Y_k\}_{k \in \mathbb{Z}}$. From the fact that the seasonality has been separated, we can assume that γ vanishes for large arguments (i.e. time lags). In this case we have the useful asymptotic property

$$n^{-1} \sum_{j=1}^{n} Y_j \xrightarrow{\text{P}} 0, \quad \text{as } n \to \infty, \tag{2.1.5}$$

which follows from Theorem 2.5.4.1 and Chebichev's inequality.

There are two approaches and two contexts or assumptions in the context of the analysis of the time series $\{X_k\}_{k \in \mathbb{Z}}$. We describe these situations as follows.

(1) *Extraction approach.* We extract $\{m_k\}_{k \in \mathbb{Z}}$ and $\{s_k\}_{k \in \mathbb{Z}}$, if present, in order to make $\{X_k\}_{k \in \mathbb{Z}}$ stationary, and we find a model for the stationary remainder $\{Y_k\}_{k \in \mathbb{Z}}$.

(2) *Differentiation approach (Box and Jenkins).* We apply repeatedly difference operators to the time series $\{X_k\}_{k \in \mathbb{Z}}$ until a stationary time series is obtained.

(a) *Presence of trend and absence of seasonality.* We assume that $\{m_k\}_{k \in \mathbb{Z}}$ is present but $\{s_k\}_{k \in \mathbb{Z}}$ is absent or that it has already be eliminated.

(b) *Presence of trend and seasonality.* We assume that $\{m_k\}_{k \in \mathbb{Z}}$ and $\{s_k\}_{k \in \mathbb{Z}}$ are present.

We start the presentation under the assumption (a), where the scope is the elimination of trend in absence of seasonality. Let us consider the sample $X_1, \ldots, X_n$ of the time series $\{X_k\}_{k \in \mathbb{Z}}$, here without seasonality. Thus the model is

$$X_k = m_k + Y_k, \quad \forall k \in \mathbb{Z}. \tag{2.1.6}$$

(a.1.i) Extraction approach in absence of seasonality: function fitting

We extract the trend $\{m_k\}_{k\in\mathbb{Z}}$ from (2.1.6) by minimizing

$$\sum_{k=1}^{n}(X_k - m_k)^2,$$

where m_k is typically considered a polynomial of the variable k. In this case the minimization is carried on w.r.t. the coefficients of the polynomial.

(a.1.ii) Extraction approach in absence of seasonality: moving average

Define the moving average of order $q \in \mathbb{N}^*$ of $\{X_k\}_{k\in\mathbb{Z}}$ by

$$W_k = \frac{1}{1+2q}\sum_{j=-q}^{q}X_{k+j} = \frac{1}{1+2q}\sum_{j=-q}^{q}m_{k+j} + \frac{1}{1+2q}\sum_{j=-q}^{q}Y_{k+j}, \ \forall k \in \mathbb{Z}.$$

According to (2.1.5), the second term on the right side of the second equality above is with high probability close to zero. Furthermore, if we assume that m_k is approximately a linear function of k, i.e., $m_k \simeq a + bk$ for some $a, b \in \mathbb{R}$, then $\hat{m}_k = W_k \simeq m_k$ provides a good estimator of m_k, $\forall k \in \mathbb{Z}$. In this situation, Viewed as an operator of $\{X_k\}_{k\in\mathbb{Z}}$, $\hat{m}_k$ is called low pass filter, because it removes high-frequency components and lets low frequency components pass through, without distortion. The more q increases, the smoother $\hat{m}_k$ becomes (unless m_k is already exactly linear). However, as q increases, the bias also increases. So the choice of the smoothing parameter q brings the usual tradeoff between bias and variance and must be carefully made.

In general,

$$\sum_{j=-\infty}^{\infty}a_jX_{k+j},$$

is called time invariant linear filter, where time invariance is due to the fact that the sequence $\{a_j\}_{j\in\mathbb{Z}}$ does not depend on the present time k. The moving average $\hat{m}_k$ is the time invariant linear filter obtained with

$$a_j = \begin{cases} \frac{1}{1+2q}, & \text{if } j \in \{-q, \dots, q\}, \\ 0, & \text{otherwise.} \end{cases}$$

(a.2) Differentiation approach in absence of seasonality

We now present the approach of Box and Jenkins (1970), which mainly consists in repeatedly differentiating the time series until stationarity becomes apparent. This approach applies linear operators of a time series. Operators of a real- or complex-valued sequence $\{x_k\}_{k\in\mathbb{Z}}$ are used in numerical analysis and are defined as follows.

Definition 2.1.12 (Operator of sequence). *We call A operator of sequence any law that allows for the computation of $y_k \in \mathbb{C}$, $\forall k \in \mathbb{Z}$, by means of $\{x_k\}_{k\in\mathbb{Z}} \in \mathbb{C}^\infty$. The value y_k depends on k and on $\{x_k\}_{k\in\mathbb{Z}}$, but the computational formula does not. We denote, for simplicity and without rigor,*

$$y_k = Ax_k, \quad \forall k \in \mathbb{Z}.$$

Thus, the operator of sequence A transforms the sequence $\{x_k\}_{k\in\mathbb{Z}}$ to another sequence $\{y_k\}_{k\in\mathbb{Z}}$.

Definition 2.1.13 (Linear operator of sequence). *The operator of sequence A is linear if, for any two sequences $\{x_k\}_{k\in\mathbb{Z}}$ and $\{y_k\}_{k\in\mathbb{Z}}$ both in $\mathbb{C}^\infty$,*

$$A(x_k + y_k) = Ax_k + Ay_k \text{ and}$$
$$A\lambda x_k = \lambda Ax_k, \quad \forall \lambda \in \mathbb{C}, \ k \in \mathbb{Z}.$$

The important operators for the construction of stationary time series are the following.

Definition 2.1.14 (Backward shift and difference operators). *Let $\{X_k\}_{k\in\mathbb{Z}}$ be a time series.*

(1) The backward shift operator is the operator B such that

$$BX_k = X_{k-1}, \quad \forall k \in \mathbb{Z}.$$

(2) The difference operator is the operator ∇ such that

$$\nabla X_k = (I - B)X_k = X_k - X_{k-1}, \quad \forall k \in \mathbb{Z},$$

where I stands for the identity operator.

(3) $\forall j \in \mathbb{N}^*$, the j-th composition of the operators B and ∇ is given by

$$B^j X_k = B(B^{j-1}X_k) \quad \text{and} \quad \nabla^j X_k = \nabla(\nabla^{j-1}X_k),$$

where $B^0 X_k = \nabla^0 X_k = X_k$, $\forall k \in \mathbb{Z}$. Thus $B^0 = \nabla^0$ is the identity operator.

Backward shift and difference operators are clearly linear.
The binomial expansion is useful in this context:

$$\forall j \in \mathbb{N}, \quad \nabla^j X_k = (I - B)^j X_k = \sum_{l=0}^{j} \binom{j}{l} (-I)^l B^l X_k, \qquad (2.1.7)$$

where $k \in \mathbb{Z}$. As usual, the binomial coefficient is

$$\binom{x}{k} = \begin{cases} \frac{[x]_k}{k!}, & \text{if } k = 1, 2, \ldots, \\ 1, & \text{if } k = 0, \\ 0, & \text{if } k = -1, -2, \ldots, \end{cases} \quad \forall x \in \mathbb{R},$$

and the descending factorial is

$$[z]_k = \begin{cases} z(z-1) \cdots (z-k+1), & \text{if } k = 1, 2, \ldots, \\ 1, & \text{if } k = 0, \end{cases} \quad \forall z \in \mathbb{C} \backslash \{0, -1, \ldots\}.$$

Let us assume that the trend m_k is a polynomial of degree $p \in \mathbb{N}$ of the variable k, i.e.

$$m_k = a_0 + a_1 k + \cdots + a_p k^p,$$

for some coefficients $a_1, \ldots, a_p \in \mathbb{R}$, with $a_p \neq 0$. Then the application of (2.1.7) shows that ∇m_k has degree $p - 1$. By iterating this result we obtain that $\forall j \in \{0, \ldots, p\}$, $\nabla^j m_k$ has degree $p - j$ and that the leading coefficient of $\nabla^j m_k$ is

$$[p]_j a_p k^{p-j}.$$

In particular, we have

$$\nabla^p m_k = p! a_p.$$

For the nonstationary time series $\{X_k\}_{k \in \mathbb{Z}}$ without seasonality but with trend m_k, we thus obtain that

$$\nabla^p X_k = p! a_p + \nabla^p Y_k, \quad \forall k \in \mathbb{Z},$$

is a stationary time series with mean $p! a_p$. Therefore the differentiation approach consists in repeated differentiation until stationarity is achieved.

We now consider Case (b) and the objective is the elimination of trend and seasonality.

(b.1) Extraction approach with seasonality

This approach uses a particular moving average in order to eliminate the seasonality. We define, for $q + 1 \leq k \leq n - q$,

$$
\tilde{m}_k = \begin{cases} \frac{1}{2q}\left(\frac{1}{2}X_{k-q} + X_{k-q+1} + \cdots + X_{k+q-1} + \frac{1}{2}X_{k+q}\right), & \text{if } d = 2q, \\ \\ \frac{1}{1+2q}\left(X_{k-q} + \cdots + X_{k+q}\right), & \text{if } d = 1 + 2q. \end{cases}
$$

It follows from $\sum_{j=1}^{d} s_j = 0$ that, for $d = 2q$,

$$
\tilde{m}_k = \frac{1}{2q}\Big(\frac{1}{2}(Y_{k-q} + m_{k-q}) + (Y_{k-q+1} + m_{k-q+1}) + \cdots
$$

$$
+ (Y_{k+q-1} + m_{k+q-1}) + \frac{1}{2}(Y_{k+q} + m_{k+q})\Big)
$$

$$
\simeq \frac{1}{2q}\Big(\frac{1}{2}m_{k-q} + m_{k-q+1} + \cdots + m_{k+q-1} + \frac{1}{2}m_{k+q}\Big),
$$

and for $d = 2q + 1$,

$$
\tilde{m}_k = \frac{1}{1 + 2q}\big((Y_{k-q} + m_{k-q}) + \cdots + (Y_{k+q} + m_{k+q})\big)
$$

$$
\simeq \frac{1}{1 + 2q}\big(m_{k-q} + \cdots + m_{k+q}\big).
$$

Thus, as we noted in (a.1.ii), if m_k is approximately a linear function w.r.t. k, then $\tilde{m}_k \simeq m_k$.

When the value of d is known, then the value of q is clearly given. But when the value of d is unknown, we can obtain a convenient value for q as follows. Assume that m_k is an approximately linear function of k. Then, by taking different values of q, we observe that $\tilde{m}_k$ is an approximately linear function when either $2q$ or $1+2q$ equals jd, for some $j \geq 1$. Therefore q must be selected as any integer such that $\tilde{m}_k$ is a function without considerable variations. In this way we also estimate the period length d.

Now that q is fixed, we define for $q + 1 \leq k \leq n - q$ the mean deviation of season k as

$$
W_k = \frac{1}{n_k} \sum_{j \,\mid\, q+1 \leq k+jd \leq n-q} X_{k+jd} - \tilde{m}_{k+jd},
$$

where n_k is the number of summands. Following previously given arguments, when m_k is approximately linear, one can show that $W_k \simeq s_k$. We could use W_k as estimator of s_k, but it is generally not true that $\sum_{k=1}^{d} W_k = 0$. This leads to the definition of the centered estimator

$$
\hat{s}_k = \begin{cases} W_k - \frac{1}{d}\sum_{j=1}^{d} W_j, & \text{if } k \in \{1, \ldots, d\}, \\ \\ \hat{s}_{k \bmod d}, & \text{otherwise.} \end{cases}
$$

Finally, we can estimate the trend according to (a.1.i) or (a.1.ii) by using the values without seasonality

$$X_k - \hat{s}_k, \quad \text{for } k = 1, \ldots, n,$$

and denote this estimation of trend by $\{\hat{m}_k\}_{k=1,\ldots,n}$. The final result is

$$\hat{Y}_k = X_k - \hat{m}_k - \hat{s}_k, \quad \text{for } k = 1, \ldots, n,$$

which is expected to verify the conditions of stationarity.

(b.2) Differentiation approach with seasonality

This approach uses the lag d difference operator defined as follows.

Definition 2.1.15 (Lag d difference operator). *The lag d difference operator is given by*

$$\nabla_d X_k = (1 - B^d) X_k = X_k - X_{k-d}, \quad \forall k \in \mathbb{Z}.$$

The lag d difference operator should not be confused with the d-th composition of the differentiation operator: $\nabla_d X_k \neq \nabla^d X_k$. Because $\{s_k\}_{k\in\mathbb{Z}}$ is a d-periodic sequence, we have

$$\nabla_d X_k = m_k - m_{k-d} + Y_k - Y_{k-d}, \quad \forall k \in \mathbb{Z}.$$

This is a time series without seasonality and with trend equal to $m_k - m_{k-d}$. Then, by applying (a.2) to $\{\nabla_d X_k\}_{k\in\mathbb{Z}}$, we obtain a stationary time series.

Remarks 2.1.16. *(1) In the first approach, the parametric function fitting (a.1.i) is useful for the prediction problem, whereas the nonparametric moving average (a.1.ii) is less convenient in this context.*

(2) The variance of $\{X_k\}_{k\in\mathbb{Z}}$ can sometimes be made constant by using variance stabilizing transforms. Typical transforms are $\sqrt{x}$, $\log x$, etc. The variance stabilizing transform is simple and effective, at least as an initial procedure.

(3) Regarding the first approach (a.1.i), if the explanatory variables $u_{1k}, \ldots, u_{lk}$ of the trend were available, for $k = 1, \ldots, n$, then one may set

$$m_k = \beta_0 + \beta_1 u_{1k} + \cdots + \beta_l u_{lk},$$

and obtain the unknown coefficients from the least-squares estimator $\hat{\beta}$ of the regression model

$$X_k = \boldsymbol{u}_k^\top \boldsymbol{\beta} + Y_k,$$

Table 2.1: Methods constructing stationary time series under different assumptions.

Assumption Methodology	(a) Presence of trend and absence of seasonality	(b) Presence of trend and seasonality
(1) Function fitting or low pass filter	(i) function fitting	seasonality removal by moving average;
	(ii) moving average	goto (a.1.i) or (a.1.ii)
(2) Differentiation	p-composed difference operator ∇^p	seasonality removal by lag d difference operator ∇_d; go to (a.2)

where $\boldsymbol{u}_k = (1, u_{1k}, \ldots, u_{lk})^\top$ and $\boldsymbol{\beta} = (\beta_0, \ldots, \beta_l)^\top$. We recall the Gauss-Markov theorem. When $\mathsf{E}[Y_k] = 0$, $\hat{\boldsymbol{\beta}}$ is an unbiased estimator of $\boldsymbol{\beta}$. It is also most efficient, among all linear unbiased estimators, when $\{Y_k\}_{k=1,\ldots,n}$ has constant variance and uncorrelated elements. With correlated elements, $\hat{\boldsymbol{\beta}}$ can become rather inefficient. Null expectation and constant variance do indeed hold, however some correlations different than zero are possible.

(4) A stationary time series with long range dependence is one for which the a.c.v.f. decays very slowly, as it happens with a periodic time series with very long period. Although the model is indeed stationary, it lies somehow at the limit of stationarity. It is indeed very difficult to distinguish a constant trend is a time series from a periodic time series with very long period, based on a sample of consecutive values.

Table 2.1 summarizes the methodology for the construction of a stationary time series that is presented in this section.

We now illustrate with two real samples of time series how stationarity can be obtained by differentiation, precisely by following (2) of Table 2.1.

Example 2.1.17 (International airline passengers data). *In this example we present the classical airline sample of Box and Jenkins (1970), which consists of monthly counts of international airline passengers, from years 1949 to 1960, both included. The sample values are given in thousands of passengers and they are rounded to the thousandth. This sample is denoted by $\{X_k\}_{k=1,\ldots,n}$, with $n = 12 \cdot 12 = 144$. The graph of $\{X_k\}_{k=1,\ldots,144}$ is given*

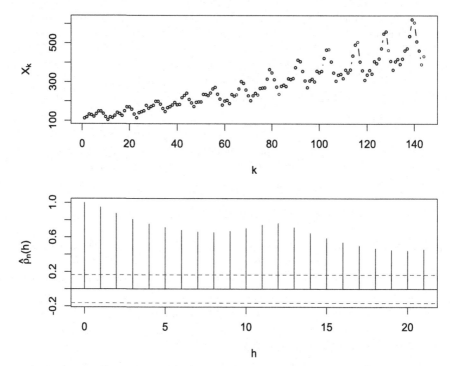

Fig. 2.1:   Airline passengers from 1949 to 1960 in thousands $\{X_k\}_{k=1,...,144}$, in upper graph, and its empirical a.c.r.f. $\hat{\rho}_n(h)$, for $h = 0, \ldots, 22$, in the lower graph.

in the upper graph of Figure 2.1 and the empirical a.c.r.f. is given in the lower graph. The empirical a.c.r.f. is given in Definition 2.1.30. The logarithmic transform of the sample is shown in the upper graph of Figure 2.2 and we see that it addresses appropriately the growth of the variance. The lower graph of Figure 2.2 shows the empirical a.c.r.f., which confirms the presence of a trend. The seasonality appears very clearly in the upper graph of Figure 2.2 and so we apply the lag $d = 12$ difference operator. Figure 2.3 shows $\{\nabla_{12} \log X_k\}_{k=13,...,144}$ after re-indexing in the upper graph and its empirical a.c.r.f. in the lower graph. The graph of the empirical a.c.r.f. shows that the time series $\{\nabla_{12} \log X_k\}_{k=13,...,144}$ does not have the yearly periodicity anymore. We also see that most values are positive. We deduce from $\nabla_{12} \log X_k = \log X_{k+12} - \log X_k > 0$ that $\log X_{k+12} > \log X_k$ and so

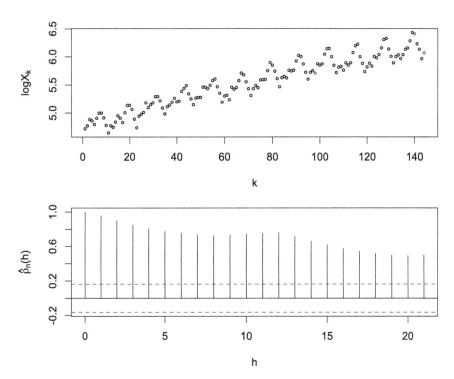

Fig. 2.2: Logarithmic airline passengers sample $\{\log X_k\}_{k=1,\ldots,144}$, in the upper graph, and its empirical a.c.r.f. $\hat{\rho}_n(h)$, for $h = 0, \ldots, 22$, in the lower graph.

that there was indeed an increasing trend before removing the seasonality. The trends remains after removing the seasonality. This is seen in graph of the empirical a.c.r.f., which is the lower graph of Figure 2.3. The empirical a.c.r.f. decays slowly until lag $h = 9$. Consequently we apply the differentiation operator to and thus consider $\{\nabla\nabla_{12} \log X_k\}_{k=14,\ldots,144}$, in order to remove the trend. The graph of $\{\nabla\nabla_{12} \log X_k\}_{k=14,\ldots,144}$ re-indexed is given in the upper graph of Figure 2.4. The time series $\{\nabla\nabla_{12} \log X_k\}_{k=14,\ldots,144}$ is spread evenly around zero and this confirms the removal of the trend. The empirical a.c.r.f. is shown in the lower graph of Figure 2.4.

The next example studies some novel and recent data.

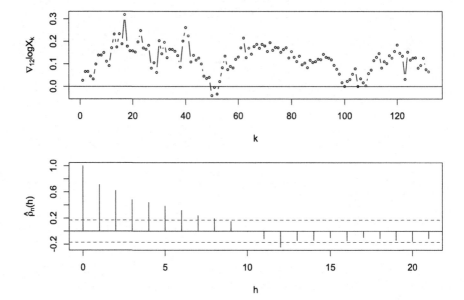

Fig. 2.3: Deseasonalized logarithmic airline passengers sample $\{\nabla_{12}$ $\log X_k\}_{k=13,\dots,144}$ re-indexed, in the upper graph, and its empirical a.c.r.f. $\hat{\rho}_n(h)$, for $h = 0, \dots, 22$, in the lower graph.

Example 2.1.18 (Swiss coronavirus data). *This example studies the sample of daily numbers of reported coronavirus (SARS-CoV-2) infections, in Switzerland and between March 29 and June 17 2021. This sample is provided by the Swiss Office of Public Health and it is reported in Table B.1. We denote it $\{X_k\}_{k=1,\dots,n}$, with $n = 81$. Its graph is given in the upper part of Figure 2.5 and the graph of the empirical a.c.r.f. is given in the lower part. This sample shows same features of the international airline passengers data: nonconstant variance, seasonality and trend. The main differences are that the seasonality has now order $d = 7$ and that the variability now decreases w.r.t. time. Consequently, we treat this sample in the same way: we stabilize the variance with the logarithmic transform, we remove the seasonality with the ∇_7 operator and we finally erase the trend with the ∇ operator. The final result of these transforms is shown in Figure 2.6.*

The following example considers a basic model that is perturbed by WN.

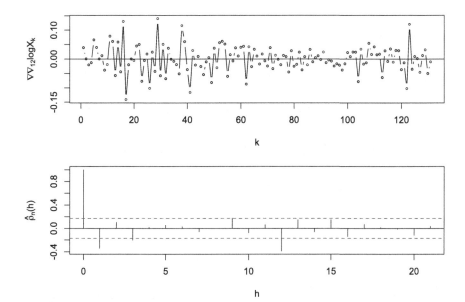

Fig. 2.4: Differentiated and deseasonalized logarithmic airline passengers sample $\{\nabla\nabla_{12}\log X_k\}_{k=14,\ldots,144}$ re-indexed, in the upper graph, and its empirical a.c.r.f. $\hat{\rho}_n(h)$, for $h = 0, \ldots, 22$, in the lower graph.

Example 2.1.19 (Random walk perturbed by WN). *Let* $\{T_k\}_{k\in\mathbb{Z}}$ *and* $\{Z_k\}_{k\in\mathbb{Z}}$ *be two independent WN time series. Let* $\sigma_T^2 = \mathsf{var}(T_1)$ *and* $\sigma_Z^2 = \mathsf{var}(Z_1)$. *We define the perturbed random walk by*

$$X_k = W_k + T_k, \quad \forall k \in \mathbb{Z},$$

where

$$W_k = W_{k-1} + Z_k, \quad \forall k \in \mathbb{Z},$$

is the standard or unperturbed random walk. In the practice the magnitude of the perturbations should be significantly smaller that the magnitude of the individual steps of the random walk, so that σ_T *should be considered substantially smaller than* σ_Z.

None of the random walk $\{W_k\}_{k\in\mathbb{Z}}$ *and perturbed random walk* $\{X_k\}_{k\in\mathbb{Z}}$ *are stationary. This claim follows from Proposition 2.2.3.2. However,* $\{\nabla X_k\}_{k\in\mathbb{Z}}$ *is a stationary time series and it possesses the a.c.v.f.*

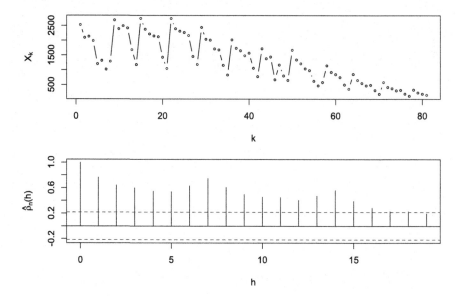

Fig. 2.5:   Daily numbers of reported coronavirus (SARS-CoV-2) infections in Switzerland between March 29 and June 17 2021 $\{X_k\}_{k=1,\ldots,81}$, in upper graph, and its empirical a.c.r.f. $\hat{\rho}_n(h)$, for $h = 0, \ldots, 22$, in lower graph.

given by

$$\gamma(h) = \begin{cases} \sigma_Z^2 + 2\sigma_T^2, & \text{if } h = 0, \\ -\sigma_T^2, & \text{if } h = \pm 1, \\ 0, & \text{if } h = \pm 2, \pm 3, \ldots. \end{cases}$$

2.1.3   *Properties of the autocovariance function*

The a.c.v.f. of a stationary time series possesses various properties that have practical importance for the analysis of the time series. This section presents most of these mathematical properties. For example, if γ denotes the a.c.v.f., the Cauchy-Schwarz inequality yields

$$|\gamma(h)| \leq \gamma(0), \quad \forall h \in \mathbb{Z},$$

telling that γ is a bounded function. It turns out that the a.c.v.f. is characterized as nonnegative definite (n.n.d.) function; cf. Definition 2.1.20 and Theorem 2.1.27. So the a.c.v.f. inherits all properties of n.n.d. functions and we thus study these properties.

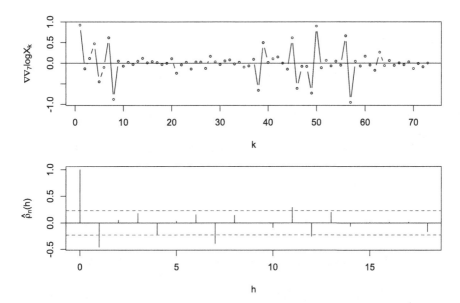

Fig. 2.6: Differentiated and deseasonalized logarithmic coronavirus infections sample $\{\nabla\nabla_7 \log X_k\}_{k=9,\dots,81}$ re-indexed, in the upper graph, and its empirical a.c.r.f. $\hat{\rho}_n(h)$, for $h = 0,\dots,22$, in the lower graph.

The class of n.n.d. functions has an important role in the context of Fourier analysis. For instance, a nonnegative Fourier series can always be obtained from a n.n.d. function; cf. Corollary 2.4.13. This class of functions is defined as follows.

Definition 2.1.20 (N.n.d. function). *The function* $f : \mathbb{R}^d \to \mathbb{C}$, *for given* $d \geq 1$, *is n.n.d. if*

$$\sum_{j=1}^{n} \sum_{k=1}^{n} c_j \overline{c_k} f(\boldsymbol{x}_j - \boldsymbol{x}_k) \geq 0,$$

$\forall \boldsymbol{x}_1,\dots,\boldsymbol{x}_n \in \mathbb{R}^d$, $c_1,\dots,c_n \in \mathbb{C}$ *and* $n \in \mathbb{N}^*$.

Note that, when $f(\boldsymbol{x}_j - \boldsymbol{x}_k)$ is replaced by $K(\boldsymbol{x}_j, \boldsymbol{x}_k)$ in Definition 2.1.20, then $K : \mathbb{R}^d \times \mathbb{R}^d \to \mathbb{C}$ is usually called n.n.d. kernel. N.n.d. kernels are important in statistical learning and in kriging.

Examples 2.1.21. *The functions $g(\boldsymbol{x}) = 1$ and $f(\boldsymbol{x}) = \mathrm{e}^{\mathrm{i}\langle \boldsymbol{a}, \boldsymbol{x} \rangle}$, $\forall \boldsymbol{x} \in \mathbb{R}^d$ and for some $\boldsymbol{a} \in \mathbb{R}^d$, are n.n.d. functions.*

Proposition 2.1.22 provides important properties of this class of functions.

Proposition 2.1.22 (Properties of n.n.d. functions). *Let $k \geq 1$ and $f, f_1, \ldots, f_k$ be n.n.d. functions. Then the following statements hold:*

(1) $\sum_{j=1}^{k} a_j f_j$ is n.n.d., $\forall a_1, \ldots, a_k \in \mathbb{R}_+$;
(2) $f(\boldsymbol{0}) \geq 0$;
(3) $f(-\boldsymbol{x}) = \overline{f(\boldsymbol{x})}$, $\forall \boldsymbol{x} \in \mathbb{R}^d$, i.e. f is an Hermitian function.
(4) $|f(\boldsymbol{x})| \leq f(\boldsymbol{0})$, $\forall \boldsymbol{x} \in \mathbb{R}^d$;
(5) $f(\boldsymbol{0}) = 0 \implies f(\boldsymbol{x}) = 0$, $\forall \boldsymbol{x} \in \mathbb{R}^d$.

Proof. 1. This first property can be trivially shown.

2. Let $n \in \mathbb{N}^*$, $\boldsymbol{x}_1 = \cdots = \boldsymbol{x}_n \in \mathbb{R}^d$ and $c_1, \ldots, c_n \in \mathbb{C}$. Then the result follows by

$$\sum_{j=1}^{n} \sum_{k=1}^{n} c_j \overline{c_k} f(\boldsymbol{0}) = \left| \sum_{j=1}^{n} c_j \right|^2 f(\boldsymbol{0}) \geq 0.$$

3. Let $n = 2$ and choose $\boldsymbol{x}_1 = \boldsymbol{0}$, $\boldsymbol{x}_2 \in \mathbb{R}^d$, $c_1 = 1$ and $c_2 \in \mathbb{C}$. Then

$$\sum_{j=1}^{2} \sum_{k=1}^{2} c_j \overline{c_k} f(\boldsymbol{x}_j - \boldsymbol{x}_k) = (1 + |c_2|^2) f(\boldsymbol{0}) + c_2 f(\boldsymbol{x}_2) + \overline{c_2} f(-\boldsymbol{x}_2) \geq 0.$$

By taking $c_2 = 1$, the previous inequality yields $\operatorname{Im} f(\boldsymbol{x}_2) = -\operatorname{Im} f(-\boldsymbol{x}_2)$. Similarly, by taking $c_2 = \mathrm{i}$, we obtain $\operatorname{Re} f(\boldsymbol{x}_2) = \operatorname{Re} f(-\boldsymbol{x}_2)$.
4. Let $n = 2$ and choose $\boldsymbol{x}_1 = \boldsymbol{0}$, $\boldsymbol{x}_2 \in \mathbb{R}^d$, $c_1 = |f(\boldsymbol{x}_2)|$ and $c_2 = -\overline{f(\boldsymbol{x}_2)}$. According to 3, we obtain

$$\sum_{j=1}^{2} \sum_{k=1}^{2} c_j \overline{c_k} f(\boldsymbol{x}_j - \boldsymbol{x}_k) = 2|f(\boldsymbol{x}_2)|^2 f(\boldsymbol{0}) - 2|f(\boldsymbol{x}_2)|^3 \geq 0.$$

If $|f(\boldsymbol{x}_2)| > 0$, then division by $|f(\boldsymbol{x}_2)|^2$ gives the desired result. If $|f(\boldsymbol{x}_2)| = 0$, then the result follows directly from 2.
5. This property follows directly from 4. $\qquad \square$

Example 2.1.23. *The function $f(x) = \cos x = (\mathrm{e}^{\mathrm{i}x} + \mathrm{e}^{-\mathrm{i}x})/2$ is n.n.d.; cf. Example 2.1.21 and Proposition 2.1.22.1.*

Theorem 2.1.24 provides an equivalent but simpler definition of a n.n.d. function, whenever the function is real-valued.

Theorem 2.1.24 (Real-valued n.n.d. function). *The function $f : \mathbb{R}^d \to \mathbb{R}$ is n.n.d. iff it is even and, $\forall n \in \mathbb{N}^*$, $\boldsymbol{x}_1, \dots, \boldsymbol{x}_n \in \mathbb{R}^d$ and $c_1, \dots, c_n \in \mathbb{R}$,*

$$\sum_{j=1}^n \sum_{k=1}^n c_j c_k f(\boldsymbol{x}_j - \boldsymbol{x}_k) \geq 0.$$

A direct consequence of Theorem 2.1.24 is the following.

Corollary 2.1.25 (N.n.d. function and matrix). *The function $f : \mathbb{Z} \to \mathbb{R}$ is n.n.d. iff, $\forall n \in \mathbb{N}^*$, the matrix $(f(j-k))_{j,k=1,\dots,n}$ is n.n.d.*

In order to show the important characterization of the a.c.v.f. in terms of n.n.d. function, which is given by Theorem 2.1.27 below, we need the theorem of Kolmogorov on the existence of a stochastic process. This is Theorem 3.1.7, which is however presented later in Section 3.1 with the theory of stochastic processes. Theorem 2.1.26 below is Theorem 3.1.7 simplified to stochastic processes with time domain $\mathbb{Z}$. It provides a practical sufficient condition that allow to justify the existence of a stochastic process with time domain $\mathbb{Z}$.

Theorem 2.1.26 (Kolmogorov existence of stochastic process with discrete time). *Let $\{F_{k_1,\dots,k_n}\}_{k_1 < \dots < k_n \in \mathbb{Z},\, n \in \mathbb{N}^*}$ be a set of d.f. satisfying the consistency condition*

$$\lim_{x_j \to \infty} F_{k_1,\dots,k_n}(x_1, \dots, x_n)$$

$$= F_{k_1,\dots,k_{j-1},k_{j+1},\dots,k_n}(x_1, \dots, x_{j-1}, x_{j+1}, \dots, x_n),$$

$\forall k_1 < \dots < k_n \in \mathbb{Z}$, $x_1, \dots, x_{j-1}, x_{j+1}, \dots, x_n \in \mathbb{R}$ and $j \leq n \in \mathbb{N}^*$. Then there exists a probability space $(\Omega, \mathcal{F}, P)$ and a stochastic process $\{X_k\}_{k \in \mathbb{Z}}$ defined over that space such that

$$\mathsf{P}[X_{k_1} \leq x_1, \dots, X_{k_n} \leq x_n] = F_{k_1,\dots,k_n}(x_1, \dots, x_n), \tag{2.1.8}$$

$\forall k_1 < \dots < k_n \in \mathbb{Z}$, $x_1, \dots, x_n \in \mathbb{R}$ and $n \in \mathbb{N}^*$.

Theorem 2.1.26 is a simplified version of Theorem 3.1.7 that holds for stochastic processes with continuous time as well. For a given stochastic process $\{X_k\}_{k \in \mathbb{Z}}$, the set of d.f. satisfying (2.1.8) is called the set of finite dimensional d.f. (f.d.d.f.) of the process.

Theorem 2.1.27 (Characterization of a.c.v.f.). *The function $\kappa : \mathbb{Z} \to \mathbb{R}$ is the a.c.v.f. of a (strictly) stationary time series $\Longleftrightarrow$ κ is n.n.d.*

Proof. ($\Leftarrow$) Let κ be the *a.c.v.f.* of a stationary time series. Let $k_1 < \cdots < k_n \in \mathbb{Z}$ and $n \in \mathbb{N}^*$. Define the random vector

$$\boldsymbol{Z} = \left(X_{k_1} - \mathsf{E}[X_{k_1}], \ldots, X_{k_n} - \mathsf{E}[X_{k_n}] \right)^{\top},$$

with values in $\mathbb{R}^n$. Then $\forall \boldsymbol{c} \in \mathbb{R}^{n \times 1}$,

$$0 \leq \mathrm{var}(\langle \boldsymbol{c}, \boldsymbol{Z} \rangle) = \boldsymbol{c}^{\top} \mathsf{E}[\boldsymbol{Z}\boldsymbol{Z}^{\top}] \boldsymbol{c} = \boldsymbol{c}^{\top} \left(\kappa(k_j - k_l) \right)_{j,l=1,\ldots n} \boldsymbol{c}.$$

The result that κ is n.n.d. follows by Corollary 2.1.25.

($\Rightarrow$) Let us now assume that κ is a n.n.d. function. Let $n \in \mathbb{N}^*$, $k_1 < \cdots < k_n \in \mathbb{Z}$ and define the vector $\boldsymbol{k} = (k_1, \ldots, k_n)$. Let $F_{\boldsymbol{k}}$ be a d.f. over $\mathbb{R}^n$ with characteristic function

$$\varphi_{\boldsymbol{k}}(\boldsymbol{v}) = \mathrm{e}^{-\frac{1}{2}\boldsymbol{v}^{\top}\boldsymbol{\Sigma}\boldsymbol{v}}, \quad \forall \boldsymbol{v} \in \mathbb{R}^{n \times 1},$$

where $\boldsymbol{\Sigma} = (\kappa(k_j - k_l))_{j,l=1,\ldots,n} \in \mathbb{R}^{n \times n}$ is a n.n.d. matrix. In this situation $F_{\boldsymbol{k}}$ is a singular normal d.f.

Let $j \in \{1, \ldots, n\}$ and $\boldsymbol{u} \in \mathbb{R}^n$, then we define

$$\boldsymbol{u}_{(j)} = (u_1, \ldots, u_{j-1}, u_{j+1}, \ldots, u_n) \in \mathbb{R}^{n-1},$$

namely the vector $\boldsymbol{u}$ after removal of its j-th element. We define $\varphi_{\boldsymbol{k}(j)}$ and $F_{\boldsymbol{k}(j)}$ respectively as $\varphi_{\boldsymbol{k}}$ and $F_{\boldsymbol{k}}$, with $\boldsymbol{k}$ replaced by $\boldsymbol{k}(j)$. It can be easily verified that, $\forall \boldsymbol{v} \in \mathbb{R}^n$ and $k_1 < \cdots < k_n \in \mathbb{Z}$,

$$\varphi_{\boldsymbol{k}(j)}(\boldsymbol{v}_{(j)}) = \lim_{v_j \to 0} \varphi_{\boldsymbol{k}}(\boldsymbol{v}).$$

This is equivalent to

$$F_{k(j)}(x_{(j)}) = \lim_{x_j \to \infty} F_k(x),$$

$\forall x \in \mathbb{R}^n$ and $k_1 < \cdots < k_n \in \mathbb{Z}$. Thus the consistency condition of Kolmogorov's existence theorem 2.1.26 is fulfilled. Therefore, there exists a stochastic process $\{X_k\}_{k \in \mathbb{Z}}$ with f.d.d.f. given by F_k, $\forall k = (k_1, \ldots, k_n)$, with $k_1 < \cdots < k_n \in \mathbb{Z}$, and $\forall n \in \mathbb{N}^*$. In particular,

$$\operatorname{cov}(X_{k_j} X_{k_l}) = \kappa(k_j - k_l), \quad \text{for } j, l = 1, \ldots, n.$$

This means that the stochastic process $\{X_k\}_{k \in \mathbb{Z}}$ is (strictly) stationary. $\square$

An extension of Theorem 2.1.27 to continuous time and to complex-valued stochastic processes is presented in Theorem 3.1.21. An heuristic application of Theorem 2.1.27 would be for checking whether a function is n.n.d. Example 2.1.4 illustrates this idea. For fixed $\theta \in [0, \pi]$, the function $\cos h\theta$, $\forall h \in \mathbb{Z}$, is n.n.d. because it is the a.c.v.f. of a stationary time series (which takes the form of the sum of two harmonics). Proposition 2.1.28 studies a basic a.c.v.f.

Proposition 2.1.28. *Let* $\gamma : \mathbb{Z} \to \mathbb{R}$ *be given by*

$$\gamma(h) = \begin{cases} 1, & \text{if } h = 0, \\ \rho, & \text{if } |h| = 1, \\ 0, & \text{if } |h| \geq 2. \end{cases}$$

Then γ *is the a.c.v.f. of a stationary time series iff* $|\rho| \leq 1/2$.

Proof. Let $|\rho| \leq 1/2$. By considering Example 2.1.8 with $\sigma^2 = 1/(1 + \theta^2)$ it follows that $\rho = \theta/(1 + \theta^2)$, i.e. $\theta = (2\rho)^{-1}(1 \pm \sqrt{1 - 4\rho^2})$, which is a real number iff $|\rho| \leq 1/2$.

Let $\rho > 1/2$. Let $n \in \mathbb{N}^*$ and define $\mathbf{\Gamma} = \big(\gamma(i - j)\big)_{i,j=1,\ldots,n} \in \mathbb{R}^{n \times n}$ and, for n even, $a_n = (1, -1, 1, -1, \ldots, 1, -1)^\top \in \mathbb{R}^{n \times 1}$. Then we find

$$a_n^\top \mathbf{\Gamma} a_n = (1, -1, 1, \ldots, 1, -1) \begin{pmatrix} 1 & \rho & & & & & \\ \rho & 1 & \rho & & & \mathbf{0} & \\ & \rho & 1 & & & & \\ & & & \ddots & \ddots & \ddots & \\ & \mathbf{0} & & & & 1 & \rho \\ & & & & & \rho & 1 \end{pmatrix} \begin{pmatrix} 1 \\ -1 \\ 1 \\ \vdots \\ 1 \\ -1 \end{pmatrix}$$

$$= 2\rho + n(1 - 2\rho),$$

which is negative iff $n > 2\rho/(2\rho - 1)$. Thus, Theorem 2.1.27 tells that there cannot exist a stationary time series.

Let $\rho < -1/2$. Similar arguments yield that γ is not n.n.d.   □

Proposition 2.1.29 tells that the matrix obtained from the first values of the a.c.v.f. or, equivalently, the covariance matrix of any vector of consecutive values of a stationary time series, is nonsingular under a rather weak condition.

Proposition 2.1.29 (Nonsingularity of the a.c.v.f. matrix). *Let γ be the a.c.v.f. of a nondegenerate and stationary time series (thus with $\gamma(0) > 0$). If $\gamma(h) \overset{h\to\infty}{\longrightarrow} 0$, then the matrix of the a.c.v.f. $\boldsymbol{\Gamma}_n = (\gamma(i-j))_{i,j=1,\ldots,n}$ is nonsingular, for $n = 1, 2, \ldots$.*

It could seem weird that an asymptotic property of a function imposes strong conditions at all finite arguments. But stationarity makes the a.c.v.f. a very particular function.

Proof. Let $\{X_k\}_{k\in\mathbb{Z}}$ be any nondegenerate and stationary time series with mean null and a.c.v.f. γ. We proceed by contradiction and assume the existence of $n \geq 2$ such that $\boldsymbol{\Gamma}_n$ singular. This is equivalent to rank $\boldsymbol{\Gamma}_n < n$. We know that $\boldsymbol{\Gamma}_n$ is n.n.d. and thus incomplete rank means that $\boldsymbol{g}^\top \boldsymbol{\Gamma}_n \boldsymbol{g} = 0$, for some vector $\boldsymbol{g} = (g_1, \ldots, g_n)^\top \neq \boldsymbol{0}$ in $\mathbb{R}^n$. Thus for that $\boldsymbol{g}$ and for $\boldsymbol{X} = (X_1, \ldots, X_n)^\top$, we have $\|\boldsymbol{g}^\top \boldsymbol{X}\|^2 = \mathsf{var}(\boldsymbol{g}^\top \boldsymbol{X}) = 0$. In other terms, $\boldsymbol{g}^\top \boldsymbol{X} = 0$ a.s.

One deduces that $\exists m \in \mathbb{N}^*$ such that

$$X_{m+1} = \sum_{j=1}^{m} g_j X_j, \text{ a.s.,} \tag{2.1.9}$$

for some $g_1, \ldots, g_m \in \mathbb{R}$, and such that $\boldsymbol{\Gamma}_m$ is nonsingular. Stationarity of $\{X_k\}_{k\in\mathbb{Z}}$ implies that (2.1.9) can be generalized to

$$X_{m+h} = \sum_{j=1}^{m} g_j X_{j+h-1}, \text{ a.s., } \text{ for } h = 1, 2, \ldots.$$

By repeated use of this identity one finds

$$X_n = \sum_{j=1}^{m} g_{nj} X_j, \text{ a.s., } \text{ for } n = m+1, m+2, \ldots,$$

for some $\boldsymbol{g}_n = (g_{n1}, \ldots, g_{nm})^\top \in \mathbb{R}^m$.

Consequently,

$$\gamma(0) = \boldsymbol{g}_n^\top \boldsymbol{\Gamma}_n \boldsymbol{g}_n = \boldsymbol{g}_n^\top \boldsymbol{P}\boldsymbol{\Lambda}\boldsymbol{P}^\top \boldsymbol{g}_n,$$

where

$$\boldsymbol{\Lambda} = \begin{pmatrix} \lambda_1 & & \boldsymbol{0} \\ & \ddots & \\ \boldsymbol{0} & & \lambda_m \end{pmatrix},$$

for some $0 < \lambda_1 \leq \ldots \leq \lambda_m$, the matrix $\boldsymbol{\Gamma}_m$ being nonsingular, and for an orthogonal matrix $\boldsymbol{P} \in \mathbb{R}^{m \times m}$. It follows from this result that

$$\gamma(0) \geq \lambda_1 \boldsymbol{g}_n^\top \boldsymbol{P}\boldsymbol{P}^\top \boldsymbol{g}_n = \lambda_1 \|\boldsymbol{g}_n\|^2. \tag{2.1.10}$$

So we find

$$0 < \gamma(0)$$

$$= \mathsf{cov}\left(X_n, \sum_{j=1}^m g_{nj} X_j \right)$$

$$= \sum_{j=1}^m g_{nj} \gamma(n-j)$$

$$\leq \sum_{j=1}^m \underbrace{|g_{nj}|}_{\substack{<\infty, \\ \text{from (2.1.10)}}} \underbrace{|\gamma(n-j)|}_{\overset{n\to\infty}{\longrightarrow}0}$$

$$\overset{n\to\infty}{\longrightarrow} 0.$$

We have thus obtained the desired contradiction. $\qquad\square$

The last main subject of this section concerns the empirical or the sample version of the a.c.v.f. and of the a.c.r.f.

Definition 2.1.30 (Empirical a.c.v.f. and a.c.r.f.). *Consider the sample of n consecutive random variables $X_1, \ldots, X_n$ of the time series $\{X_k\}_{k\in\mathbb{Z}}$. The empirical or sample a.c.v.f. is given by*

$$\hat{\gamma}_n(h) = \begin{cases} \frac{1}{n} \sum_{j=1}^{n-h} (X_{j+h} - M_n)(X_j - M_n), & \text{if } 0 \leq h < n, \\ \frac{1}{n} \sum_{j=1}^{n+h} (X_{j-h} - M_n)(X_j - M_n), & \text{if } -n < h < 0, \end{cases}$$

where $M_n = n^{-1} \sum_{j=1}^n X_j$ is the sample mean.
The empirical or sample a.c.r.f. is given by

$$\hat{\rho}_n(h) = \frac{\hat{\gamma}_n(h)}{\hat{\gamma}_n(0)}, \quad \text{if } |h| < n,$$

provided $\hat{\gamma}_n(0) \neq 0$.

Just like the a.c.v.f γ, the empirical a.c.v.f. $\hat{\gamma}_n$ is n.n.d.

Proposition 2.1.31 (N.n.d. empirical a.c.v.f. and a.c.r.f.). *The empirical a.c.v.f. $\hat{\gamma}_n$ and a.c.r.f. $\hat{\rho}_n$ based on the sample $X_1, \ldots, X_n$ of the time series $\{X_k\}_{k \in \mathbb{Z}}$ based on the sample $X_1, \ldots, X_n$ are n.n.d. functions.*

Proof. According to Corollary 2.1.25, we show that the matrix of empirical a.c.v.f.

$$\hat{\boldsymbol{\Gamma}}_n = \big(\hat{\gamma}_n(|j-k|)\big)_{j,k=1,\ldots,n}$$

is n.n.d.

Consider the centered sample $Y_j = X_j - M_n$, for $j = 1, \ldots, n$, and define the matrix $\boldsymbol{A} \in \mathbb{R}^{n \times 2n}$ by

$$\boldsymbol{A} = \begin{pmatrix} 0 & 0 & 0 & \ldots & 0 & 0 & Y_1 & \ldots & Y_{n-1} & Y_n \\ 0 & 0 & 0 & \ldots & 0 & Y_1 & Y_2 & \ldots & Y_n & 0 \\ \vdots & \ddots & & & & & & & \ddots & \vdots \\ 0 & Y_1 & Y_2 & \ldots & Y_{n-1} & Y_n & 0 & \ldots & 0 & 0 \end{pmatrix}.$$

We find that $\boldsymbol{A}\boldsymbol{A}^\top = n\hat{\boldsymbol{\Gamma}}_n$ and that

$$\boldsymbol{a}^\top \hat{\boldsymbol{\Gamma}}_n \boldsymbol{a} = \frac{1}{n} \boldsymbol{a}^\top \boldsymbol{A}\boldsymbol{A}^\top \boldsymbol{a} = \frac{1}{n} \big(\boldsymbol{A}^\top \boldsymbol{a}\big)^\top \boldsymbol{A}^\top \boldsymbol{a} \geq 0, \quad \forall \boldsymbol{a} \in \mathbb{R}^n,$$

which means that the matrix $\hat{\boldsymbol{\Gamma}}_n$ is n.n.d.

Provided $\hat{\gamma}_n(0) \neq 0$, which happens a.s., $\hat{\boldsymbol{\Gamma}}_n / \hat{\gamma}_n(0)$ is also n.n.d. □

In absence of information beyond the sample $X_1, \ldots, X_n$ of the stationary time series $\{X_k\}_{k \in \mathbb{Z}}$, it is not possible to estimate its a.c.v.f. $\gamma(h)$ and its a.c.r.f. $\rho(h)$ for time lags $h \geq n$. Moreover, for k slightly smaller than n, we should not expect accuracy from the empirical a.c.v.f. or a.c.r.f., because only few pairs (X_{j+h}, X_j) would be available in this situation (and only one when $h = n - 1$).

Empirical a.c.v.f. and a.c.r.f. are important for the statistical analysis of a time series. They can be applied to a time series that may not be stationary. A trend can be detected by slow decay of these two functions. A seasonality is identified by repeated peaks at a given periodicity.

2.2   ARMA time series

This section presents the most important model for stationary time series, which is the ARMA model. The definition and the motivations are given in Section 2.2.1. The general ARMA time series is only implicitly defined through a difference equation and Section 2.2.2 provides an explicit representation or solution in terms of a random series. This solution or random can be simplified in two different ways, depending on whether the series is causal or invertible. These two notions are defined in Section 2.2.3.

2.2.1 *Definitions and motivations*

This section presents the most important model for stationary time series: the autoregressive and moving average (ARMA) model. The ARMA model provides a simple and flexible representation of a stationary time series. It is composed by two polynomials: one for the autoregressive (AR) part and the one for the moving average (MA) part. The MA part is a model for the error term, precisely for the white noise (WN), in terms of a linear combination of present and past values of errors. The AR part is the linear regression of the present value of the time series on its past values. ARMA models allow for prediction of future values of the time series. The origin of ARMA models or time series can be traced back to Whittle (1951), whereas the first important general reference is Box and Jenkins (1970). An important reference, which is followed in in this section, is Brockwell and Davis (1991).

We begin by reminding some important notation for this section. We denote by $\langle \cdot, \cdot \rangle$ the scalar product over $\mathcal{L}_2(\Omega, \mathcal{F}, \mathsf{P})$ or shortly $\mathcal{L}_2$. Precisely, for any two random variables $X, Y \in \mathcal{L}_2$,

$$\langle X, Y \rangle = \mathsf{E}[XY].$$

The induced norm is denoted $\|X\| = \sqrt{\langle X, X \rangle}$ the induced norm. For a given sequence of complex number $\{\xi_j\}_{j \in \mathbb{Z}}$, we denote by $\xi(z)$ the series

$$\xi(z) = \sum_{j=-\infty}^{\infty} \xi_j z^j,$$

$\forall z \in \mathbb{C}$ where the series is convergent. $\forall \varepsilon \geq 0$ and $r > 1$, we denote the centered disk and annulus over the complex plane respectively by

$$\mathcal{D}_\varepsilon = \{z \in \mathbb{C} \,|\, |z| \leq 1 + \varepsilon\}, \tag{2.2.1}$$

and

$$\mathcal{A}_r = \left\{z \in \mathbb{C} \,\middle|\, \frac{1}{r} \leq |z| \leq r\right\}. \tag{2.2.2}$$

The central definition of this chapter is the following.

Definition 2.2.1 (ARMA(p,q) time series). *Let $\{Z_k\}_{k\in\mathbb{Z}}$ be WN(σ^2). Given $p, q \in \mathbb{N}^*$, the stationary time series $\{X_k\}_{k\in\mathbb{Z}}$ is called:*

- *moving average time series of order q, denoted* MA(q), *when*

$$X_k = Z_k + \theta_1 Z_{k-1} + \cdots + \theta_q Z_{k-q}, \quad \forall k \in \mathbb{Z}, \qquad (2.2.3)$$

  for some $\theta_1, \ldots, \theta_q \in \mathbb{R}$, with $\theta_q \neq 0$;
- *autoregressive time series of order p, denoted* AR(p), *when*

$$X_k = \varphi_1 X_{k-1} + \cdots + \varphi_p X_{k-p} + Z_k, \quad \forall k \in \mathbb{Z}, \qquad (2.2.4)$$

  for some $\varphi_1, \ldots, \varphi_p \in \mathbb{R}$, with $\varphi_p \neq 0$;
- *autoregressive and moving average of orders p and q, denoted* ARMA(p,q), *when*

$$X_k - \varphi_1 X_{k-1} - \cdots - \varphi_p X_{k-p} = Z_k + \theta_1 Z_{k-1} + \cdots + \theta_q Z_{k-q}, \ \forall k \in \mathbb{Z}, \qquad (2.2.5)$$

  for some $\varphi_1, \ldots, \varphi_p, \theta_1, \ldots, \theta_q \in \mathbb{R}$, with $\varphi_p, \theta_q \neq 0$.

The MA(q) *time series with $q = 0$ and the* AR(p) *time series with $p = 0$ are given by $X_k = Z_k$, $\forall k \in \mathbb{Z}$. We use the following notation:* ARMA($p, 0$) $=$ AR(p) *and* ARMA($0, q$) $=$ MA(q).

We note that, as written in Definition 2.2.1, the ARMA(p, q) time series is stationary by assumption. Although this convention is not universal, it is convenient to impose stationarity because only stationary time series that satisfy (2.2.5) have practical interest.

Let φ and θ be the polynomials

$$\varphi(z) = \varphi_0 - \varphi_1 z - \cdots - \varphi_p z^p \text{ and } \theta(z) = \theta_0 + \theta_1 z + \cdots + \theta_q z^q, \ \forall z \in \mathbb{C},$$

where $\theta_0 = \varphi_0 = 1$ and $\varphi_1, \ldots, \varphi_p, \theta_1, \ldots, \theta_q \in \mathbb{R}$. Then, the ARMA($p, q$) equation can be expressed in terms of these two polynomials applied to the backward shift operator B, i.e.

$$\varphi(B) = \varphi_0 - \varphi_1 B - \cdots - \varphi_p B^p \text{ and } \theta(B) = \theta_0 + \theta_1 B + \cdots + \theta_q B^q.$$

Thus we obtain the compact representation of (2.2.5) given by

$$\varphi(B)(X_k) = \theta(B)(Z_k), \quad \forall k \in \mathbb{Z}. \qquad (2.2.6)$$

As a matter of fact, the MA equation, i.e. (2.2.3), always yields a stationary time series and its a.c.v.f. is given by the following proposition.

Proposition 2.2.2 (A.c.v.f. of MA(q)). *Let* $q \in \mathbb{N}^*$. *There exists an unique solution to the* MA(q) *equation (2.2.3), it is stationary, with mean zero and with a.c.v.f. given by*

$$\gamma(h) = \begin{cases} \sigma^2 \sum_{j=0}^{q-|h|} \theta_j \theta_{|h|+j}, & \text{if } |h| \leq q, \\ 0, & \text{if } |h| \geq q+1. \end{cases}$$

The justification of this proposition is rather direct.

We conclude this introductory section with some remarks of general order. Because of their simplicity and high modelling flexibility, ARMA models provide the most practical models for stationary time series. In particular, they provide a formula for the covariance structure of the time series MA time series are applied when various random phenomena of current and near past periods sum up to a present value of some other phenomenon of interest. For example, an economic indicator depends on inflation, interest rates, etc. AR time series are applied when the present value of a phenomenon of interest is the sum its past values plus another random phenomenon. For example, the velocity of wind in a given direction depends on past values of the velocity plus a random term. An ARMA model typically requires fewer parameters than a MA or an AR would require individually. In this sense ARMA models provide an important level of flexibility while remaining simple, or at least simpler than individual MA or AR models. This situation can be compared with the classical problem of numerical analysis, where for a given amount of computational burden, a rational approximation (of a function over some finite interval) has typically smaller error than a polynomial approximation.

2.2.2 Explicit representations of ARMA time series

In this section we study the stationary solutions to the ARMA equations or, in other terms, the explicit representations of ARMA time series, which are stationary by definition. Before starting any formal analysis, we should mention that an AR(p) time series is mathematically more sophisticated than linear regression on p independent explanatory variables: we now have a difference equation whose stationary solution may or may not exist. Let $k \in \mathbb{Z}$, then the AR equation (2.2.4) tells that

$$\mathsf{E}[X_k | X_{k-1}, \ldots, X_{k-p}] = \varphi_1 X_{k-1} + \cdots + \varphi_p X_{k-p},$$

i.e. that the conditional mean of X_k is a linear function of its p past values. There is a priori no reason for which a solution $\{X_k\}_{k \in \mathbb{Z}}$ does exist.

This solution should not be confused with a particular nonstationary solution $\{X_k\}_{k\in\mathbb{Z}}$ that is obtained when arbitrary initial values are given. For example, when $p = 1$ a particular solution can be obtained by taking for X_0 any specified random variable of $\mathcal{L}_2$ that is uncorrelated with $\{Z_k\}_{k\in\mathbb{Z}}$. But that particular solution may not be stationary.

For simplicity we begin with the analysis of the AR(1).

Proposition 2.2.3 (Stationary solution of AR(1) equation and a.c.v.f.). *Let $\varphi_1 \in \mathbb{R}$ and consider the AR(1) equation*

$$X_k = \varphi_1 X_{k-1} + Z_k, \quad \forall k \in \mathbb{Z}, \qquad (2.2.7)$$

where $\{Z_k\}_{k\in\mathbb{Z}}$ is WN(σ^2), *for some $\sigma > 0$.*
1. If $|\varphi_1| \neq 1$, then (2.2.7) admits an unique stationary solution. This solution is given by

$$X_k = \begin{cases} \sum_{j=0}^{\infty} \varphi_1^j Z_{k-j}, & \text{if } |\varphi_1| < 1, \\ -\sum_{j=0}^{\infty} \varphi_1^{-j-1} Z_{k+j+1} = -\sum_{j=-1}^{-\infty} \varphi_1^j Z_{k-j}, & \text{if } |\varphi_1| > 1, \end{cases} \quad \forall k \in \mathbb{Z},$$

where the two series above converge in $\mathcal{L}_2$ and a.s.

When $|\varphi_1| < 1$, the above stationary solution has mean zero and a.c.v.f. given by

$$\gamma(h) = \sigma^2 \frac{\varphi_1^{|h|}}{1 - \varphi_1^2}, \quad \forall h \in \mathbb{Z}. \qquad (2.2.8)$$

2. When $|\varphi_1| = 1$, the AR(1) equation (2.2.7) does not admit stationary solutions.

Proof. 1. Let $\varphi_1 \in \mathbb{R}$ such that $|\varphi_1| < 1$, let $k \in \mathbb{Z}$ and $l \in \mathbb{N}$.

$$\begin{aligned} X_k &= \varphi_1 X_{k-1} + Z_k \\ &= \varphi_1(\varphi_1 X_{k-2} + Z_{k-1}) + Z_k \\ &= \varphi_1^2 X_{k-2} + \varphi_1 Z_{k-1} + Z_k \\ &= \cdots \\ &= \varphi_1^{l+1} X_{k-l-1} + \varphi_1^l Z_{k-l} + \cdots + \varphi_1 Z_{k-1} + Z_k. \end{aligned} \qquad (2.2.9)$$

It follows from the assumed stationarity of $\{X_k\}_{k\in\mathbb{Z}}$ that $\|X_k\|^2$ does not depend on k, so that

$$\left\| X_k - \sum_{j=0}^{l} \varphi_1^j Z_{k-j} \right\|^2 = \varphi_1^{2(l+1)} \|X_{k-l-1}\|^2 \xrightarrow{l\to\infty} 0. \qquad (2.2.10)$$

Because $\sum_{j=0}^{l} \varphi_1^j Z_{k-j}$, $\forall l \in \mathbb{N}$, is a Cauchy sequence of the complete space $\mathcal{L}_2$, it converges towards an unique element of $\mathcal{L}_2$, which must be X_k, from (2.2.10).

Furthermore, monotone convergence yields

$$\mathsf{E}\left[\sum_{j=0}^{\infty} |\varphi_1^j Z_{k-j}|\right] = \sum_{j=0}^{\infty} \mathsf{E}[|\varphi_1^j Z_{k-j}|] \leq \mathsf{E}[|Z_{k-j}|] \sum_{j=0}^{\infty} |\varphi_1|^j < \infty.$$

Consequently, $\sum_{j=0}^{\infty} |\varphi_1^j Z_{k-j}| < \infty$ a.s. and

$$X_k = \sum_{j=0}^{\infty} \varphi_1^j Z_{k-j} \quad \text{a.s.} \tag{2.2.11}$$

The equality between $\mathcal{L}_2$ and a.s. limits is given by Lemma A.3.11. Also, it can be easily verified that $\sum_{j=0}^{\infty} \varphi_1^j Z_{k-j}$ solves the AR(1) equation (2.2.7).

Now we show that such solution is indeed stationary. It follows from Fubini-Tonelli's theorem A.5.6 that

$$\mathsf{E}[X_k] = \mathsf{E}\left[\sum_{j=0}^{\infty} \varphi_1^j Z_{k-j}\right] = \sum_{j=0}^{\infty} \varphi_1^j \mathsf{E}[Z_{k-j}] = 0.$$

The continuity of the scalar product in $\mathcal{L}_2$ (cf. Lemma A.1.5) yields, $\forall h \in \mathbb{N}$,

$$\begin{aligned}
\mathsf{cov}(X_k, X_{k+h}) \\
&= \mathsf{E}[X_k X_{k+h}] \\
&= \mathsf{E}\left[\sum_{j=0}^{\infty} \varphi_1^j Z_{k-j} \sum_{i=0}^{\infty} \varphi_1^i Z_{k+h-i}\right] \\
&= \sum_{j=0}^{\infty} \varphi_1^j \varphi_1^{h+j} \mathsf{E}[Z_{k-j}^2] \quad (-j = h - i \Leftrightarrow i = h + j; h \geq 0 \Rightarrow i \geq 0) \\
&= \varphi_1^h \sum_{j=0}^{\infty} \varphi_1^{2j} \sigma^2 \\
&= \sigma^2 \frac{\varphi_1^h}{1 - \varphi_1^2},
\end{aligned}$$

which does not depend on k. The independence of k of this result leads to (2.2.8).

Regarding the unicity of the solution, let us assume the existence of another solution $\{Y_k\}_{k \in \mathbb{Z}}$. Let $k \in \mathbb{Z}$. Then (2.2.9) must hold with X_k

replaced by Y_k and to that equation we subtract (2.2.11). This leads to

$$Y_k - X_k = \varphi_1^{l+1} Y_{k-l-1} - \sum_{j=l+1}^{\infty} \varphi_1^j Z_{k-j}, \quad \forall l \in \mathbb{N}.$$

Denote $U_l = \varphi_1^{l+1} Y_{k-l-1}$ and $V_l = \sum_{j=l+1}^{\infty} \varphi_1^j Z_{k-j}$, $\forall l \in \mathbb{N}$. Assume

$$\sum_{l=0}^{\infty} \mathsf{P}[|U_l - V_l| > \varepsilon] < \infty, \quad \forall \varepsilon > 0. \qquad (2.2.12)$$

In this situation Borel-Cantelli's Lemma A.3.1 yields:

$$\mathsf{P}[\{|U_l - V_l| > \varepsilon\}, \ l \geq 0, \ \text{i.o.}] = 0, \ \forall \varepsilon > 0$$
$$\implies \mathsf{P}[\{|U_l - V_l| > \varepsilon\} \ \text{for at most finitely many } l \geq 0] = 1, \ \forall \varepsilon > 0$$
$$\implies \mathsf{P}[U_l = V_l, \ \forall l \in \mathbb{N}] = 1$$
$$\implies \mathsf{P}[X_k = Y_k] = 1.$$

Let us now show (2.2.12). Let $k \in \mathbb{Z}, l \in \mathbb{N}$ and $\varepsilon > 0$. We know that $\mathsf{P}[|U_l - V_l| > \varepsilon] \leq \mathsf{P}[|U_l| > \varepsilon/2] + \mathsf{P}[|V_l| > \varepsilon/2]$. With Markov's inequality we obtain that for some $c > 0$,

$$\mathsf{P}\left[|U_l| > \frac{\varepsilon}{2}\right] = \mathsf{P}\left[|\varphi_1^{l+1} Y_{k-l+1}| > \frac{\varepsilon}{2}\right] \leq \frac{c}{\varepsilon} |\varphi_1|^{l+1},$$

and

$$\mathsf{P}\left[|V_l| > \frac{\varepsilon}{2}\right] = \mathsf{P}\left[\left|\varphi_1^{l+1} \sum_{j=0}^{\infty} \varphi_1^j Z_{k-j-l-1}\right| > \frac{\varepsilon}{2}\right] \leq \frac{c}{\varepsilon} |\varphi_1|^{l+1}.$$

Thus

$$\sum_{l=0}^{\infty} \mathsf{P}[|U_l - V_l| > \varepsilon] \leq 2\frac{c}{\varepsilon} \sum_{l=0}^{\infty} |\varphi_1|^{l+1} < \infty.$$

Let us now consider the case $|\varphi_1| > 1$. Let $k \in \mathbb{Z}$ and $l \in \mathbb{N}$, then

$$X_k = -\varphi_1^{-1} Z_{k+1} + \varphi_1^{-1} X_{k+1}$$
$$= -\varphi_1^{-1} Z_{k+1} + \varphi_1^{-1}(-\varphi_1^{-1} Z_{k+2} + \varphi_1^{-1} X_{k+2})$$
$$= \cdots$$
$$= -\varphi_1^{-1} Z_{k+1} - \varphi_1^{-2} Z_{k+2} - \cdots - \varphi_1^{-l-1} Z_{k+l+1} + \varphi_1^{-l-1} X_{k+l+1}.$$

It follows from the assumed stationarity of $\{X_k\}_{k \in \mathbb{Z}}$ that $\|X_k\|^2$ does not depend on k, so that

$$\left\|X_k + \sum_{j=1}^{l+1} \varphi_1^{-j} Z_{k+j}\right\|^2 = \varphi_1^{-2(l+1)} \|X_{k+l+1}\|^2 \xrightarrow{l \to \infty} 0.$$

Therefore we obtain that

$$X_k = -\sum_{j=0}^{\infty} \varphi_1^{-j-1} Z_{k+j+1},$$

is the unique stationary solution in $\mathcal{L}_2$. By applying analogue arguments to the ones of the previous paragraphs, we can also justify a.s. convergence and stationarity of this solution.

2. Let $n \in \{2, 4, \ldots\}$. When $|\varphi_1| = 1$, (2.2.9) gives

$$X_k - X_{k-n} = \sum_{j=0}^{n-1} \varphi_1^j Z_{k-j}. \tag{2.2.13}$$

The variance of the right side of (2.2.13) is $n\sigma^2$. The variance of the left side of (2.2.13) is

$$\mathsf{var}(X_k - X_{k-n}) = \mathsf{var}(X_k) - 2\mathsf{cov}(X_k, X_{k-n}) + \mathsf{var}(X_{k-n}) = 2\{\gamma(0) - \gamma(n)\}.$$

Hence

$$2\{\gamma(0) - \gamma(n)\} = n\sigma^2. \tag{2.2.14}$$

Recall that $|\gamma(n)| \leq \gamma(0)$, so that the left side of (2.2.14) is bounded. However the right side diverges as $n \to \infty$. This is a contradiction, therefore $\{X_k\}_{k \in \mathbb{Z}}$ cannot be stationary. $\qquad\square$

Remark 2.2.4. *When $|\varphi_1| \neq 1$, the AR(1) time series has an unique stationary solution that depends only on the WN. Precisely, let $k \in \mathbb{Z}$. Then we have that:*

- *if $|\varphi_1| < 1$, then the time series depends on present and past values of the WN, i.e., X_k is a function of $\{Z_k, Z_{k-1}, Z_{k-2}, \ldots\}$;*
- *if $|\varphi_1| > 1$, then the time series depends on future values of the WN, i.e., X_k is a function of $\{Z_{k+1}, Z_{k+2}, \ldots\}$.*

ARMA time series depending on present and past values of the WN are called causal. We refer to Definition 2.2.10. In the practice, causal time series are more relevant than non-causal ones.

If $\{\psi_j\}_{j \in \mathbb{Z}}$ is a sequence of complex numbers, then we call $\psi(B) = \sum_{j=-\infty}^{\infty} \psi_j B^j$ the MA operator. Proposition 2.2.5 and 2.2.6 provide two general results on the convergence of $\psi(B)X_k$, where $k \in \mathbb{Z}$ and $\{X_k\}_{k \in \mathbb{Z}}$ is a time series.

Proposition 2.2.5 (Convergence of MA transform of time series). *Let $\{X_k\}_{k \in \mathbb{Z}}$ be a time series and let $\{\psi_j\}_{j \in \mathbb{Z}}$ be a sequence of real numbers such that $\sum_{j=-\infty}^{\infty} |\psi_j| < \infty$. Then the two following results hold.*

(1) If

$$\sup_{k \in \mathbb{Z}} \mathsf{E}[|X_k|] < \infty,$$

then the series

$$\psi(B)X_k = \sum_{j=-\infty}^{\infty} \psi_j B^j X_k = \sum_{j=-\infty}^{\infty} \psi_j X_{k-j}, \qquad (2.2.15)$$

converges absolutely a.s., $\forall k \in \mathbb{Z}$.

(2) If

$$\sup_{k \in \mathbb{Z}} \mathsf{E}[|X_k|^2] < \infty,$$

then the series $\psi(B)X_k$ converges in $\mathcal{L}_2$ and it converges in $\mathcal{L}_2$ to the a.s. limit (or sum), $\forall k \in \mathbb{Z}$.

We remind that, if $\sum a_j$ is a series of complex numbers that converges absolutely, then any one of the rearrangements of $\sum a_j$ converges and all of them converge to the same value. In particular, $\sum_{j=-\infty}^{\infty} a_j = \lim_{n \to \infty} \sum_{j=-n}^{n} a_j$.

Proof. 1. Monotone convergence theorem A.5.1 implies that

$$\mathsf{E}\left[\sum_{j=-\infty}^{\infty} |\psi_j X_{k-j}| \right] = \lim_{n \to \infty} \mathsf{E}\left[\sum_{j=-n}^{n} |\psi_j X_{k-j}| \right]$$

$$\leq \lim_{n \to \infty} \sum_{j=-n}^{n} |\psi_j| \sup_{k \in \mathbb{Z}} \mathsf{E}[|X_k|] < \infty,$$

from $\sup_{k \in \mathbb{Z}} \mathsf{E}[|X_k|] < \infty$. Thus $\sum_{j=-\infty}^{\infty} |\psi_j X_{k-j}| < \infty$ a.s. and so $\sum_{j=-\infty}^{\infty} \psi_j X_{k-j}$ converges a.s. Moreover, absolute convergence a.s. allows us to write

$$\sum_{j=-\infty}^{\infty} \psi_j X_{k-j} = \lim_{n \to \infty} \sum_{j=-n}^{n} \psi_j X_{k-j} \quad \text{a.s.}$$

2. If $\sup_{k \in \mathbb{Z}} \mathsf{E}[|X_k|^2] = \sup_{k \in \mathbb{Z}} \|X_k\|^2 < \infty$, then $\sum_{j=-n}^{n} \psi_j X_{k-j}$, $\forall n \in \mathbb{N}$, is a Cauchy sequence in $\mathcal{L}_2$, because, $\forall m < n \in \mathbb{N}$,

$$\left\| \sum_{j=-n}^{n} \psi_j X_{k-j} - \sum_{j=-m}^{m} \psi_j X_{k-j} \right\| \leq \sum_{j=-m}^{m} \|\psi_j X_{k-j}\| + \sum_{j=-n}^{-n} \|\psi_j X_{k-j}\|$$

$$\leq \sup_{k \in \mathbb{Z}} \|X_k\| \left(\sum_{j=m}^{n} |\psi_j| + \sum_{j=-n}^{-m} |\psi_j| \right)$$

$$\xrightarrow{m,n \to \infty} 0.$$

Therefore $\sum_{j=-n}^{n} \psi_j X_{k-j}$ converges in $\mathcal{L}_2$ to $\sum_{j=-\infty}^{\infty} \psi_j X_{k-j}$, as $n \to \infty$.
    The equality between $\mathcal{L}_2$ and a.s. limits is obtained by applying Lemma A.3.11. □

For stationary time series we have the following result.

Proposition 2.2.6 (Convergence of MA transform of stationary time series). *Let $\{X_k\}_{k\in\mathbb{Z}}$ be a stationary time series with a.c.v.f. γ and let $\{\psi_j\}_{j\in\mathbb{Z}}$ be a sequence of real numbers such that $\sum_{j=-\infty}^{\infty} |\psi_j| < \infty$. Then the two following results hold.*

(1) The random series

$$\psi(B)X_k = \sum_{j=-\infty}^{\infty} \psi_j X_{k-j},$$

converges in $\mathcal{L}_2$, a.s. and to the same limit (or sum), $\forall k \in \mathbb{Z}$.
(2) Let Y_k be the a.s. or $\mathcal{L}_2$ limit (or sum) of the above series, i.e. $Y_k = \psi(B)X_k$, $\forall k \in \mathbb{Z}$. Then $\{Y_k\}_{k\in\mathbb{Z}}$ is a stationary time series with mean

$$\mathsf{E}[X_1] \sum_{j=-\infty}^{\infty} \psi_j,$$

and with a.c.v.f.

$$\gamma_Y(h) = \sum_{j=-\infty}^{\infty} \sum_{l=-\infty}^{\infty} \psi_j \psi_l \gamma(h - j + l), \quad \forall h \in \mathbb{Z}.$$

Proof. 1. It follows from stationarity of $\{X_k\}_{k\in\mathbb{Z}}$ that $\sup_{k\in\mathbb{Z}} \mathsf{E}[|X_k|^2] = \mathsf{E}[|X_1|^2] < \infty$ and this part follows from Proposition 2.2.5.2.
2. The continuity of the scalar product (cf. Lemma A.1.5) yields

$$\mathsf{E}[Y_k] = \mathsf{E}\left[\sum_{j=-\infty}^{\infty} \psi_j X_{k-j} \right]$$

$$= \sum_{j=-\infty}^{\infty} \psi_j \mathsf{E}[X_{k-j}] = \mathsf{E}[X_1] \sum_{j=-\infty}^{\infty} \psi_j, \quad \forall k \in \mathbb{Z}.$$

The continuity of the scalar product yields $\forall k, h \in \mathbb{Z}$,

$$\mathsf{E}[Y_k Y_{k+h}]$$

$$= \mathsf{E}\left[\sum_{j=-\infty}^{\infty} \psi_j X_{k-j} \sum_{l=-\infty}^{\infty} \psi_l X_{k+h-l} \right]$$

$$= \sum_{j=-\infty}^{\infty} \sum_{l=-\infty}^{\infty} \psi_j \psi_l \mathsf{E}[X_{k-j} X_{k+h-l}]$$

$$= \sum_{j=-\infty}^{\infty} \sum_{l=-\infty}^{\infty} \psi_j \psi_l \Big(\mathsf{E}[X_{k-j} X_{k+h-l}] - \mathsf{E}[X_{k-j}]\mathsf{E}[X_{k+h-l}]$$

$$+ \mathsf{E}[X_{k-j}]\mathsf{E}[X_{k+h-l}] \Big)$$

$$= \sum_{j=-\infty}^{\infty} \sum_{l=-\infty}^{\infty} \psi_j \psi_l \big(\gamma(h+j-l) + \mathsf{E}[X_{k-j}]\mathsf{E}[X_{k+h-l}] \big)$$

$$= \sum_{j=-\infty}^{\infty} \sum_{l=-\infty}^{\infty} \psi_j \psi_l \gamma(h+j-l) + \sum_{j=-\infty}^{\infty} \sum_{l=-\infty}^{\infty} \psi_j \psi_l \mathsf{E}[X_{k-j}]\mathsf{E}[X_{k+h-l}]$$

$$= \sum_{j=-\infty}^{\infty} \sum_{l=-\infty}^{\infty} \psi_j \psi_l \gamma(h+j-l) + \mathsf{E}[Y_k]\mathsf{E}[Y_{k+h}].$$

$\square$

It follows from Propositions 2.2.5 and 2.2.6 that the MA operator $\psi(B) = \sum_{j=-\infty}^{\infty} \psi_j B^j$ is meaningful whenever $\sum_{j=-\infty}^{\infty} |\psi_j| < \infty$. Roughly stated, it inherits the properties of power series. For example, the MA operator is commutative: for any two MA operators $\alpha(B)$ and $\beta(B)$, we have

$$\alpha(B)\{\beta(B)X_k\} = \beta(B)\{\alpha(B)X_k\} = (\alpha\beta)(B)X_k = (\beta\alpha)(B)X_k, \ \forall k \in \mathbb{Z},$$

where $(\alpha\beta)(B) = (\beta\alpha)(B)$ denotes the convolution or the product. This is restated in Proposition 2.2.7 and shown thereafter. Yet, these properties are not obvious consequences of the linearity of the MA operator: linear operators are generally not commutative. For instance, matrix multiplication is not commutative.

Proposition 2.2.7 (Commutativity MA operators). *Let* $\{X_k\}_{k\in\mathbb{Z}}$ *be a time series for which*

$$\sup_{k\in\mathbb{Z}} \mathsf{E}[|X_k|] < \infty,$$

is satisfied. Let $\{\alpha_j\}_{j\in\mathbb{Z}}$ and $\{\beta_j\}_{j\in\mathbb{Z}}$ be two sequences of complex numbers such that

$$\sum_{j=-\infty}^{\infty} |\alpha_j| < \infty \quad \text{and} \quad \sum_{j=-\infty}^{\infty} |\beta_j| < \infty. \tag{2.2.16}$$

Then, $\forall k \in \mathbb{Z}$,

$$\alpha(B)\beta(B)X_k = \beta(B)\alpha(B)X_k = \psi(B)X_k = \sum_{j=-\infty}^{\infty} \psi_j X_{k-j} < \infty, \quad \text{a.s.,} \tag{2.2.17}$$

where

$$\psi_j = \sum_{l=-\infty}^{\infty} \alpha_l \beta_{j-l} = \sum_{l=-\infty}^{\infty} \alpha_{j-l}\beta_l, \ \forall j \in \mathbb{Z}, \text{ and } \sum_{j=-\infty}^{\infty} |\psi_j| < \infty. \tag{2.2.18}$$

Thus $\psi = \alpha\beta = \beta\alpha$.

Proof. We first note that (2.2.16) implies

$$\sum_{j=-\infty}^{\infty} |\psi_j| = \sum_{j=-\infty}^{\infty} \left| \sum_{l=-\infty}^{\infty} \alpha_l \beta_{j-l} \right| \leq \sum_{j=-\infty}^{\infty} \sum_{l=-\infty}^{\infty} |\alpha_l \beta_{j-l}|$$

$$= \sum_{j=-\infty}^{\infty} |\alpha_j| \sum_{j=-\infty}^{\infty} |\beta_j| < \infty,$$

given the remark on the rearrangement of the sum at p. 50. In fact that remark tells that (2.2.18) holds as well. Therefore Proposition 2.2.5.1 yields a.s. convergence of $\psi(B)X_k$, $\forall k \in \mathbb{Z}$.

Next, let $k \in \mathbb{Z}$. Then,

$$\alpha(B)\beta(B)X_k = \sum_{j=-\infty}^{\infty} \sum_{l=-\infty}^{\infty} \alpha_j \beta_l B^{j+l} X_k = \sum_{n=-\infty}^{\infty} \underbrace{\sum_{l=-\infty}^{\infty} \alpha_{n-l}\beta_l}_{=\psi_n} B^n X_k$$

$$= \sum_{n=-\infty}^{\infty} \underbrace{\sum_{l=-\infty}^{\infty} \alpha_l \beta_{n-l}}_{=\psi_n} B^n X_k = \sum_{j=-\infty}^{\infty} \sum_{l=-\infty}^{\infty} \alpha_l \beta_j B^{j+l} X_k$$

$$= \beta(B)\alpha(B)X_k.$$

Repeated application of Proposition 2.2.5.1 tells that the above series converges a.s. We have thus shown (2.2.17). □

Theorem 2.2.8 provides the explicit form of the stationary solution of the ARMA equation, in terms of MA of WN.

Theorem 2.2.8 (Explicit solution of ARMA(p, q)). *Let $\{X_k\}_{k \in \mathbb{Z}}$ satisfy the ARMA(p, q) equation*

$$\varphi(B)X_k = \theta(B)Z_k, \quad \forall k \in \mathbb{Z}, \qquad (2.2.19)$$

where $\{Z_k\}_{k \in \mathbb{Z}}$ is WN(σ^2). Assume that $\varphi(z) \neq 0$, $\forall z \in \mathbb{C}$ such that $|z| = 1$, i.e. $\forall z \in \mathcal{A}_1$. Then the two following results hold.

(1) The unique stationary solution of (2.2.19) is given by

$$X_k = \sum_{j=-\infty}^{\infty} \psi_j Z_{k-j}, \quad \forall k \in \mathbb{Z},$$

where $\dots, \psi_{-1}, \psi_0, \psi_1, \dots \in \mathbb{R}$ are the coefficients of the Laurent series of

$$\psi(z) = \frac{\theta(z)}{\varphi(z)},$$

namely of

$$\psi(z) = \sum_{j=-\infty}^{\infty} \psi_j z^j,$$

which is convergent $\forall z \in \mathcal{A}_r$, namely in the annulus (2.2.2), for some $r > 1$. Moreover $\sum_{j=-\infty}^{\infty} |\psi_j| < \infty$.

(2) The mean of $\{X_k\}_{k \in \mathbb{Z}}$ is null and its a.c.v.f. is given by

$$\gamma(h) = \sigma^2 \sum_{j=-\infty}^{\infty} \psi_j \psi_{|h|+j}, \quad \forall h \in \mathbb{Z}.$$

Proof. 1. Because φ is a polynomial, it is continuous and so the assumption that it is nonnull over the circumference of the centered unit circle can be extended to some small annulus containing the circle. In other terms,

$$\exists r > 1 \text{ such that } \varphi(z) \neq 0, \quad \forall z \in \mathcal{A}_r.$$

Consequently, $\xi(z) = 1/\varphi(z)$ is continuous and also differentiable at any $z \in \mathcal{A}_r^\circ$, which is the interior of the annulus. So $\xi(z)$ admits the Laurent expansion at $z = 0$ given by

$$\xi(z) = \sum_{j=-\infty}^{\infty} \xi_j z^j,$$

which converges $\forall z \in \mathcal{A}_r^\circ$. We refer to Theorem A.7.3 for the Laurent expansion.

Let $1/r < \alpha < 1 < \beta < r$. Then $\xi_j \beta^j \to 0$, as $j \to \infty$, i.e.

$$\forall \delta > 0, \exists j_0 \in \mathbb{N} \text{ such that } j \geq j_0 \Rightarrow |\xi_j \beta^j| < \delta \text{ i.e. } |\xi_j| \leq \frac{\delta}{\beta^j}.$$

From this we obtain $\sum_{j=0}^\infty \delta/\beta^j < \infty$, which gives us $\sum_{j=0}^\infty |\xi_j| < \infty$.
Similarly, $\xi_j \alpha^j \to 0$, as $j \to -\infty$, and so,

$$\forall \delta > 0, \exists j_0 \in \mathbb{N} \text{ such that } j \leq -j_0 \Longrightarrow |\xi_j \alpha^j| < \delta \text{ i.e. } |\xi_j| \leq \frac{\delta}{\alpha^j}.$$

Thus $\sum_{j=0}^\infty \delta \alpha^j < \infty$ yields $\sum_{j=-\infty}^0 |\xi_j| < \infty$.

Thus $\sum_{j=-\infty}^\infty |\xi_j| < \infty$ and the stationarity of $\{X_k\}_{k \in \mathbb{Z}}$ implies that $\{X_k\}_{k \in \mathbb{Z}}$ is bounded in $\mathcal{L}_1$. Consequently, Proposition 2.2.7 applies and it yields

$$X_k = \xi(B)\varphi(B)X_k = \xi(B)\theta(B)Z_k = \psi(B)Z_k, \quad \forall k \in \mathbb{Z}, \qquad (2.2.20)$$

where $\psi(z) = \xi(z)\theta(z) = \theta(z)/\varphi(z)$ and $\sum_{j=-\infty}^\infty |\psi_j| < \infty$. Because $\theta(z)$ and $\xi(z)$ have real coefficients, as shown in Proposition A.7.6, we obtain that $\{\psi_j\}_{j \in \mathbb{Z}}$ is a real-valued sequence. It follows from Proposition 2.2.6 that the time series $\{\psi(B)Z_k\}_{k \in \mathbb{Z}}$ is indeed stationary. Thus (2.2.20) tells that $\{\psi(B)Z_k\}_{k \in \mathbb{Z}}$ is the only stationary solution.
2. Let $\mu \in \mathbb{R}$ be the mean of $\{X_k\}_{k \in \mathbb{Z}}$. Then by taking the expectation on both sides of (2.2.19), we obtain

$$\mu\varphi(1) = 0.$$

The claim follows from the assumption $\varphi(1) \neq 0$.

Given that $\{X_k\}_{k \in \mathbb{Z}}$ takes the form of the random series, Proposition 2.2.6.2 can be applied to the case where $\{Z_k\}_{k \in \mathbb{Z}}$ is the initial stationary time series. Thus the a.c.v.f. of $\{X_k\}_{k \in \mathbb{Z}}$ is given by

$$\gamma(h) = \sum_{j=-\infty}^\infty \sum_{l=-\infty}^\infty \psi_j \psi_l \sigma^2 \mathbb{1}\{h - j + l = 0\}$$

$$= \sigma^2 \sum_{j=-\infty}^\infty \psi_j \psi_{j-h}, \quad \forall h \in \mathbb{Z}.$$

Because $\gamma(h) = \gamma(-h)$, $\forall h \in \mathbb{Z}$, we obtain the desired result. $\square$

Remark 2.2.9 (Absolute convergence of Laurent series). *It follows from the proof of Theorem 2.2.8 that, for some $r > 0$, the Laurent series $\sum_{j=-\infty}^\infty \psi_j z^j$ converges absolutely $\forall z \in \mathcal{A}_r^\circ$.*

Indeed, by using the notation of that proof, we have that, $\forall z \in \mathcal{A}_r^\circ$ *with* $|z| > 1$, $\exists \beta \in (|z|, r)$ *and* $j_0 \in \mathbb{N}$ *such that*

$$\sum_{j=-\infty}^{\infty} |\xi_j z^j| \leq \sum_{j=-\infty}^{j_0} |\xi_j z^j| + \delta \sum_{j=j_0+1}^{\infty} \left(\frac{|z|}{\beta}\right)^j < \infty.$$

Similar arguments can be given for the case $|z| < 1$.

2.2.3 Causality and invertibility

As before we denote by $\{Z_k\}_{k \in \mathbb{Z}}$ any $\mathrm{WN}(\sigma^2)$ time series, where $\sigma \in (0, \infty)$, and by $\varphi(z)$ and $\theta(z)$ the polynomials given by

$$\varphi(z) = \varphi_0 - \varphi_1 z - \cdots - \varphi_p z^p \quad \text{and} \quad \theta(z) = \theta_0 + \theta_1 z + \cdots + \theta_q z^q,$$

$\forall z \in \mathbb{C}$, where $\theta_0 = \varphi_0 = 1$ and $\varphi_1, \ldots, \varphi_p, \theta_1, \ldots, \theta_q \in \mathbb{R}$. Under the assumption of Theorem 2.2.8, the explicit form of an $\mathrm{ARMA}(p, q)$ time series takes the form of a random series. However, there is a practical and important subclass of ARMA time series in which this random series can be substantially simplified: the class of causal time series.

Definition 2.2.10 (Causal ARMA). *The* $\mathrm{ARMA}(p, q)$ *time series* $\{X_k\}_{k \in \mathbb{Z}}$ *defined by*

$$\varphi(B)X_k = \theta(B)Z_k, \quad \forall z \in \mathbb{Z},$$

is a causal function of $\{Z_k\}_{k \in \mathbb{Z}}$, *or simply causal, if there exists a sequence of real numbers* $\{\psi_j\}_{j \in \mathbb{N}}$ *such that* $\sum_{j=0}^{\infty} |\psi_j| < \infty$ *and*

$$X_k = \sum_{j=0}^{\infty} \psi_j Z_{k-j}, \quad \forall k \in \mathbb{Z}.$$

In relation with Theorem 2.2.8, a causal time series is one for which the Laurent series of $\psi(z) = \theta(z)/\varphi(z)$ has no principal part and is thus limited to its analytic part.

The next remark shows that the causal AR time series is Markovian.

Remark 2.2.11 (Causal AR(p) and Markov property). *Consider the causal* AR(p) *time series* $\{X_k\}_{k \in \mathbb{Z}}$ *defined by* $\varphi(B)X_k = Z_k$, $\forall k \in \mathbb{Z}$. *Let* $x \in \mathbb{R}$, *then*

$$\mathsf{P}[X_k \leq x | X_{k-1}, X_{k-2}, \ldots]$$
$$= \mathsf{P}[Z_k \leq x - \varphi_1 X_{k-1} - \cdots - \varphi_p X_{k-p} | X_{k-1}, X_{k-2}, \ldots]$$
$$= \mathsf{P}[Z_k \leq x - \varphi_1 X_{k-1} - \cdots - \varphi_p X_{k-p} | X_{k-1}, \ldots, X_{k-p}]$$
$$= F_k(x - \varphi_1 X_{k-1} - \cdots - \varphi_p X_{k-p}),$$

where F_k denotes the d.f. of Z_k. The second equality is a consequence of causality: Z_k is independent of $X_{k-p-1}, X_{k-p-2}, \ldots$. In this sense, the causal AR(p) time series satisfies the Markov property of order p, where the Markov property of order one is given in Definition 3.1.11. Markovian stochastic processes are useful in many applied sciences.

One of the most important results of this chapter is provided by Theorem 2.2.12. It provides sufficient conditions for causality and the explicit solution to the causal ARMA(p, q) time series. In this sense it completes Theorem 2.2.8.

Theorem 2.2.12 (Causal ARMA(p, q) and explicit solution). *Let $\{X_k\}_{k \in \mathbb{Z}}$ be the ARMA(p, q) time series with equation*

$$\varphi(B)X_k = \theta(B)Z_k, \quad \forall z \in \mathbb{Z}, \tag{2.2.21}$$

where $\{Z_k\}_{k \in \mathbb{Z}}$ is a WN. Assume that the polynomials φ and θ possess no common root over the disk $\mathcal{D}_0$; cf. (2.2.1). Then, the following statements hold.

(1) $\{X_k\}_{k \in \mathbb{Z}}$ is causal $\iff \varphi(z) \neq 0, \forall z \in \mathcal{D}_0$.
(2) If $\{X_k\}_{k \in \mathbb{Z}}$ is causal, then

$$X_k = \sum_{j=0}^{\infty} \psi_j Z_{k-j}, \quad \forall k \in \mathbb{Z},$$

where $\psi_0, \psi_1, \ldots \in \mathbb{R}$ are the coefficients of the Taylor series of

$$\psi(z) = \frac{\theta(z)}{\varphi(z)},$$

namely of

$$\psi(z) = \sum_{j=0}^{\infty} \psi_j z^j,$$

which is convergent $\forall z \in \mathcal{D}_0$, namely the disk (2.2.1). Moreover, $\sum_{j=0}^{\infty} |\psi_j| < \infty$.

Proof. 1. ($\Leftarrow$) Assume first $\varphi(z) \neq 0, \forall z \in \mathcal{D}_0$. Because φ is a polynomial, it is continuous and thus $\exists \varepsilon > 0$ such that $\varphi(z) \neq 0, \forall z \in \mathcal{D}_\varepsilon$. Thus, $\xi(z) = 1/\varphi(z)$ is continuous and also differentiable over $\mathcal{D}_\varepsilon^\circ$. Therefore, it admits a Taylor expansion at $z = 0$. That is,

$$\xi(z) = \sum_{j=0}^{\infty} \xi_j z^j,$$

where $\xi_0, \xi_1, \ldots \in \mathbb{R}$ and where the series converges $\forall z \in \mathcal{D}_\varepsilon^\circ$. We refer to Proposition A.7.7 for a detailed justification. Thus,

$$\xi_j \left(1 + \frac{\varepsilon}{2}\right)^j \longrightarrow 0, \quad \text{as } j \to \infty.$$

Consequently, $\forall \delta > 0, \exists j_0 \in \mathbb{N}$ such that

$$j \geq j_0 \implies \left| \xi_j \left(1 + \frac{\varepsilon}{2}\right)^j \right| < \delta \quad \text{i.e. } |\xi_j| \leq \delta \left(\frac{1}{1 + \frac{\varepsilon}{2}}\right)^j.$$

Therefore $\sum_{j=0}^{\infty} 1/(1 + \varepsilon/2)^j < \infty$ yields $\sum_{j=0}^{\infty} |\xi_j| < \infty$.

Because $\{X_k\}_{k \in \mathbb{Z}}$ is stationary, it is bounded in $\mathcal{L}_1$, so that Proposition 2.2.7 yields

$$X_k = \xi(B)\varphi(B)X_k = \xi(B)\theta(B)Z_k = \psi(B)Z_k, \quad \forall k \in \mathbb{Z},$$

where

$$\psi(z) = \xi(z)\theta(z) = \frac{\theta(z)}{\varphi(z)},$$

and $\sum_{j=0}^{\infty} |\psi_j| < \infty$. We deduce from the fact that the Taylor series $\xi(z)$ and the polynomial $\theta(z)$ have real coefficients that the coefficients of the series expansion of $\psi(z)$ are real, i.e. $\{\psi_j\}_{j \in \mathbb{N}} \in \mathbb{R}^\infty$. Clearly, Proposition 2.2.7 tells that $\{\psi(B)Z_k\}_{k \in \mathbb{Z}}$ is indeed a solution of the ARMA equation (2.2.21) and thus $\{X_k\}_{k \in \mathbb{Z}}$ is causal.

($\Rightarrow$) Let now assume that $\{X_k\}_{k \in \mathbb{Z}}$ is causal. Then,

$$X_k = \sum_{j=0}^{\infty} \psi_j Z_{k-j}, \quad \forall k \in \mathbb{Z},$$

where $\{\psi_j\}_{j \in \mathbb{N}}$ is a real-valued sequence such that $\sum_{j=0}^{\infty} |\psi_j| < \infty$. Let $\alpha(z) = \varphi(z)\psi(z) = \sum_{j=0}^{\infty} \alpha_j z^j$. It follows from Proposition 2.2.7 that $\sum_{j=0}^{\infty} |\alpha_j| < \infty$ and that

$$\theta(B)Z_k = \varphi(B)X_k = \varphi(B)\psi(B)Z_k = \alpha(B)Z_k, \quad \forall k \in \mathbb{Z},$$

which is equivalent to

$$\sum_{j=0}^{q} \theta_j Z_{k-j} = \sum_{j=0}^{\infty} \alpha_j Z_{k-j}, \quad \forall k \in \mathbb{Z}.$$

Therefore

$$\left\langle Z_{k-l}, \sum_{j=0}^{q} \theta_j Z_{k-j} \right\rangle = \left\langle Z_{k-l}, \sum_{j=0}^{\infty} \alpha_j Z_{k-j} \right\rangle, \quad \forall l \in \mathbb{N}.$$

From the continuity of the scalar product, this holds iff

$$\alpha_l = \begin{cases} \theta_l, & \text{for } l = 0, \ldots, q, \\ 0, & \text{for } l = q+1, q+2, \ldots. \end{cases}$$

Thus, $\theta(z) = \alpha(z)$, that is, $\theta(z) = \varphi(z)\psi(z)$, $\forall z \in \mathcal{D}_0$. In particular, if $\varphi(z) = 0$ for some $z \in \mathcal{D}_0$, then $\theta(z) = 0$ must hold as well. But this contradicts the hypothesis that $\varphi(z)$ and $\theta(z)$ have no common roots in $\mathcal{D}_0$. Consequently it must hold that $\varphi(z) \neq 0$, $\forall z \in \mathcal{D}_0$.

2. It follows from the part 1 ($\Leftarrow$) of this proof, that if $\{X_k\}_{k \in \mathbb{Z}}$ is causal, then the only possible solution is given by $X_k = \sum_{j=0}^{\infty} \psi_j Z_{k-j}$, $\forall k \in \mathbb{Z}$. $\square$

Another subclass of ARMA time series is obtained by the invertible time series, which is defined as follows.

Definition 2.2.13 (Invertible ARMA). *The* ARMA(p,q) *time series* $\{X_k\}_{k \in \mathbb{Z}}$ *defined by*

$$\varphi(B)X_k = \theta(B)Z_k, \quad \forall k \in \mathbb{Z}, \tag{2.2.22}$$

is invertible, if there exists a sequence of real numbers $\{\tau_j\}_{j \in \mathbb{N}}$ *such that* $\sum_{j=0}^{\infty} |\tau_j| < \infty$ *and*

$$Z_k = \sum_{j=0}^{\infty} \tau_j X_{k-j}, \quad \forall k \in \mathbb{Z}.$$

The analogue version of Theorem 2.2.12 for invertible time series is the following.

Theorem 2.2.14 (Invertible ARMA(p,q) and explicit solution). *Let* $\{X_k\}_{k \in \mathbb{Z}}$ *be the* ARMA(p,q) *time series with equation*

$$\varphi(B)X_k = \theta(B)Z_k, \quad \forall k \in \mathbb{Z}, \tag{2.2.23}$$

where $\{Z_k\}_{k \in \mathbb{Z}}$ *is a WN. Assume that the polynomials* φ *and* θ *possess no common root over the disk* $\mathcal{D}_0$. *Then, the following statements hold.*

(1) $\{X_k\}_{k \in \mathbb{Z}}$ *is invertible* $\iff \theta(z) \neq 0$, $\forall z \in \mathcal{D}_0$.

(2) If $\{X_k\}_{k \in \mathbb{Z}}$ *is invertible, then*

$$Z_k = \sum_{j=0}^{\infty} \tau_j X_{k-j}, \quad \forall k \in \mathbb{Z},$$

where $\tau_0, \tau_1, \ldots \in \mathbb{R}$, *are the coefficients of the Taylor series of*

$$\tau(z) = \frac{\varphi(z)}{\theta(z)},$$

namely of

$$\tau(z) = \sum_{j=0}^{\infty} \tau_j z^j,$$

which is convergent $\forall z \in \mathcal{D}_0$. *Moreover* $\sum_{j=0}^{\infty} |\tau_j| < \infty$.

Proof. 1. ($\Leftarrow$) Assume $\theta(z) \neq 0$, $\forall z \in \mathcal{D}_0$. Because θ is continuous, $\exists \varepsilon > 0$ such that $\theta(z) \neq 0$, $\forall z \in \mathcal{D}_\varepsilon$. Thus, $\eta(z) = 1/\theta(z)$ is continuous and differentiable over $\mathcal{D}_\varepsilon^\circ$. Consequently, it admits a Taylor expansion at $z = 0$, cf. Proposition A.7.7, which is given by

$$\eta(z) = \sum_{j=0}^{\infty} \eta_j z^j,$$

where $\eta_0, \eta_1, \ldots \in \mathbb{R}$ and where the series converges $\forall z \in \mathcal{D}_\varepsilon^\circ$. Then

$$\eta_j \left(1 + \frac{\varepsilon}{2}\right)^j \longrightarrow 0, \quad \text{as } j \to \infty.$$

So, $\forall \delta > 0$, $\exists j_0 \in \mathbb{N}$ such that

$$j \geq j_0 \implies \left| \eta_j \left(1 + \frac{\varepsilon}{2}\right)^j \right| < \delta \text{ i.e. } |\eta_j| \leq \delta \left(\frac{1}{1 + \frac{\varepsilon}{2}}\right)^j.$$

Therefore $\sum_{j=0}^{\infty} 1/(1 + \varepsilon/2)^j < \infty$ yields $\sum_{j=0}^{\infty} |\eta_j| < \infty$. By Proposition 2.2.7 we obtain

$$Z_k = \eta(B)\theta(B)Z_k = \eta(B)\varphi(B)X_k = \tau(B)X_k, \quad \forall k \in \mathbb{Z},$$

where

$$\tau(z) = \eta(z)\varphi(z) = \frac{\varphi(z)}{\theta(z)},$$

and $\sum_{j=0}^{\infty} |\tau_j| < \infty$. It follows from the fact that the Taylor series $\eta(z)$ and the polynomial $\varphi(z)$ have real coefficients that $\{\tau_j\}_{j \in \mathbb{N}} \in \mathbb{R}^\infty$. Thus $\{X_k\}_{k \in \mathbb{Z}}$ is invertible.

($\Rightarrow$) Let us now assume that $\{X_k\}_{k \in \mathbb{Z}}$ is invertible. Then, $\forall k \in \mathbb{Z}$,

$$Z_k = \sum_{j=0}^{\infty} \tau_j X_{k-j},$$

where $\{\tau_j\}_{j \in \mathbb{N}}$ is a real-valued sequence such that $\sum_{j=0}^{\infty} |\tau_j| < \infty$. Let $\alpha(z) = \tau(z)\theta(z) = \sum_{j=0}^{\infty} \alpha_j z^j$. It follows from Proposition 2.2.7 that $\sum_{j=0}^{\infty} |\alpha_j| < \infty$ and that

$$\varphi(B)Z_k = \varphi(B)\tau(B)X_k = \tau(B)\varphi(B)X_k = \tau(B)\theta(B)Z_k = \alpha(B)Z_k,$$

which is equivalent to

$$\sum_{j=0}^{q} \varphi_j Z_{k-j} = \sum_{j=0}^{\infty} \alpha_j Z_{k-j}, \quad \forall k \in \mathbb{Z}.$$

Therefore

$$\left\langle Z_{k-l}, \sum_{j=0}^{q} \varphi_j Z_{k-j} \right\rangle = \left\langle Z_{k-l}, \sum_{j=0}^{\infty} \alpha_j Z_{k-j} \right\rangle, \quad \forall l \in \mathbb{N}.$$

From the continuity of the scalar product, this holds iff

$$\alpha_l = \begin{cases} \varphi_l, & \text{for } l = 0, \ldots, q, \\ 0, & \text{for } l = q+1, q+2, \ldots. \end{cases}$$

Thus, $\varphi(z) = \alpha(z)$, that is, $\varphi(z) = \tau(z)\theta(z)$, $\forall z \in \mathcal{D}_0$. In particular, if $\theta(z) = 0$ for some $z \in \mathcal{D}_0$, then $\varphi(z) = 0$ must be true. This is in contradiction with the hypothesis and so $\theta(z) \neq 0$, $\forall z \in \mathcal{D}_0$.

2. It follows from the part 1 ($\Leftarrow$) of this proof, that if $\{X_k\}_{k\in\mathbb{Z}}$ is invertible, then the only possible series expression of Z_k in terms of $\{X_k\}_{k\in\mathbb{Z}}$ is given by $Z_k = \sum_{j=0}^{\infty} \tau_j X_{k-j}$, $\forall k \in \mathbb{Z}$. $\qquad \square$

With the previous results, we can provide the series representation of a particular ARMA time series.

Example 2.2.15 (Explicit solution of ARMA$(2,1)$). *Consider the* ARMA$(2,1)$ *time series with coefficients* $\theta_1 = 1$, $\varphi_1 = 5/2$ *and* $\varphi_2 = -1$. *Let us first see if this time series is causal or invertible. The AR and MA polynomials are respectively given by*

$$\varphi(z) = z^2 - \frac{5}{2}z + 1 = \left(z - \frac{1}{2}\right)(z - 2) \text{ and } \theta(z) = z + 1.$$

Because $\varphi(1/2) = 0$ *and* $\theta(-1) = 0$, *this time series is neither causal nor invertible; cf. Theorems 2.2.12.1 and 2.2.14.1.*

In this situation, we can however provide an explicit solution in terms of a series depending on all values of the WN; cf. Theorem 2.2.8. For this purpose, we have to compute the Laurent series of the rational function

$$\psi(z) = \frac{\varphi(z)}{\theta(z)} = \frac{1+z}{\left(z - \frac{1}{2}\right)(z - 2)}.$$

This can be done by partial fraction decomposition. We must find constants c_1, c_2 *such that the equation*

$$\frac{1+z}{\left(z - \frac{1}{2}\right)(z - 2)} = \frac{c_1}{z - \frac{1}{2}} + \frac{c_2}{z - 2}, \tag{2.2.24}$$

holds. By multiplying (2.2.24) by $z - 1/2$ and by setting $z = 1/2$, we obtain $c_1 = -1$. By multiplying (2.2.24) by $z - 2$ and by setting $z = 2$, we obtain $c_2 = 2$. Since $\varphi(z) \neq 0$, $\forall z \in \mathbb{C}$ such that with $|z| = 1$, we know that ψ admits an absolutely convergent Laurent series in a region that includes the Annulus $\mathcal{A}_r^{\circ}$, for some $r > 1$. This Laurent series is given by

$$\psi(z) = \frac{1 + z}{\left(z - \frac{1}{2}\right)(z - 2)} = \frac{-1}{z - \frac{1}{2}} + \frac{2}{z - 2}$$

$$= -\frac{1}{z}\frac{1}{1 - \frac{1}{2z}} - \frac{1}{1 - \frac{z}{2}} = -\frac{1}{z}\sum_{j=0}^{\infty}\left(\frac{1}{2z}\right)^j - \sum_{j=0}^{\infty}\left(\frac{z}{2}\right)^j$$

$$= \sum_{j=1}^{\infty} -2^{-j+1}z^{-j} - \sum_{j=0}^{\infty}(2)^{-j}z^j.$$

Hence we obtain the coefficients $\psi_j = 2^{-j}$, for $j = 0, 1, \ldots$, and $\psi_{-j} = -2^{-j+1}$ for, $j = 1, 2, \ldots$. The series representation of X_k is is thus given by

$$X_k = \sum_{j=-\infty}^{\infty} \psi_j Z_{k-j} = \sum_{j=0}^{\infty} 2^{-j} Z_{k-j} - 2\sum_{j=1}^{\infty} 2^{-j} Z_{k+j}, \quad \forall k \in \mathbb{Z}.$$

An obvious corollary of the last Theorems 2.2.12 and 2.2.14 is the following.

Corollary 2.2.16 (Causal and invertible ARMA). *Let $\{X_k\}_{k\in\mathbb{Z}}$ be the ARMA(p, q) time series with equation*

$$\varphi(B)X_k = \theta(B)Z_k, \quad \forall k \in \mathbb{Z},$$

where $\{Z_k\}_{k\in\mathbb{Z}}$ is WN. Assume that the polynomials φ and θ possess no root over $\mathcal{D}_0$. Then the time series $\{X_k\}_{k\in\mathbb{Z}}$ is causal and invertible.

Definition 2.2.17 (AR(∞)). *The time series $\{X_k\}_{k\in\mathbb{Z}}$ is called AR(∞), if there exists a WN time series $\{Z_k\}_{k\in\mathbb{Z}}$ and a sequence of real numbers $\{\tau_j\}_{j\in\mathbb{N}}$ such that $\sum_{j=0}^{\infty}|\tau_j| < \infty$ and*

$$Z_k = \sum_{j=0}^{\infty} \tau_j X_{k-j}, \quad \forall k \in \mathbb{Z}.$$

Definition 2.2.18 (MA(∞)). *The time series $\{X_k\}_{k\in\mathbb{Z}}$ is called MA(∞), if there exists a WN time series $\{Z_k\}_{k\in\mathbb{Z}}$ and a sequence of real numbers $\{\psi_j\}_{j\in\mathbb{N}}$ such that $\sum_{j=0}^{\infty}|\psi_j| < \infty$ and*

$$X_k = \sum_{j=0}^{\infty} \psi_j Z_{k-j}, \quad \forall k \in \mathbb{Z}.$$

Examples 2.2.19 (Three MA(∞)).

(1) The MA(q) *is trivially* MA(∞) *with*

$$\psi_j = \begin{cases} \theta_j, & \text{if } j = 1, \ldots, q, \\ 0, & \text{otherwise.} \end{cases}$$

(2) The AR(1) *with* $|\varphi_1| < 1$ *is* MA(∞) *with*

$$\psi_j = \begin{cases} \varphi_1^j, & \text{if } j = 0, 1, \ldots, \\ 0, & \text{if } j = -1, -2, \ldots. \end{cases}$$

We refer to Proposition 2.2.3 for this result. The AR(1) *is causal iff* $\varphi(z) = 1 - \varphi_1 z$ *has no root over* $\mathcal{D}_0$ *or, equivalently,* $|\varphi_1| < 1$.
The AR(1) *with* $|\varphi_1| > 1$ *is not* MA(∞). *It takes the form of* $\sum_{j=-\infty}^{\infty} \psi_j Z_{k-j}$ *with*

$$\psi_j = \begin{cases} 0, & \text{if } j = 0, 1, \ldots, \\ -\varphi_1^{-j}, & \text{if } j = -1, -2, \ldots. \end{cases}$$

We refer to Proposition 2.2.3 for this result. We note that $\varphi(z) \neq 0$, $\forall z \in \mathcal{A}_1$, *which is the condition of Theorem 2.2.8. This* AR(1) *is not causal.*

(3) The causal ARMA(p, q), *whose autoregressive polynomial* φ *possess no roots over the unit disk* $\mathcal{D}_0$, *is* MA(∞). *This is shown in the proof of Theorem 2.2.12.1, part* ($\Leftarrow$). *Note that Proposition 2.3.16 states that there exists a WN that converts a non-causal* ARMA(p, q) *to a causal one, allowing thus for the* MA(∞) *representation.*

Proposition 2.2.2 provides the a.c.v.f. of the MA(q) time series and we see that it vanishes after a finite time lag. In fact, an a.c.v.f. with this feature must be the one of a MA(q).

Proposition 2.2.20 (Vanishing a.c.v.f. and MA(q)). *Let* $\{X_k\}_{k\in\mathbb{Z}}$ *be a stationary time series with zero mean. If its a.c.v.f.* γ *satisfies*

$$\gamma(\pm q) \neq 0 \text{ and } \gamma(h) = 0, \ \forall h \in \mathbb{Z} \ \text{ such that } |h| \geq q + 1,$$

for some $q \geq 1$, *then* $\{X_k\}_{k\in\mathbb{Z}}$ *must be a* MA(q) *time series.*

The proof makes use of some mathematical definitions related to Hilbert spaces that are given in Section A.1.

Proof of Proposition 2.2.20. Let $j < k \in \mathbb{Z}$, let $H_k = \overline{\mathrm{sp}}\{X_j\}_{j \leq k}$ in $\mathcal{L}_2$ and let

$$Z_k = \left(I - \mathrm{pro}_{H_{k-1}}\right) X_k = X_k - \mathrm{pro}_{H_{k-1}} X_k,$$

be the residual of the projection. As stated in Definition A.1.8, the expectation in $\mathcal{L}_2$ is the orthogonal projection onto the space of constant random variables. Therefore the iterativity of the orthogonal projection gives

$$\mathsf{E}\,[Z_k] = 0. \tag{2.2.25}$$

It follows from $Z_j \in H_j \subset H_{k-1}$ and $Z_k \in H_{k-1}^{\perp}$ that

$$\mathsf{E}\,[Z_j Z_k] = \langle Z_j, Z_k \rangle = 0. \tag{2.2.26}$$

Let $n \in \mathbb{N}^*$, then it could be shown that

$$\mathrm{pro}_{\overline{\mathrm{sp}}\{X_j\}_{j=k-n,\ldots,k-1}} X_k \xrightarrow{\mathcal{L}_2} \mathrm{pro}_{H_{k-1}} X_k.$$

This last result together with stationarity and with continuity of the $\mathcal{L}_2$-norm give

$$
\begin{aligned}
\|Z_{k+1}\|^2 &= \left\| \left(I - \mathrm{pro}_{H_k}\right) X_{k+1} \right\|^2 \\
&= \left\| X_{k+1} - \mathrm{pro}_{\overline{\mathrm{sp}}\{X_j\}_{j=k-n,\ldots,k}} X_{k+1} \right\|^2 \\
&\quad - \underbrace{\left\| \mathrm{pro}_{H_k} X_{k+1} - \mathrm{pro}_{\overline{\mathrm{sp}}\{X_j\}_{j=k-n,\ldots,k}} X_{k+1} \right\|^2}_{\xrightarrow{n \to \infty} 0} \\
&= \lim_{n \to \infty} \left\| X_{k+1} - \mathrm{pro}_{\overline{\mathrm{sp}}\{X_j\}_{j=k-n,\ldots,k}} X_{k+1} \right\|^2 \\
&= \lim_{n \to \infty} \left\| X_k - \mathrm{pro}_{\overline{\mathrm{sp}}\{X_j\}_{j=k-n-1,\ldots,k-1}} X_k \right\|^2 \\
&= \left\| X_k - \mathrm{pro}_{H_{k-1}} X_k \right\|^2 \\
&= \|Z_k\|^2. \tag{2.2.27}
\end{aligned}
$$

Therefore (2.2.25), (2.2.26) and (2.2.27) imply that $\{Z_k\}_{k \in \mathbb{Z}}$ is $\mathrm{WN}(\sigma^2)$, where $\sigma^2 = \|Z_k\|^2 = \mathsf{E}\left[Z_k^2\right]$. We thus have

$$
\begin{aligned}
H_{k-1} &= H_{k-2} \oplus \overline{\mathrm{sp}}\,\{Z_{k-1}\} \\
&= H_{k-3} \oplus \overline{\mathrm{sp}}\,\{Z_{k-2}\} \oplus \overline{\mathrm{sp}}\,\{Z_{k-1}\} \\
&= H_{k-3} \oplus \overline{\mathrm{sp}}\,\{Z_{k-2}, Z_{k-1}\} \\
&= \cdots \\
&= H_{k-q-1} \oplus \overline{\mathrm{sp}}\,\{Z_{k-q}, \ldots, Z_{k-1}\},
\end{aligned}
$$

where $\oplus$ denotes the direct sum between orthogonal subspaces of $\mathcal{L}_2$; cf. Section A.1. This decomposition, the assumption of the theorem and the normal equations give us

$$\text{pro}_{H_{k-1}} X_k = \underbrace{\text{pro}_{H_{k-q-1}} X_k}_{\substack{=0, \text{ because} \\ \gamma(h)=0, \text{ for } h \geq q+1}} + \text{pro}_{\overline{\text{sp}}\{Z_{k-q},\ldots,Z_{k-1}\}} X_k$$

$$= \frac{\langle Z_{k-1}, X_k \rangle}{\|Z_{k-1}\|^2} Z_{k-1} + \cdots + \frac{\langle Z_{k-q}, X_k \rangle}{\|Z_{k-q}\|^2} Z_{k-q}.$$

By denoting

$$\theta_j = \frac{\langle Z_{k-j}, X_k \rangle}{\|Z_{k-j}\|^2}, \quad \text{for } j = 1, \ldots, q,$$

we have obtained the desired MA(q) equation

$$Z_k = \left(I - \text{pro}_{H_{k-1}} \right) X_k = X_k - \theta_1 Z_{k-1} - \cdots - \theta_q Z_{k-q}.$$

$\square$

Proposition 2.2.20 can be re-expressed in terms of h-correlation, given Definition 2.1.9.1.

Proposition 2.2.21 (MA(q) and q-correlation). *Let $\{X_k\}_{k\in\mathbb{Z}}$ be a stationary time series with mean zero and let $q \in \mathbb{N}^*$. Then $\{X_k\}_{k\in\mathbb{Z}}$ is MA(q) iff it is q.*

Example 2.2.22 (Unicity of ARMA(p,q)). *Let $\{X_k\}_{k\in\mathbb{Z}}$ and $\{Y_k\}_{k\in\mathbb{Z}}$ be stationary time series with mean null and same a.c.v.f. Assume that $\{Y_k\}_{k\in\mathbb{Z}}$ is ARMA(p,q) with AR polynomial φ. In this situation $\{X_k\}_{k\in\mathbb{Z}}$ is necessarily ARMA(p,q) and with same AR polynomial φ.*

In order to prove this claim, we first show that

$$U_k = X_k - \varphi_1 X_{k-1} - \cdots - \varphi_p X_{k-p}, \quad \forall k \in \mathbb{Z},$$

has null a.c.v.f. at time lags larger that q. We have that, $\forall k, h \in \mathbb{Z}$,

$$\text{cov}(U_{k+h}, U_k) = \text{cov}\left(\varphi(B) X_{k+h}, \varphi(B) X_k \right) = \text{cov}\left(\varphi(B) Y_{k+h}, \varphi(B) Y_k \right).$$

$$(2.2.28)$$

Since $\{\varphi(B) Y_k\}_{k\in\mathbb{Z}}$ is MA(q), the right side of (2.2.28) is independent of k and null $\forall h \in \mathbb{Z}$ such that $|h| > q$; cf. Proposition 2.2.2. Thus the a.c.v.f. of $\{U_k\}_{k\in\mathbb{Z}}$ satisfies these properties and $\{U_k\}_{k\in\mathbb{Z}}$ is stationary. The conditions of Proposition 2.2.20 being satisfied, $\{U_k\}_{k\in\mathbb{Z}}$ is MA(q). So there exist $\{Z_k\}_{k\in\mathbb{Z}}$ WN and $\theta_1, \ldots, \theta_q \in \mathbb{R}$ such that, $\forall k \in \mathbb{Z}$,

$$X_k - \varphi_1 X_{k-1} - \cdots - \varphi_p X_{k-p} = U_k = Z_k + \theta_1 Z_{k-1} + \cdots + \theta_q Z_{k-q}.$$

In the practice, Proposition 2.2.20 can be used for model identification, as illustrated by the next example.

Example 2.2.23 (Swiss coronavirus data). *Consider the Swiss coronavirus data presented in Example 2.1.18. The sample concerns daily numbers of reported coronavirus infections in Switzerland between March 29 and June 17 2021 and it is given in Table B.1. We consider the differentiated and deseasonalized logarithmic sample* $\{\nabla\nabla_7 \log X_k\}_{k=9,\ldots,81}$, *which is redenoted* $\{Y_k\}_{k=1,\ldots,72}$. *The* $n = 72$ *sample values are displayed in the upper graph of Figure 2.6. The empirical a.c.r.f. is provided by lower graph of Figure 2.6. The last value of the empirical a.c.r.f. that appears clearly significant, namely clearly outside the dashed band, is obtained at time lag* $h = 7$. *Thus, the* MA(7) *appears as a reasonable candidate for this sample. The Gaussian m.l.e. estimators of the* MA(7) *coefficients together with their estimated standard deviations are given in Table 2.2. Precisely, we find: the estimator of the j-the coefficient, denoted* $\hat\theta_{jn}$, *its standard deviations, denoted* $\hat\sigma(\hat\theta_{jn})$, *and its approximate confidence interval at level 95%, which is computed by* $\hat\theta_{jn} \pm 2\hat\sigma(\hat\theta_{jn})$, *denoted* $(l_n(\theta_j), u_n(\theta_j))$, *for* $j = 1, \ldots, 7$. *Because none of these intervals contain zero, all estimated coefficients can be considered significant. The estimator of the WN variance is* $\hat\sigma_n^2 = 203084$ *and thus* $\hat\sigma_n = 450.6484$.

Theorem 2.2.24 tells that the a.c.v.f. of the MA(∞) is simply the limit of the a.c.v.f. of the MA(q), which is given in Proposition 2.2.2, as $q \to \infty$.

Table 2.2:   Estimation of MA(7) of differentiated and deseasonalized logarithmic coronavirus infections sample $\{Y_k\}_{k=1,\ldots,72}$: estimators of coefficients, $\hat\theta_{\cdot n}$, their standard deviations, $\hat\sigma(\hat\theta_{\cdot n})$, and their confidence interval at level 95%, $(l_n(\theta_\cdot), u_n(\theta_\cdot))$.

	θ_1	θ_2	θ_3	θ_4	θ_5	θ_6	θ_7
$\hat\theta_{\cdot n}$	1.1177	1.2878	1.4727	1.0235	0.6206	0.4338	0.4603
$\hat\sigma(\hat\theta_{\cdot n})$	0.1480	0.2250	0.3040	0.3038	0.2486	0.1509	0.1189
$l_n(\theta_\cdot)$	1.4137	1.7379	2.0806	1.6311	1.1179	0.7356	0.6981
$u_n(\theta_\cdot)$	0.8217	0.8378	0.8647	0.4159	0.1234	0.1319	0.2226

Theorem 2.2.24 (A.c.v.f. of MA(∞)). *The MA(∞) time series is stationary and its a.c.v.f. is given by*

$$\gamma(h) = \sigma^2 \sum_{j=0}^{\infty} \psi_j \psi_{|h|+j}, \quad \forall h \in \mathbb{Z},$$

where σ^2 is the variance of the WN.

Proof. Given that $\{X_k\}_{k\in\mathbb{Z}}$ is MA(∞), there exists a WN(σ^2) time series $\{Z_k\}_{k\in\mathbb{Z}}$ and a sequence of real numbers $\{\psi_j\}_{j\in\mathbb{N}}$, such that $\sum_{j=0}^{\infty} |\psi_j| < \infty$ and $X_k = \sum_{j=0}^{\infty} \psi_j Z_{k-j}$, $\forall k \in \mathbb{Z}$. Then, Proposition 2.2.6 applied to $\{Z_k\}_{k\in\mathbb{Z}}$ and to $\{\psi_j\}_{j\in\mathbb{Z}}$, where $\psi_j = 0$, $\forall j < 0$, implies that $\{X_k\}_{k\in\mathbb{Z}}$ is stationary with a.c.v.f. given by

$$\gamma(h) = \sum_{j=0}^{\infty} \sum_{l=0}^{\infty} \psi_j \psi_l \sigma^2 \mathbb{I}\{h - j + l = 0\} = \sigma^2 \sum_{j=0}^{\infty} \psi_j \psi_{j-h}, \quad \forall h \in \mathbb{Z}.$$

Because $\gamma(h) = \gamma(-h)$, $\forall h \in \mathbb{Z}$, we obtain the desired result. $\square$

According to Example 2.2.19.3, the MA(∞) time series is the stationary solution to the causal ARMA(p, q) equation. We thus have the a.c.v.f. of the causal ARMA(p, q).

Corollary 2.2.25 (A.c.v.f. of causal ARMA(p, q)). *Let $\{X_k\}_{k\in\mathbb{Z}}$ be the ARMA(p, q) time series with equation*

$$\varphi(B)X_k = \theta(B)Z_k, \quad \forall k \in \mathbb{Z}, \tag{2.2.29}$$

where $\{Z_k\}_{k\in\mathbb{Z}}$ is a WN(σ^2), for some $\sigma > 0$. Assume that the polynomial φ possesses no root over the disk $\mathcal{D}_0$. Let $\psi_0, \psi_1, \ldots \in \mathbb{R}$ be the coefficients of the Taylor series of $\psi(z) = \theta(z)/\varphi(z)$, namely of

$$\psi(z) = \sum_{j=0}^{\infty} \psi_j z^j,$$

which is convergent $\forall z \in \mathcal{D}_0$.

Then the mean of $\{X_k\}_{k\in\mathbb{Z}}$ is null and the a.c.v.f. is given by

$$\gamma(h) = \sigma^2 \sum_{j=0}^{\infty} \psi_j \psi_{j+|h|}, \quad \forall h \in \mathbb{Z}.$$

Proof. The a.c.v.f. is obtained by joining Theorems 2.2.12 and 2.2.24 together.

Denote by $\mu \in \mathbb{R}$ be the mean of $\{X_k\}_{k\in\mathbb{Z}}$. Then by taking the expectation on both sides of (2.2.29), we obtain

$$\mu\varphi(1) = 0.$$

The claim follows from the assumption $\varphi(1) \neq 0$. $\square$

Example 2.2.26 (A.c.v.f. of causal ARMA(1,1)). *Let $\{X_k\}_{k \in \mathbb{Z}}$ be the ARMA(1,1) time series*

$$X_k - \varphi_1 X_{k-1} = Z_k + \theta_1 Z_{k-1}, \quad \forall k \in \mathbb{Z},$$

where $|\varphi_1|, |\theta_1| < 1$ and with $\{Z_k\}_{k \in \mathbb{Z}}$ is WN(σ^2). The root of $\varphi(z) = 1 - \varphi_1 z$ is $1/\varphi_1$ and the root of $\theta(z) = 1 + \theta_1 z$ is $-1/\theta_1$. Both roots lie outside the unit disc $\mathcal{D}_0$ and so this time series is causal and invertible whenever $\varphi_1 \neq -\theta_1$. Thus it admits the representation $X_k = \sum_{j=0}^{\infty} \psi_j Z_{k-j}$, $\forall k \in \mathbb{Z}$, with $\psi(z) = \theta(z)/\varphi(z)$, $\forall z \in \mathcal{D}_0$. Consider $z \in \mathcal{D}_0$, then we find the convergent series

$$\frac{1}{\varphi(z)} = \frac{1}{1 - \varphi_1 z} = \sum_{j=0}^{\infty} (\varphi_1 z)^j,$$

and thus

$$\frac{\theta(z)}{\varphi(z)} = (1 + \theta_1 z) \sum_{j=0}^{\infty} \varphi_1^j z^j = 1 + (\varphi_1 + \theta_1) \sum_{j=1}^{\infty} \varphi_1^{j-1} z^j.$$

This leads to

$$\psi_0 = 1 \text{ and } \psi_j = (\varphi_1 + \theta_1)\varphi_1^{j-1}, \text{ for } j = 1, 2, \dots. \quad (2.2.30)$$

It follows from Corollary 2.2.25 that

$$\gamma(h) = \sigma^2 \sum_{j=0}^{\infty} \psi_j \psi_{|h|+j}$$

$$= \sigma^2 \left(\psi_{|h|} + \sum_{j=1}^{\infty} \psi_j \psi_{|h|+j} \right)$$

$$= \sigma^2 \left((\varphi_1 + \theta_1)\varphi_1^{|h|-1} + (\varphi_1 + \theta_1)^2 \sum_{j=0}^{\infty} (\varphi_1^2)^j \varphi_1^{|h|} \right)$$

$$= \sigma^2 \frac{(\varphi_1 + \theta_1)(1 + \varphi_1 \theta_1)}{1 - \varphi_1^2} \varphi_1^{|h|-1}, \quad \forall h \in \mathbb{Z}^*. \quad (2.2.31)$$

We then find

$$\gamma(0) = \sigma^2 \frac{1 + \theta_1^2 + 2\varphi_1 \theta_1}{1 - \varphi_1^2}. \quad (2.2.32)$$

Note that, although the ARMA(1,1) time series assumes $\varphi_1 \neq 0$ and $\theta_1 \neq 1$, the results (2.2.31) and (2.2.32) with $\varphi_1 = 0$ yield the a.c.v.f. of the MA(1), given in Proposition 2.2.2, and these results with $\theta_1 = 0$ yield the a.c.v.f. of the AR(1), given in (2.2.8).

2.3 Autocovariance and related functions

The central topics of this section are covariance and correlation. Section 2.3.1 gives some comments on the computation of the a.c.v.f., Section 2.3.2 introduces the partial autocorrelation function (p.a.c.r.f.) and Section 2.3.3 introduces the autocovariance generating function (a.c.v.g.f.).

2.3.1 *Computation of autocovariance function*

This section is purely algorithmic. It first presents a recursive method for the computation of the coefficients of the series of the causal ARMA(p,q) time series. This allows to obtain the a.c.v.f. Then, particular Yule-Walker method for the causal AR(p) is introduced. This method is also used for estimating the parameters of the AR(p).

Causal ARMA(p,q) *time series*

According to the proof of Theorem 2.2.12.1, the ARMA(p,q) time series with equation

$$\varphi(B)X_k = \theta(B)Z_k, \quad \forall z \in \mathbb{Z},$$

is causal if $\varphi(z) \neq 0$, $\forall z \in \mathcal{D}_0$. (This implication does not require that that the polynomials φ and θ possess no common root over the disk $\mathcal{D}_0$.) Under this condition, Corollary 2.2.25 tells that the a.c.v.f. can be obtained from the coefficients of the series

$$\psi(z) = \frac{\theta(z)}{\varphi(z)} = \sum_{j=0}^{\infty} \psi_j z^j, \quad \forall z \in \mathcal{D}_0.$$

This leads us to solve w.r.t. $\psi_0, \psi_1, \ldots$ the equation

$$(1-\varphi_1 z-\varphi_2 z^2-\cdots-\varphi_p z^p)(\psi_0+\psi_1 z+\psi_2 z^2+\cdots) = 1+\theta_1 z+\theta_2 z^2+\cdots+\theta_q z^q.$$

This equation leads to the following recursive system of equations,

$$\psi_k - \sum_{j=1}^{k} \psi_{k-j}\varphi_j = \theta_k, \quad \text{if } k = 0,\ldots,q,$$

$$\psi_k - \sum_{j=1}^{p} \psi_{k-j}\varphi_j = 0, \quad \text{if } k = q+1, q+2, \ldots,$$

with $\theta_0 = 1$, $\psi_j = 0$, for $j < 0$, and $\varphi_j = 0$, for $j > p$, and with the convention $\sum_{j=1}^{0} = 0$. This is compactly re-expressed as

$$\psi_k - \sum_{j=1}^{k \wedge p} \psi_{k-j}\varphi_j = \begin{cases} \theta_k, & \text{if } k = 0,\ldots,q, \\ 0, & \text{if } k = q+1, q+2, \ldots. \end{cases} \quad (2.3.1)$$

Example 2.3.1 (MA(q)). *We consider the* MA(q) *time series. In this case, the recursive system of equations (2.3.1) simplifies to*

$$\psi_k = \begin{cases} \theta_k, & \text{for } k = 0, \dots, q, \\ 0, & \text{for } k = q+1, q+2, \dots. \end{cases}$$

This leads to the a.c.v.f. given in Proposition 2.2.2.

Example 2.3.2 (Causal AR(p)). *We consider the causal* AR(p) *time series, for $p = 1, 2, \dots$. The system of equations (2.3.1) simplifies to*

$$\psi_k = \begin{cases} 1, & \text{for } k = 0, \\ \sum_{j=1}^{k \wedge p} \psi_{k-j} \varphi_j, & \text{for } k = 1, 2, \dots. \end{cases} \tag{2.3.2}$$

Consider the alternative notation $\psi_{p,k}$ for the coefficient ψ_k belonging to the causal AR(p) *model, for $k = 0, 1, \dots$. It follows from (2.3.2) that*

$$\psi_{p,k} = \psi_{p+1,k} = \psi_{p+2,k} = \dots, \text{ for } k = 0, \dots, p.$$

Thus we have the following cases.

(1) *With $p = 1$, causality is equivalent to $|\varphi_1| < 1$. The recursive system of equations (2.3.2) simplifies to $\psi_0 = 1$ and $\psi_k = \psi_{k-1} \varphi_1$, for $k = 1, 2, \dots$. We thus obtain:*

$$\psi_0 = 1, \ \psi_1 = \varphi_1, \ \psi_2 = \varphi_1^2, \dots.$$

In particular, this leads to the a.c.v.f. (2.2.8) given in Proposition 2.2.3.

(2) *With $p = 2$, we obtain form (2.3.2):*

$$\psi_0 = 1, \ \psi_1 = \varphi_1,$$
$$\psi_2 - \psi_1 \varphi_1 - \psi_0 \varphi_2 = 0 \iff \psi_2 = \varphi_1^2 + \varphi_2,$$
$$\psi_3 - \psi_2 \varphi_1 - \psi_1 \varphi_2 = 0 \iff \psi_3 = \varphi_1^3 + 2\varphi_1 \varphi_2,$$
$$\psi_4 - \psi_3 \varphi_1 - \psi_2 \varphi_2 = 0 \iff \psi_4 = \varphi_1^4 + 3\varphi_1^2 \varphi_2 + \varphi_2^2,$$
$$\dots.$$

(3) *With $p = 3$, we obtain form (2.3.2):*

$$\psi_0 = 1, \ \psi_1 = \varphi_1, \ \psi_2 = \varphi_1^2 + \varphi_2,$$
$$\psi_3 - \psi_2 \varphi_1 - \psi_1 \varphi_2 - \psi_0 \varphi_3 = 0 \iff \psi_3 = \varphi_1^3 + 2\varphi_1 \varphi_2 + \varphi_3,$$
$$\psi_4 - \psi_3 \varphi_1 - \psi_2 \varphi_2 - \psi_1 \varphi_3 = 0 \iff \psi_4 = \varphi_1^4 + 3\varphi_1^2 \varphi_2 + \varphi_2^2 + 2\varphi_1 \varphi_3,$$
$$\dots.$$

Causal AR(p) *and Yule-Walker method*

Consider the causal AR(p) time series $\{X_k\}_{k\in\mathbb{Z}}$ and its sample $X_1,\dots,X_n$, where $n > p$. Let $k \in \mathbb{Z}$, then for $j = 0,\dots,p$, we have

$$X_{k-j}X_k - \varphi_1 X_{k-j}X_{k-1} - \cdots - \varphi_p X_{k-j}X_{k-p}$$
$$= X_{k-j}Z_k = \sum_{l=0}^{\infty} \psi_l Z_{k-j-l}Z_k, \tag{2.3.3}$$

where the second equality refers to the causal solution $X_k = \sum_{l=0}^{\infty} \psi_l Z_{k-l}$, $\{Z_k\}_{k\in\mathbb{Z}}$ being WN(σ^2). By applying the expectation to (2.3.3), we obtain

$$\gamma(j) - \varphi_1\gamma(j-1) - \cdots - \varphi_p\gamma(j-p) = \begin{cases} \sigma^2\psi_0, & \text{for } j = 0, \\ 0, & \text{for } j = 1,\dots,p. \end{cases} \tag{2.3.4}$$

Let

$$\boldsymbol{\Gamma} = \Big(\gamma(i-j)\Big)_{i,j=1,\dots,p}, \quad \boldsymbol{\varphi} = (\varphi_1,\dots,\varphi_p)^{\top} \text{ and } \boldsymbol{\gamma} = (\gamma(1),\dots,\gamma(p))^{\top}.$$

We have obtained

$$\boldsymbol{\Gamma}\boldsymbol{\varphi} = \boldsymbol{\gamma} \text{ and } \boldsymbol{\gamma}^{\top}\boldsymbol{\varphi} = \gamma(0) - \sigma^2. \tag{2.3.5}$$

These are the Yule-Walker equations and they allow us to determine $\gamma(0),\dots,\gamma(p)$ from $\varphi_1,\dots,\varphi_p$ and σ^2.

Alternatively, we can replace the a.c.v.f. γ by the empirical a.c.v.f. $\hat{\gamma}_n$ and define

$$\hat{\boldsymbol{\Gamma}}_n = \Big(\hat{\gamma}_n(i-j)\Big)_{i,j=1,\dots,p} \text{ and } \hat{\boldsymbol{\gamma}}_n = (\hat{\gamma}_n(1),\dots,\hat{\gamma}_n(p))^{\top}.$$

By inserting these quantities in (2.3.5), we obtain

$$\hat{\boldsymbol{\varphi}}_n = \hat{\boldsymbol{\Gamma}}_n^{-1}\hat{\boldsymbol{\gamma}}_n \text{ and } \hat{\sigma}_n^2 = \hat{\gamma}_n(0) - \hat{\boldsymbol{\gamma}}_n^{\top}\hat{\boldsymbol{\Gamma}}_n^{-1}\hat{\boldsymbol{\gamma}}_n,$$

that are called Yule-Walker estimators of $\varphi_1,\dots,\varphi_p$ and σ^2. Note that invertibility of $\hat{\boldsymbol{\Gamma}}_n$ holds when $\hat{\gamma}_n(0) > 0$.

Example 2.3.3 (Causal AR(2)). *For the causal* AR(2) *and a sample of size* $n > 2$ *we have:* $\hat{\gamma}_n(0) = 950$, $\hat{\gamma}_n(1) = 603$ *and* $\hat{\gamma}_n(2) = 290$. *The Yule-Walker estimators* $\hat{\varphi}_{1n}, \hat{\varphi}_{2n}$ *and* $\hat{\sigma}_n^2$ *are obtained as follows:*

$$\hat{\boldsymbol{\varphi}}_n = \begin{pmatrix} \hat{\varphi}_{1n} \\ \hat{\varphi}_{2n} \end{pmatrix} = \hat{\boldsymbol{\Gamma}}_n^{-1}\boldsymbol{\gamma}_n$$

$$= \frac{1}{\hat{\gamma}_n^2(0) - \hat{\gamma}_n^2(1)} \begin{pmatrix} \hat{\gamma}_n(0) & -\hat{\gamma}_n(1) \\ -\hat{\gamma}_n(1) & \hat{\gamma}_n(0) \end{pmatrix} \begin{pmatrix} \hat{\gamma}_n(1) \\ \hat{\gamma}_n(2) \end{pmatrix}$$

$$= \frac{1}{\hat{\gamma}_n^2(0) - \hat{\gamma}_n^2(1)} \begin{pmatrix} \hat{\gamma}_n(1)\{\hat{\gamma}_n(0) - \hat{\gamma}_n(2)\} \\ -\hat{\gamma}_n^2(1) + \hat{\gamma}_n(0)\hat{\gamma}_n(2) \end{pmatrix},$$

and

$$\hat{\sigma}_n^2 = \hat{\gamma}_n(0) - \left(\hat{\gamma}_n(1), \hat{\gamma}_n(2)\right) \begin{pmatrix} \hat{\varphi}_{1n} \\ \hat{\varphi}_{2n} \end{pmatrix}$$

$$= \hat{\gamma}_n(0) - \hat{\varphi}_{1n}\hat{\gamma}_n(1) - \hat{\varphi}_{2n}\hat{\gamma}_n(2).$$

This yields: $\hat{\varphi}_{1n} = 0.74$, $\hat{\varphi}_{1n} = -0.16$ *and* $\hat{\sigma}_n^2 = 552.09$.

Yule-Walker equations (2.3.4) hold for $j > p$ as well and thus they constitute a linear difference equation of order p, which may be solved directly. We illustrate this procedure in the following example.

Example 2.3.4 (A.c.r.f. of causal AR(2)). *Consider the causal* AR(2) *time series*

$$X_k - \frac{1}{3}X_{k-1} - \frac{2}{9}X_{k-2} = Z_k, \quad \forall k \in \mathbb{Z}.$$

We show that the Yule-Walker equations allow to obtain its a.c.r.f. in the following form,

$$\rho(h) = \frac{16}{21}\left(\frac{2}{3}\right)^{|h|} + \frac{5}{21}\left(-\frac{1}{3}\right)^{|h|}, \quad \forall h \in \mathbb{Z}. \tag{2.3.6}$$

Indeed, the Yule-Walker equations (2.3.4) lead to

$$\rho(j) - \frac{1}{3}\rho(j-1) - \frac{2}{9}\rho(j-2) = 0, \text{ for } j = 1, 2, \ldots. \tag{2.3.7}$$

This is a second order linear difference equation. According to Theorem A.9.2, the solution takes the form

$$\rho(j) = a_1 r_1^j + a_2 r_2^j,$$

whenever r_1 and r_2 are distinct roots of the characteristic equation

$$r^2 - \frac{1}{3}r - \frac{2}{9} = 0.$$

We find indeed $r_1 r_2 = -2/9$ and $r_1 + r_2 = 1/3$, viz. $r_1 = 2/3$ and $r_2 = -1/3$. Further, we know $\rho(0) = a_1 + a_2 = 1$ and (2.3.4) with $j = 1$ leads to $\rho(1) = 1/3 + 2/9\rho(1)$, thus $\rho(1) = 3/7$. We deduce $3/7 = a_1 2/3 + a_2(-1/3)$. Thus $a_1 = 16/21$ and $a_2 = 5/21$ and symmetry leads to (2.3.6).

2.3.2 Partial autocorrelation function

In statistics, the partial correlation coefficient between two random variables X and Y, for a given a set of $k \geq 1$ explanatory random variables $Z_1, \ldots, Z_k$, is the correlation between the residuals of the linear regression of X and Y on $Z_1, \ldots, Z_k$. The application of this notion to time series leads to the p.a.c.r.f. The operator pro denotes the orthogonal projection and $\overline{\text{sp}}$ the closed span, both defined in Section A.1.

Definition 2.3.5 (Partial autocorrelation function). *The partial autocorrelation function (p.a.c.r.f.) of the stationary time series* $\{X_k\}_{k \in \mathbb{Z}}$ *is given by*

$$\beta(k) = \begin{cases} \text{corr}(X_2, X_1), & \text{if } k = 1, \\ \text{corr}\big(X_{k+1} - \text{pro}_{\overline{\text{sp}}\{1, X_2, \ldots, X_k\}} X_{k+1}, X_1 - \text{pro}_{\overline{\text{sp}}\{1, X_2, \ldots, X_k\}} X_1\big), \\ \qquad \text{if } k = 2, 3, \ldots, \end{cases}$$

where the symbol 1 appearing in $\overline{\text{sp}}$ *denotes the constant random variable taking value one.*

Note that the part $k = 1$ of Definition 2.3.5 is justified by

$$\text{corr}\big(X_2 - \text{pro}_{\overline{\text{sp}}\{1\}} X_2, X_1 - \text{pro}_{\overline{\text{sp}}\{1\}} X_1\big) = \text{corr}(X_2 - \mathsf{E}[X_2], X_1 - \mathsf{E}[X_1])$$
$$= \text{corr}(X_2, X_1).$$

The p.a.c.r.f. $\beta(k)$ gives the correlation between the two residuals that are obtained from the orthogonal projection onto $\overline{\text{sp}}\{1, X_2, \ldots, X_k\}$ of X_{k+1} and of X_1, where $k = 2, 3, \ldots$. Heuristically, It measures the correlation between X_{k+1} and X_1 that subsists after removing all information of these two random variables that can be expressed in terms of the $k - 1$ intermediate random variables $X_2, \ldots, X_k$.

We can also give the following interpretation to the p.a.c.r.f.

- If $\beta(k) \simeq 0$, then the two residuals of the regression of X_{k+1} and of X_1 on the explanatory variables $1, X_2, \ldots, X_k$ are uncorrelated. This fact is considered as a validation for the regression model: because $1, X_2, \ldots, X_k$ appear as appropriate explanatory variables for X_{k+1} and X_1, the correlation lag can be considered $k - 1$, or perhaps less. This means that the $k - 1$ intermediate random variables, or perhaps less, contain sufficient information on X_1 and X_{k+1}.
- If $\beta(k) \simeq \pm 1$, then these two residuals are correlated. The regression model with $X_2, \ldots, X_k$ as explanatory variables cannot be validated:

because it should be improved by adding further explanatory viz. intermediate random variables $X_{k+1}, X_{k+2}, \ldots$, the correlation lag can be considered k or more. Thus, for some $j \geq 1$, $k + j - 1$ intermediate random variables are required in order to retain sufficient information on X_1 and X_{k+j+1}.

Example 2.3.6 (P.a.c.r.f. of h-correlated Gaussian time series). *Let us consider a the h-correlated and Gaussian time series $\{X_k\}_{k \in \mathbb{Z}}$. Refer to Definition 2.1.9 for h-correlation and h-dependence, which are equivalent in this example because the time series is Gaussian. Thus $\{X_k\}_{k \in \mathbb{Z}}$ is assumed stationary and we consider $h = k - 1$, for some $k \geq 2$. We consequently obtain*

$$\mathsf{corr}(X_1, X_{1+l}) \begin{cases} \neq 0, & \text{for } l = k - 1, \\ = 0, & \text{for } l = k,\, k+1, \ldots. \end{cases}$$

Define the residuals of projection

$$R_{k+1} = X_{k+1} - \mathsf{pro}_{\overline{\mathrm{sp}}\{1, X_2, \ldots, X_k\}} X_{k+1} \text{ and } R_1 = X_1 - \mathsf{pro}_{\overline{\mathrm{sp}}\{1, X_2, \ldots, X_k\}} X_1.$$

Thus R_{k+1} is a linear function of $X_{k+1}, X_2, \ldots, X_k$, R_1 is linear function of $X_1, X_2, \ldots, X_k$ and both R_{k+1} and R_1 are uncorrelated with $X_2, \ldots, X_k$. Thus R_{k+1} and R_1 are also Gaussian random variables. Through these remarks we obtain

$$\beta(k) = \mathsf{corr}\,(R_{k+1}, R_1) = \mathsf{corr}\,(R_{k+1}, R_1 \mid X_2, \ldots, X_k) = 0,$$

where the second equality above is due to independence and the third equality follows from $(k-1)$-correlation.

Let us now derive a computational formula for the p.a.c.r.f. Denote by $\{X_k\}_{k \in \mathbb{Z}}$ a stationary time series and let $k \in \{2, 3, \ldots\}$, then it holds from some $\beta_1, \ldots, \beta_k \in \mathbb{R}$ that

$$\mathsf{pro}_{\overline{\mathrm{sp}}\{1, X_2, \ldots, X_k\}} X_{k+1} = \beta_1 + \beta_2 X_2 + \cdots + \beta_k X_k.$$

The coefficients $\beta_1, \ldots, \beta_k$ are obtained from the normal equations

$$\left\langle X_{k+1} - \beta_1 - \sum_{j=2}^{k} \beta_j X_j,\, 1 \right\rangle = 0,$$

$$\left\langle X_{k+1} - \beta_1 - \sum_{j=2}^{k} \beta_j X_j,\, X_l \right\rangle = 0, \quad \text{for } l = 2, \ldots, k,$$

or, equivalently, from

$$
\begin{pmatrix}
1 & \mathsf{E}[X_2] & \dots & \mathsf{E}[X_k] \\
\mathsf{E}[X_2] & \mathsf{E}[X_2^2] & \dots & \mathsf{E}[X_2 X_k] \\
\vdots & \vdots & \ddots & \vdots \\
\mathsf{E}[X_k] & \mathsf{E}[X_k X_2] & \dots & \mathsf{E}[X_k^2]
\end{pmatrix}
\begin{pmatrix}
\beta_1 \\ \beta_2 \\ \vdots \\ \beta_k
\end{pmatrix}
=
\begin{pmatrix}
\mathsf{E}[X_{k+1}] \\ \mathsf{E}[X_2 X_{k+1}] \\ \vdots \\ \mathsf{E}[X_k X_{k+1}]
\end{pmatrix}.
$$

But $\{X_k\}_{k\in\mathbb{Z}}$ is stationary and therefore $\beta_1, \dots, \beta_k \in \mathbb{R}$ solve the equation

$$
\begin{pmatrix}
1 & \mathsf{E}[X_1] & \dots & \mathsf{E}[X_1] \\
\mathsf{E}[X_1] & \gamma(0) + \mathsf{E}^2[X_1] & \dots & \gamma(k-2) + \mathsf{E}^2[X_1] \\
\vdots & \vdots & \ddots & \vdots \\
\mathsf{E}[X_1] & \gamma(k-2) + \mathsf{E}^2[X_1] & \dots & \gamma(0) + \mathsf{E}^2[X_1]
\end{pmatrix}
\begin{pmatrix}
\beta_1 \\ \beta_2 \\ \vdots \\ \beta_k
\end{pmatrix}
$$

$$
=
\begin{pmatrix}
\mathsf{E}[X_1] \\ \gamma(k-1) + \mathsf{E}^2[X_1] \\ \vdots \\ \gamma(1) + \mathsf{E}^2[X_1]
\end{pmatrix}. \tag{2.3.8}
$$

We obtain in a similar way that

$$
\mathrm{pro}_{\overline{\mathrm{sp}}\{1,X_2,\dots,X_k\}} X_1 = \alpha_1 + \alpha_2 X_2 + \dots + \alpha_k X_k,
$$

where $\alpha_1, \dots, \alpha_k \in \mathbb{R}$ solve the equation

$$
\begin{pmatrix}
1 & \mathsf{E}[X_1] & \dots & \mathsf{E}[X_1] \\
\mathsf{E}[X_1] & \gamma(0) + \mathsf{E}^2[X_1] & \dots & \gamma(k-2) + \mathsf{E}^2[X_1] \\
\vdots & \vdots & \ddots & \vdots \\
\mathsf{E}[X_1] & \gamma(k-2) + \mathsf{E}^2[X_1] & \dots & \gamma(0) + \mathsf{E}^2[X_1]
\end{pmatrix}
\begin{pmatrix}
\alpha_1 \\ \alpha_2 \\ \vdots \\ \alpha_k
\end{pmatrix} \tag{2.3.9}
$$

$$
=
\begin{pmatrix}
\mathsf{E}[X_1] \\ \gamma(1) + \mathsf{E}^2[X_1] \\ \vdots \\ \gamma(k-1) + \mathsf{E}^2[X_1]
\end{pmatrix}.
$$

Thus, the two vectors of coefficients $(\beta_1, \dots, \beta_k)^\top$ and $(\alpha_1, \dots, \alpha_k)^\top \in \mathbb{R}^k$ that solve (2.3.8) and (2.3.9) only depend on the mean and on the a.c.v.f. at time lags $h = 0, \dots, k-1$. This implies that the same two vectors of coefficients give, for $l = k+1, k+2, \dots$,

$$
\mathrm{pro}_{\overline{\mathrm{sp}}\{1,X_{l-k+1},\dots,X_{l-1}\}} X_l = \beta_1 + \beta_2 X_{l-k+1} + \dots + \beta_k X_{l-1}
$$

and

$$
\mathrm{pro}_{\overline{\mathrm{sp}}\{1,X_{l-k+1},\dots,X_{l-1}\}} X_{l-k} = \alpha_1 + \alpha_2 X_{l-k+1} + \dots + \alpha_k X_{l-1}.
$$

Therefore for $k = 2, 3, \ldots$ and for $l = k + 1, k + 2, \ldots$, we have

$$\beta(k)$$
$$= \mathrm{corr}\Big(X_l - \mathrm{pro}_{\overline{\mathrm{sp}}\{1, X_{l-k+1}, \ldots, X_{l-1}\}} X_l, X_{l-k} - \mathrm{pro}_{\overline{\mathrm{sp}}\{1, X_{l-k+1}, \ldots, X_{l-1}\}} X_{l-k}\Big).$$

In other terms, stationarity implies the above shift invariance property of the p.a.c.r.f. This makes Definition 2.3.5 of the p.a.c.r.f. meaningful.

Example 2.3.7 (Causal AR(1)). *Let $\{X_k\}_{k \in \mathbb{Z}}$ be the time series given by*

$$X_k = 0.9 X_{k-1} + Z_k, \quad \forall k \in \mathbb{Z},$$

where $\{Z_k\}_{k \in \mathbb{Z}}$ is $\mathrm{WN}(\sigma^2)$. Since $\rho(z) = 1 - 0.9z$ has not root in $\mathcal{D}_0$, the time series $\{X_k\}_{k \in \mathbb{Z}}$ is causal. Proposition 2.2.3 implies that the time series is stationary with zero mean and a.c.v.f. given by (2.2.8). Thus the a.c.r.f. is

$$\rho(h) = 0.9^{|h|}, \quad \forall h \in \mathbb{Z}.$$

Let $k \in \{2, 3, \ldots\}$. Because the mean is null, (2.3.8) becomes

$$\begin{pmatrix} 1 & 0 & \ldots & 0 \\ 0 & \gamma(0) & \ldots & \gamma(k-2) \\ \vdots & \vdots & \ddots & \vdots \\ 0 & \gamma(k-2) & \ldots & \gamma(0) \end{pmatrix} \begin{pmatrix} \beta_1 \\ \beta_2 \\ \vdots \\ \beta_k \end{pmatrix} = \begin{pmatrix} 0 \\ \gamma(k-1) \\ \vdots \\ \gamma(1) \end{pmatrix}.$$

Then $\beta_1 = 0$ and in this case the operators $\mathrm{pro}_{\overline{\mathrm{sp}}\{1, X_2, \ldots, X_k\}}$ and $\mathrm{pro}_{\overline{\mathrm{sp}}\{X_2, \ldots, X_k\}}$ become equivalent. The equivalence of these operators when the mean is null can be precisely justified from the linearity, the idempotence and the iterativity of the linear projection, that are given in Proposition A.1.2. We can thus reduce the above equation to

$$\begin{pmatrix} \gamma(0) & \ldots & \gamma(k-2) \\ \vdots & \ddots & \vdots \\ \gamma(k-2) & \ldots & \gamma(0) \end{pmatrix} \begin{pmatrix} \beta_2 \\ \vdots \\ \beta_k \end{pmatrix} = \begin{pmatrix} \gamma(k-1) \\ \vdots \\ \gamma(1) \end{pmatrix}.$$

and, by dividing both sides by $\gamma(0)$ we obtain,

$$\begin{pmatrix} 1 & 0.9 & 0.9^2 & \ldots & 0.9^{k-2} \\ 0.9 & 1 & 0.9 & \ldots & 0.9^{k-3} \\ 0.9^2 & 0.9 & 1 & \ldots & 0.9^{k-4} \\ \vdots & \vdots & \vdots & \ddots & \vdots \\ 0.9^{k-2} & 0.9^{k-3} & 0.9^{k-4} & \ldots & 1 \end{pmatrix} \begin{pmatrix} \beta_2 \\ \beta_3 \\ \beta_4 \\ \vdots \\ \beta_k \end{pmatrix} = \begin{pmatrix} 0.9^{k-1} \\ 0.9^{k-2} \\ 0.9^{k-3} \\ \vdots \\ 0.9 \end{pmatrix}.$$

We have a constant diagonal matrix, also called Toeplitz matrix. The system is solved by $\beta_k = 0.9$ and $\beta_2 = \cdots = \beta_{k-1} = 0$, and so,

$\mathsf{pro}_{\overline{\mathrm{sp}}\{1,X_2,\ldots,X_k\}}X_k = 0.9X_k.$ *Similarly,* $\mathsf{pro}_{\overline{\mathrm{sp}}\{1,X_2,\ldots,X_k\}}X_1 = 0.9X_2,$ *because the system becomes*

$$
\begin{pmatrix}
1 & 0.9 & 0.9^2 & \ldots 0.9^{k-2} \\
0.9 & 1 & 0.9 & \ldots 0.9^{k-3} \\
0.9^2 & 0.9 & 1 & \ldots 0.9^{k-4} \\
\vdots & \vdots & \vdots & \ddots & \vdots \\
0.9^{k-2} & 0.9^{k-3} & 0.9^{k-4} & \ldots & 1
\end{pmatrix}
\begin{pmatrix}
\alpha_2 \\ \alpha_3 \\ \alpha_4 \\ \vdots \\ \alpha_k
\end{pmatrix}
=
\begin{pmatrix}
0.9 \\ 0.9^2 \\ 0.9^3 \\ \vdots \\ 0.9^{k-1}
\end{pmatrix},
$$

which yields $\alpha_2 = 0.9$ *and* $\alpha_3 = \cdots = \alpha_k = 0.$ *So for* $k = 2, 3, \ldots,$

$$\beta(k) = \mathsf{corr}(X_{k+1} - 0.9X_k, X_1 - 0.9X_2) = \mathsf{corr}(Z_{k+1}, X_1 - 0.9X_2) = 0.$$

In particular $\beta(2) = \mathsf{corr}(X_3 - 0.9X_2, X_1 - 0.9X_2) = 0.$ *Thus, if the causal* AR(1) *model holds, then a plot of* $(X_{k-2} - 0.9X_{k-1}, X_k - 0.9X_{k-1})$ *should not display any particular structure.*

Example 2.3.8 (Causal AR(p)). *Let* $\{X_k\}_{k\in\mathbb{Z}}$ *be any causal* AR(p) *time series. Let* $k \geq p+1$ *i.e.* $k + 1 - p \geq 2.$ *Then we find that*

$$\mathsf{pro}_{\overline{\mathrm{sp}}\{1,X_2,\ldots,X_k\}}X_{k+1} = \mathsf{pro}_{\overline{\mathrm{sp}}\{X_2,\ldots,X_k\}}X_{k+1} = \sum_{j=1}^{p}\varphi_j X_{k+1-j},$$

because the mean is null and because of the following explanation:

$$Y \in \overline{\mathrm{sp}}\{X_2,\ldots,X_k\}, \ X_l = \sum_{j=0}^{\infty}\psi_j Z_{l-j}, \ \forall l \in \mathbb{Z} \Longrightarrow Y \in \overline{\mathrm{sp}}\{Z_j\}_{j\leq k}$$

$$\Longrightarrow \left\langle X_{k+1} - \sum_{j=1}^{p}\varphi_j X_{k+1-j}, Y \right\rangle = \langle Z_{k+1}, Y \rangle = 0,$$

where $\{Z_k\}_{k\in\mathbb{Z}}$ *is WN and* $\{\psi_k\}_{k\in\mathbb{Z}} \in \mathbb{R}^{\infty}.$ *Therefore,*

$$\beta(k) = \mathsf{corr}\left(X_{k+1} - \sum_{j=1}^{p}\varphi_j X_{k+1-j}, X_1 - \mathsf{pro}_{\overline{\mathrm{sp}}\{X_2,\ldots,X_k\}}X_1\right)$$

$$= \mathsf{corr}(Z_{k+1}, Y) = 0,$$

where $Y \in \overline{\mathrm{sp}}\{X_1,\ldots,X_k\}.$

The computation for $k \leq p$ *can be obtained from the next proposition and* $\beta(k) \neq 0,$ *generally.*

Let $\{X_k\}_{k\in\mathbb{Z}}$ be a stationary time series. Let $k \in \mathbb{N}^*$ and let $\varphi_{k,1}, \ldots, \varphi_{k,k} \in \mathbb{R}$ satisfy

$$\mathsf{pro}_{\overline{\mathrm{sp}}\{X_1,X_2,\ldots,X_k\}}X_{k+1} = \varphi_{k,k}X_1 + \varphi_{k,k-1}X_2 + \cdots + \varphi_{k,1}X_k$$

$$= \sum_{l=1}^{k}\varphi_{k,l}X_{k+1-l}.$$

Then for $j = 1, \ldots, k$,

$$\left\langle X_{k+1} - \sum_{l=1}^{k} \varphi_{k,l} X_{k+1-l}, X_j \right\rangle = 0 \iff \gamma(k+1-j) = \sum_{l=1}^{k} \varphi_{k,l} \gamma(k+1-j-l).$$

By dividing both sides of the last equality by $\gamma(0)$, we obtain that $(\varphi_{k,1}, \ldots, \varphi_{k,k}) \in \mathbb{R}^k$ solves

$$\begin{pmatrix} \rho(0) & \rho(1) & \ldots & \rho(k-1) \\ \rho(1) & \rho(0) & \ldots & \rho(k-2) \\ \vdots & \vdots & \ddots & \vdots \\ \rho(k-1) & \rho(k-2) & \cdots & \varphi(0) \end{pmatrix} \begin{pmatrix} \varphi_{k,1} \\ \varphi_{k,2} \\ \vdots \\ \varphi_{k,k} \end{pmatrix} = \begin{pmatrix} \rho(1) \\ \rho(2) \\ \vdots \\ \rho(k) \end{pmatrix}. \tag{2.3.10}$$

Proposition 2.3.9. *Let $\{X_k\}_{k\in\mathbb{Z}}$ be a stationary time series with zero mean and a.c.v.f. γ, such that $\gamma(h) \to 0$, as $h \to \infty$. Then its p.a.c.r.f. is given by*

$$\beta(k) = \varphi_{k,k}, \ \forall k \in \mathbb{N}^*.$$

where $\varphi_{k,k}$ is implicitly defined in (2.3.10).

This proposition is given without proof and provides an equivalent definition of the p.a.c.r.f. The sample p.a.c.r.f. is obtained by replacing φ by $\hat{\varphi}_n$.

2.3.3 *Autocovariance generating function*

This section introduces a the generating function of the a.c.v.f., which is called autocovariance generating function (a.c.v.g.f.). The a.c.v.g.f. of ARMA time series is provided. With the help of this the a.c.v.g.f. it is shown that under a weak condition, an ARMA(p, q) time series can be transformed to a causal and invertible ARMA(p, q) time series, by appropriate choices AR and MA polynomials and of WN.

Definition 2.3.10 (A.c.v.g.f.). *Consider a stationary time series with a.c.v.f. γ, then its a.c.v.g.f. is given by*

$$G(z) = \sum_{k=-\infty}^{\infty} \gamma(k) z^k,$$

whenever there exists $r > 1$ such that $G(z)$ is convergent $\forall z \in \mathcal{A}_r^\circ$, which is the interior of the closed annulus delimited by the radii $1/r$ and r.

Example 2.3.11 (A.c.v.g.f. of WN). *A stationary time series is WN(σ^2) iff its a.c.v.g.f. is $G(z) = \sigma^2$, $\forall z \in \mathbb{C}$.*

Proposition 2.3.12 (A.c.v.g.f. of stationary time series). *Let $\{Z_k\}_{k\in\mathbb{Z}}$ be* $\mathrm{WN}(\sigma^2)$ *and let $\{\psi_j\}_{j\in\mathbb{Z}}$ be a sequence of real numbers such that,*

$$\sum_{j=-\infty}^{\infty} |\psi_j z^j| < \infty, \quad \forall z \in \mathcal{A}_r^\circ,$$

for some $r > 1$. Then the a.c.v.g.f. of the time series given by

$$X_k = \sum_{j=-\infty}^{\infty} \psi_j Z_{k-j}, \quad \forall k \in \mathbb{Z},$$

is given by

$$G(z) = \sigma^2 \psi(z)\psi(z^{-1}), \quad \forall z \in \mathcal{A}_r^\circ.$$

Proof. It follows from Proposition 2.2.6 that $\{X_k\}_{k\in\mathbb{Z}}$ is a stationary time series with mean null and a.c.v.f. given by, $\forall h \in \mathbb{Z}$,

$$\gamma(h) = \sum_{j=-\infty}^{\infty} \sum_{l=-\infty}^{\infty} \psi_j \psi_l \gamma_Z(h - j + l)$$

$$= \sigma^2 \sum_{j=-\infty}^{\infty} \psi_j \psi_{j-h},$$

where $\gamma_Z(h) = \sigma^2 \mathbb{I}\{h = 0\}$, $\forall h \in \mathbb{Z}$, is the a.c.v.f. of $\mathrm{WN}(\sigma^2)$. Consequently,

$$G(z) = \sigma^2 \sum_{k=-\infty}^{\infty} \sum_{j=-\infty}^{\infty} \psi_j \psi_{j-k} z^k$$

$$= \sigma^2 \sum_{j=-\infty}^{\infty} \psi_j z^j \sum_{k=-\infty}^{\infty} \psi_{j-k} z^{-(j-k)}$$

$$= \sigma^2 \psi(z)\psi(z^{-1}),$$

which is by assumption convergent $\forall z \in \mathcal{A}_r^\circ$. (Refer to the remark at p. 50 for the rearrangement of the sum.) $\qquad\square$

The direct application of Theorem 2.2.8 and Proposition 2.3.12 to the ARMA time series yields Corollary 2.3.13.

Corollary 2.3.13 (A.c.v.g.f. of ARMA(p,q) time series). *Let $\{X_k\}_{k\in\mathbb{Z}}$ be the ARMA(p,q) time series with equation*

$$\varphi(B)X_k = \theta(B)Z_k, \quad \forall k \in \mathbb{Z},$$

where $\{Z_k\}_{k\in\mathbb{Z}}$ is $\mathrm{WN}(\sigma^2)$. *Assume that* $\varphi(z)\neq 0$, $\forall z\in\mathbb{C}$ *such that* $|z|=1$. *Then the a.c.v.g.f. of* $\{X_k\}_{k\in\mathbb{Z}}$ *is given by*

$$G(z)=\sigma^2\psi(z)\psi(z^{-1})=\sigma^2\frac{\theta(z)\theta(z^{-1})}{\varphi(z)\varphi(z^{-1})},\quad \forall z\in\mathcal{A}_r^\circ,$$

for some $r>1$, is the a.c.v.g.f. of $\{X_k\}_{k\in\mathbb{Z}}$, where $\ldots,\psi_{-1},\psi_0,\psi_1,\ldots\in\mathbb{R}$ are the coefficients of the Laurent series of

$$\psi(z)=\frac{\theta(z)}{\varphi(z)},\text{ i.e. of } \psi(z)=\sum_{j=-\infty}^{\infty}\psi_j z^j,\quad \forall z\in\mathcal{A}_r^\circ.$$

Proof. We know from Theorem 2.2.8 that $\sum_{j=-\infty}^{\infty}|\psi_j|<\infty$ and that

$$X_k=\sum_{j=-\infty}^{\infty}\psi_j Z_{k-j},\quad \forall k\in\mathbb{Z},$$

where $\ldots,\psi_{-1},\psi_0,\psi_1,\ldots\in\mathbb{R}$ are the coefficients of the above Laurent series. Let $z\in\mathbb{C}$ such that $|z|=1$. We have

$$\sum_{j=-\infty}^{\infty}|\psi_j z^j|<\infty.$$

Because of continuity, this inequality holds $\forall z\in\mathcal{A}_r^\circ$, for some $r>1$. Then Proposition 2.3.12 gives

$$G(z)=\sigma^2\psi(z)\psi(z^{-1})=\sigma^2\frac{\theta(z)\theta(z^{-1})}{\varphi(z)\varphi(z^{-1})},\quad \forall z\in\mathcal{A}_r^\circ.$$

$\square$

Example 2.3.14 (MA(2)). *The a.c.v.g.f. of the MA(2) time series is*

$$G(z)=\sum_{k=-\infty}^{\infty}\gamma(k)z^k=\sigma^2\big(1+\theta_1 z+\theta_2 z^2\big)\big(1+\theta_1 z^{-1}+\theta_2 z^{-2}\big),$$

which implies

$$\gamma(0)=\sigma^2(1+\theta_1^2+\theta_2^2),\ \gamma(-1)=\gamma(1)=\sigma^2\theta_1(1+\theta_2),\ \gamma(-2)=\gamma(2)=\sigma^2\theta_2,$$

and $\gamma(h)=0$, for $h=\pm 3,\pm 4,\ldots.$ This result is also obtained in Proposition 2.2.2.

Example 2.3.15 (Conversion to invertible MA(1)). *Let $\{Z_k\}_{k\in\mathbb{Z}}$ be* $\mathrm{WN}(\sigma^2)$ *and let $\{X_k\}_{k\in\mathbb{Z}}$ be the MA(1) time series with equation*

$$X_k=Z_k-2Z_{k-1},\quad \forall k\in\mathbb{Z}.$$

The time series $\{X_k\}_{k\in\mathbb{Z}}$ is not invertible, because the polynomial $\theta(z) = 1 - 2z$ vanishes at $z = 1/2 \in \mathcal{D}_0$; cf. Theorem 2.2.14.1. Let $\varphi(z) = 1 - z/2$ and let $\{Z_k^*\}_{k\in\mathbb{Z}}$ be the ARMA(1, 1) time series satisfying

$$\varphi(B)Z_k^* = \theta(B)Z_k, \quad \forall k \in \mathbb{Z}.$$

The a.c.v.g.f. of $\{Z_k^*\}_{k\in\mathbb{Z}}$ is given by

$$G(z) = \sigma^2 \frac{\theta(z)\theta(z^{-1})}{\varphi(z)\varphi(z^{-1})} = \sigma^2 \frac{(1-2z)\left(1-\frac{2}{z}\right)}{\left(1-\frac{z}{2}\right)\left(1-\frac{1}{2z}\right)} = 4\sigma^2 \frac{(1-2z)\left(1-\frac{2}{z}\right)z}{(2-z)\left(2-\frac{1}{z}\right)z} = 4\sigma^2.$$

Thus $\{Z_k^*\}_{k\in\mathbb{Z}}$ is WN($4\sigma^2$). Moreover,

$$X_k = \theta(B)Z_k = \varphi(B)Z_k^*, \quad \forall k \in \mathbb{Z},$$

which is invertible w.r.t. the WN $\{Z_k^*\}_{k\in\mathbb{Z}}$, because $\varphi(z) = 0$ iff $z = 2$ and $2 \notin \mathcal{D}_0$. Thus, with formal notation we obtain

$$Z_k^* = \frac{X_k}{\varphi(B)} = \frac{1}{1-\frac{B}{2}}X_k = \sum_{j=0}^{\infty}\left(\frac{B}{2}\right)^j X_k = \sum_{j=0}^{\infty}\frac{1}{2^j}X_{k-j}, \quad \forall k \in \mathbb{Z}.$$

The next proposition generalizes this example, by converting an ARMA(p, q) to another causal and invertible ARMA(p, q).

Proposition 2.3.16 (Conversion to causal and invertible ARMA(p, q)). *Let $\{X_k\}_{k\in\mathbb{Z}}$ be the ARMA(p, q) time series with equation*

$$\varphi(B)X_k = \theta(B)Z_k, \quad \forall k \in \mathbb{Z},$$

where $\{Z_k\}_{k\in\mathbb{Z}}$ is WN(σ^2). Assume that $\varphi(z) \neq 0$ and $\theta(z) \neq 0$, $\forall z \in \mathbb{C}$ such that $|z| = 1$. Then there exist two polynomials $\tilde{\varphi}$ and $\tilde{\theta}$, of respective orders p and q, and a WN time series $\{Z_k^*\}_{k\in\mathbb{Z}}$ such that $\tilde{\varphi}(z) \neq 0$ and $\tilde{\theta}(z) \neq 0$, $\forall z \in \mathcal{D}_0$, and

$$\tilde{\varphi}(B)X_k = \tilde{\theta}(B)Z_k^*, \quad \forall k \in \mathbb{Z}.$$

Thus $\{X_k\}_{k\in\mathbb{Z}}$ is causal and invertible ARMA(p, q) w.r.t. the WN $\{Z_k^*\}_{k\in\mathbb{Z}}$.

Proof. Let $a_{r+1}, \ldots, a_p$ and $b_{s+1}, \ldots, b_q$ be the roots of φ and θ, respectively, that belong to $\mathcal{D}_0^\circ$. Define

$$\tilde{\varphi}(z) = \varphi(z) \prod_{j=r+1}^{p} \frac{1 - a_j z}{1 - \frac{z}{a_j}} \quad \text{and} \quad \tilde{\theta}(z) = \theta(z) \prod_{j=s+1}^{q} \frac{1 - b_j z}{1 - \frac{z}{b_j}},$$

with the convention that if $p = 0$, then $\prod_{j=r+1}^{p} = 1$, and if $q = 0$, then $\prod_{j=s+1}^{q} = 1$. We note that $\tilde{\varphi}$ and $\tilde{\theta}$ are polynomials with real coefficients[2] and of degrees p and q, respectively, and that their roots lie outside the unit disk $\mathcal{D}_0$. In particular, these roots have modulus different than one and so the Laurent series of $\tilde{\theta}(z)/\tilde{\varphi}(z)$ converges absolutely $\forall z \in \mathcal{A}_r^\circ$, for some $r > 0$. This fact and Proposition 2.2.5 allow us to define, with formal notation,

$$Z_k^* = \frac{\tilde{\varphi}(B)}{\tilde{\theta}(B)} X_k, \quad \forall k \in \mathbb{Z},$$

as a time series in $\mathcal{L}_2$. Let

$$\chi(B) = \frac{\prod_{j=r+1}^{p} \frac{1-a_j B}{1-\frac{B}{a_j}}}{\prod_{j=s+1}^{q} \frac{1-b_j B}{1-\frac{B}{b_j}}} \cdot$$

Then we have

$$Z_k^* = \frac{\tilde{\varphi}(B)}{\tilde{\theta}(B)} X_k = \chi(B) \frac{\varphi(B)}{\theta(B)} X_k = \chi(B) Z_k, \quad \forall k \in \mathbb{Z},$$

because of the commutativity of MA operators of Proposition 2.2.7.

So $\{Z_k^*\}_{k \in \mathbb{Z}}$ is the MA transform of a stationary time series and thus itself stationary (cf. Proposition 2.2.6). Its a.c.v.g.f. is given by

$G(z)$
$$= \sigma^2 \chi(x) \chi(z^{-1})$$
$$= \sigma^2 \frac{\prod_{j=r+1}^{p}(1 - a_j z)\prod_{j=s+1}^{q}(1 - \frac{z}{b_j})\prod_{j=r+1}^{p}(1 - \frac{a_j}{z})\prod_{j=s+1}^{q}(1 - \frac{1}{b_j z})}{\prod_{j=r+1}^{p}(1 - \frac{z}{a_j})\prod_{j=s+1}^{q}(1 - b_j z)\prod_{j=r+1}^{p}(1 - \frac{1}{a_j z})\prod_{j=s+1}^{q}(1 - \frac{b_j}{z})}$$
$$= \sigma^2 \prod_{j=r+1}^{p} \underbrace{\frac{1 - a_j z}{1 - \frac{1}{a_j z}}}_{=-a_j z} \underbrace{\frac{1 - \frac{a_j}{z}}{1 - \frac{z}{a_j}}}_{=-\frac{a_j}{z}} \prod_{j=s+1}^{q} \underbrace{\frac{1 - \frac{z}{b_j}}{1 - \frac{b_j}{z}}}_{=-\frac{z}{b_j}} \underbrace{\frac{1 - \frac{1}{b_j z}}{1 - b_j z}}_{=-\frac{1}{b_j z}}$$
$$= \sigma^2 \frac{\prod_{j=r+1}^{p} |a_j|^2}{\prod_{j=s+1}^{q} |b_j|^2},$$

---

[2]If the coefficients of a polynomial are real, then its purely imaginary roots appear in conjugate pairs. So we have

$$(z - a_j)(z - \overline{a_j}) \frac{1 - a_j z}{1 - \frac{z}{a_j}} \frac{1 - \overline{a_j} z}{1 - \frac{z}{\overline{a_j}}} = |a_j|^2 (|a_j|^2 z^2 - 2\text{Re}\,(a_j) z + 1).$$

$\forall z \in \mathcal{A}_r^\circ$ and for some $r > 0$. Thus $\{Z_k^*\}_{k \in \mathbb{Z}}$ is WN. Because $\tilde{\varphi}(z) \neq 0$ and $\tilde{\theta}(z) \neq 0$, $\forall z \in \mathcal{D}_0$, and because

$$\tilde{\varphi}(B)X_k = \tilde{\theta}(B)Z_k^*, \quad \forall k \in \mathbb{Z},$$

the time series $\{X_k\}_{k \in \mathbb{Z}}$ is indeed causal and invertible w.r.t. $\{Z_k^*\}_{k \in \mathbb{Z}}$, as given by Corollary 2.2.16.

Note that if $\operatorname{Im} a_j > 0$ for some $j = r + 1, \ldots, p$, then $\overline{a}_j$ also is a root of φ, belonging also to $\mathcal{D}_0$. Similar argument holds for the roots of θ.

This implies that $\tilde{\varphi}$ and $\tilde{\theta}$ have real coefficients and G is a positive and constant function. Thus $\{Z_k^*\}_{k \in \mathbb{Z}}$ is WN and, because $\tilde{\varphi}(z) \neq 0$ and $\tilde{\theta}(z) \neq 0$, $\forall z \in \mathcal{D}_0$, the time series $\{X_k\}_{k \in \mathbb{Z}}$ is causal and invertible w.r.t. $\{Z_k^*\}_{k \in \mathbb{Z}}$. $\qquad \square$

2.4 Analysis in frequency domain

The methods for the analysis of stationary time series presented so far act on the time domain. An alternative approach is provided by the so-called frequency domain analysis, where the spectral distribution takes the central role that the a.c.v.f. has in the time domain analysis. quantity of the time series. The stationary time series, viewed as a random signal over time, is the sum of various harmonics, possessing several frequencies with random amplitudes. The frequency approach offers various advantages that make it more convenient for various applications, in particular in engineering. Consider for example the reliability control of an engine: a faulty gear wheel of particular type or size can be detected by the identification of its particular frequency in the analysis of the sound frequencies. This section considers complex-valued instead of real-valued time series. This generalization is obtained conceptual difficulties.

Section 2.4.1 starts by presenting complex-valued time series and gives then the spectral decomposition of a time series. Section 2.4.2 introduces Herglotz's theorem, which relates the a.c.v.f. to the spectral distribution, and provides further results on the spectral distribution. Section 2.4.3 presents the spectral density of ARMA time series and introduces the theory of linear filters. Section 2.4.4 provides the important theoretical results that stationary time series can be arbitrarily well approximated by AR and MA time series. Section 2.4.5 introduces the periodogram, which is a nonparametric estimator of the spectral density.

2.4.1 Complex-valued time series and spectral decomposition

In this section we illustrate the essential features of the spectral decomposition with arbitrary stationary and complex-valued time series. It may be more direct to consider complex-valued instead of real-valued time series. Besides this heuristic justification, complex-valued time series provide sometimes preferred representations of bivariate signals, mainly because their compact formulation. They have been applied in various technical domains, such as digital signal processing. magnetic resonance imaging, cf. e.g. Rowe (2005), or oceanography, cf. e.g. Gonella (1972).

A complex-valued random variable is a straightforward generalization of the real-valued random variable: the possible values it can take are complex numbers. Thus any complex-valued random variable is a pair of real-valued random variables. The distribution of a complex random variable can be represented by the joint distribution of two real random variables.

The Hilbert space of complex-valued and square-integrable random variables is rigorously defined in Section A.1.2. In this paragraph we merely report some important facts. A random variable taking values in $\mathbb{C}$ has the form

$$X = U + \mathrm{i}V,$$

where U and V are real-valued random variables defined over the probability space $(\Omega, \mathcal{F}, \mathsf{P})$.

Whenever $\mathsf{E}[U]$ and $\mathsf{E}[V]$ exist, we define the expectation of X as

$$\mathsf{E}[X] = \mathsf{E}[U] + \mathrm{i}\mathsf{E}[V].$$

All complex-valued random variables X and Y satisfying $\mathsf{E}[|X|^2] < \infty$ and $\mathsf{E}[|Y|^2] < \infty$ constitute the Hilbert space $\mathcal{L}_2(\Omega, \mathcal{F}, \mathsf{P})$ with scalar product

$$\langle X, Y \rangle = \mathsf{E}\left[X\overline{Y}\right],$$

and seminorm[3]

$$\|X\| = \langle X, X \rangle^{\frac{1}{2}} = \mathsf{E}^{\frac{1}{2}}\left[|X|^2\right].$$

Definition 2.4.1 (Variance, covariance and pseudo-covariance of complex random variables). *Let $X, Y \in \mathcal{L}_2$ be complex-valued random variables. Their covariance is given by*

$$\mathsf{cov}(X, Y) = \mathsf{E}\left[(X - \mathsf{E}[X])\overline{(Y - \mathsf{E}[Y])}\right],$$

---

[3]Cf. p. 343.

and their pseudo-covariance is given by

$$\text{cov}^*(X,Y) = \text{E}[(X - \text{E}[X])(Y - \text{E}[Y])].$$

Also, the variance of X is given by

$$\text{var}(X) = \text{cov}(X,X) = \text{E}\left[|X - \text{E}[X]|^2\right].$$

Note the equivalent and practical formulas for covariance

$$\text{cov}(X,Y) = \text{E}\left[X\overline{Y}\right] - \text{E}[X]\text{E}\left[\overline{Y}\right],$$

and variance

$$\text{var}(X) = \text{cov}(X,X) = \text{E}\left[|X|^2\right] - |\text{E}[X]|^2.$$

Two basic formulae are

$$\text{var}(X) = \text{var}(U) + \text{var}(V),$$

and

$$\text{var}(X + Y) = \text{var}(X) + \text{var}(Y) + \text{cov}(X,Y) + \overline{\text{cov}(X,Y)}.$$

The following example shows that with complex-valued random variables, the concepts of independence and covariance null (i.e. orthogonality, when means are null) can differ substantially.

Example 2.4.2 (Total dependence and null covariance). *Let $X = U + iV$, for real-valued random variables U, V satisfying $\text{E}[U] = \text{E}[V] = 0$, $\sigma^2 = \text{var}(U) = \text{var}(V) < \infty$ and $\text{cov}(U,V) = 0$. In this situation, we find*

$$\text{cov}(X,\overline{X}) = \text{E}[XX] = \text{E}[(U + iV)^2] = \text{E}[U^2] - \text{E}[V^2] + 2i\,\text{E}[UV] = 0.$$

Therefore, although X and $\overline{X}$ are totally dependent, they have null covariance or, in other terms, they are orthogonal. Thus null covariance or orthogonality in the complex case cannot always be understood as approximate independence, like in the real case.

We note however that the pseudo-covariance is nonnull,

$$\text{cov}^*(X,\overline{X}) = \text{E}[X\overline{X}] = \text{E}[(U + iV)(U - iV)] = \text{E}[U^2] + \text{E}[V^2] = 2\sigma^2 > 0.$$

This example motivates the following specific definition of uncorrelation for complex-valued random variables.

Definition 2.4.3 (Uncorrelated complex-valued random variables). *Two complex-valued random variables $X, Y \in \mathcal{L}_2$ are uncorrelated if*

$$\text{cov}(X,Y) = \text{cov}^*(X,Y) = 0.$$

The a.c.v.f. of a complex-valued time series is the following.

Definition 2.4.4 (A.c.v.f. of complex-valued time series). *The a.c.v.f. of the complex-valued time series* $\{X_k\}_{k \in \mathbb{Z}}$ *in* $\mathcal{L}_2$ *is the function*

$$\gamma : \quad \mathbb{Z} \times \mathbb{Z} \to \mathbb{C}(k+h, k)$$
$$\mapsto \mathsf{cov}(X_{k+h}, X_k) = \mathsf{E}\left[X_{k+h}\overline{X_k}\right] - \mathsf{E}[X_{k+h}]\mathsf{E}\left[\overline{X_k}\right].$$

According to Example 2.4.2, a null or small value of the complex a.c.v.f. does not necessarily indicate approximate independence. It does neither indicate null or small correlation, in the sense of Definition 2.4.3.

This function simplifies with a stationary time series.

Definition 2.4.5 (Stationary complex-valued time series). *The complex-valued time series* $\{X_k\}_{k \in \mathbb{Z}}$ *in* $\mathcal{L}_2$, *i.e.* $\mathsf{E}[|X_k|^2] < \infty$, $\forall k \in \mathbb{Z}$, *is stationary if*

(1) $\mathsf{E}[X_k]$ *does not depend on* k, $\forall k \in \mathbb{Z}$, *and*
(2) $\gamma(h) = \gamma(h, 0) = \gamma(k+h, k)$, $\forall h, k \in \mathbb{Z}$.

The properties of the complex-valued a.c.v.f. correspond to those of the n.n.d. function presented in Proposition 2.1.22. For example, we have:

- $\gamma(0) \geq 0$,
- $|\gamma(h)| \leq \gamma(0)$, $\forall h \in \mathbb{Z}$, and
- $\gamma(-h) = \overline{\gamma(h)}$, $\forall h \in \mathbb{Z}$.

Theorem 2.4.6 below characterizes the a.c.v.f. of a complex-valued stationary time series. It generalizes Theorem 2.1.27, which is restricted to real-valued time series. We do not give the proof of Theorem 2.4.6, instead of it we give the proof of the similar Theorem 3.1.21 that holds for complex-valued and continuous time stationary processes.

Theorem 2.4.6 (Characterization of the a.c.v.f.). *The function* $\kappa : \mathbb{Z} \to \mathbb{C}$ *is the a.c.v.f. of a stationary time series iff* κ *is n.n.d.*

We now introduce the spectral decomposition of a stationary time series, which is the central topic.

Finite number of frequencies in $(-\pi, \pi]$

For simplicity, let us first consider the simple complex-valued harmonic sum introduced by Rice (1944, 1945) and given by

$$X_k = \sum_{j=1}^{n} A(\alpha_j)\mathrm{e}^{\mathrm{i}k\alpha_j} = \sum_{j=1}^{n} A(\alpha_j)(\cos k\alpha_j + \mathrm{i}\sin k\alpha_j), \quad \forall k \in \mathbb{Z}, \quad (2.4.1)$$

where $n \in \mathbb{N}^*$,

- $-\pi < \alpha_1 < \alpha_2 < \cdots < \alpha_n \le \pi$ are angles called (angular) frequencies[4] and
- $A(\alpha_1), \ldots, A(\alpha_n)$ are complex-valued and uncorrelated random variables such that

$$\mathsf{E}[A(\alpha_j)] = 0 \quad \text{and} \quad \sigma_j^2 = \mathsf{E}\left[|A(\alpha_j)|^2\right] < \infty, \quad \text{for } j = 1, \ldots, n.$$

The random variable or coefficient $A(\alpha_j)$ represents the complex amplitude of the frequency α_j, for $j = 1, \ldots, n$.

The time series $\{X_k\}_{k \in \mathbb{Z}}$ is indeed stationary: it has zero mean and

$$\gamma(h) = \mathsf{E}\left[X_{k+h}\overline{X_k}\right] = \sum_{j=1}^{n} \sigma_j^2 e^{ih\alpha_j}, \quad \forall h, k \in \mathbb{Z}.$$

The a.c.v.f. of $\{X_k\}_{k \in \mathbb{Z}}$ can be re-expressed in terms of the Riemann-Stieltjes integral

$$\gamma(h) = \int_{(-\pi, \pi]} e^{ih\alpha} dF(\alpha), \tag{2.4.2}$$

where the function F is defined by

$$F(\alpha) = \sum_{j=1}^{n} \sigma_j^2 \mathsf{I}\{\alpha_j \le \alpha\} = \sum_{j=1}^{n} \mathsf{E}\left[|A(\alpha_j)|^2\right] \mathsf{I}\{\alpha_j \le \alpha\}, \forall \alpha \in [-\pi, \pi].$$

Formula (2.4.2) is called spectral decomposition of the a.c.v.f. and the function F is a d.f., called spectral d.f. of the time series $\{X_k\}_{k \in \mathbb{Z}}$. In this simple example, F is the step function that assigns mass σ_j^2 to the frequency α_j, for $j = 1, \ldots, n$. We also note that

$$F(-\pi) = 0 \quad \text{and} \quad F(\pi) = \sum_{j=1}^{n} \sigma_j^2 = \gamma(0) = \mathsf{E}\left[|X_k|^2\right], \quad \forall k \in \mathbb{Z},$$

is the total spectral mass. It is important to note that the distribution of the time series $\{X_k\}_{k \in \mathbb{Z}}$ is fixed neither by the spectral d.f. F nor by the a.c.v.f. γ. Indeed, no complete assumption is made on the distribution of the amplitudes $A(\alpha_1), \ldots, A(\alpha_n)$, besides those on moments of orders one and two. In this sense $\{X_k\}_{k \in \mathbb{Z}}$ is not completely determined. Perhaps the most important case is obtained by considering Gaussian amplitudes, yielding a Gaussian time series. The Gaussian case is studied in Section 3.2.1.

---

[4]The angular frequency α is the number of radians per unit of time, whereas the frequency $\alpha/(2\pi)$ is the number of cycles per unit of time. For convenience, we will not mention that the frequency is of angular type whenever it is clear from the context. The common measurement unit of the frequency $\alpha/(2\pi)$ is the Hertz [Hz].

All frequencies of $(-\pi, \pi]$

The simple situation in which only a finite number of frequencies are present can be generalized to the case in which all frequencies in $(-\pi, \pi]$ appear. This is the general situation. Every stationary time series with mean null, $\{X_k\}_{k \in \mathbb{Z}}$, admits the spectral decomposition

$$X_k = \int_{(-\pi, \pi]} e^{ik\alpha} dZ_\alpha, \quad \forall k \in \mathbb{Z}, \tag{2.4.3}$$

where the integral is a stochastic integral w.r.t. to the stochastic process with orthogonal increments $\{Z_\alpha\}_{\alpha \in [-\pi, \pi]}$, which is in this context called spectral process. The process with orthogonal increments $\{Z_\alpha\}_{\alpha \in [-\pi, \pi]}$ satisfies $\mathsf{E}[Z_\alpha] = 0$, $\forall \alpha \in [-\pi, \pi]$, $\mathsf{E}[|Z_\alpha|^2] < \infty$ and does not depend on $\alpha \in [-\pi, \pi]$ and the increments are uncorrelated, i.e.

$$\mathsf{E}\left[(Z_{\alpha_2} - Z_{\alpha_1})\overline{(Z_{\alpha_4} - Z_{\alpha_3})}\right] = 0, \ \forall -\pi < \alpha_1 < \alpha_2 < \alpha_3 < \alpha_4 \leq \pi.$$

The complete definition of the spectral process is given later, in Definition 3.6.1.

An important stochastic process with orthogonal increment over $[-\pi, \pi]$ can be obtained from the Wiener process over $\mathbb{R}$ that is restricted to $[-\pi, \pi]$. Let $\left\{W_t^{(1)}\right\}_{t \in \mathbb{R}_+}$ and $\left\{W_t^{(2)}\right\}_{t \in \mathbb{R}_+}$ be two independent Wiener processes with time domain $\mathbb{R}_+$. Then

$$W_t = \begin{cases} W_t^{(1)}, & \text{if } t \geq 0, \\ W_{-t}^{(2)}, & \text{if } t < 0, \end{cases}$$

is a Wiener process over $\mathbb{R}$, which can be restricted to $[-\pi, \pi]$. One can verify that it is a stochastic process with orthogonal increments. We refer to Definition 3.2.12 and Proposition 3.2.13 for a more complete explanation.

The a.c.v.f. of $\{X_k\}_{k \in \mathbb{Z}}$ can be expressed in the integral form (2.4.2), in which the function F satisfies $F(\pi) = \gamma(0) = \mathsf{E}[|X_k|^2]$, $\forall k \in \mathbb{Z}$. Precisely, F is the d.f. given by Definition 2.4.7.

Definition 2.4.7 (Spectral d.f. and density of time series). *A spectral d.f. is any d.f. F satisfying $F(-\pi) = 0$ and $F(\pi) < \infty$. When F is absolutely continuous w.r.t. the Lebesgue measure, its density f is called spectral density.*

Simplifications arise when $\{X_k\}_{k \in \mathbb{Z}}$ is $\mathbb{R}$-valued. First, $\gamma(h) \in \mathbb{R}$, $\forall h \in \mathbb{Z}$, and consequently we have the simpler formula

$$\gamma(h) = \int_{(-\pi, \pi]} \cos h\alpha \, dF(\alpha), \quad \forall h \in \mathbb{Z}.$$

However, in this case the spectral d.f. F is symmetric around zero and so it is enough to consider nonnegative frequencies. If the symmetric d.f. F admits the density f, then f is symmetric around zero. Indeed,

$$0 = \int_{(-\pi,\pi]} \sin h\alpha \, dF(\alpha) = \int_{(0,\pi]} \sin h\alpha \, \{f(\alpha) - f(-\alpha)\} d\alpha, \quad \forall h \in \mathbb{Z},$$

which holds iff $f(\alpha) = f(-\alpha)$, for $\alpha \in (0,\pi]$ a.e. This implies trivially that

$$\gamma(h) = 2 \int_0^\pi \cos h\alpha f(\alpha) d\alpha, \quad \forall h \in \mathbb{Z}.$$

Example 2.4.8 (Triangular spectral density). *Define the triangular probability density*

$$p(x) = \begin{cases} 1 - |x|, & \text{if } |x| \le 1, \\ 0, & \text{if } |x| > 1. \end{cases} \tag{2.4.4}$$

We want to determine the a.c.v.f. of the stationary time series with spectral density

$$f(\alpha) = \frac{1}{\pi} p\left(\frac{\alpha}{\pi}\right) = \frac{\pi - |\alpha|}{\pi^2}, \quad \forall \alpha \in [-\pi, \pi].$$

We have

$$\gamma(h) = \frac{2}{\pi^2} \int_0^\pi \cos h\alpha(\pi - \alpha) \, d\alpha$$

$$= \frac{2(1 - \cos h\pi)}{(\pi h)^2}, \quad \forall h \in \mathbb{Z}^*.$$

We also have $\gamma(0) = 1$. So we obtain the a.c.v.f.

$$\gamma(h) = \begin{cases} 1, & \text{if } h = 0, \\ \frac{4}{(\pi h)^2}, & \text{if } h = \pm 1, \pm 3, \ldots, \\ 0, & \text{if } h = \pm 2, \pm 4, \ldots. \end{cases}$$

Remark 2.4.9 (One-sided spectral d.f.). *Sometimes one defines the one-sided spectral d.f. by doubling the original spectral mass over $\mathbb{R}_+$ and by correcting for leaving the total mass unchanged, namely equal to $\gamma(0)$. The precise formula appears in Definition 3.5.7 and particular properties are given in Remarks 3.5.8, in the context of continuous time stationary processes.*

The spectral decomposition (2.4.3) is one of major results of the analysis of stationary time series. It was introduced by Cramér (1942) and Loève (1948) and it can be extended to stationary processes with continuous time. A detailed presentation of the spectral decomposition with continuous time is given in Section 3.6.2.

2.4.2 *Spectral distribution and Herglotz's theorem*

By Theorem 2.4.6 we know that γ is the a.c.v.f. of a stationary time series iff γ is n.n.d. A second important characterization of the a.c.v.f. of stationary time series can be directly deduced from Theorem 2.4.10 of Herglotz.

Theorem 2.4.10 (Herglotz). *The function $\kappa : \mathbb{Z} \to \mathbb{C}$ is n.n.d.* $\Longleftrightarrow$

$$\kappa(h) = \int_{(-\pi,\pi]} e^{ih\alpha} dF(\alpha), \quad \forall h \in \mathbb{Z},$$

for some d.f. F over $[-\pi, \pi]$, with finite mass and with $F(-\pi) = 0$.

Thus a n.n.d. function takes the form of a Fourier coefficient and so does the a.c.v.f. of a stationary time series. The version of Herglotz's theorem for $\kappa : \mathbb{R} \to \mathbb{C}$, i.e. for the a.c.v.f. a continuous time stationary process, is given in Theorem 3.5.1 and it is called Bochner's theorem.

Many useful properties of the a.c.v.f. can be obtained from Theorem 2.4.10. For example, the product of two a.c.v.f. has the spectral distribution given by the convolution of the two original spectral distributions. As a consequence, the product of two a.c.v.f. becomes another a.c.v.f.

An important component of the proof of the Theorem 2.4.10 is Helly's theorem A.3.19 that is given in the appendix.

Proof of Theorem 2.4.10. ($\Leftarrow$) Given $\kappa(h) = \int_{(-\pi,\pi]} e^{ih\alpha} dF(\alpha)$, $\forall h \in \mathbb{Z}$, we have that, $\forall a_1, \ldots, a_n \in \mathbb{C}$ and $n \in \mathbb{N}^*$,

$$\sum_{j=1}^{n} \sum_{k=1}^{n} a_j \overline{a_k} \kappa(j-k) = \int_{(-\pi,\pi]} \sum_{j=1}^{n} a_j e^{ij\alpha} \sum_{k=1}^{n} \overline{a_k} e^{-ik\alpha} dF(\alpha)$$

$$= \int_{(-\pi,\pi]} \left| \sum_{j=1}^{n} a_j e^{ij\alpha} \right|^2 dF(\alpha)$$

$$\geq 0.$$

($\Rightarrow$) Let κ be a n.n.d. function and let $n \in \mathbb{N}^*$. We have

$$f_n(\alpha) = \frac{1}{2\pi n} \sum_{k=1}^{n} \sum_{l=1}^{n} e^{-ik\alpha} \kappa(k-l) e^{il\alpha}$$

$$= \frac{1}{2\pi n} \sum_{j=-(n-1)}^{n-1} (n - |j|) e^{-ij\alpha} \kappa(j)$$

$$= \frac{1}{2\pi} \sum_{j=-(n-1)}^{n-1} p\left(\frac{j}{n}\right) e^{-ij\alpha} \kappa(j)$$

$$\geq 0, \quad \forall \alpha \in [-\pi, \pi], \tag{2.4.5}$$

where p denotes the triangular probability density (2.4.4). The function

$$F_n(\alpha) = \int_{-\pi}^{\alpha} f_n(\beta)\mathrm{d}\beta, \quad \forall \alpha \in [-\pi, \pi],$$

is a d.f. over $[-\pi, \pi]$, it has finite total mass and it satisfies $F_n(-\pi) = 0$. We then obtain

$$\int_{(-\pi,\pi]} \mathrm{e}^{\mathrm{i}h\alpha}\mathrm{d}F_n(\alpha) = \sum_{j=-(n-1)}^{n-1} p\left(\frac{j}{n}\right)\kappa(j)\frac{1}{2\pi}\int_{-\pi}^{\pi} \mathrm{e}^{-\mathrm{i}(j-h)\alpha}\mathrm{d}\alpha$$

$$= p\left(\frac{h}{n}\right)\kappa(h), \quad \forall h \in \mathbb{Z}. \tag{2.4.6}$$

We check directly that $F_n(\pi) = \kappa(0) \in [0, \infty)$, $\forall n \in \mathbb{N}^*$, and we can exclude the trivial case $\kappa(0) = 0$. This fact allows for the application of Helly's Theorem A.3.19: there exists a subsequence $\{F_{n_k}\}_{k\in\mathbb{N}^*}$ and a d.f. F, with $F(-\pi) = 0$ and $F(\pi) = \kappa(0)$, such that

$$F_{n_k}(x) \overset{k\to\infty}{\longrightarrow} F(x),$$

$\forall x \in \mathbb{R}$ continuity point of F. This can be re-expressed as weak convergence, i.e. $F_{n_k}/\kappa(0) \overset{\mathrm{w}}{\longrightarrow} F/\kappa(0)$, and the characterization of weak convergence given in Lemma A.3.15 yields that, for any function g that is bounded, continuous and that satisfies $g(-\pi) = g(\pi)$,

$$\int_{(-\pi,\pi]} g(\alpha)\mathrm{d}F_{n_k}(\alpha) \overset{k\to\infty}{\longrightarrow} \int_{(-\pi,\pi]} g(\alpha)\mathrm{d}F(\alpha),$$

holds. Therefore we have

$$p\left(\frac{h}{n_k}\right)\kappa(h) = \int_{(-\pi,\pi]} \mathrm{e}^{\mathrm{i}h\alpha}\mathrm{d}F_{n_k}(\alpha) \overset{k\to\infty}{\longrightarrow} \int_{(-\pi,\pi]} \mathrm{e}^{\mathrm{i}h\alpha}\mathrm{d}F(\alpha), \quad \forall h \in \mathbb{Z}.$$

$\square$

Remarks 2.4.11.

(1) Concerning part ($\Rightarrow$) of the proof of Herglotzs theorem 2.4.10, we note that if we could always take the limit as $n \to \infty$ of (2.4.5) after inserting the limit inside the infinite sum, then the proof would be essentially finished. Indeed, we would obtain a spectral density in the form of a Fourier series whose coefficients are $\kappa(j)$, $\forall j \in \mathbb{Z}$. Taking the limit in this way is however not allowed without absolute summability of κ. These arguments are explained in detail in Theorem 2.4.12 and Corollary 2.4.13 that follow.

(2) *The d.f. F with finite mass and $F(-\pi) = 0$ is uniquely determined by $\kappa(h) = \int_{(-\pi,\pi]} e^{ih\alpha} dF(\alpha)$, $\forall h \in \mathbb{Z}$. This follows from the fact that all trigonometric polynomials form a dense space in the space of continuous functions g over $[-\pi, \pi]$ that satisfy the condition $g(\pi) = g(-\pi)$; cf. e.g. Rudin (1964) p. 190. Let F and G be two d.f. with finite mass, with $F(-\pi) = G(-\pi) = 0$ and satisfying*

$$\kappa(h) = \int_{(-\pi,\pi]} e^{ih\alpha} dF(\alpha) = \int_{(-\pi,\pi]} e^{ih\alpha} dG(\alpha), \quad \forall h \in \mathbb{Z}.$$

Thus we have

$$\int_{(-\pi,\pi]} g(\alpha) dF(\alpha) = \int_{(-\pi,\pi]} g(\alpha) dG(\alpha),$$

for all continuous functions g over $[-\pi, \pi]$ satisfying $g(\pi) = g(-\pi)$. Hence $F = G$ a.e. over $[-\pi, \pi]$.

(3) *In view of Theorem 2.4.6, Herglotzs theorem 2.4.10 states that the a.c.v.f. is the Fourier coefficient of the spectral distribution. Some practical properties of the a.c.v.f. can be deduced from this fact. For example, the product of two a.c.v.f. has the spectral distribution given by the convolution of the two original spectral distributions. Consequently, the product of two a.c.v.f. is another a.c.v.f.*

Theorem 2.4.12 is a standard result of Fourier analysis.

Theorem 2.4.12 (Inversion of Fourier series). *Let $\kappa : \mathbb{Z} \to \mathbb{C}$ be a function such that*

$$\sum_{j=-\infty}^{\infty} |\kappa(j)| < \infty,$$

and let

$$f(\alpha) = \frac{1}{2\pi} \sum_{j=-\infty}^{\infty} e^{-ij\alpha} \kappa(j), \quad \forall \alpha \in \mathbb{R}.$$

Then

$$\kappa(j) = \int_{-\pi}^{\pi} e^{ij\alpha} f(\alpha) d\alpha, \quad \forall j \in \mathbb{Z}.$$

Proof. We have, $\forall j \in \mathbb{Z}$,

$$\int_{-\pi}^{\pi} e^{ij\alpha} f(\alpha) d\alpha = \frac{1}{2\pi} \sum_{l=-\infty}^{\infty} \kappa(l) \int_{-\pi}^{\pi} e^{i(j-l)\alpha} d\alpha = \kappa(j),$$

where the application of Fubini-Tonelli's theorem is justified by

$$\int_{-\pi}^{\pi} \sum_{l=-\infty}^{\infty} \left| e^{i(j-l)\alpha} \kappa(l) \right| d\alpha < \infty.$$

$\square$

In fact, the absolute summability of the a.c.v.f. implies the existence of the spectral density.

Corollary 2.4.13 (Inversion formula for spectral density of time series). *Let $\kappa : \mathbb{Z} \to \mathbb{C}$ be a function such that*

$$\sum_{j=-\infty}^{\infty} |\kappa(j)| < \infty.$$

Then, κ is the a.c.v.f. of a stationary time series $\iff$

$$f(\alpha) = \frac{1}{2\pi} \sum_{j=-\infty}^{\infty} e^{-ij\alpha} \kappa(j) \geq 0, \quad \forall \alpha \in [-\pi, \pi],$$

and $\int_{-\pi}^{\pi} f(\alpha) d\alpha < \infty$, making f the spectral density of the stationary time series with a.c.v.f. κ.

Proof. ($\Rightarrow$) By Theorem 2.4.6, κ is n.n.d. and so we have

$$0 \leq \frac{1}{2\pi n} \sum_{j=1}^{n} \sum_{k=1}^{n} e^{-ij\alpha} \kappa(j-k) e^{ik\alpha}$$

$$= \frac{1}{2\pi n} \sum_{l=-(n-1)}^{n-1} (n - |l|) e^{-il\alpha} \kappa(l)$$

$$= \frac{1}{2\pi} \sum_{l=-\infty}^{\infty} p\left(\frac{l}{n}\right) e^{-il\alpha} \kappa(l)$$

$$\xrightarrow{n \to \infty} \frac{1}{2\pi} \sum_{l=-\infty}^{\infty} e^{-il\alpha} \kappa(l)$$

$$= f(\alpha), \quad \forall \alpha \in [-\pi, \pi],$$

where the limit follows from Dominated convergence theorem and where p is defined in (2.4.4). Besides being nonnegative, f has finite mass because

$$\left| \int_{-\pi}^{\pi} f(\alpha) d\alpha \right| \leq \int_{-\pi}^{\pi} \frac{1}{2\pi} \sum_{j=-\infty}^{\infty} \left| e^{-ij\alpha} \kappa(j) \right| d\alpha < \infty.$$

Theorems 2.4.12 and 2.4.10 imply that f is the spectral density of the stationary time series with a.c.v.f. κ.

($\Leftarrow$) Because κ is absolutely summable, Theorem 2.4.12 gives $\kappa(j) = \int_{-\pi}^{\pi} e^{ij\alpha} f(\alpha) d\alpha$, $\forall j \in \mathbb{Z}$. But f is nonnegative and so, from Herglotzs theorem 2.4.10 and Theorem 2.4.6, it is the spectral density of a stationary time series with a.c.v.f. $\kappa(h) = \int_{-\pi}^{\pi} e^{ih\alpha} f(\alpha) d\alpha$, $\forall h \in \mathbb{Z}$. $\square$

Example 2.4.14. *Consider the time series $X_k = \sum_{j=-\infty}^{\infty} \psi_j Z_{k-j}$, with $\{Z_k\}_{k\in\mathbb{Z}}$ WN(1) and $\sum_{j=-\infty}^{\infty} |\psi_j| < \infty$. Then $\sum_{h=-\infty}^{\infty} |\gamma(h)| < \infty$, where γ is the a.c.v.f. of $\{X_k\}_{k\in\mathbb{Z}}$. Thus Corollary 2.4.13 tells that there exists a spectral density and it can be expressed as a Fourier series. We have indeed, $\forall k, h \in \mathbb{Z}$,*

$$\gamma(h) = \mathsf{cov}(X_{k+h}, X_k)$$

$$= \mathsf{cov}\left(\sum_{j=-\infty}^{\infty} \psi_j Z_{k+h-j}, \sum_{j=-\infty}^{\infty} \psi_j Z_{k-j} \right) = \sum_{j=-\infty}^{\infty} \psi_j \psi_{j-h}.$$

Hence

$$\sum_{h=-\infty}^{\infty} |\gamma(h)| = \sum_{h=-\infty}^{\infty} \left| \sum_{j=-\infty}^{\infty} \psi_j \psi_{j-h} \right| \leq \sum_{h=-\infty}^{\infty} \sum_{j=-\infty}^{\infty} |\psi_j||\psi_{j-h}|$$

$$= \sum_{j=-\infty}^{\infty} \sum_{h=-\infty}^{\infty} |\psi_j||\psi_{j-h}| = \sum_{j=-\infty}^{\infty} |\psi_j| \sum_{h=-\infty}^{\infty} |\psi_{j-h}| < \infty,$$

where the change of order of summation order is justified by absolute summability. This time series is the solution of the general ARMA(p, q) (cf. Theorem 2.2.8) and a simple form of the spectral density is given in Theorem 2.4.30.

Theorem 2.4.15 provides the Fourier inversion formula of the a.c.v.f. for the spectral d.f. It is given without proof.

Theorem 2.4.15 (Inversion formula for spectral d.f. of time series). *Let γ be the a.c.v.f. of a stationary time series and let F be the corresponding spectral d.f. Define $F^*(\alpha) = \{F(\alpha) + F(\alpha-)\}/2$, $\forall \alpha \in (-\pi, \pi]$. Then, $\forall \alpha < \beta \in (-\pi, \pi]$,*

$$F^*(\beta) - F^*(\alpha) = \frac{1}{2\pi} \gamma(0)(\beta - \alpha)$$

$$+ \lim_{n \to \infty} \frac{1}{2\pi} \sum_{\substack{j=-n \\ j \neq 0}}^{n} \frac{\exp\{-i\beta j\} - \exp\{-i\alpha j\}}{-ij} \gamma(j).$$

Two basic examples are the following.

Examples 2.4.16 (Two basic spectral densities).

(1) (WN) Any $\mathrm{WN}(\sigma^2)$ *time series has a.c.v.f.* $\gamma(h) = \sigma^2 \mathbb{1}\{h = 0\}$, $\forall h \in \mathbb{Z}$.
Thus the last corollary gives the spectral density

$$f(\alpha) = \frac{\sigma^2}{2\pi}, \quad \forall \alpha \in [-\pi, \pi].$$

This spectral density is an intuitive characteristic of the WN: all frequencies are equally represented.
(2) (MA(1)) Consider the a.c.v.f. γ *defined by*

$$\gamma(h) = \begin{cases} 1, & \text{if } h = 0, \\ \rho & \text{if } |h| = 1, \\ 0, & \text{if } |h| \geq 2. \end{cases}$$

Obviously, γ *is absolutely summable and we have*

$$f(\alpha) = \frac{1}{2\pi} \sum_{j=-\infty}^{\infty} \mathrm{e}^{-ij\alpha} \gamma(j) = \frac{1}{2\pi}(1 + 2\rho \cos \alpha), \ \forall \alpha \in [-\pi, \pi], \quad (2.4.7)$$

which is nonnegative iff $|\rho| \leq 1/2$. *Referring to Proposition 2.1.28,* γ *is the a.c.v.f. of a stationary time series iff* $|\rho| \leq 1/2$.

Example 2.4.17. *We apply Corollary 2.4.13 in order to show that a given function cannot be the a.c.v.f. of a stationary time series. Consider the candidate a.c.v.f.*

$$\kappa(h) = \begin{cases} 1, & \text{if } h = 0, \\ -\frac{1}{2}, & \text{if } h = \pm 2, \\ -\frac{1}{4}, & \text{if } h = \pm 3, \\ 0, & \text{otherwise.} \end{cases}$$

We have $\sum_{h=-\infty}^{\infty} |\kappa(h)| < \infty$ *and so Corollary 2.4.13 tells that* κ *is the a.c.v.f. of a stationary time series iff the spectral density* f *obtained by inversion satisfies* $f(\alpha) \geq 0$, $\forall \alpha \in [-\pi, \pi]$. *The inversion of* κ *gives*

$$f(\alpha) = \frac{1}{2\pi}\left(1 - \cos 2\alpha - 2 \cos 3\alpha\right), \quad \forall \alpha \in [-\pi, \pi].$$

In particular, $f(0) = -1/\pi$. *Therefore* κ *cannot be the a.c.v.f. of a stationary time series.*

Remark 2.4.18 (Cardioid and other circular distributions). *For $0 \le \rho \le 1/2$, the spectral density f of the* MA(1) *time series given in (2.4.7) is the probability density of an important circular distribution and it is called cardioid density. The cardioid curve over the circle can be obtained from the trajectory of a point on the perimeter of a circle, which rolls around another circle with same radius. This trajectory is called cardioid because it takes the shape of a heart. Let θ be a random angle, also called circular random variable, with circular density f, and let $e^{i\theta}$ be the random unit vector associated to θ. Note that $e^{i\theta}$ has the interpretation of a random planar direction. Then the so-called first trigonometric moment of θ is defined by*

$$
\begin{aligned}
\mathsf{E}\left[e^{i\theta}\right] &= \int_{-\pi}^{\pi} e^{i\alpha} f(\alpha) d\alpha \\
&= \frac{1}{2\pi} \int_{-\pi}^{\pi} (\cos\alpha + i \sin\alpha)(1 + 2\rho\cos\alpha) d\alpha \\
&= \frac{\rho}{\pi} \int_{-\pi}^{\pi} \cos^2\alpha d\alpha = \rho.
\end{aligned}
$$

In fact any spectral distribution can be a circular distribution and conversely. There is thus a close relation between circular probability distributions and spectral distributions of time series and more details about this relation are given in Section 4.9. There has been a considerable amount of research on circular distributions during the last few decades. Two monographs on circular statistics are Mardia and Jupp (2000) and Jammalamadaka and SenGupta (2001) and a introductory article on this topic is Gatto and Jammalamadaka (2015).

2.4.3   Linear filters and spectral distribution of ARMA time series

A linear operator applied to stationary data, that is to a stationary signal observed at equidistant epochs, is a specific weighted sum of the data that follows some particular objectives, like for example the elimination of rapid variations. The theoretical analysis of filtered data is more convenient in the frequency domain, where different frequency components of the signal can be easily identified. The spectral analysis of stationary time series is the ideal approach for studying linear filters, that are defined as follows.

Definition 2.4.19 (Linear filter and time invariant linear filter). *The linear filter that is identified to the complex double sequence $\{c_{k,j}\}_{k,j \in \mathbb{Z}}$*

is the operator that maps the time series $\{Y_k\}_{k \in \mathbb{Z}}$ *to the filtered time series* $\{X_k\}_{k \in \mathbb{Z}}$ *defined by*

$$X_k = \sum_{j=-\infty}^{\infty} c_{k,j} Y_j, \quad \forall k \in \mathbb{Z}.$$

This linear filter is time invariant if, $\forall j, k \in \mathbb{Z}$, $c_{k,k-j}$ *is independent of* k *and thus* $c_{k,k-j} = \psi_j$, *for some* $\psi_j \in \mathbb{C}$.

Note that with any time invariant linear filter we have

$$X_k = \sum_{j=-\infty}^{\infty} c_{k,j} Y_j = \sum_{j=-\infty}^{\infty} c_{k,k-j} Y_{k-j} = \sum_{j=-\infty}^{\infty} \psi_j Y_{k-j} = \psi(B) Y_k, \; \forall k \in \mathbb{Z}.$$

$$(2.4.8)$$

With continuous time stochastic processes, the analogue of the sequence $\{\psi_j\}_{j \in \mathbb{Z}}$ is called the impulse response function (see Definition 3.9.1).

In what follows we identify the time invariant linear filter either with its associated sequence of coefficients $\{\psi_j\}_{j \in \mathbb{Z}}$ or with $\psi(z) = \sum_{j=-\infty}^{\infty} \psi_j z^j$ or with $\psi(B)$.

Two basic examples are the following.

Example 2.4.20. *For some* $a \in \mathbb{R}^*$, $X_k = a Y_{-k}$, $\forall k \in \mathbb{Z}$, *is the output of the linear filter*

$$c_{k,k-j} = \begin{cases} a, & \text{if } j = 2k, \\ 0, & \text{if } j \neq 2k, \end{cases}$$

which is not time invariant.

Example 2.4.21 (Moving average filter). *The moving average filter is the sample mean of the current, the* q *past and the* q *next random variables the time series:*

$$X_k = \frac{1}{1+2q} \sum_{j=-q}^{q} Y_{k-j}, \quad \forall k \in \mathbb{Z},$$

where $q \in \mathbb{N}^*$ *is the smoothing parameter. This filter is time invariant. It will be shown, in Example 2.4.27, that it is a particular low pass linear filter, in the sense that it lets the low frequencies pass and its cancels the high frequencies, of the signal* $\{Y_k\}_{k \in \mathbb{Z}}$.

Theorem 2.4.22 below shows that the spectral d.f. of the output of a time invariant linear filter is obtained from the spectral d.f. of the input stationary time series.

Theorem 2.4.22 (Spectral distribution of filtered time series). *Let* $\{Y_k\}_{k \in \mathbb{Z}}$ *be a complex-valued stationary time series with spectral d.f.* F_Y *and let*

$$X_k = \sum_{j=-\infty}^{\infty} \psi_j Y_{k-j}, \quad \forall k \in \mathbb{Z},$$

where $\{\psi_j\}_{j \in \mathbb{Z}}$ *is a complex sequence such that* $\sum_{j=-\infty}^{\infty} |\psi_j| < \infty$. *Then* $\{X_k\}_{k \in \mathbb{Z}}$ *is stationary with spectral d.f.*

$$F_X(\alpha) = \int_{(-\pi, \alpha]} |\psi(e^{-i\beta})|^2 dF_Y(\beta), \quad \forall \alpha \in [-\pi, \pi].$$

In particular, if $\{Y_k\}_{k \in \mathbb{Z}}$ *has spectral density* f_Y, *then* $\{X_k\}_{k \in \mathbb{Z}}$ *has spectral density*

$$f_X(\alpha) = |\psi(e^{-i\alpha})|^2 f_Y(\alpha), \quad \forall \alpha \in [-\pi, \pi].$$

Proof. A suitable version of Proposition 2.2.6 for complex-valued time series implies that $\{X_k\}_{k \in \mathbb{Z}}$ is a stationary time series and that it has a.c.v.f.

$$\gamma_X(h) = \sum_{j=-\infty}^{\infty} \sum_{l=-\infty}^{\infty} \psi_j \overline{\psi_l} \gamma_Y(h - j + l), \quad \forall h \in \mathbb{Z},$$

where γ_Y denotes the a.c.v.f. of $\{Y_k\}_{k \in \mathbb{Z}}$. It follows from Herglotzs theorem 2.4.10 that there exists the spectral d.f. F_X and F_Y such that

$$
\begin{aligned}
\gamma_X(h) &= \sum_{j=-\infty}^{\infty} \sum_{l=-\infty}^{\infty} \psi_j \overline{\psi_l} \int_{(-\pi, \pi]} e^{i(h-j+l)\alpha} dF_Y(\alpha) \\
&= \int_{(-\pi, \pi]} e^{ih\alpha} \sum_{j=-\infty}^{\infty} \psi_j e^{-ij\alpha} \sum_{l=-\infty}^{\infty} \overline{\psi_l} e^{il\alpha} dF_Y(\alpha) \\
&= \int_{(-\pi, \pi]} e^{ih\alpha} \left| \sum_{j=-\infty}^{\infty} \psi_j e^{-ij\alpha} \right|^2 dF_Y(\alpha) \\
&= \int_{(-\pi, \pi]} e^{ih\alpha} dF_X(\alpha), \quad \forall h \in \mathbb{Z}.
\end{aligned}
$$

Consequently, if $\{Y_k\}_{k \in \mathbb{Z}}$ has spectral density f_Y, then $\{X_k\}_{k \in \mathbb{Z}}$ has spectral density given by

$$f_X(\alpha) = \left| \sum_{j=-\infty}^{\infty} \psi_j e^{-ij\alpha} \right|^2 f_Y(\alpha) = \left| \psi(e^{-i\alpha}) \right|^2 f_Y(\alpha), \quad \forall \alpha \in [-\pi, \pi],$$

where $\psi(z) = \sum_{j=-\infty}^{\infty} \psi_j z^j$. $\qquad \square$

We now give some terminology.

Definition 2.4.23 (Transfer and power transfer functions). *Let $\psi(B)$ be a time invariant linear filter satisfying $\sum_{j=-\infty}^{\infty} |\psi_j| < \infty$.*

(1) The function

$$\psi(\mathrm{e}^{-\mathrm{i}\alpha}) = \sum_{j=-\infty}^{\infty} \mathrm{e}^{-\mathrm{i}j\alpha}\psi_j, \quad \forall \alpha \in [-\pi, \pi], \tag{2.4.9}$$

is the transfer function of the filter $\psi(B)$.
(2) The function $|\psi(\mathrm{e}^{-\mathrm{i}\alpha})|^2$, $\forall \alpha \in [-\pi, \pi]$, is the power transfer function of the filter $\psi(B)$.

We clearly have $|\psi(\mathrm{e}^{-\mathrm{i}\alpha})| \leq \sum_{j=-\infty}^{\infty} |\psi_j| < \infty$, $\forall \alpha \in [-\pi, \pi]$. In order to understand how the filter works, we note that $\psi(B)$ totally removes all the frequencies $\alpha \in [-\pi, \pi]$ that solve

$$\psi(\mathrm{e}^{-\mathrm{i}\alpha}) = 0,$$

and it attenuates frequencies α for which this holds approximately.

Example 2.4.24 (Filter of lag d operator). *Let $\{Y_k\}_{k\in\mathbb{Z}}$ be a stationary time series with spectral density f_Y. The lag d difference operator ∇_d, for some $d \in \mathbb{N}^*$ (cf. Definition 2.1.15), is a time invariant and causal linear filter. It removes the d-periodic component, which is a seasonal effect. Let*

$$X_k = \nabla_d Y_k = Y_k - Y_{k-d} = \psi(B)X_k, \quad \forall k \in \mathbb{Z},$$

for the filter $\psi(B) = I - B^d$. Thus $\psi_0 = 1$, $\psi_d = -1$ and $\psi_j = 0$, $\forall j \notin \{0, d\}$. So $|1 - \mathrm{e}^{-\mathrm{i}d\alpha}|^2$ is the power transfer function and

$$f_X(\alpha) = |1 - \mathrm{e}^{-\mathrm{i}d\alpha}|^2 f_Y(\alpha), \quad \forall \alpha \in [-\pi, \pi],$$

is the spectral density of $\{X_k\}_{k\in\mathbb{Z}}$. The filter $\psi(B)$ removes all frequencies $\alpha \in [-\pi, \pi]$ that solve

$$1 - \mathrm{e}^{-\mathrm{i}d\alpha} = 0,$$

that is the frequencies

$$\alpha = \frac{2k\pi}{d}, \quad \text{for } k = -\left\lfloor \frac{d-1}{2} \right\rfloor, \dots, \left\lfloor \frac{d}{2} \right\rfloor.$$

Thus, all frequencies integer multiples of $2\pi/d$ are removed, i.e. the component of period d is removed. Indeed, the period $d \in \{1, 2, \dots\}$ corresponds to the frequency $1/d \in (0, 1]$ and to the angular frequency $2\pi/d \in (0, 2\pi]$, which is called fundamental.

The Dirichlet kernel is given in Definition 2.4.26 and it is used Example 2.4.27. Its definition is based on the following elementary result.

Lemma 2.4.25. *Let $\alpha \in \mathbb{R}^*$ and $n \in \mathbb{N}^*$, then the following formula holds,*

$$1 + 2 \sum_{k=1}^{n} \cos k\alpha = \frac{\sin\left(\frac{1}{2} + n\right)\alpha}{\sin\frac{\alpha}{2}}.$$

Proof. It follows from the identity $\cos\alpha\sin\beta = \{\sin(\alpha+\beta) - \sin(\alpha-\beta)\}/2$ that

$$\cos k\alpha \sin\frac{\alpha}{2} = \frac{1}{2}\left\{\sin\left(k+\frac{1}{2}\right)\alpha - \sin\left(k-\frac{1}{2}\right)\alpha\right\}.$$

Then we have

$$\left(\frac{1}{2} + \sum_{k=1}^{n} \cos k\alpha\right)\sin\frac{\alpha}{2} = \frac{1}{2}\sin\frac{\alpha}{2} + \frac{1}{2}\sum_{k=1}^{n}\sin\left(k+\frac{1}{2}\right)\alpha - \sin\left(k-\frac{1}{2}\right)\alpha$$

$$= \frac{1}{2}\sin\frac{\alpha}{2} + \frac{1}{2}\left\{\sin\left(\frac{1}{2}+n\right)\alpha - \sin\frac{\alpha}{2}\right\}$$

$$= \frac{1}{2}\sin\left(\frac{1}{2}+n\right)\alpha.$$

$\square$

Definition 2.4.26 (Dirichlet kernel)**.** *The Dirichlet kernel is the function*

$$D_n(\alpha) = \frac{1}{1+2n}\sum_{k=-n}^{n} e^{ik\alpha} = \begin{cases} \frac{\sin\left(\frac{1}{2}+n\right)\frac{\alpha}{2}}{\left(\frac{1}{2}+n\right)\sin\frac{\alpha}{2}}, & \text{if } \alpha \neq 0, \\ 1, & \text{if } \alpha = 0, \end{cases} \quad \forall \alpha \in [-\pi, \pi],$$

where $n \in \mathbb{N}$.

Figure 2.7 shows the graphs of the Dirichlet kernel for $n = 4$ and 20. The Dirichlet kernel is an important function in the analysis of Fourier series because its convolution with a 2π-periodic function f yields the function $(1 + 2n)/(2\pi)f_n$, where f_n is the n-th Fourier series approximation to f, given by (1.2.3). The Fourier series inherits various properties of the Dirichlet kernel.

Example 2.4.27 (Moving average filter and Dirichlet kernel)**.** *Let*

$$\psi_j = \begin{cases} \frac{1}{1+2n}, & \text{if } j = -n, \ldots, n, \\ 0, & \text{otherwise,} \end{cases}$$

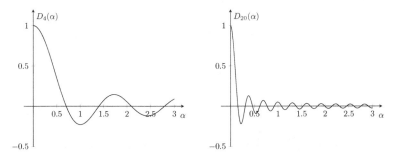

Fig. 2.7:   Dirichlet kernel D_n with $n = 4$ (left) and $n = 20$ (right).

where $n \in \mathbb{N}^*$. With this choice of coefficients, the transfer function is given by

$$\psi(e^{-i\alpha}) = \frac{1}{1 + 2n} \sum_{j=-n}^{n} e^{-ij\alpha}$$

$$= \frac{1}{1 + 2n} \sum_{j=-n}^{n} \cos j\alpha$$

$$= \frac{1}{1 + 2n} \left(1 + 2 \sum_{j=1}^{n} \cos j\alpha \right), \quad \forall \alpha \in [-\pi, \pi].$$

This can be re-expressed in terms of the Dirichlet kernel as

$$\psi(e^{-i\alpha}) = D_n(\alpha), \quad \forall \alpha \in [-\pi, \pi],$$

and so we refer to Figure 2.7 for the visualization of this filter. We see that with a large value of the index n, the main effect of the moving average filter is to temper substantially high frequencies and to leave low frequencies almost unaltered.

Example 2.4.28 (Best low pass filter). *The best low pass filter would have transfer function*

$$\psi(e^{-i\alpha}) = \begin{cases} 1, & \text{if } -a \le \alpha \le a, \\ 0, & \text{otherwise,} \end{cases}$$

for some $a \in (0, \pi)$. The coefficients $\{\psi_j\}_{j \in \mathbb{Z}}$ are not directly given, as it was in the two previous examples. The series expansion of transfer function *is a Fourier series, $\psi(e^{-i\alpha}) = \sum_{j=-\infty}^{\infty} \psi_j e^{-ij\alpha}$, $\forall \alpha \in [-\pi, \pi]$. So these coefficients are Fourier coefficients and thus we have*

$$\psi_j = \frac{1}{2\pi} \int_{-\pi}^{\pi} \psi(e^{-i\alpha}) e^{ij\alpha} d\alpha$$

$$= \frac{1}{2\pi} \int_{-a}^{a} e^{ij\alpha} d\alpha$$

$$= \frac{a}{\pi} \mathrm{sinc}\, ja, \; \forall j \in \mathbb{Z}.$$

For the sinc function refer to Definition A.8.1.

In practice this filter must be approximated: the term $X_k = \sum_{j=-\infty}^{\infty} \psi_j Y_{k-j}$ is computed by $\sum_{j=-n}^{n} \psi_j Y_{k-j}$, $\forall k \in \mathbb{Z}$, where n is a possibly large integer. The suitability of this truncation should be investigated through the truncated transfer function. The index n should be chosen in such a way that $\psi^{(n)}(e^{-i\alpha}) = \sum_{j=-n}^{n} \psi_j e^{-ij\alpha}$ is sufficiently close to $\psi(e^{-i\alpha})$, $\forall \alpha \in [0, \pi]$. This situation is illustrated in Figure 2.8, where the best low pass filter with $a = \pi/8$ is approximated by $\psi^{(4)}(e^{-i\alpha})$, drawn with dashed line, and by $\psi^{(20)}(e^{-i\alpha})$, drawn with dashed and dotted line. We see that $\psi^{(20)}(e^{-i\alpha})$ is a substantially better approximation than $\psi^{(4)}(e^{-i\alpha})$, but both of them perform poorly for α close to the point of jump $a = \pi/8$. This is due to Dirichlet's theorem (1.2.7), which tells that the Fourier approximations at point $a = \pi/8$ converge to $1/2$ (the mean of left of right limits), instead of 1. Generally, whenever a transfer function possesses a large jump, a high order Fourier approximation should be preferred.

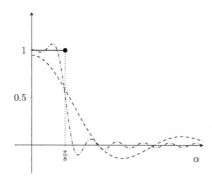

Fig. 2.8:   Best low pass filter $\psi(e^{-i\alpha})$ with $a = \pi/8$ (solid line) together with approximations $\psi^{(4)}(e^{-i\alpha})$ (dashed line) and $\psi^{(20)}(e^{-i\alpha})$ (dashed and dotted line).

Example 2.4.29 (Narrow banded WN). *By applying the best low pass filter of Example 2.4.28 to* WN(σ^2)*, we obtain the narrow banded WN whose spectral density is given by*

$$f(\alpha) = |\psi(e^{-i\alpha})|^2 \frac{\sigma^2}{2\pi} = \frac{\sigma^2}{2\pi} \mathrm{I}\{|\alpha| \le a\}, \quad \forall \alpha \in [-\pi, \pi].$$

The corresponding a.c.v.f. is given by

$$\gamma(h) = \frac{\sigma^2}{2\pi} \int_{-a}^{a} e^{ih\alpha} d\alpha = \frac{\sigma^2 a}{\pi} \mathrm{sinc}\, ah, \quad \forall h \in \mathbb{Z}.$$

Theorem 2.4.30 is an important result for the analysis of ARMA time series.

Theorem 2.4.30 (Spectral density of ARMA time series). *Let* $\{X_k\}_{k \in \mathbb{Z}}$ *be the* ARMA(p, q) *time series with equation*

$$\varphi(B)X_k = \theta(B)Z_k, \quad \forall k \in \mathbb{Z},$$

where $\{Z_k\}_{k \in \mathbb{Z}}$ *is* WN(σ^2)*. Assume that* $\varphi(z) \ne 0$, $\forall z \in \mathbb{C}$ *such that* $|z| = 1$. *Then* $\{X_k\}_{k \in \mathbb{Z}}$ *has spectral density*

$$f(\alpha) = \frac{\sigma^2}{2\pi} \frac{|\theta(e^{-i\alpha})|^2}{|\varphi(e^{-i\alpha})|^2}, \quad \forall \alpha \in [-\pi, \pi]. \tag{2.4.10}$$

The spectral density (2.4.10) is referred to as rational.

Proof. Theorem 2.2.8 tells that the unique stationary solution to the ARMA(p, q) equation is $X_k = \sum_{j=-\infty}^{\infty} \psi_j Z_{k-j}$, $\forall k \in \mathbb{Z}$, where $\psi(z) = \sum_{j=-\infty}^{\infty} \psi_j z^j$ is the Laurent series of $\theta(z)/\varphi(z)$, $\forall z \in \mathcal{A}_r$, for some $r > 1$. We obtain from Corollary 2.4.13 that $\{Z_k\}_{k \in \mathbb{Z}}$ has spectral density

$$f_Z(\alpha) = \frac{1}{2\pi} \sum_{j=-\infty}^{\infty} e^{-ij\alpha} \sigma^2 \mathrm{I}\{j = 0\} = \frac{\sigma^2}{2\pi}, \quad \forall \alpha \in [-\pi, \pi].$$

Then Theorem 2.4.22 implies that the spectral density of $\{X_k\}_{k \in \mathbb{Z}}$ equals

$$f(\alpha) = |\psi(e^{-i\alpha})|^2 f_Z(\alpha) = \frac{\sigma^2}{2\pi} \frac{|\theta(e^{-i\alpha})|^2}{|\varphi(e^{-i\alpha})|^2},$$

given that $|e^{-i\alpha}| = 1$, $\forall \alpha \in [-\pi, \pi]$. $\qquad \square$

The direct application of Theorem 2.4.30 gives the two next examples.

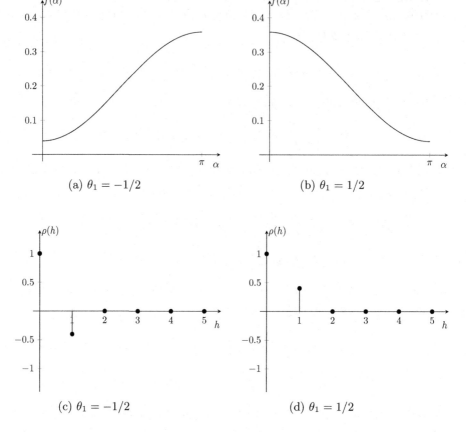

Fig. 2.9:   Spectral densities and a.c.r.f. of MA(1) for $\theta_1 = -1/2$, in (a) and (c), and for $\theta_1 = 1/2$, in (b) and (d).

Example 2.4.31 (MA(1)). *The* MA(1) *time series has the spectral density*

$$f(\alpha) = \frac{\sigma^2}{2\pi}|1 + \theta_1 e^{-i\alpha}|^2 = \frac{\sigma^2}{2\pi}(1 + \theta_1^2 + 2\theta_1\cos\alpha), \quad \forall\alpha \in [-\pi, \pi].$$

For the MA(1) *time series, two spectral densities and their corresponding a.c.r.f. are drawn in Figure 2.9: the case with $\theta_1 = -1/2$ is shown in the graphs (a) and (c) and the case with $\theta_1 = 1/2$ is shown in the graphs (b) and (d).*

Example 2.4.32 (AR(1)). *The* AR(1) *time series, with* $|\varphi_1| \neq 1$*, has spectral density*

$$f(\alpha) = \frac{\sigma^2}{2\pi} \frac{1}{|1 - \varphi_1 e^{-i\alpha}|^2} = \frac{\sigma^2}{2\pi}(1 + \varphi_1^2 - 2\varphi_1 \cos\alpha)^{-1}, \quad \forall \alpha \in [-\pi, \pi].$$

For the AR(1) *time series, two spectral densities and the corresponding a.c.r.f. are drawn in Figure 2.10: the case with* $\varphi_1 = -1/2$ *is shown in the graphs (a) and (c) and the case with* $\varphi_1 = 1/2$ *is shown in the graphs (b) and (d).*

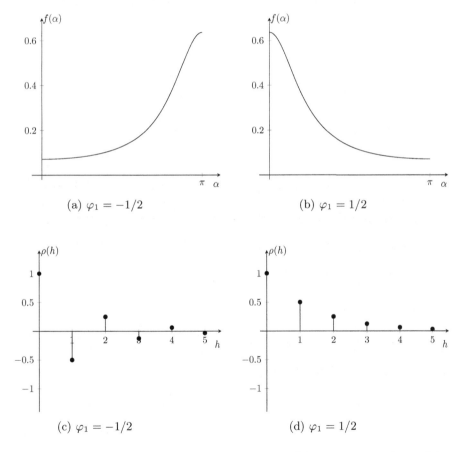

(a) $\varphi_1 = -1/2$ (b) $\varphi_1 = 1/2$

(c) $\varphi_1 = -1/2$ (d) $\varphi_1 = 1/2$

Fig. 2.10: Spectral densities and a.c.r.f. of AR(1) for $\varphi_1 = -1/2$, in (a) and (c), and for $\varphi_1 = 1/2$, in (b) and (d).

Remark 2.4.33 (Wrapped Cauchy distribution). *It is explained in Remark 2.4.18 that the spectral distribution of the* MA(1) *time series leads to the well-known cardioid circular distribution. Similarly, the spectral density of the* AR(1) *corresponds to another important circular density, namely the wrapped Cauchy. It is obtained by wrapping the Cauchy distribution with density*

$$g(x) = \frac{1}{\pi} \frac{\tau}{\tau^2 + (x - \mu)^2}, \quad \forall x \in \mathbb{R}, \tag{2.4.11}$$

where $\mu \in \mathbb{R}$ and $\tau > 0$, around the unit circle. The wrapped Cauchy density admits the simple form

$$g^\circ(\alpha) = \sum_{k=-\infty}^{\infty} g(\alpha + 2k\pi) = \frac{1}{2\pi} \frac{1 - \rho^2}{1 + \rho^2 - 2\rho\cos(\alpha - \mu)}, \forall \alpha \in [-\pi, \pi],$$

$$\tag{2.4.12}$$

where $\rho = e^{-\tau}$ is called concentration parameter. The second equality in (2.4.12) can be verified as follows. First note that, $\forall \alpha \in [-\pi, \pi]$,

$$\operatorname{Re} \sum_{k=1}^{\infty} \rho^k e^{-ik(\alpha - \mu)} = \operatorname{Re} \frac{\rho e^{-i(\alpha - \mu)}}{1 - \rho e^{-i(\alpha - \mu)}} = \frac{\rho \cos(\alpha - \mu) - \rho^2}{1 - 2\rho \cos(\alpha - \mu) + \rho^2}.$$

$$\tag{2.4.13}$$

A general result is that the Fourier coefficients of a wrapped probability density, also called trigonometric moments, correspond to the characteristic function φ of the original or linear density at integer values. Let us denote by G the linear d.f. having density g, then the wrapped d.f. is given by

$$G^\circ(\alpha) = \sum_{k=-\infty}^{\infty} G(\alpha + 2k\pi) - G((2k - 1)\pi), \ \forall \alpha \in [-\pi, \pi].$$

We find that the j-th trigonometric moment of G° is given by

$$\varphi_j = \int_{(-\pi, \pi]} e^{ij\alpha} dG^\circ(\alpha)$$

$$= \int_{(-\pi, \pi]} e^{ij\alpha} d\left\{ \sum_{k=-\infty}^{\infty} G(\alpha + 2k\pi) \right\}$$

$$= \sum_{k=-\infty}^{\infty} \int_{(-\pi, \pi]} e^{ij\alpha} dG(\alpha + 2k\pi)$$

$$= \sum_{k=-\infty}^{\infty} \int_{((2k-1)\pi, (2k+1)\pi]} e^{ij(\alpha - 2\pi k)} dG(\alpha)$$

$$= \int_{\mathbb{R}} e^{ijx} dG(x)$$

$$= \varphi(j), \ \forall j \in \mathbb{Z}.$$

This fact and the formula of the characteristic function of the Cauchy distribution

$$\varphi(v) = e^{i\mu v - \tau|v|}, \quad \forall v \in \mathbb{R}, \tag{2.4.14}$$

yield

$$g^\circ(\alpha) = \frac{1}{2\pi}\left\{1 + 2\operatorname{Re}\sum_{k=1}^{\infty}\rho^k e^{-ik(\alpha-\mu)}\right\}. \tag{2.4.15}$$

Then (2.4.13) and (2.4.15) lead to the wrapped Cauchy density (2.4.12). This circular density is often used for statistical modelling of planar directions, like e.g. directions of winds or sea waves. It is circularly unimodal and symmetric around its mode μ. An important generalization of the wrapped Cauchy distribution is the wrapped α-stable distribution, which is analyzed in Gatto and Jammalamadaka (2003). It is obtained by wrapping an α-stable distribution around the unit circle and the special case with stability index $\alpha = 1$ yields the wrapped Cauchy.

We end this section with the study of the spectral distribution of the sum of two stationary time series.

Example 2.4.34 (Sum of AR(1) and stationary time series). *Let $\{X_k\}_{k\in\mathbb{Z}}$ be an AR(1) time series with $|\varphi_1| \neq 1$ and with $\mathrm{WN}(\sigma^2)$ given by $\{Z_k\}_{k\in\mathbb{Z}}$, for some $\sigma > 0$. Let $\{T_k\}_{k\in\mathbb{Z}}$ be a stationary time series that is uncorrelated with $\{Z_k\}_{k\in\mathbb{Z}}$. The time series of interest is given by*

$$Y_k = X_k + T_k, \quad \forall k \in \mathbb{Z}.$$

Let us derive the spectral density of $\{Y_k\}_{k\in\mathbb{Z}}$. It follows from $|\varphi_1| \neq 1$ that there exists a sequence $\{\psi_j\}_{j\in\mathbb{Z}}$ such that the expansion $X_k = \sum_{j=-\infty}^{\infty}\psi_j Z_{k-j}$, $\forall k \in \mathbb{Z}$, holds; cf. Theorem 2.2.8. Consequently we obtain

$$\langle X_k, T_l\rangle = \left\langle \sum_{k=-\infty}^{\infty}\psi_j Z_{k-j}, T_l\right\rangle = \sum_{k=-\infty}^{\infty}\psi_j\langle Z_{k-j}, T_l\rangle = 0, \quad \forall k, l \in \mathbb{Z},$$

by using the continuity of the scalar product. Therefore the time series $\{X_k\}_{k\in\mathbb{Z}}$ and $\{T_k\}_{k\in\mathbb{Z}}$ are uncorrelated. It follows from this fact that the a.c.v.f. of $\{Y_k\}_{k\in\mathbb{Z}}$ is given by

$$\begin{aligned}\gamma_Y(h) &= \gamma_{X+T}(h) \\ &= \gamma_X(h) + \gamma_T(h) \\ &= \int_{(-\pi,\pi]} e^{ih\alpha}\{f_X(\alpha) + f_T(\alpha)\}\,d\alpha, \quad \forall h \in \mathbb{Z}.\end{aligned}$$

The spectral density of $\{Y_k\}_{k\in\mathbb{Z}}$ is thus given by $f_Y = f_X + f_T$, where f_T is undetermined and where

$$f_X(\alpha) = \frac{\sigma^2}{2\pi} \frac{1}{1 + \varphi_1^2 - 2\varphi_1 \cos\alpha}, \quad \forall \alpha \in [-\pi, \pi];$$

cf. Example 2.4.32.

Now we consider that case where $\{T_k\}_{k\in\mathbb{Z}}$ is another $\mathrm{WN}(\sigma^2)$ time series. Then the formula of f_Y can be re-expressed in terms of the spectral density of an $\mathrm{ARMA}(p,q)$, for some $p, q \geq 1$. In order to see this, note first that the spectral density of $\{T_k\}_{k\in\mathbb{Z}}$ is constant at $\sigma^2/(2\pi)$. Therefore

$$f_Y(\alpha) = f_X(\alpha) + f_T(\alpha) = \frac{\sigma^2}{2\pi} \frac{2 + \varphi_1^2 - 2\varphi_1 \cos\alpha}{1 + \varphi_1^2 - 2\varphi_1 \cos\alpha}, \quad \forall \alpha \in [-\pi, \pi].$$

The spectral density of the $\mathrm{ARMA}(1,1)$ time series with MA coefficient $-\theta$, AR coefficient φ and WN variance τ^2 is given by

$$f(\alpha) = \frac{\tau^2}{2\pi} \frac{1 + \theta^2 - 2\theta \cos\alpha}{1 + \varphi^2 - 2\varphi \cos\alpha}, \quad \forall \alpha \in [-\pi, \pi];$$

cf. Theorem 2.4.30. We can rewrite the spectral density f_Y as

$$f_Y(\alpha) = \frac{\tau^2}{2\pi} \frac{c(2 + \varphi_1^2) - 2c\varphi_1 \cos\alpha}{1 + \varphi_1^2 - 2\varphi_1 \cos\alpha}, \quad \forall \alpha \in [-\pi, \pi],$$

where $c = \sigma^2/\tau^2 > 0$. Consequently, f_Y is of the same type as f if

$$c(2 + \varphi_1^2) = 1 + \theta^2, \quad -c\varphi_1 = \theta \quad \text{and} \quad \varphi_1 = \varphi.$$

The first two equations above imply

$$2c + \frac{\theta^2}{c} = 1 + \theta^2 \quad \text{and thus} \quad \theta = \sqrt{c\frac{2c - 1}{c - 1}}.$$

Thus θ is a non-zero real number iff

$$c \in (0, 1/2) \cup (1, \infty) \quad \text{and thus iff} \quad \tau^2 \in (0, \sigma^2) \cup (2\sigma^2, \infty).$$

Thus, f_Y can be written as the spectral density of an $\mathrm{ARMA}(1,1)$ time series with AR coefficient φ_1, with WN variance $\tau^2 \in (0, \sigma^2) \cup (2\sigma^2, \infty)$ and with MA coefficient $-\sqrt{c(2c-1)/(c-1)}$, where $c = (\sigma/\tau)^2$.

Let us now analyze the stationarity of the two time series $\{Y_k\}_{k\in\mathbb{Z}}$ and $\{\nabla Y_k\}_{k\in\mathbb{Z}}$ when $\varphi_1 = 1$ and when $\{T_k\}_{k\in\mathbb{Z}}$ is $\mathrm{WN}(\sigma^2)$, as before. We obtain the random walk perturbed by WN, which is introduced in Example 2.1.19. Example 2.1.19 tells that $\{Y_k\}_{k\in\mathbb{Z}}$ is not stationary but $U_k = \nabla Y_K$, $\forall k \in \mathbb{Z}$, is stationary and it has the a.c.v.f.

$$\gamma_U(h) = \begin{cases} 3\sigma^2, & h = 0, \\ -\sigma^2, & |h| = 1, \\ 0, & h = \pm 1, \pm 2, \dots. \end{cases}$$

Thus

$$f_U(\alpha) = \frac{1}{2\pi} \sum_{j=-\infty}^{\infty} \gamma_U(j) e^{-ij\alpha}$$

$$= \frac{1}{2\pi} \{3\sigma^2 - \sigma^2 (e^{-i\alpha} + e^{i\alpha})\}$$

$$= \frac{\sigma^2}{2\pi} (3 - 2\cos\alpha), \quad \forall \alpha \in [-\pi, \pi].$$

The spectral d.f. is given by

$$F_U(\alpha) = \frac{\sigma^2}{2\pi} \{3(\alpha + \pi) - 2\sin\alpha\}, \quad \forall \alpha \in [-\pi, \pi].$$

We have indeed $F_U(\pi) = 3\sigma^2 = \gamma(0)$.

2.4.4 AR and MA approximations to stationary time series

This section presents two important theoretical properties of AR and MA time series: for any real-valued and stationary time series with continuous spectral density f, there exists a casual AR(p) and an invertible MA(q) time series whose spectral densities are arbitrarily close to the target spectral density f.

Proposition 2.4.35 (Approximation of symmetric spectral density). *Let f be a symmetric and continuous spectral density over $[-\pi, \pi]$ and let $\varepsilon > 0$. Then there exists a polynomial of degree $p \geq 0$ that has either the trivial form $\zeta(z) = 1$ or, when $p > 0$, the form*

$$\zeta(z) = \prod_{j=1}^{p} \left(1 - \frac{z}{\eta_j}\right) = 1 + \zeta_1 z + \cdots + \zeta_p z^p, \quad \forall z \in \mathbb{Z},$$

with $\eta_1, \ldots, \eta_p \in \mathbb{C}$ satisfying $|\eta_1|, \ldots, |\eta_p| > 1$ and with $\zeta_1, \ldots, \zeta_p \in \mathbb{R}$, such that

$$\left| a |\zeta(e^{-i\alpha})|^2 - f(\alpha) \right| < \varepsilon, \quad \forall \alpha \in [-\pi, \pi],$$

where $a = \left\{2\pi(1 + \zeta_1^2 + \cdots + \zeta_p^2)\right\}^{-1} \int_{-\pi}^{\pi} f(\alpha) d\alpha > 0$.

Remarks 2.4.36.

(1) *Thus $a|\zeta(e^{-i\cdot})|^2$ provides a smooth approximation that is uniformly close to f. Moreover, we can control directly that $a|\zeta(e^{-i\cdot})|^2$ is symmetric and that it possesses same integral as f over $[-\pi, \pi)$. According to Section 1.2.1, $\zeta(e^{-i\cdot})$ is a linear combination of the elements $e_0, \ldots, e_p$ of the orthonormal basis $\{e_j\}_{j \in \mathbb{Z}}$ of $\mathcal{L}_2([-\pi, \pi))$, which is defined in (1.2.1).*

(2) We also note that $\zeta(z) \neq 0$, $\forall z \in \mathcal{D}_0$, which has two important practical consequences. If $\zeta = \theta$, the polynomial of the MA(p) time series, then we have indeed an invertible MA(p). If $\zeta = \varphi$, the polynomial of the AR(p) time series, then we are in presence of a causal AR(p). This second claim can be heuristically deduced from the replacement of f by $1/f$ in Proposition 2.4.35. Precise statements and justification are given in Theorems 2.4.37 and 2.4.38 and their proofs.

Proof of Proposition 2.4.35. Let $\varepsilon > 0$, $s = \sup_{-\pi \leq \alpha \leq \pi} f(\alpha) > 0$,

$$\delta = \min \left\{ s, 1 + 2\varepsilon \left(\frac{2\pi s}{\int_{-\pi}^{\pi} f(\alpha)\mathrm{d}\alpha} \right)^{-1} \right\},$$

and $f_\delta(\alpha) = \max\{f(\alpha), \delta\}$, which is the lower truncation of the spectral density f. Thus

$$0 \leq f_\delta(\alpha) - f(\alpha) \leq \delta, \quad \forall \alpha \in [-\pi, \pi].$$

It then follows, from uniform convergence of Cèsàro sum of Fourier approximations to the continuous function f_δ, which is mentioned in (1.2.8), that $\exists n \in \mathbb{N}^*$ such that

$$\left| \frac{1}{n} \sum_{j=0}^{n-1} \sum_{k=-j}^{j} b_k \mathrm{e}^{-\mathrm{i}k\alpha} - f_\delta(\alpha) \right| < \delta, \quad \forall \alpha \in [-\pi, \pi], \qquad (2.4.16)$$

where $b_k = 1/(2\pi) \int_{-\pi}^{\pi} \mathrm{e}^{\mathrm{i}kv} f_\delta(\alpha)\mathrm{d}\alpha$, $\forall k \in \mathbb{Z}$, are the Fourier coefficients. This leads to

$$\frac{1}{n} \sum_{j=0}^{n-1} \sum_{k=-j}^{j} b_k \mathrm{e}^{-\mathrm{i}k\alpha} = \frac{1}{n} \sum_{k=-(n-1)}^{n-1} (n - |k|) b_k \mathrm{e}^{-\mathrm{i}k\alpha}$$

$$= \sum_{k=-(n-1)}^{n-1} \left(1 - \frac{|k|}{n} \right) b_k \mathrm{e}^{-\mathrm{i}k\alpha} \in \mathbb{R}, \quad \forall \alpha \in [-\pi, \pi],$$

because $b_{-k} = b_k$. Let

$$g(z) = \sum_{k=-(n-1)}^{n-1} p\left(\frac{k}{n} \right) b_k z^k, \qquad (2.4.17)$$

with $z \in \mathbb{C}$, where p is the triangular probability density (2.4.4), and let $p = \max\{1 \leq k \leq n - 1 | b_k \neq 0\}$. Then

$$z^p g(z) = c_1 \prod_{j=1}^{p} \left(1 - \eta_j^{-1}z \right) (1 - \eta_j z),$$

is a polynomial of degree $2p$, for some $c_1 \in \mathbb{R}$ and $\eta_1, \ldots, \eta_p \in \mathbb{C}$ such that $|\eta_j| > 1$, for $j = 1, \ldots, p$. This form can be justified as follows. First note that $g(z) = g(z^{-1})$, which follows from $b_{-k} = b_k$. Thus if $\eta \neq 0$ is a root of g, then η^{-1} is another one. We see from (2.4.16) that η satisfying $|\eta| = 1$ cannot be a root of g. We finally note that the coefficients of the monomials of degrees 0 and $2p$ of $z^p g(z)$ must coincide. Thus we have

$$
\begin{aligned}
g(z) &= c_1 \prod_{j=1}^{p} \left(1 - \eta_j^{-1} z\right) z^{-1} \left(1 - \eta_j z\right) \\
&= c_1 \prod_{j=1}^{p} \left(1 - \eta_j^{-1} z\right) (-\eta_j) \left(1 - \eta_j^{-1} z^{-1}\right) \\
&= c_2 \zeta(z) \zeta(z^{-1}),
\end{aligned} \tag{2.4.18}
$$

where $c_2 = (-1)^p c_1 \prod_{j=1}^{p} \eta_j$. By equating constant terms of (2.4.17) and (2.4.18) we obtain $b_0 = c_2 \left(1 + \zeta_1^2 + \cdots + \zeta_p^2\right)$, which implies

$$
c_2 = \frac{1}{2\pi \left(1 + \zeta_1^2 + \cdots + \zeta_p^2\right)} \int_{-\pi}^{\pi} f_\delta(\alpha) d\alpha.
$$

So we obtain

$$
\begin{aligned}
(2.4.16) &\iff \left| g\left(e^{-i\alpha}\right) - f_\delta(\alpha) \right| < \delta, \quad \forall \alpha \in [-\pi, \pi] \\
&\iff \left| c_2 \zeta\left(e^{-i\alpha}\right) \zeta\left(e^{i\alpha}\right) - f_\delta(\alpha) \right| < \delta, \quad \forall \alpha \in [-\pi, \pi] \\
&\iff \left| c_2 \left| \zeta\left(e^{-i\alpha}\right) \right|^2 - f_\delta(\alpha) \right| < \delta, \quad \forall \alpha \in [-\pi, \pi].
\end{aligned} \tag{2.4.19}
$$

Further,

$$
\begin{aligned}
(ref3, 2) &\implies c_2 \left| \zeta(e^{-i\alpha}) \right|^2 - f_\delta(\alpha) < \delta, \; \forall \alpha \in [-\pi, \pi] \\
&\iff \frac{\left| \zeta\left(e^{-i\alpha}\right) \right|^2}{1 + \zeta_1^2 + \cdots + \zeta_p^2} < 2\pi \{ \underbrace{\delta}_{\leq s} + \underbrace{f_\delta(\alpha)}_{\leq s} \} \left(\int_{-\pi}^{\pi} f_\delta(\alpha) d\alpha \right)^{-1}, \\
& \qquad\qquad \forall \alpha \in [-\pi, \pi] \\
&\iff \frac{\left| \zeta\left(e^{-i\alpha}\right) \right|^2}{1 + \zeta_1^2 + \cdots + \zeta_p^2} < 4\pi s \left(\int_{-\pi}^{\pi} f(\alpha) d\alpha \right)^{-1}, \quad \forall \alpha \in [-\pi, \pi].
\end{aligned}
$$

This gives

$$\left| a \left| \zeta \left(e^{-i\alpha} \right) \right|^2 - c_2 \left| \zeta \left(e^{-i\alpha} \right) \right|^2 \right|$$

$$= \frac{\left| \zeta \left(e^{-i\alpha} \right) \right|^2}{1 + \zeta_1^2 + \cdots + \zeta_p^2} \frac{1}{2\pi} \int_{-\pi}^{\pi} \{ f_\delta(\beta) - f(\beta) \} \, d\beta$$

$$< 4\pi s \left(\int_{-\pi}^{\pi} f(\beta) d\beta \right)^{-1} \frac{1}{2\pi} \int_{-\pi}^{\pi} \underbrace{\{ f_\delta(\beta) - f(\beta) \} d\beta}_{\leq \delta}$$

$$\leq 4\pi s \delta \left(\int_{-\pi}^{\pi} f(\beta) d\beta \right)^{-1}, \quad \forall \alpha \in [-\pi, \pi]. \tag{2.4.20}$$

This last inequality together with (2.4.19) give

$$\left| a \left| \zeta \left(e^{-i\alpha} \right) \right|^2 - f(\alpha) \right|$$

$$\leq \left| a \left| \zeta \left(e^{-i\alpha} \right) \right|^2 - c_2 \left| \zeta \left(e^{-i\alpha} \right) \right|^2 \right| + \left| c_2 \left| \zeta \left(e^{-i\alpha} \right) \right|^2 - f_\delta(\alpha) \right|$$

$$+ \left| f_\delta(\alpha) - f(\alpha) \right|$$

$$< 4\pi s \delta \left(\int_{-\pi}^{\pi} f(\beta) d\beta \right)^{-1} + \delta + \delta$$

$$< \delta \left\{ 4\pi s \left(\int_{-\pi}^{\pi} f(\beta) d\beta \right)^{-1} + 2 \right\}$$

$$< \varepsilon, \quad \forall \alpha \in [-\pi, \pi].$$

$\square$

Theorem 2.4.37 (Approximation by AR time series). *For any symmetric and continuous spectral density f over $[-\pi, \pi]$ and $\varepsilon > 0$, there exists a causal AR(p) time series with spectral density f_X such that*

$$|f_X(\alpha) - f(\alpha)| < \varepsilon, \quad \forall \alpha \in [-\pi, \pi].$$

Proof. Let $\varepsilon > 0$ and define

$$f_\varepsilon(\alpha) = \max \left\{ f(\alpha), \frac{\varepsilon}{2} \right\}, \quad \forall \alpha \in [-\pi, \pi],$$

$$m = \max_{\alpha \in [-\pi, \pi]} f_\varepsilon(\alpha) \quad \text{and} \quad \delta = \min \left\{ \frac{\varepsilon}{(2m)^2}, \frac{1}{2m} \right\}.$$

Theorem 2.4.35 applied to (the spectral density) $1/f_\varepsilon$ yields the existence of a polynomial with real coefficients $\zeta(z) = 1 + \zeta_1 z + \cdots + \zeta_p z^p$ satisfying $\zeta(z) \neq 0$, $\forall z \in \mathcal{D}_0$, and the existence of $a > 0$ such that

$$\left| a |\zeta(e^{-i\alpha})|^2 - \frac{1}{f_\varepsilon(\alpha)} \right| < \delta, \quad \forall \alpha \in [-\pi, \pi].$$

Consequently and $\forall \alpha \in [-\pi, \pi]$ we have,

$$\frac{1}{f_\varepsilon(\alpha)} - \delta < a|\zeta(e^{-i\alpha})|^2,$$

which implies

$$\frac{f_\varepsilon(\alpha)}{1 - \delta f_\varepsilon(\alpha)} > \frac{1}{a|\zeta(e^{-i\alpha})|^2}.$$

We can also directly check that

$$\frac{f_\varepsilon(\alpha)}{1 - \delta f_\varepsilon(\alpha)} < \frac{m}{1 - \delta m} \leq 2m.$$

These two last results together yield $a^{-1}|\zeta(e^{-i\alpha})|^{-2} < 2m$. Therefore we find

$$\left| a^{-1}|\zeta(e^{-i\alpha})|^{-2} - f_\varepsilon(\alpha) \right| = a^{-1}|\zeta(e^{-i\alpha})|^{-2} f_\varepsilon(\alpha) \left| \frac{1}{f_\varepsilon(\alpha)} - a|\zeta(e^{-i\alpha})|^2 \right|$$

$$\leq 2m^2 \delta$$

$$< \frac{\varepsilon}{2}, \quad \forall \alpha \in [-\pi, \pi].$$

The latter inequality and the triangular inequality imply

$$\left| a^{-1}|\zeta(e^{-i\alpha})|^{-2} - f(\alpha) \right| < \delta, \quad \forall \alpha \in [-\pi, \pi].$$

Theorem 2.4.30 tells that the causal $AR(p)$ time series given by $\zeta(B)X_k = Z_k$, $\forall k \in \mathbb{Z}$, with $\{Z_k\}_{k \in \mathbb{Z}}$ WN$(2\pi/a)$, has spectral density $f_X(\alpha) = a^{-1}|\zeta(e^{-i\alpha})|^{-2}$, $\forall \alpha \in [-\pi, \pi]$. $\qquad \square$

Theorem 2.4.38 (Approximation by MA time series). *For any symmetric and continuous spectral density f over $[-\pi, \pi]$ and $\varepsilon > 0$, there exists an invertible* MA(q) *time series with spectral density f_X such that*

$$|f_X(\alpha) - f(\alpha)| < \varepsilon, \quad \forall \alpha \in [-\pi, \pi].$$

Besides their theoretical relevance, Theorems 2.4.37 and 2.4.38 are two useful results for the simulation of any stationary time series with continuous spectral density. One can approximate this density by the one of an $AR(p)$ or by the one of an $MA(q)$ time series and use the model equations for generating the desired stationary time series.

2.4.5 *Periodogram and estimation of spectral density*

The periodogram is an important transform of a sample of consecutive values of a time series. It provides an estimator of the spectral density. Because of its nonparametric nature, the periodogram is well-suited for detecting special features of the time series, such as periodicities, that can usually not be identified by parametric estimators of the spectral density.

Discrete Fourier transform and periodogram

In order to define the periodogram, we need to introduce the discrete Fourier transform. Let $n \in \mathbb{N}^*$ and consider the set of n indices around the origin given by

$$\mathcal{F}_n = \left\{ -\left\lfloor \frac{n-1}{2} \right\rfloor, \ldots, \left\lfloor \frac{n}{2} \right\rfloor \right\}.$$

We define the Fourier frequencies by

$$\omega_j = \frac{2\pi j}{n}, \quad \forall j \in \mathcal{F}_n.$$

The Fourier frequencies are multiple of the so-called fundamental frequency $2\pi/n$ and satisfy

$$\omega_j \in (-\pi, \pi], \quad \forall j \in \mathcal{F}_n.$$

As mentioned, we have card $\mathcal{F}_n = n$ and so n Fourier frequencies.

In order to define the periodogram, consider the vector space $\mathbb{C}^n$ with scalar product $\langle \boldsymbol{x}, \boldsymbol{y} \rangle = \sum_{j=1}^{n} x_j \overline{y}_j, \ \forall \boldsymbol{x} = (x_1, \ldots, x_n)^\top, \boldsymbol{y} = (y_1, \ldots, y_n)^\top \in \mathbb{C}^n$. Let $\{\boldsymbol{e}_j\}_{j \in \mathcal{F}_n}$ be the vectors of $\mathbb{C}^n$ defined by

$$\boldsymbol{e}_j = n^{-\frac{1}{2}} (e^{i\omega_j}, \ldots, e^{in\omega_j})^\top, \quad \forall j \in \mathcal{F}_n.$$

Then $\{\boldsymbol{e}_j\}_{j \in \mathcal{F}_n}$ form an orthonormal basis of $\mathbb{C}^n$, because

$$\langle \boldsymbol{e}_j, \boldsymbol{e}_k \rangle = n^{-1} \sum_{l=1}^{n} e^{il(\omega_j - \omega_k)}$$

$$= \begin{cases} 1, & \text{if } j = k \in \mathcal{F}_n, \\ n^{-1} e^{i(\omega_j - \omega_k)} \frac{1 - e^{in(\omega_j - \omega_k)}}{1 - e^{i(\omega_j - \omega_k)}} = 0, & \text{if } j \neq k \in \mathcal{F}_n. \end{cases}$$

Thus, each vector $\boldsymbol{x} = (x_1, \ldots, x_n)^\top \in \mathbb{C}^n$ can be expressed w.r.t. the basis $\{\boldsymbol{e}_j\}_{j \in \mathcal{F}_n}$ as

$$\boldsymbol{x} = \sum_{j \in \mathcal{F}_n} a_j \boldsymbol{e}_j, \tag{2.4.21}$$

for some coefficients $a_1, \ldots, a_n \in \mathbb{C}$. That is,

$$x_k = n^{-\frac{1}{2}} \sum_{j \in \mathcal{F}_n} a_j e^{ik\omega_j} = \langle \boldsymbol{a}, \overline{\boldsymbol{e}_j} \rangle, \quad \text{for } k = 1, \ldots, n, \tag{2.4.22}$$

where $\boldsymbol{a} = (a_{\lfloor (n-1)/2 \rfloor}, \ldots a_{\lfloor n/2 \rfloor})^\top$. The two expressions (2.4.21) and (2.4.22) are called inverse discrete Fourier transform of the vector of coefficients $\boldsymbol{a}$.

Furthermore, we have

$$a_j = \langle \boldsymbol{x}, \boldsymbol{e}_j \rangle = n^{-\frac{1}{2}} \sum_{k=1}^{n} x_k \mathrm{e}^{-\mathrm{i}k\omega_j}, \quad \forall j \in \mathcal{F}_n. \tag{2.4.23}$$

The expression (2.4.23) is called discrete Fourier transform of $\boldsymbol{x}$, which represents data obtained from n complex (or two-dimensional) measurements or observations. It is the analogue of the Fourier transform of a function (1.2.9). In particular we have,

$$a_0 = n^{-\frac{1}{2}} \sum_{k=1}^{n} x_k,$$

and

$$\|x\|^2 = \langle \boldsymbol{x}, \boldsymbol{x} \rangle \sum_{j \in \mathcal{F}_n} a_j \langle \boldsymbol{e}_j, \boldsymbol{x} \rangle = \sum_{j \in \mathcal{F}_n} a_j \overline{a_j} = |a_j|^2.$$

The periodogram is the following transform of a sample of a time series.

Definition 2.4.39 (Periodogram). *The periodogram of the realization $\boldsymbol{x} = (x_1, \ldots, x_n)^\top \in \mathbb{C}^n$, of a sample of n consecutive values of a time series, is the function of the Fourier frequencies given by*

$$\Lambda_n(\omega_j) = \left| n^{-\frac{1}{2}} \sum_{k=1}^{n} x_k \mathrm{e}^{-\mathrm{i}k\omega_j} \right|^2 = |a_j|^2, \quad \forall j \in \mathcal{F}_n.$$

The value $\Lambda_n(\omega_j)$ is the contribution of the j-th Fourier frequency to $\|x\|^2$, where $j \in \mathcal{F}_n$. In particular we have

$$\Lambda_n(0) = \Lambda_n(\omega_0) = \frac{1}{n} \left| \sum_{k=1}^{n} x_k \right|^2. \tag{2.4.24}$$

Simplifications occur for real-valued time series. Let $j \in \mathcal{F}_n$ be such that $-\omega_j = \omega_{-j} \in (-\pi, \pi)$, i.e. $j \neq n/2$ whenever n is even. Then $\boldsymbol{e}_{-j} = \overline{\boldsymbol{e}_j}$ and so $a_j = \langle \boldsymbol{x}, \overline{\boldsymbol{e}_{-j}} \rangle = \overline{\langle \overline{\boldsymbol{x}}, \boldsymbol{e}_{-j} \rangle}$. Therefore when $\boldsymbol{x} \in \mathbb{R}^n$ we obtain

$$a_j = \overline{a_{-j}} \quad \text{and so} \quad \Lambda_n(\omega_j) = \Lambda_n(\omega_{-j}).$$

The real-case gives also the simplification

$$\boldsymbol{x} = a_0 \boldsymbol{e}_0 + \sum_{j=1}^{\lfloor \frac{n-1}{2} \rfloor} (a_j \boldsymbol{e}_j + \overline{a_j \boldsymbol{e}_j}) + a_{\frac{n}{2}} \boldsymbol{e}_{\frac{n}{2}} \mathsf{I}\{n \text{ even}\}.$$

For $j \in \mathcal{F}_n$, we define

$$a_j = r_j e^{i\theta_j},$$

$$c_j = \sqrt{\frac{2}{n}}(\cos\omega_j, \cos2\omega_j, \ldots, \cos n\omega_j)^\top \text{ and}$$

$$s_j = \sqrt{\frac{2}{n}}(\sin\omega_j, \sin2\omega_j, \ldots, \sin n\omega_j)^\top.$$

We then obtain, $\forall x \in \mathbb{R}^n$,

$$x = a_0 e_0 + \frac{1}{\sqrt{2}} \sum_{j=1}^{\lfloor\frac{n-1}{2}\rfloor} r_j\{e^{i\theta_j}(c_j + is_j) + e^{-i\theta_j}(c_j - is_j)\} + a_{\frac{n}{2}} e_{\frac{n}{2}} \mathsf{I}\{n \text{ even}\}$$

$$= \frac{a_0 c_0}{\sqrt{2}} + \sqrt{2} \sum_{j=1}^{\lfloor\frac{n-1}{2}\rfloor} r_j(\cos\theta_j c_j - \sin\theta_j s_j) + a_{\frac{n}{2}} e_{\frac{n}{2}} \mathsf{I}\{n \text{ even}\}.$$

One could show that

$$\left\{ \frac{c_0}{\sqrt{2}}, c_1, s_1, \ldots, c_{\lfloor\frac{n-1}{2}\rfloor}, s_{\lfloor\frac{n-1}{2}\rfloor}, e_{\frac{n}{2}} \mathsf{I}\{n \text{ is even}\} \right\},$$

is an orthonormal basis of $\mathbb{R}^n$. Thus for $j = 1, \ldots, \lfloor(n-1)/2\rfloor$, one obtains $\langle x, c_j \rangle = \sqrt{2}r_j\cos\theta_j$, $\langle x, s_j \rangle = \sqrt{2}r_j\sin\theta_j$ and

$$\langle x, c_j \rangle^2 + \langle x, s_j \rangle^2 = 2r_j^2 = 2|a_j|^2 = 2\Lambda_n(\omega_j) = \Lambda_n(\omega_j) + \Lambda_n(\omega_{-j}),$$

which is either the contribution of the j-th or the $-j$-th harmonics to $\|x\|^2$ or twice the contribution of j-th harmonic, in the one-sided spectrum over $[0, \pi]$. Furthermore we have

$$\left\langle x, \frac{c_0}{\sqrt{2}} \right\rangle^2 = a_0^2 = \Lambda_n(\omega_0) = \Lambda_n(0),$$

and, for n even,

$$\langle x, e_{\frac{n}{2}} \rangle^2 = a_{\frac{n}{2}}^2 = \Lambda_n(\omega_{\frac{n}{2}}) = \Lambda_n(\pi).$$

Estimation of the spectral density

Proposition 2.4.40 below shows that the periodogram takes the form of the discrete Fourier transform of the empirical a.c.v.f.

Proposition 2.4.40 (Periodogram in terms of empirical a.c.v.f.). *Consider the realization $x = (x_1, \ldots, x_n)^\top \in \mathbb{C}^n$ of a sample of n consecutive values* *of a complex-valued time series, define the mean by* $m_n = n^{-1}\sum_{k=1}^{n} x_k$ *and the empirical a.c.v.f. by*

$$\gamma_n(h) = \frac{1}{n}\sum_{j=1}^{n-h}(x_{j+h} - m_n)(\overline{x_j} - \overline{m_n}) \text{ and } \gamma_n(-h) = \overline{\gamma_n(h)},$$

for $h = 0, \ldots, n-1$ *(extending thus Definition 2.1.30 of the real case). Then,* $\forall j \in \mathcal{F}_n\backslash\{0\}$, *we have*

$$\Lambda_n(\omega_j) = \sum_{h=-(n-1)}^{n-1} \gamma_n(h)e^{-ih\omega_j}.$$

Proof. Let $j \in \mathcal{F}_n\backslash\{0\}$. Then we have

$$\sum_{k=1}^{n} e^{ik\omega_j} = e^{i\omega_j}\frac{1 - e^{in\omega_j}}{1 - e^{i\omega_j}} = 0.$$

Therefore we find

$$\Lambda_n(\omega_j) = \left| n^{-\frac{1}{2}}\sum_{k=1}^{n} x_k e^{-ik\omega_j} \right|^2$$

$$= \frac{1}{n}\sum_{l=1}^{n} x_l e^{-il\omega_j}\sum_{k=1}^{n}\overline{x_k}e^{ik\omega_j}$$

$$= \frac{1}{n}\sum_{l=1}^{n}\sum_{k=1}^{n}(x_l - m_n)(\overline{x_k} - \overline{m_n})e^{-i(l-k)\omega_j}.$$

We note the last expression corresponds to the sum of all elements of an Hermitian matrix A. By applying diagonal summation and by exploiting the Hermitian property $A^\top = \overline{A}$, i.e. $a_{kl} = \overline{a_{lk}}$, for $l, k = 1, \ldots, n$, we have

$$\sum_{k=1}^{n}\sum_{l=1}^{n} a_{kl} = \sum_{k=1}^{n} a_{kk} + \sum_{k=1}^{n-1}\sum_{l=1}^{n-k} a_{l,l+k} + \sum_{k=1}^{n-1}\sum_{l=1}^{n-k}\overline{a_{l,l+k}}.$$

We thus continue the previous computation as follows,

$$\Lambda_n(\omega_j) = \frac{1}{n}\sum_{k=1}^{n}(x_k - m_n)(\overline{x_k} - \overline{m_n}) + \sum_{k=1}^{n-1}\underbrace{\frac{1}{n}\sum_{l=1}^{n-k}(x_l - m_n)(\overline{x_{l+k}} - \overline{m_n})}_{=\overline{\gamma_n(k)}}e^{ik\omega_j}$$

$$+ \sum_{k=1}^{n-1}\underbrace{\frac{1}{n}\sum_{l=1}^{n-l}(\overline{x_l} - \overline{m_n})(x_{l+k} - m_n)}_{=\gamma_n(k)}e^{-ik\omega_j}$$

$$= \gamma_n(0) + \sum_{k=1}^{n-1}\gamma_n(-k)e^{-i(-k)\omega_j} + \sum_{k=1}^{n-1}\gamma_n(k)e^{-ik\omega_j}.$$

$\square$

We remember from Corollary 2.4.13 that if the a.c.v.f. γ satisfies $\sum_{k=-\infty}^{\infty} |\gamma(k)| < \infty$, then $f(\alpha) = (2\pi)^{-1} \sum_{k=-\infty}^{\infty} \gamma(k) e^{-ik\alpha} \geq 0$, $\forall \alpha \in [-\pi, \pi]$, is the spectral density of a stationary time series with a.c.v.f. γ. This last formula is similar to the one of the periodogram of Proposition 2.4.40.

Assume that the sample $\boldsymbol{X} = (X_1, \ldots, X_n)^{\top}$ is generated from a stationary time series with spectral density f. Thus, if $\hat{\Lambda}_n$ is constructed by extending, by piecewise constant interpolation, the periodogram of $\boldsymbol{X}$ from the non-null Fourier frequencies to $(-\pi, \pi] \backslash \{0\}$, then

$$\frac{1}{2\pi} \hat{\Lambda}_n(\alpha), \tag{2.4.25}$$

provides an estimator to $f(\alpha)$, $\forall \alpha \in (-\pi, \pi] \backslash \{0\}$. In this way the spectral mass estimated at the $n-1$ Fourier frequencies ω_j, $\forall j \in \mathcal{F}_n \backslash \{0\}$, is spread over $(-\pi, \pi] \backslash \{0\}$. One can certainly consider other estimators and the most direct one would be obtained by inserting any $\alpha \in (-\pi, \pi] \backslash \{0\}$ directly in the formula of the periodogram (thus without restricting to non-null Fourier frequencies).

According to (2.4.24), we define $\hat{\Lambda}_n(\omega_0) = \hat{\Lambda}_n(0) = n|M_n|^2$, where $M_n = n^{-1} \sum_{j=1}^{n} X_j$. Under the assumption of absolute summability of the a.c.v.f., Theorem 2.5.4.2 yields

$$\mathsf{E}\left[\hat{\Lambda}_n(\omega_0)\right] - n\mu = n\mathsf{var}(M_n) \overset{n\to\infty}{\longrightarrow} \sum_{h=-\infty}^{\infty} \gamma(h) = 2\pi f(0),$$

where $\mu = \mathsf{E}[X_1]$. We can deduce from this limit that an estimator of $f(0)$ is given by

$$\frac{1}{2\pi} \left\{ \hat{\Lambda}_n(\omega_0) - n\mu \right\}.$$

We should however mention that any estimator of type (2.4.25) possesses two major drawbacks. The first one is that its graph is typically irregular. In order to understand this irregularity, we remember that the periodogram at a Fourier frequency is the squared modulus of the corresponding amplitude; cf. Definition 2.4.39. But because the amplitudes of a stationary time series are uncorrelated, the plot of these consecutive values, namely of the periodogram, is unlikely to follow a regular curve. This irregularity can be addressed by the application of smoothing techniques. Another problem is that, for given $\alpha \in (-\pi, \pi] \backslash \{0\}$, (2.4.25) does not provide a consistent estimation of $f(\alpha)$. We avoid the precise proof of this statement and instead provide an intuitive justification. As the sample size

n increases, the number of Fourier frequencies n and thus the number of basic quantities to estimate increase simultaneously. This means that the asymptotics $n \to \infty$ does not seem to improve the precision of estimation.

One can nevertheless construct a consistent estimator of the spectral density $f(\alpha)$, at $\alpha \in (-\pi, \pi] \backslash \{0\}$, by the following procedure:

(1) select a neighborhood of α that vanishes, as $n \to \infty$, and that simultaneously contains a diverging number of Fourier frequencies, as $n \to \infty$;
(2) compute the mean of $\hat{\Lambda}_n(\alpha)/(2\pi)$ with α equal to any of the Fourier frequencies in the selected neighborhood.

We illustrate this procedure with the following example.

Example 2.4.41 (Consistent estimator of spectral density). *Let $m_n = \lfloor \sqrt{n} \rfloor$ and $\omega_n(\alpha)$ be the closest multiple of $2\pi/n$ to α, for some $\alpha \in (-\pi, \pi] \backslash \{0\}$. Then the above procedure yields the estimator of the spectral density at α, $f(\alpha)$, given by*

$$\hat{f}_n(\alpha) = \frac{1}{2\pi(2m_n + 1)} \sum_{j=-m_n}^{m_n} \hat{\Lambda}_n\left(\omega_n(\alpha) + \frac{2\pi j}{n}\right).$$

This estimator is consistent. We note that there are $2m_n + 1 \overset{n\to\infty}{\longrightarrow} \infty$ Fourier frequencies averaged over a neighborhood of vanishing length $2\pi/n 2m_n = 4\pi\{\sqrt{n}/n - (\sqrt{n} - \lfloor\sqrt{n}\rfloor)/n\} = 4\pi n^{-1/2}\{1 + O(n^{-1/2})\} = O(n^{-1/2})$, as $n \to \infty$.

The following general formula provides a spectral density estimator that consistent for ARMA time series,

$$\hat{f}_n(\alpha) = \frac{1}{2\pi} \sum_{j=-m_n}^{m_n} u_{jn} \hat{\Lambda}_n\left(\omega_n(\alpha) + \frac{2\pi j}{n}\right),$$

where $m_n \to \infty$, $m_n/n \to 0$, as $n \to \infty$, and where the weights u_{jn}, for $j = -m_n, \ldots, m_n$, satisfy $u_{jn} = u_{-jn} \geq 0$,

$$\sum_{j=-m_n}^{m_n} u_{jn} = 1, \ \forall n \geq 1, \quad \text{and} \quad \sum_{j=-m_n}^{m_n} u_{jn}^2 \to 0, \ \text{as } n \to \infty.$$

Precisely, this result holds for time series that take the form $X_k = \sum_{j=-\infty}^{\infty} \psi_j Z_{k-j}$, $\forall k \in \mathbb{Z}$, where $\sum_{j=-\infty}^{\infty} |\psi_j| |j|^{\frac{1}{2}} < \infty$ and $\{Z_k\}_{k\in\mathbb{Z}}$ is a sequence of i.i.d. random variables with $\mathsf{E}[Z_1^4] < \infty$.

2.5   Further classical topics on time series

This last section of introduces some more important topics. of the analysis
of real-valued time series. These topics are the following: the problem of
prediction, in Section 2.5.1, the estimation of the mean and the a.c.v.f., in
Section 2.5.2, the analysis of nonstationary integrated ARMA (ARIMA)
time series, in Section 2.5.3, the determination of the two orders p and q of
ARMA(p, q) time series, in Section 2.5.4, and finally the state space model
and the Kalman filter, in Section 2.5.5.

2.5.1   *Prediction with stationary time series*

The problem of making predictions with a probabilistic model is an important topic of statistical inference. Temporal prediction, that is prediction
for measurements or observation at future epochs, is also referred as forecasting in statistics. Thus, the scope of forecasting is to estimate values of a
time series at some epochs in the near future, when present and past values
of the time series are known.

Linear prediction

Consider stationary time series $\{X_k\}_{k\in\mathbb{N}^*}$ of the Hilbert space $\mathcal{L}_2$. We
consider the problem of prediction or estimation of the future values
$X_{n+1}, X_{n+2}, \ldots$ from the past and present values $X_1, \ldots, X_n$, where $n \in
\mathbb{N}^*$. We search for the best linear predictor of X_{n+h}, that is for the orthog-
onal projection

$$\mathrm{pro}_{\overline{\mathrm{sp}}\{1, X_1, \ldots, X_n\}} X_{n+h}.$$

This best linear predictor would correspond to the conditional expectation
$\mathsf{E}[X_{n+h}|\ X_1, \ldots, X_n]$, where $h \in \mathbb{N}^*$, if normality was assumed.

   If μ is the mean and γ the a.c.v.f. of the time series $\{X_k\}_{k\in\mathbb{N}^*}$, then the
time series $Y_k = X_k - \mu$, for $k = 1, 2, \ldots$, is stationary with mean 0, a.c.v.f.
γ and we find

$$
\begin{aligned}
\mathrm{pro}_{\overline{\mathrm{sp}}\{1, X_1, \ldots, X_n\}} X_{n+h} &= \mu + \mathrm{pro}_{\overline{\mathrm{sp}}\{1, X_1, \ldots, X_n\}} (X_{n+h} - \mu) \\
&= \mu + \mathrm{pro}_{\overline{\mathrm{sp}}\{1, X_1-\mu, \ldots, X_n-\mu\}} (X_{n+h} - \mu) \\
&= \mu + \mathrm{pro}_{\overline{\mathrm{sp}}\{1, Y_1, \ldots, Y_n\}} Y_{n+h} \\
&= \mu + \mathrm{pro}_{\overline{\mathrm{sp}}\{Y_1, \ldots, Y_n\}} Y_{n+h},
\end{aligned}
$$

from the linearity, the idempotence and the iterativity of the orthogonal
projection, given in Proposition A.1.2, and because the mean of $\{Y_k\}_{k\in\mathbb{N}^*}$

is null. So we will directly assume $\mu = 0$ and so, for $h = 1, 2, \ldots$, define by

$$\hat{X}_{n+h} = \mathrm{pro}_{\overline{\mathrm{sp}}\{X_1, \ldots, X_n\}} X_{n+h}, \tag{2.5.1}$$

the best linear predictor of X_{n+h}.

Consider $h = 1$ and re-express (2.5.1) as

$$\hat{X}_{n+1} = \varphi_{n1} X_n + \cdots + \varphi_{nn} X_1, \tag{2.5.2}$$

where

$$\left\langle X_{n+1} - \sum_{i=1}^{n} \varphi_{ni} X_{n-i+1}, X_{n-j+1} \right\rangle = 0, \quad \text{for } j = 1, \ldots, n,$$

i.e.

$$\sum_{i=1}^{n} \varphi_{ni} \gamma(j - i) = \gamma(j), \quad \text{for } j = 1, \ldots, n.$$

This can be re-expressed as

$$\boldsymbol{\Gamma}_n \boldsymbol{\varphi}_n = \boldsymbol{\gamma}_n, \tag{2.5.3}$$

where

$$\boldsymbol{\Gamma}_n = (\gamma(j-i))_{i,j=1,\ldots,n}, \ \boldsymbol{\gamma}_n = (\gamma(1), \ldots, \gamma(n))^\top \text{ and } \boldsymbol{\varphi}_n = (\varphi_{n1}, \ldots, \varphi_{nn})^\top.$$

The existence of the orthogonal projection tells that $\hat{X}_{n+1}$ must have the form (2.5.2), for some coefficients $\varphi_{n1}, \ldots, \varphi_{nn}$. This is due to the assumption of null mean. Thus (2.5.3) must have at least one solution $\boldsymbol{\varphi}_n$. The unicity of the orthogonal projection tells that each solution $\boldsymbol{\varphi}_n$ must yield the same $\hat{X}_{n+1}$ through (2.5.2). If $\boldsymbol{\Gamma}_n$ is nonsingular, then we have

$$\boldsymbol{\varphi}_n = \boldsymbol{\Gamma}_n^{-1} \boldsymbol{\gamma}_n,$$

as unique solution.

According to Proposition 2.1.29, if $\gamma(h) \xrightarrow{h \to \infty} 0$ (and $\gamma(0) > 0$), then $\boldsymbol{\Gamma}_n$ nonsingular, for $n = 1, 2, \ldots$. In this situation, the best linear predictor is given by

$$\hat{X}_{n+1} = \sum_{j=1}^{n} \varphi_{nj} X_{n-j+1} = \boldsymbol{\varphi}_n^\top \begin{pmatrix} X_n \\ \vdots \\ X_1 \end{pmatrix} = \boldsymbol{\gamma}_n^\top \boldsymbol{\Gamma}_n^{-1} \begin{pmatrix} X_n \\ \vdots \\ X_1 \end{pmatrix}.$$

The mean square prediction error is the square $\mathcal{L}_2$-norm of the projection or prediction residual $R_n = \hat{X}_{n+1} - X_{n+1}$. When $\boldsymbol{\Gamma}_n$ is nonsingular, it is given by

$$
\begin{aligned}
v_n &= \|R_n\|^2 \\
&= \mathsf{E}\left[(\hat{X}_{n+1} - X_{n+1})^2\right] \\
&= \mathsf{E}\left[\hat{X}_{n+1}^2\right] + \mathsf{E}\left[X_{n+1}^2\right] - 2\mathrm{cov}(\hat{X}_{n+1}, X_{n+1}) \\
&= \boldsymbol{\gamma}_n^\top \boldsymbol{\Gamma}_n^{-1} \boldsymbol{\Gamma}_n \boldsymbol{\Gamma}_n^{-1} \boldsymbol{\gamma}_n + \gamma(0) - 2\boldsymbol{\gamma}_n^\top \boldsymbol{\Gamma}_n^{-1}\mathrm{cov}\left(\begin{pmatrix} X_n \\ \vdots \\ X_1 \end{pmatrix}, X_{n+1}\right) \\
&= \gamma(0) - \boldsymbol{\gamma}_n^\top \boldsymbol{\Gamma}_n^{-1} \boldsymbol{\gamma}_n.
\end{aligned}
\tag{2.5.4}
$$

We now turn to the best linear h-step predictor of X_{n+h} that is given by (2.5.1). It takes the form of

$$
\hat{X}_{n+h} = \varphi_{n1}^{(h)} X_n + \cdots + \varphi_{nn}^{(h)} X_1,
$$

where $\varphi_{n1}^{(h)}, \ldots, \varphi_{nn}^{(h)}$ solve

$$
\boldsymbol{\Gamma}_n \boldsymbol{\varphi}_n^{(h)} = \boldsymbol{\gamma}_n^{(h)},
$$

for

$$
\boldsymbol{\varphi}_n^{(h)} = (\varphi_{n1}^{(h)}, \ldots, \varphi_{nn}^{(h)})^\top \quad \text{and} \quad \boldsymbol{\gamma}_n^{(h)} = (\gamma(h), \ldots, \gamma(h+n-1))^\top.
$$

By assuming $\boldsymbol{\Gamma}_n$ invertible, we obtain

$$
\hat{X}_{n+h} = \boldsymbol{\gamma}_n^{(h)\top} \boldsymbol{\Gamma}_n^{-1} \begin{pmatrix} X_n \\ \vdots \\ X_1 \end{pmatrix}.
$$

Prediction with Gaussian time series

Assume now that $\{X_k\}_{k\in\mathbb{N}^*}$ is a Gaussian stationary time series. Then we have

$$
\hat{X}_{n+h} = \mathsf{E}[X_{n+h} \mid X_1, \ldots, X_n].
$$

This follows directly from the linear form of the conditional expectation of jointly Gaussian random variables, which is precisely given in Lemma A.1.9.

Define the h-step prediction residual

$$
R_n^{(h)} = X_{n+h} - \hat{X}_{n+h}.
$$

Because the time series is Gaussian, we have

$$
R_n^{(h)} \sim \mathcal{N}\left(0, v_n^{(h)}\right),
$$

where $v_n^{(h)} = \mathsf{E}[R_n^{(h)2}]$ is the mean square error of the h-step prediction. By following (2.5.4), we obtain

$$v_n^{(h)} = \gamma(0) - \gamma_n^{(h)\top} \mathbf{\Gamma}_n^{-1} \gamma_n^{(h)}.$$

This leads to the prediction interval with coverage level $1 - \alpha$, for some $\alpha \in (0,1)$ small, given by

$$1 - \alpha = \mathsf{P}\left[z_{\frac{\alpha}{2}} \leq \frac{R_n^{(h)}}{\sqrt{v_n^{(h)}}} \leq z_{-\frac{\alpha}{2}} \right]$$

$$= \mathsf{P}\left[\hat{X}_{n+h} + z_{\frac{\alpha}{2}} \sqrt{v_n^{(h)}} \leq X_{n+h} \leq \hat{X}_{n+h} - z_{\frac{\alpha}{2}} \sqrt{v_n^{(h)}} \right],$$

where $z_{\alpha/2} = \Phi^{-1}(\alpha/2)$.

Prediction with causal and invertible ARMA time series

Assume that the time series $\{X_k\}_{k \in \mathbb{Z}}$ is a causal and invertible ARMA(p,q) time series. Denote $H_n = \overline{\mathrm{sp}}\{X_j\}_{j \leq n}$, for simplicity, and

$$\hat{X}_k = \mathrm{pro}_{H_n} X_k$$

the best linear predictor of X_k in terms of $\{X_j\}_{j \leq n}$, $\forall k, n \in \mathbb{Z}$. We note that $\hat{X}_k = X_k$, if $k \leq n$. The best linear predictor and its mean square in this situation are given by Theorem 2.5.1.

Theorem 2.5.1 (Prediction with ARMA time series). *Let $\{X_k\}_{k \in \mathbb{Z}}$ be an ARMA(p,q) time series satisfying $\varphi(z) \neq 0$ and $\theta(z) \neq 0, \forall z \in \mathcal{D}_0$. Then we have the best linear predictor*

$$\hat{X}_{n+h} = \mathrm{pro}_{H_n} X_{n+h} = -\sum_{j=1}^{\infty} \tau_j \hat{X}_{n+h-j} = \sum_{j=h}^{\infty} \psi_j Z_{n+h-j},$$

and the mean square prediction error

$$\mathsf{E}\left[\left(X_{n+h} - \hat{X}_{n+h} \right)^2 \right] = \sigma^2 \sum_{j=0}^{h-1} \psi_j^2,$$

$\forall h \in \mathbb{N}^*, n \in \mathbb{Z}$, *where*

$$\tau(z) = \frac{\varphi(z)}{\theta(z)} = \sum_{j=0}^{\infty} \tau_j z^j, \ \psi(z) = \frac{\theta(z)}{\varphi(z)} = \sum_{j=0}^{\infty} \psi_j z^j, \ \forall z \in \mathcal{D}_0,$$

$\sum_{j=0}^{\infty} |\tau_j| < \infty$ *and* $\sum_{j=0}^{\infty} |\psi_j| < \infty$.

Note that $\psi_0 = \psi(0) = \theta(0)/\varphi(0) = 1$, giving the simple formula $\mathsf{E}[(X_{n+1} - \hat{X}_{n+1})^2] = \sigma^2$.

Proof. Let $h \in \mathbb{N}^*$ and $n \in \mathbb{Z}$. Corollary 2.2.16 tells that under the given assumptions, $\{X_k\}_{k \in \mathbb{Z}}$ is causal and invertible. Thus we have

$$Z_{n+h} = \sum_{j=0}^{\infty} \tau_j X_{n+h-j} = X_{n+h} + \sum_{j=1}^{\infty} \tau_j X_{n+h-j}, \qquad (2.5.5)$$

and

$$X_{n+h} = \sum_{j=0}^{\infty} \psi_j Z_{n+h-j}, \qquad (2.5.6)$$

because $\tau_0 = \tau(0) = \varphi(0)/\theta(0) = 1$. Therefore, $\forall \alpha_n, \alpha_{n-1}, \ldots \in \mathbb{R}$, (2.5.6) gives

$$\langle Z_{n+h}, \alpha_n X_n + \alpha_{n-1} X_{n-1} + \cdots \rangle$$

$$= \alpha_n \langle Z_{n+h}, X_n \rangle + \alpha_{n-1} \langle Z_{n+h}, X_{n-1} \rangle + \cdots$$

$$= \alpha_n \left\langle Z_{n+h}, \sum_{j=0}^{\infty} \psi_j Z_{n-j} \right\rangle + \alpha_{n-1} \left\langle Z_{n+h}, \sum_{j=0}^{\infty} \psi_j Z_{n-1-j} \right\rangle + \cdots$$

$$= \alpha_n \sum_{j=0}^{\infty} \psi_j \langle Z_{n+h}, Z_{n-j} \rangle + \alpha_{n-1} \sum_{j=0}^{\infty} \psi_j \langle Z_{n+h}, Z_{n-1-j} \rangle + \cdots$$

$$= 0.$$

Thus Z_{n+h} is orthogonal to H_n and so it follows from (2.5.5) that

$$\mathsf{pro}_{H_n} Z_{n+h} = \mathsf{pro}_{H_n} X_{n+h} + \sum_{j=1}^{\infty} \tau_j \mathsf{pro}_{H_n} X_{n+h-j} = 0, \qquad (2.5.7)$$

which yields the first prediction formula to show,

$$\hat{X}_{n+h} = -\sum_{j=1}^{\infty} \tau_j \hat{X}_{n+h-j}.$$

Then (2.5.5) gives, for $h - j \leq 0$,

$$Z_{n+h-j} = X_{n+h-j} + \tau_1 X_{n+h-j-1} + \cdots \in H_{n+h-j} \subset H_n. \qquad (2.5.8)$$

One obtains from (2.5.6) that

$$\mathsf{pro}_{H_n} X_{n+h} = \sum_{j=0}^{\infty} \psi_j \mathsf{pro}_{H_n} Z_{n+h-j}$$

$$= \sum_{j=0}^{h-1} \psi_j \underbrace{\mathsf{pro}_{H_n} Z_{n+h-j}}_{\substack{=0, \\ \text{from } (2.5.7)}} + \sum_{j=h}^{\infty} \psi_j \underbrace{\mathsf{pro}_{H_n} Z_{n+h-j}}_{\substack{=Z_{n+h-j}, \\ \text{from } (2.5.8)}}.$$

This is equivalent to the second prediction formula to show, which is

$$\hat{X}_{n+h} = \sum_{j=h}^{\infty} \psi_j Z_{n+h-j}. \tag{2.5.9}$$

By subtracting (2.5.9) from (2.5.6) one obtains

$$X_{n+h} - \hat{X}_{n+h} = \sum_{j=0}^{h-1} \psi_j Z_{n+h-j},$$

which justifies the formula for the mean square prediction error. □

The prediction formula of Theorem 2.5.1 can be decomposed in three parts,

$$\hat{X}_{n+h}$$

$$= - \underbrace{\sum_{j=1}^{h-1} \tau_j \hat{X}_{n+h-j}}_{\substack{\text{depends on predictions} \\ \hat{X}_{n+1},\dots,\hat{X}_{n+h-1}}} - \underbrace{\sum_{j=h}^{n+h-1} \tau_j X_{n+h-j}}_{\substack{\text{depends on sample} \\ X_1,\dots,X_n}} - \underbrace{\sum_{j=n+h}^{\infty} \tau_j X_{n+h-j}}_{\substack{\text{depends on past} \\ X_0,X_{-1},\dots}}, \tag{2.5.10}$$

by assuming $h \geq 2$ and $n \geq 1$. If $h = 1$, then the first of the three parts above vanishes.

The prediction formula (2.5.10) can be computed recursively as follows:

$$\hat{X}_{n+1} = -\sum_{j=1}^{\infty} \tau_j \hat{X}_{n+1-j} = -\sum_{j=1}^{\infty} \tau_j X_{n+1-j},$$

$$\hat{X}_{n+2} = -\tau_1 \hat{X}_{n+1} - \sum_{j=2}^{\infty} \tau_j X_{n+2-j} = \tau_1 \sum_{j=1}^{\infty} \tau_j X_{n+1-j} - \sum_{j=2}^{\infty} \tau_j X_{n+2-j},$$

$$\hat{X}_{n+3} = -\tau_1 \hat{X}_{n+2} - \tau_2 \hat{X}_{n+1} - \sum_{j=3}^{\infty} \tau_j X_{n+3-j} = \dots,$$

. . . .

One can obtain an approximation to the predictor by setting $\sum_{j=n+h}^{\infty} \tau_j X_{n+h-j} = 0$ in (2.5.10). This is approximately correct for large values of the sample size n large, because $\sum_{j=0}^{\infty} \tau_j$ is absolutely convergent. Based on this approximation, we can define the truncated prediction at time n by

$$\hat{X}_{n+h}^{\dagger} = \underbrace{-\tau_1 \hat{X}_{n+h-1}^{\dagger} - \dots - \tau_{h-1} \hat{X}_{n+1}^{\dagger}}_{\substack{\text{depends on predictions} \\ \hat{X}_{n+1}^{\dagger},\dots,\hat{X}_{n+h-1}^{\dagger}}} \underbrace{-\tau_h X_n - \dots - \tau_{n+h-1} X_1}_{\substack{\text{depends on sample} \\ X_1,\dots,X_n}},$$

if $h \geq 2$ If $h = 1$, then the first part of the above formula vanishes. Thus $\hat{X}_{n+h}^{\dagger}$ provides a computable approximation to $\hat{X}_{n+h}$, which can be obtained recursively as follows:

$$\hat{X}_{n+1}^{\dagger} = -\sum_{j=1}^{n} \tau_j X_{n+1-j},$$

$$\hat{X}_{n+2}^{\dagger} = -\tau_1 \hat{X}_{n+1}^{\dagger} - \sum_{j=2}^{n+1} \tau_j X_{n+2-j} = \tau_1 \sum_{j=1}^{n} \tau_j X_{n+1-j} - \sum_{j=2}^{n+1} \tau_j X_{n+2-j},$$

$$\hat{X}_{n+3}^{\dagger} = -\tau_1 \hat{X}_{n+2}^{\dagger} - \tau_2 \hat{X}_{n+1}^{\dagger} - \sum_{j=3}^{n+2} \tau_j X_{n+3-j} = \cdots,$$

$$\cdots.$$

Two basic applications of Theorem 2.5.1 are the following.

Example 2.5.2 (Causal AR(p)). *Consider the AR(p) time series with $|\varphi_1| < 1$. The conditions of Theorem 2.5.1 are satisfied and thus we have $\tau(z) = \varphi(z)$ and $\hat{X}_{n+1} = -\tau_1 \hat{X}_n - \cdots - \tau_p \hat{X}_{n-p+1}$. It also holds that $\hat{X}_j = \text{pro}_{H_n} X_j = X_j$, for $j = n - p + 1, \ldots, n$, and therefore we obtain*

$$\hat{X}_{n+1} = \varphi_1 X_n + \cdots + \varphi_p X_{n-p+1}.$$

Note the related analysis of the AR(p) given in Example 2.3.8. We also have

$$\mathsf{E}\left[(X_{n+1} - \hat{X}_{n+1})^2\right] = \mathsf{E}\left[Z_{n+1}^2\right] = \sigma^2.$$

Example 2.5.3 (MA(1)). *Consider the MA(1) time series with $|\theta_1| < 1$. Then*

$$\tau(z) = \frac{\varphi(z)}{\theta(z)} = \frac{1}{1 - (-\theta_1)z} = \sum_{j=0}^{\infty} (-\theta_1)^j z^j, \quad \forall z \in \mathcal{D}_0.$$

So with $\tau_j = (-\theta_1)^j$, for $j = 0, 1, \ldots$, and with $\hat{X}_n = X_n, \hat{X}_{n-1} = X_{n-1}, \ldots$, Theorem 2.5.1 yields

$$\hat{X}_{n+1} = -\sum_{j=1}^{\infty} (-\theta_1)^j X_{n+1-j} = \sum_{j=1}^{\infty} (-1)^{j+1} \theta_1^j X_{n+1-j},$$

and

$$\mathsf{E}\left[(X_{n+1} - \hat{X}_{n+1})^2\right] = \sigma^2 \psi_0^2 = \sigma^2.$$

The truncation

$$\hat{X}_{n+1}^{\dagger} = -\sum_{j=1}^{n} (-\theta_1)^j X_{n+1-j},$$

provides a poor approximation to the predictor $\hat{X}_{n+1}$, whenever the value of $|\theta_1|$ is close to 1.

2.5.2 Estimation and asymptotic inference for mean and autocovariance function

This section presents the problem of estimation of the mean value and of the a.c.v.f. of a real-valued stationary time series and it provides asymptotic results for inference on these quantities.

Estimation and inference for mean

We consider a real-valued $\{X_k\}_{k \in \mathbb{Z}}$ stationary time series with mean $\mu = \mathsf{E}[X_1]$ and with a.c.v.f. γ. Let $n \in \mathbb{N}^*$ and define the mean of $X_1, \ldots, X_n$ as

$$M_n = \frac{1}{n} \sum_{k=1}^{n} X_k, \tag{2.5.11}$$

which provides an unbiased estimator of μ. Theorem 2.5.4 gives two conditions under which $M_n \xrightarrow{\text{P}} \mu$ or, in other terms, under which it is a weakly consistent estimator of μ. Any one of these two conditions with Markov's inequality A.2.5 give weak consistency.

The variance of M_n is given by

$$n \,\mathsf{var}(M_n) = \frac{1}{n} \sum_{i=1}^{n} \sum_{j=1}^{n} \mathsf{cov}(X_i, X_j)$$

$$= \frac{1}{n} \sum_{h=-(n-1)}^{n-1} (n - |h|) \gamma(h)$$

$$= \sum_{h=-(n-1)}^{n-1} p\left(\frac{h}{n}\right) \gamma(h)$$

$$= \gamma(0) + 2 \sum_{h=1}^{n-1} p\left(\frac{h}{n}\right) \gamma(h), \tag{2.5.12}$$

with p denoting the triangular density (2.4.4), and its asymptotic value is given in Theorem 2.5.4.2.

Theorem 2.5.4 (Asymptotic value of variance). *Let $\{X_k\}_{k \in \mathbb{Z}}$ be a real-valued stationary time series with mean μ and a.c.v.f. γ. Then the following results hold.*

(1) $\gamma(h) \xrightarrow{h \to \infty} 0 \implies \mathsf{var}(M_n) \xrightarrow{n \to \infty} 0.$

(2) $\sum_{h=0}^{\infty} |\gamma(h)| < \infty \implies n \,\mathsf{var}(M_n) \xrightarrow{n \to \infty} \sum_{n=-\infty}^{\infty} \gamma(h) = \gamma(0) + 2 \sum_{h=1}^{\infty} \gamma(h).$

Proof. 1. We deduce from (2.5.12) that

$$n\mathrm{var}(M_n) \leq \sum_{h=-(n-1)}^{n-1} |\gamma(h)|.$$

Therefore, $\gamma(h) \xrightarrow{h\to\infty} 0$ leads to

$$\lim_{n\to\infty} \mathrm{var}(M_n) \leq \lim_{n\to\infty} \frac{1}{n} \sum_{h=-(n-1)}^{n-1} |\gamma(h)|$$

$$= \lim_{n\to\infty} \frac{\gamma(0)}{n} + 2 \lim_{n\to\infty} \sum_{h=1}^{n-1} \frac{|\gamma(h)|}{n}$$

$$= 0,$$

where the last equality uses the fact that convergence implies Cesàro summability; cf. p. 8.

2. By applying Dominated convergence theorem to (2.5.12) we obtain

$$\lim_{n\to\infty} n\mathrm{var}(M_n) = \lim_{n\to\infty} \sum_{h=-\infty}^{\infty} p\left(\frac{h}{n}\right)\gamma(h) = \sum_{h=-\infty}^{\infty} \gamma(h).$$

The symmetry of the a.c.v.f. γ leads to the desired result. □

Remarks 2.5.5.

(1) *Theorem 2.5.4.1 or 2.5.4.2 imply the simple ergodic result $M_n \xrightarrow{P} \mu$.
It is ergodic in the sense that a characteristic of the time series at
any fixed time (which is here the expectation μ) is obtained from the
analogue characteristic on the infinite time horizon.*

(2) *We can re-express Theorem 2.5.4.2 in terms of the spectral den-
sity. If the a.c.v.f. γ satisfies $\sum_{j=-\infty}^{\infty} |\gamma(j)| < \infty$, then Corol-
lary 2.4.13 provides the spectral density in the form $f(\alpha) =
1/(2\pi) \sum_{j=-\infty}^{\infty} e^{-ij\alpha}\gamma(j) \geq 0, \forall\alpha \in [-\pi, \pi]$. Thus*

$$n\,\mathrm{var}(M_n) \xrightarrow{n\to\infty} 2\pi f(0).$$

Example 2.5.6 (ARMA). *Let $\{X_k\}_{k\in\mathbb{Z}}$ be an ARMA(p, q) time series with
autoregressive polynomial φ satisfying $\varphi(z) \neq 0, \forall z \in \mathbb{Z}$ such that $|z| = 1$.
Theorem 2.2.8.1 provides the MA representation $X_k = \sum_{j=-\infty}^{\infty} \psi_j Z_{k-j}$,
$\forall k \in \mathbb{Z}$, where $\sum_{j=-\infty}^{\infty} |\psi_j| < \infty$ and where $\{Z_k\}_{k\in\mathbb{Z}}$ is WN(σ^2). Theorem
2.2.8.2 tells that the mean is null and that the a.c.v.f. is given by $\gamma(h) =
\sigma^2 \sum_{j=0}^{\infty} \psi_j \psi_{|h|+j}, \forall h \in \mathbb{Z}$. We deduce that*

$$\sum_{h=-\infty}^{\infty} |\gamma(h)| \leq \sigma^2 \sum_{h=-\infty}^{\infty} \sum_{j=-\infty}^{\infty} |\psi_j| |\psi_{|h|+j}| \sum_{j=-\infty}^{\infty} |\psi_j| \sum_{h=-\infty}^{\infty} |\psi_{|h|+j}| < \infty.$$

Theorem 2.4.30 gives the spectral density $f(\alpha) = \sigma^2/(2\pi)$ $|\sum_{j=-\infty}^{\infty} \psi_j e^{-ij\alpha}|^2$, $\forall \alpha \in [-\pi, \pi]$, which together with Remark 2.5.5.2 yields

$$n \operatorname{var}(M_n) \xrightarrow{n \to \infty} 2\pi f(0) = \left(\sigma \sum_{j=-\infty}^{\infty} \psi_j \right)^2.$$

The mean M_n is asymptotically normal.

Theorem 2.5.7 (Asymptotic normality of mean of stationary time series).
Let $\{X_k\}_{k \in \mathbb{Z}}$ be a stationary time series with mean μ and MA representation $X_k - \mu = \sum_{j=-\infty}^{\infty} \psi_j Z_{k-j}$, $\forall k \in \mathbb{Z}$, where $\sum_{j=-\infty}^{\infty} |\psi_j| < \infty$, $\sum_{j=-\infty}^{\infty} \psi_j \neq 0$ and where the elements of $\{Z_k\}_{k \in \mathbb{Z}}$ are i.i.d. with mean null and variance σ^2. Then

$$\sqrt{n}(M_n - \mu) \xrightarrow{d} \mathcal{N}\left(0, \xi^2\right),$$

where

$$\xi^2 = \sum_{h=-\infty}^{\infty} \gamma(h) = \left(\sigma \sum_{j=-\infty}^{\infty} \psi_j \right)^2.$$

The proof of Theorem 2.5.7 can be found in Brockwell and Davis (1991), Section 7.3. Note that the asymptotic normal approximation to the distribution of the mean M_n can be very inaccurate in various circumstances, in particular when small tail probability are desired or when the sample size n is not very large. One can indeed show that the relative error is unbounded. A superior asymptotic method is obtained from the theory large deviations, which is shortly introduced in Section 4.8 and applied to the AR(1). In contrast to the normal approximation, the large deviations approximation requires the precise distribution of the WN. Note however that if $\{X_k\}_{k \in \mathbb{Z}}$ is a stationary Gaussian time series, then the normal approximation holds exactly, namely

$$M_n \sim \mathcal{N}\left(\mu, \frac{1}{n} \sum_{h=-(n-1)}^{n-1} p\left(\frac{h}{n}\right) \gamma(h)\right), \quad \forall n \in \mathbb{N}^*,$$

where p is the triangular density (2.4.4).

Examples 2.5.8 (Asymptotic distributions of means of some ARMA).

(1) Consider the MA(q) time series whose a.c.v.f. is given by Proposition 2.2.2. Thus we have

$$\xi^2 = \sum_{h=-\infty}^{\infty} \gamma(h) = \sum_{h=-\infty}^{\infty} \sigma^2 \sum_{j=0}^{q-|h|} \theta_j \theta_{|h|+j} I\{|h| \le q\}$$

$$= \sigma^2 \left(\sum_{j=0}^{q} \theta_j^2 + 2\sum_{h=1}^{q}\sum_{j=0}^{q-h} \theta_j \theta_{h+j} \right) = \sigma^2 \left(\sum_{j=0}^{q} \theta_j^2 + 2\sum_{j=0}^{q-1}\sum_{h=j+1}^{q} \theta_j \theta_h \right)$$

$$= \{\sigma(1 + \theta_1 + \cdots + \theta_q)\}^2 .$$

Given $\psi_j = \theta_j$, for $j = 0, \ldots, q$, and $\psi_j = 0$, for $j = q+1, q+2, \ldots$, we obtain

$$\xi^2 = \left(\sigma \sum_{j=-\infty}^{\infty} \psi_j \right)^2 = \{\sigma(1 + \theta_1 + \cdots + \theta_q)\}^2 .$$

We thus obtain

$$\sqrt{n} M_n \xrightarrow{d} \mathcal{N}\left(0, \{\sigma(1 + \theta_1 + \cdots + \theta_q)\}^2 \right). \tag{2.5.13}$$

(2) Consider the causal AR(1) time series, thus with $|\varphi_1| < 1$. Its a.c.v.f. is given by (2.2.8) and so we have

$$\xi^2 = \sum_{h=-\infty}^{\infty} \gamma(h) = \sum_{h=-\infty}^{\infty} \sigma^2 \frac{\varphi_1^{|h|}}{1 - \varphi_1^2}$$

$$= \frac{\sigma^2}{1 - \varphi_1^2} \left(2\sum_{h=0}^{\infty} \varphi_1^h - 1 \right) = \left(\frac{\sigma}{1 - \varphi_1} \right)^2 .$$

Given $\psi_j = \varphi_1^j$, for $j = 0, 1, \ldots$, and $\psi_j = 0$, for $j = -1, -2, \ldots$, we obtain

$$\xi^2 = \left(\sigma \sum_{j=-\infty}^{\infty} \psi_j \right)^2 = \left(\sigma \sum_{j=0}^{\infty} \varphi_1^j \right)^2 = \left(\frac{\sigma}{1 - \varphi_1} \right)^2 .$$

We thus obtain

$$\sqrt{n} M_n \xrightarrow{d} \mathcal{N}\left(0, \left(\frac{\sigma}{1 - \varphi_1} \right)^2 \right). \tag{2.5.14}$$

(3) *Consider the causal and invertible* ARMA$(1,1)$ *of Example 2.2.26. Given* $\psi_0 = 1$ *and* $\psi_j = (\varphi_1 + \theta_1)\varphi_1^{j-1}$, *for* $j = 1, 2, \ldots$, *and* $\psi_j = 0$, *for* $j = -1, -2, \ldots$, *as obtained in (2.2.30), we easily find*

$$\xi^2 = \left(\sigma \sum_{j=-\infty}^{\infty} \psi_j\right)^2 = \left(\sigma\left[1 + (\varphi_1 + \theta_1)\sum_{j=0}^{\infty} \varphi_1^j\right]\right)^2 = \left(\sigma \frac{1+\theta_1}{1-\varphi_1}\right)^2.$$

The a.c.v.f. of the ARMA$(1,1)$ *is given by (2.2.31) and (2.2.32) and, with some perseverance, we can find*

$$\xi^2 = \sum_{h=-\infty}^{\infty} \gamma(h) = \left(\sigma \frac{1+\theta_1}{1-\varphi_1}\right)^2.$$

We thus obtain

$$\sqrt{n}M_n \xrightarrow{\mathrm{d}} \mathcal{N}\left(0, \left(\sigma \frac{1+\theta_1}{1-\varphi_1}\right)^2\right). \qquad (2.5.15)$$

Note that (2.5.15) with $\varphi_1 = 0$ *yields (2.5.13) with* $q = 1$. *Moreover, (2.5.15) with* $\theta_1 = 0$ *yields and (2.5.14).*

Estimation and inference for a.c.v.f.

Consider as before the stationary time series $\{X_k\}_{k \in \mathbb{Z}}$ and the sample $X_1, \ldots, X_n$. We remind that the empirical a.c.v.f. at time lags $h = 0, \ldots, n-1$ is given by

$$\hat{\gamma}_n(h) = \frac{1}{n} \sum_{j=1}^{n-h} \left(X_{j+h} - M_n\right)\left(X_j - M_n\right), \qquad (2.5.16)$$

and the empirical a.c.r.f. by

$$\hat{\rho}_n(h) = \frac{\hat{\gamma}_n(h)}{\hat{\gamma}_n(0)}.$$

In this context, we remind Proposition 2.1.31: the empirical a.c.v.f. is n.n.d. We understand that $\gamma(h)$ and $\rho(h)$ cannot be estimated for $h \geq n$. But also in the case $h \leq n-1$ and close to $n-1$, we cannot expect good accuracy because of the too small number of summands in (2.5.16). The next theorem gives the asymptotic distribution of the empirical a.c.r.f.

Theorem 2.5.9 (Asymptotic normality of empirical a.c.r.f.). *Let* $\{X_k\}_{k \in \mathbb{Z}}$ *be a time series with mean* μ, *a.c.v.f.* γ *and a.c.r.f.* ρ. *Assume the MA representation* $X_k - \mu = \sum_{j=-\infty}^{\infty} \psi_j Z_{k-j}$, $\forall k \in \mathbb{Z}$, *where* $\sum_{j=-\infty}^{\infty} |\psi_j| < \infty$, *where the random variables $\{Z_k\}_{k\in\mathbb{Z}}$ are i.i.d. with mean 0, variance σ^2 and with $\mathsf{E}[Z_1^4] < \infty$. Then, it holds for $h = 0, \ldots, n-1$ and as $n \to \infty$,*

$$\sqrt{n}(\hat{\boldsymbol{\rho}}_{hn} - \boldsymbol{\rho}_h) \xrightarrow{\text{d}} \mathcal{N}(0, \boldsymbol{W}),$$

where $\hat{\boldsymbol{\rho}}_{hn} = \big(\hat{\rho}_n(1), \ldots, \hat{\rho}_n(h)\big)^\top$, $\boldsymbol{\rho}_{hn} = \big(\rho_n(1), \ldots, \rho_n(h)\big)^\top$, $\boldsymbol{W} = \big(w_{ij}\big)_{i,j=1,\ldots,h}$, with

$$w_{ij} = \sum_{k=1}^{\infty} \big\{\rho(k+i) + \rho(k-i) - 2\rho(i)\rho(k)\big\}\big\{\rho(k+j) + \rho(k-j) - 2\rho(j)\rho(k)\big\},$$

for $i, j = 1, \ldots, h$ and $h = 1, \ldots, n-1$.

The condition of finiteness of the fourth moment of the i.i.d. sequence restricts to some rather light-tailed distributions. It is certainly satisfied by Gaussian WN. We refer to Brockwell and Davis (1991), Section 7.3, for the proof of Theorem 2.5.9.

We now provide two simple illustrations.

Example 2.5.10 (I.i.d. sequence). *Consider the time series $\{X_k\}_{k\in\mathbb{Z}}$ of i.i.d. random variables with mean 0 and finite moment of order four. Thus $\{X_k\}_{k\in\mathbb{Z}} = \{Z_k\}_{k\in\mathbb{Z}}$, for $\psi_j = \mathsf{I}\{j = 0\}$, $\forall j \in \mathbb{Z}$. Then $\rho(h) = 0$, for $h \neq 0$, and*

$$\omega_{ij} = \begin{cases} \rho^2(0) = 1, & \text{if } i = j, \\ 0, & \text{if } i \neq j. \end{cases}$$

Thus $\hat{\rho}_n(1), \ldots, \hat{\rho}_n(h)$ are approximately independent random variables and distributed as $\mathcal{N}(0, 1/n)$, for $h = 1, \ldots, n-1$. A graph of the points $(h, \hat{\rho}_n(h))$, for $h = 1, \ldots, n-1$, should show approximately $1 - \alpha$ of these points inside the band

$$\Big(n^{-\frac{1}{2}} z_{\frac{\alpha}{2}}, \, -n^{-\frac{1}{2}} z_{\frac{\alpha}{2}}\Big),$$

with $\alpha \in (0,1)$ small and z_β denoting the β-th quantile of the standard normal distribution, for $\beta \in (0,1)$.

Example 2.5.11 (MA(q)). *The MA(q) has equation $X_k = \theta(B)Z_k$, $\forall k \in \mathbb{Z}$, with $\{Z_k\}_{k\in\mathbb{Z}}$ WN(σ^2). The a.c.v.f. is given by Proposition 2.2.2.*

In particular $\gamma(0) = \sigma^2 \sum_{j=0}^{q} \theta_j^2$ *and thus, for* $h = q+1, q+2, \ldots$, *we have*

$$\omega_{hh} = \sum_{k=1}^{\infty} \{\rho(k+h) + \rho(k-h) - 2\rho(h)\rho(k)\}^2$$

$$= \sum_{k=h-q}^{h+q} \rho^2(k-h)$$

$$= \rho^2(-q) + \cdots + \rho^2(q)$$

$$= 1 + 2\rho^2(1) + \cdots + \rho^2(q).$$

Thus for the MA(1) *time series and for* $h = 2, \ldots, n-1$, *we find* $\omega_{hh} = 1 + 2\rho^2(1)$ *and*

$$\sqrt{n}\{\hat{\rho}_n(h) - \rho(h)\} \xrightarrow{\text{d}} \mathcal{N}(0, 1 + 2\rho^2(1)).$$

A graph of the points $(h, \hat{\rho}_n(h))$, *for* $h = 2, \ldots, n-1$, *should have approximately* $1 - \alpha$ *of these points within the band*

$$\left(z_{\frac{\alpha}{2}} n^{-\frac{1}{2}} \sqrt{1 + 2\rho^2(1)}, \ -z_{\frac{\alpha}{2}} n^{-\frac{1}{2}} \sqrt{1 + 2\rho^2(1)}\right),$$

with $\alpha \in (0,1)$ *small. We can also compute* $\rho(1) = \theta_1/(1 + \theta_1^2)$, $\rho(2) = \rho(3) = \cdots = 0$. *It is only* $\hat{\rho}_n(1)$ *that allows to distinguish the* MA(1) *from* WN.

Example 2.5.12 (Causal AR(1)). *We consider the causal* AR(1) *time series given by* $X_k - \varphi_1 X_{k-1} = Z_k$, $\forall k \in \mathbb{Z}$, *with* $\{Z_k\}_{k \in \mathbb{Z}}$ WN(σ^2) *and with* $|\varphi_1| < 1$. *We have* $\gamma(h) = \varphi_1^{|h|} \sigma^2/(1 - \varphi_1^2)$ *and* $\rho(h) = \varphi_1^{|h|}$, $\forall h \in \mathbb{Z}$. *Therefore, for* $h = 1, 2, \ldots$, *we have*

$$\omega_{hh} = \sum_{k=1}^{\infty} \{\rho(k+h) + \rho(k-h) - 2\rho(h)\rho(k)\}^2$$

$$= \sum_{k=1}^{\infty} \left(\varphi_1^{k+h} + \varphi_1^{|k-h|} - 2\varphi_1^{h+k}\right)^2$$

$$= \sum_{k=1}^{h} \left(\varphi_1^{h-k} - \varphi_1^{k+h}\right)^2 + \sum_{k=h+1}^{\infty} \left(\varphi_1^{k-h} - \varphi_1^{k+h}\right)^2$$

$$= \varphi_1^{2h} \sum_{k=1}^{h} \left(\varphi_1^{-k} - \varphi_1^{k}\right)^2 + \left(\varphi_1^{-h} - \varphi_1^{h}\right)^2 \sum_{k=h+1}^{\infty} \varphi_1^{2k}$$

$$= \varphi_1^{2h} \left(\sum_{k=0}^{h} \varphi_1^{-2k} + \sum_{k=0}^{h} \varphi_1^{2k} - 4\right) + \left(\varphi_1^{-2h} + \varphi_1^{2h} - 2\right) \varphi_1^{2(h+1)} \sum_{k=0}^{\infty} \varphi_1^{2k}$$

$$= \varphi_1^{2h} \left(\frac{\varphi_1^{-2(h+1)} - 1}{\varphi_1^{-2} - 1} + \frac{\varphi_1^{2(h+1)} - 1}{\varphi_1^2 - 1} - 4 \right)$$

$$+ \left(\varphi_1^{-2h} + \varphi_1^{2h} - 2 \right) \varphi_1^{2(h+1)} \frac{1}{1 - \varphi_1^2}$$

$$\longrightarrow \frac{1 + \varphi_1^2}{1 - \varphi_1^2}, \text{ as } h \to \infty.$$

Thus

$$\sqrt{n}\{\hat{\rho}(h) - \rho(h)\} \sim \mathcal{N}\left(0, \frac{1 + \varphi_1^2}{1 - \varphi_1^2}\right),$$

approximately, for $h = 1, \ldots, n-1$. Note that the vanishing p.a.c.r.f., precisely $\beta(h) = 0$, $\forall h \geq p+1$, is an important way of identifying the AR(p) time series.

2.5.3   Nonstationary ARIMA time series

In this section we introduce a new and closely related class of time series that is generally nonstationary but that can be transformed to an ARMA and thus to a stationary time series, by applying the difference operator $\nabla = 1 - B$, one or more times. Section 4.4 introduces a more general type of nonstationarity for time series and for continuous time stochastic processes: the intrinsic stationarity, which is given in Definition 4.4.2. ARIMA time series are thus particular intrinsic stationary time series and the precise definition is the following.

Definition 2.5.13 (ARIMA time series). *The time series $\{X_k\}_{k \in \mathbb{Z}}$ is an autoregressive integrated moving average (ARIMA) time series with parameters $p, d, q \in \mathbb{N}$ denoted ARIMA(p, d, q), if the time series $Y_k = \nabla^d X_k$, $\forall k \in \mathbb{Z}$, is a causal ARMA(p, q) time series.*

Consider the ARIMA(p, d, q) time series $\{X_k\}_{k \in \mathbb{Z}}$ and $\alpha_1, \ldots, \alpha_d \in \mathbb{R}$, for some $d \geq 1$. Then

$$X_k + \alpha_1 k^{d-1} + \cdots + \alpha_{d-1} k + \alpha_d, \quad \forall k \in \mathbb{Z},$$

is also an ARIMA(p, d, q) time series and precisely with same polynomials φ and θ and same WN. This means that an ARIMA(p, d, q) time series is rather a class of time series, which is defined modulo a polynomial of degree $d - 1$. This polynomial is the trend and it is removable by application of the operator ∇^d. We refer to Section 2.1.2. The related causal ARMA(p, q)

time series $\{Y_k\}_{k\in\mathbb{Z}}$ appears as follows,

$$\varphi(B)\underbrace{(1-B)^d X_k}_{=Y_k} = \theta(B)Z_k, \quad \forall k \in \mathbb{Z}.$$

In order to guarantee causality of $\{Y_k\}_{k\in\mathbb{Z}}$, we assume that φ and θ have no common roots over $\mathcal{D}_0$ and that $\varphi(z) \neq 0, \forall z \in \mathcal{D}_0$. We refer to Theorem 2.2.12.1 for these assumptions.

Define

$$\tilde{\varphi}(z) = \varphi(z)(1-z)^d, \quad \forall z \in \mathbb{C}.$$

Thus

$$\tilde{\varphi}(B)X_k = \theta(B)Z_k, \quad \forall k \in \mathbb{Z}, \tag{2.5.17}$$

defines the ARIMA time series and $\tilde{\varphi}$ has the root 1 with multiplicity d. Theorem 2.2.8 gives $\tilde{\varphi}(z) \neq 0, \forall z \in \mathbb{C}$ such that $|z| = 1$, as sufficient condition for (2.5.17) to have a unique stationary solution. This tells that if $\tilde{\varphi}(z)$ possesses other roots than $z = 1$ over the unit circle, then nonstationarities of other types than the one generated by a polynomial trend, added to the ARMA time series, are possible. However, in the practice, any root that is close to the unit circle indicates nonstationarity, as the following example illustrates.

Examples 2.5.14 (Nonstationary time series).

(1) Consider the time series $\{X_k\}_{k\in\mathbb{Z}}$ satisfying the AR(1) equation $X_k + 0.99X_{k-1} = Z_k, \forall k \in \mathbb{Z}$. Then $\varphi(z) = 1 + 0.99\,z = 0$ implies $z = -1/0.99 \simeq -1$, thus a root over the unit circle.

(2) Consider the time series $\{X_k\}_{k\in\mathbb{Z}}$ satisfying the AR(2) equation $X_k - X_{k-1}+0.99X_{k-2} = Z_k, \forall k \in \mathbb{Z}$. Then $\varphi(z) = 1-z+0.99z^2 = 0$ implies

$$z_1 z_2 = \frac{1}{0.99} \simeq 1, \tag{2.5.18}$$

and

$$z_1 + z_2 = \frac{1}{0.99} \simeq 1. \tag{2.5.19}$$

Let $z_1 = \rho_1 e^{i\theta_1}$ and $z_2 = \rho_2 e^{i\theta_2}$, then it follows from (2.5.18) that $\theta_1 = -\theta_2$ and $\rho_1 = \rho_2^{-1}$. It follows from these constraints and from (2.5.19) that $\theta_1 = \pi/3$, $\theta_2 = -\pi/3$ and $\rho_1 = \rho_2 = 1$.

We now provide some numerical examples obtained by simulation.

Example 2.5.15 (Simulation of ARIMA(1, 1, 0)). *We illustrate numerically the ARIMA(1, 1, 0) time series, which is given by*

$$(1 - \varphi_1 B)\nabla^1 X_k = Z_k, \quad \forall k \in \mathbb{Z},$$

where $\{Y_k\}_{k\in\mathbb{Z}} = \{\nabla^1 X_k\}_{k\in\mathbb{Z}}$ is a causal AR(1) and $\{Z_k\}_{k\in\mathbb{Z}}$ is WN(σ^2). The values of the parameters are $\varphi_1 = 0.8$, $\sigma^2 = 1$ and we consider $X_0 = 0$ and the times $k = 1, \ldots, n$, for $n = 200$.

We begin with the study of $\{X_k\}_{k\in\mathbb{Z}}$. Figure 2.11 shows the Monte Carlo simulation $\{X_k\}_{k=1,\ldots,200}$. Figure 2.12 shows the corresponding empirical

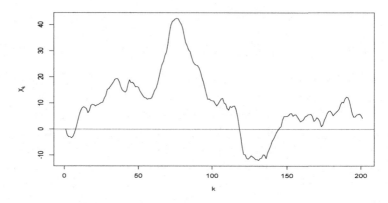

Fig. 2.11: Simulated sample path of ARIMA(1, 1, 0) $\{X_k\}_{k=1,\ldots,200}$ with $X_0 = 0$.

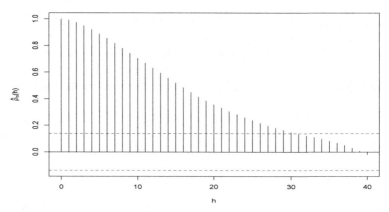

Fig. 2.12: Empirical a.c.r.f. $\hat{\rho}_n(h)$, for $h = 0, \ldots, 40$, of simulated ARIMA(1, 1, 0) $\{X_k\}_{k=1,\ldots,200}$.

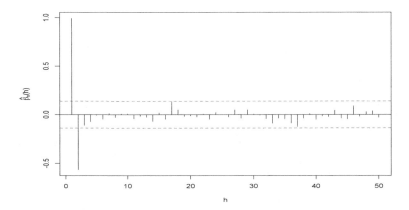

Fig. 2.13: Empirical p.a.c.r.f. $\hat{\beta}_n(h)$, for $h = 0, \ldots, 50$, of simulated ARIMA$(1, 1, 0)$ $\{X_k\}_{k=1,\ldots,200}$.

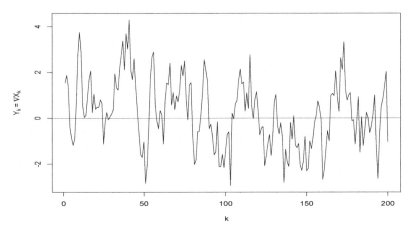

Fig. 2.14: Simulated sample path of AR(1) $\{Y_k\}_{k=1,\ldots,200}$ with $Y_0 = 0$.

a.c.r.f. $\hat{\rho}_n(h)$ at time lags $h = 0, \ldots, 40$. We can find a slow decay of this empirical a.c.r.f. It indicates the presence of a trend in the time series $\{X_k\}_{k=1,\ldots,200}$ and thus nonstationarity. In Figure 2.13 we find the corresponding empirical p.a.c.r.f. $\hat{\beta}_n(h)$, for $h = 0, \ldots, 50$. For the causal AR(p) time series it was shown in Example 2.3.8 that $\beta(h) = 0$, for $h = p + 1, p + 2, \ldots$. We however see, in Figure 2.13, that $\hat{\beta}_n(2) \neq 0$ and so we conclude that we should exclude the AR(1) time series. The general conclusion is that there is enough evidence against stationarity of $\{X_k\}_{k=1,\ldots,200}$.

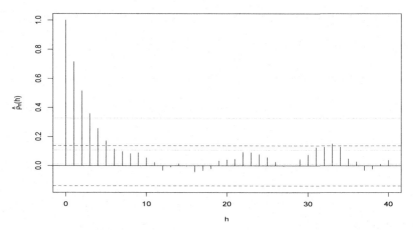

Fig. 2.15: Empirical a.c.r.f. $\hat{\rho}_n(h)$, for $h = 0, \ldots, 50$, of simulated AR(1) $\{Y_k\}_{k=1,\ldots,200}$; levels $0.8^5 = 0.328$, $0.8^{10} = 0.107$ and $0.8^{20} = 0.011$ (thin lines).

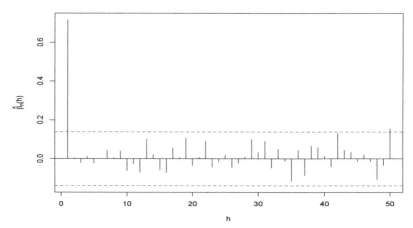

Fig. 2.16: Empirical p.a.c.r.f. $\hat{\beta}_n(h)$, for $h = 0, \ldots, 50$, of simulated AR(1) $\{Y_k\}_{k=1,\ldots,200}$.

We now turn to the analysis of $\{Y_k\}_{k \in \mathbb{Z}}$. Figure 2.14 shows the simulation of $\{Y_k\}_{k=1,\ldots,200}$ obtained from the simulation of $\{X_k\}_{k=1,\ldots,200}$. Figure 2.15 gives the corresponding empirical a.c.r.f. We observe a fast decay of the empirical a.c.r.f. Because the causal AR(1) time series has a.c.r.f. $\rho(h) = \varphi_1^{|h|}$, the levels $0.8^5 = 0.328$, $0.8^{10} = 0.107$ and $0.8^{20} = 0.011$ are marked by fine lines in Figure 2.15. Figure 2.16 gives the empirical p.a.c.r.f.

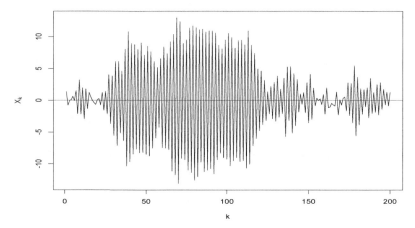

Fig. 2.17: Simulated sample path of AR(1) $\{X_k\}_{k=1,\ldots,200}$ with $X_0 = 0$.

of $\{Y_k\}_{k\in\mathbb{Z}}$*. One observes that the values* $\hat{\beta}_n(h)$*, for* $h = 2, 3, \ldots$*, are not significant. This supports the causal* AR(1) *and thus stationarity. These last three pictures provide substantial support for stationarity of* $\{Y_k\}_{k=1,\ldots,200}$.

Example 2.5.16 (Simulation of AR(1)). *We study numerically the* AR(1) *time series*

$$(1 - \varphi_1 B)X_k = Z_k, \quad \forall k \in \mathbb{Z},$$

where $\{Z_k\}_{k\in\mathbb{Z}}$ *is* WN(σ^2)*. The values of the parameters are* $\varphi_1 = -0.99$*,* $\sigma^2 = 1$ *and we consider* $X_0 = 0$ *and the times* $k = 1, \ldots, n$*, for* $n = 200$*. We have the situation of Example 2.5.14.1, where* $\varphi(-1) \simeq 0$.

Figure 2.17 shows the simulation of $\{X_k\}_{k=0,\ldots,200}$*. Figure 2.18 shows the corresponding empirical a.c.r.f. We observe that the a.c.r.f. has a very slow decay which indicates the nonstationarity. Figure 2.19 shows the empirical p.a.c.r.f. We see that* $\hat{\beta}_n(7)$*,* $\hat{\beta}_n(25)$*,* $\hat{\beta}_n(45) \neq 0$*. This contradicts the property of the causal* AR(1) *that* $\beta(h) = 0$*, for* $h = 2, 3, \ldots$.

The next example studies the AR(2) time series.

Example 2.5.17 (Simulation of AR(2)). *We now study numerically the* AR(2) *time series*

$$(1 - \varphi_1 B - \varphi_2 B^2)X_k = Z_k, \quad \forall k \in \mathbb{Z},$$

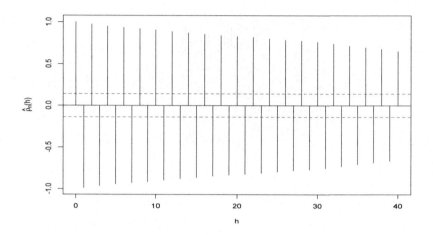

Fig. 2.18: Empirical a.c.r.f. $\hat{\rho}_n(h)$, for $h = 0, \ldots, 50$, of simulated AR(1) $\{X_k\}_{k=1,\ldots,200}$.

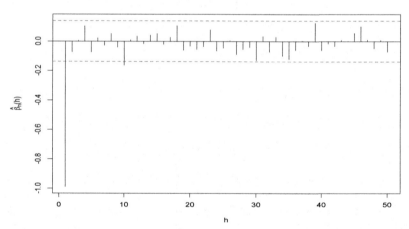

Fig. 2.19: Empirical p.a.c.r.f. $\hat{\beta}_n(h)$, for $h = 0, \ldots, 50$, of simulated AR(1) $\{X_k\}_{k=1,\ldots,200}$.

where $\{Z_k\}_{k\in\mathbb{Z}}$ is WN(σ^2). *The values of the parameters are* $\varphi_1 = 1$, $\varphi_2 = -0.99$, $\sigma^2 = 1$ *and we consider* $X_0 = 0$ *and the times* $k = 1, \ldots, n$, *for* $n = 200$. *We are in the setting of Example 2.5.14.2, where the approximate roots of φ are along the unit circle, precisely* $\varphi(e^{i\pi/3}) \simeq 0$ *and* $\varphi(e^{-i\pi/3}) \simeq 0$.

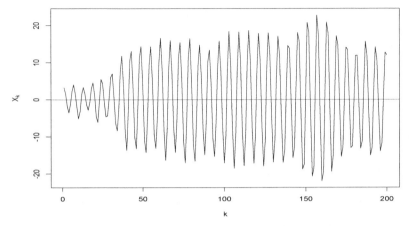

Fig. 2.20: Simulated sample path of AR(2) $\{X_k\}_{k=1,\ldots,200}$ with $X_0 = 0$.

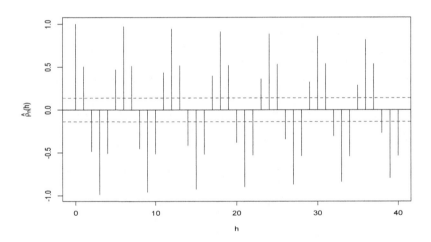

Fig. 2.21: Empirical a.c.r.f. $\hat{\rho}_n(h)$, for $h = 0, \ldots, 40$, of simulated AR(2) $\{X_k\}_{k=1,\ldots,200}$.

Figure 2.20 shows a simulation of $\{X_k\}_{k=0,\ldots,200}$. Figure 2.21 the corresponding empirical a.c.r.f. We observe that the a.c.r.f. has a very slow decay, which indicates nonstationarity. Figure 2.22 the corresponding empirical p.a.c.r.f. We see that $\hat{\alpha}_n(3) \neq 0$, which is evidence against the causal AR(2), for which $\beta(h) = 0$, for $h = 3, 4, \ldots$.

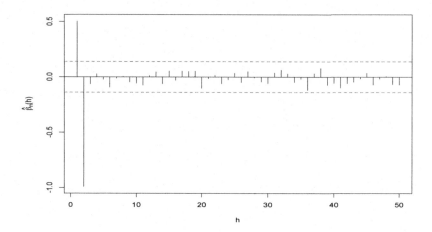

Fig. 2.22: Empirical p.a.c.r.f. $\hat{\beta}_n(h)$, for $h = 0, \ldots, 50$, of simulated AR(2) $\{X_k\}_{k=1,\ldots,200}$.

2.5.4 *Determination of order of AR time series*

If the sample does not appear stationary, then we can differentiate it until reaching stationarity. The differentiated sample may appear as a preliminary ARMA(p, q) time series. In this case, the original or integrated sample is an ARIMA(p, d, q) time series, where d is the number of differentiations applies to read stationarity. The preliminary ARMA(p, q) model can be improved by reselecting p and q according to various criteria available. One can also control the suitability of the selected ARMA(p, q) model by a control of residuals that are obtained from difference between observations and predicted values from the estimated ARMA(p, q) model. We limit the presentation of such data analytic methods to a brief presentation of the selection of the order p of the AR(p) time series.

The final prediction error (f.p.e.) is an asymptotically unbiased estimator of the mean square prediction error of the causal AR(p) model. The f.p.e. provides a practical selection rule according to which, the order p for which the AR(p) has the smallest f.p.e. is retained. Let $X_1, \ldots, X_n$ and $Y_1, \ldots, Y_n$ be two independent samples of the same causal AR(p), for some order $p < n$. Let $\hat{\varphi}_{1n}, \ldots, \hat{\varphi}_{pn}$ be the Gaussian maximum estimators (m.l.e.) of $\varphi_1, \ldots, \varphi_p$, based on the sample $X_1, \ldots, X_n$. The mean square prediction error is given by

$$\varepsilon_p = \mathsf{E}\left[(Y_{n+1} - \hat{\varphi}_{1n}Y_n - \cdots - \hat{\varphi}_{pn}Y_{n-p+1})^2\right].$$

An estimator of ε is the f.p.e., which is defined by

$$\hat{\varepsilon}_{pn} = \hat{\sigma}_n^2 \frac{n+p}{n-p},$$

where $\hat{\sigma}_n^2$ is the m.l.e. of σ^2.

The f.p.e. is an asymptotically unbiased estimator of ε_p. The main arguments of the proof of asymptotic unbiasness are the following. From causality we have

$$
\begin{aligned}
\mathsf{E}&\big[(Y_{n+1} - \hat{\varphi}_{1n}Y_n - \cdots - \hat{\varphi}_{pn}Y_{n-p+1})^2\big] \\
&= \mathsf{E}\big[\{Y_{n+1} - \varphi_1 Y_n - \cdots - \varphi_p Y_{n-p+1} + (\varphi_1 - \hat{\varphi}_{1n})Y_n + \cdots \\
&\qquad + (\varphi_p - \hat{\varphi}_{pn})Y_{n-p+1}\}^2\big] \\
&= \mathsf{E}\big[Z_{n+1}^2\big] + \mathsf{E}\Big[(\hat{\varphi}_n - \varphi)^\top \begin{pmatrix} Y_n \\ \vdots \\ Y_{n-p+1} \end{pmatrix} (Y_n \cdots Y_{n-p+1})(\hat{\varphi}_n - \varphi)\Big] \\
&= \sigma^2 + \mathsf{E}\big[(\hat{\varphi}_n - \varphi)^\top \boldsymbol{\Gamma}(\hat{\varphi}_n - \varphi)\big],
\end{aligned}
$$

where $\varphi = (\varphi_1, \ldots, \varphi_p)^\top$, $\hat{\varphi}_n = (\varphi_{1n}, \ldots, \varphi_{pn})^\top$ and

$$
\boldsymbol{\Gamma} = \mathsf{E}\left[\begin{pmatrix} Y_n Y_n & \cdots & Y_n Y_{n-p+1} \\ \vdots & \ddots & \vdots \\ Y_{n-p+1}Y_n & \cdots & Y_{n-p+1}Y_{n-p+1} \end{pmatrix}\right] = \begin{pmatrix} \mathsf{E}[Y_1 Y_1] & \cdots & \mathsf{E}[Y_1 Y_p] \\ \vdots & \ddots & \vdots \\ \mathsf{E}[Y_1 Y_p] & \cdots & \mathsf{E}[Y_1 Y_1] \end{pmatrix}.
$$

One can show that

$$\sqrt{n}(\hat{\varphi}_n - \varphi) \xrightarrow{\mathrm{d}} \mathcal{N}(0, \sigma^2 \boldsymbol{\Gamma}^{-1}).$$

Consequently, we have

$$
\varepsilon_p = \sigma^2 + \frac{\sigma^2}{n} \mathsf{E}\Big[\underbrace{\Big\{\sqrt{n}\frac{\boldsymbol{\Gamma}^{\frac{1}{2}}}{\sigma}(\hat{\varphi}_n - \varphi)\Big\}^\top \sqrt{n}\frac{\boldsymbol{\Gamma}^{\frac{1}{2}}}{\sigma}(\hat{\varphi}_n - \varphi)}_{\xrightarrow{\mathrm{d}} \chi_p^2}\Big]
$$

$$
= \sigma^2 \Big\{1 + \frac{p}{n} + \mathrm{o}(n^{-1})\Big\}, \quad \text{as } n \to \infty.
$$

One can show that

$$n\frac{\hat{\sigma}_n^2}{\sigma^2} \sim \chi_{n-p}^2,$$

approximately for large n. Thus

$$\frac{n}{n-p}\hat{\sigma}_n^2,$$

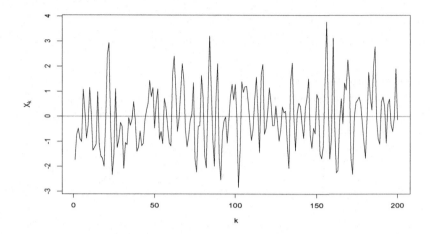

Fig. 2.23:   Simulated sample path of AR(2) $\{X_k\}_{k=1,\ldots,200}$ with $X_0 = 0$.

is an asymptotically unbiased estimator of σ^2. Consequently,

$$\hat{\varepsilon}_{pn} = \frac{n}{n-p}\hat{\sigma}_n^2\left(1 + \frac{p}{n}\right) = \frac{n+p}{n-p}\hat{\sigma}_n^2,$$

is an asymptotically unbiased estimator of ε_p, i.e. of the f.p.e.

We have obtained $\varepsilon_p \sim \sigma^2(1 + p/n)$, as $n \to \infty$, which does increase w.r.t. p. This could be understood as the cumulation of errors resulting from each one of the estimators $\hat{\varphi}_{1n}, \ldots, \hat{\varphi}_{pn}$. In the opposite way, $\hat{\sigma}_n^2$ decreases with p. The f.p.e. inherits the composition of these two effects and it turns out that typically attains a unique minimum at moderate values of p. Thus, it has the desirable effect of penalizing for large values of p. Thus, a practical rule is to select the AR(p) model that minimizes the f.p.e., i.e. $\hat{\varepsilon}_{pn}$. We now give a numerical illustration of the f.p.e.

Example 2.5.18 (F.p.e. of simulated AR(2)). *We consider the* AR(2) *time series* $\{X_k\}_{k\in\mathbb{Z}}$ *given by*

$$X_k - \varphi_1 X_{k-1} + \varphi_2 X_{k-2} = Z_k, \quad \forall k \in \mathbb{Z},$$

where $\{Z_k\}_{k\in\mathbb{Z}}$ *is* WN(σ^2). *We consider* $\varphi_1 = 1/2$, $\varphi_2 = -1/2$ *and* $\sigma^2 = 1$. *We generate* $\{X_k\}_{k=1,\ldots,200}$ *from this model with* $X_0 = 0$. *Figure 2.23 shows the path obtained with the* $n = 200$ *simulated values, which appears stationary. The corresponding empirical a.c.r.f. is given in Figure 2.24. The decay of the a.c.r.f. appears reasonably fast and it gives some support for stationarity. The corresponding empirical p.a.c.r.f. is given in Figure 2.25. For the causal* AR(p) *time series, Example 2.3.8 shows that* $\beta(h) = 0$, *for*

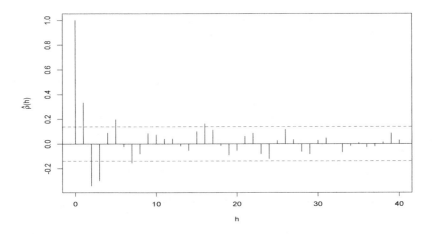

Fig. 2.24: Empirical a.c.r.f. $\hat{\rho}_n(h)$, for $h = 0, \ldots, 40$, of simulated AR(2) $\{X_k\}_{k=1,\ldots,200}$.

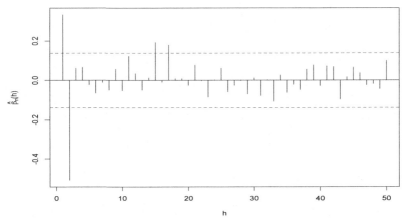

Fig. 2.25: Empirical p.a.c.r.f. $\hat{\beta}_n(h)$, for $h = 0, \ldots, 50$, of simulated AR(2) $\{X_k\}_{k=1,\ldots,200}$.

$h = p + 1, p + 2, \ldots$. *So both* AR(1) *and* AR(2) *appear admissible, according to the p.a.c.r.f. Table 2.3 shows the Gaussian m.l.e. of the coefficients* $\varphi_1, \ldots, \varphi_p$, *denoted* $\hat{\varphi}_{1n}, \ldots, \hat{\varphi}_{pn}$, *the Gaussian m.l.e. of the WN variance* σ^2, *denoted* $\hat{\sigma}_n^2$, *and the f.p.e., denoted* $\hat{\varepsilon}_p$, *for* $p = 1, \ldots, 5$. *We can observe that the f.p.e. is the smallest with* $p = 2$. *This corresponds indeed to the*

Table 2.3: Estimators of AR(p) coefficients, of WN variance and f.p.e. at different orders p, obtained from simulated AR(2) $\{X_k\}_{k=1,\ldots,200}$.

p	$\hat{\varphi}_{1n}$	$\hat{\varphi}_{2n}$	$\hat{\varphi}_{3n}$	$\hat{\varphi}_{4n}$	$\hat{\varphi}_{5n}$	$\hat{\sigma}_n^2$	$\hat{\varepsilon}_{pn}$
1	0.335					1.348	1.362
2	0.507	−0.516				0.991	1.011
3	0.544	−0.552	0.071			0.986	1.016
4	0.539	−0.516	0.036	0.064		0.982	1.022
5	0.541	−0.515	0.023	0.077	−0.024	0.982	1.032

true time series model. When $p = 2$, we note that the estimated coefficients are close to the true values.

2.5.5   State space model and Kalman filter

The filter of Kalman (1960) is a technique that provides recursive prediction formulae for the so-called state space model. This section is a simple description of the Kalman filter.

The state space model consists of two equations. The first one, called state space equation, is given by

$$X_{k+1} = \varphi X_k + Z_{k+1}, \quad \text{for } k = 1, 2, \ldots,$$

where $-1 < \varphi < 1$, $Z_2, Z_3, \ldots$ are independent and $\mathcal{N}(0, \sigma_Z^2)$, for some $\sigma_Z \in (0, \infty)$. The initial point X_1 is either a Gaussian random variable or a constant. The second equation, called observation equation, is given by

$$Y_k = \beta X_k + U_k, \quad \text{for } k = 1, 2, \ldots,$$

where $\beta \in \mathbb{R}$, $U_1, U_2, \ldots$ are independent, $\mathcal{N}(0, \sigma_U^2)$, for some $\sigma_U \in (0, \infty)$, and they are independent of $Z_2, Z_3, \ldots$. We note that the state space equation is a causal AR(1) time series which admits an unique stationary solution; cf. Proposition 2.2.3 and Remark 2.2.4. It is supposed that the states $X_2, X_3, \ldots$ are non-observable random variables and only $Y_1, Y_2, \ldots$ are observable. Denote $\mathcal{F}_k = \sigma(Y_1, \ldots, Y_k)$, i.e. the σ-algebra generated by $Y_1, \ldots, Y_k$, for $k = 1, 2, \ldots$. Thus $\{\mathcal{F}_k\}_{k \geq 1}$ is a filtration, in the sense that $\mathcal{F}_k \subset \mathcal{F}_{k+1}$, and $\mathcal{F}_k$ represents the observable information available up to time $k = 1, 2, \ldots$.

The Kalman filter is an iterative procedure for computing the predictors of future states from available information, that are given in (2.5.22) that follows and it is based on the formulae for conditional expectation and variance of Gaussian random variables, that are given in Proposition A.8.4.4. These standard formulae lead to

$$\mathsf{E}[X_{k+1} \mid \mathcal{F}_{k+1}] =$$

$$E[X_{k+1} \mid \mathcal{F}_k] + \frac{\mathsf{cov}(X_{k+1}, Y_{k+1} \mid \mathcal{F}_k)}{\mathsf{var}(Y_{k+1} \mid \mathcal{F}_k)} (Y_{k+1} - \mathsf{E}[Y_{k+1} \mid \mathcal{F}_k]),$$

$$(2.5.20)$$

and

$$\mathsf{var}(X_{k+1} \mid \mathcal{F}_{k+1}) = \mathsf{var}(X_{k+1} \mid \mathcal{F}_k) - \frac{\mathsf{cov}^2(X_{k+1}, Y_{k+1} \mid \mathcal{F}_k)}{\mathsf{var}(Y_{k+1} \mid \mathcal{F}_k)}, \quad (2.5.21)$$

for $k = 1, 2, \ldots$. Thus, (2.5.20) provides the conditional expectation and (2.5.21) provides the conditional variance, updated at any time $k \geq 2$, in terms of the same objects but updated at time $k - 1$. Note that the update of the conditional expectation at time $k + 1$, given in (2.5.20), is also a function of Y_{k+1}, whereas the same update of conditional variance, given in (2.5.21), is not. This particularity of the conditional expectation and the conditional variance is is explained just after Proposition A.8.4.

The predictors of the states, for which a computational formula is desired, are the random variables

$$\hat{X}_{k,k+j} = \mathsf{E}[X_{k+j} \mid \mathcal{F}_k], \quad \text{for } k = 1, 2, \ldots \text{ and } j = 0, 1, \ldots. \quad (2.5.22)$$

The computation of (2.5.22) is our main objective. We define the predictors of the observations as

$$\hat{Y}_{k,k+1} = \mathsf{E}[Y_{k+1} \mid \mathcal{F}_k],$$

and we define the coefficients

$$a_{k+1} = \frac{\mathsf{cov}(X_{k+1}, Y_{k+1} \mid \mathcal{F}_k)}{\mathsf{var}(Y_{k+1} \mid \mathcal{F}_k)}, \quad \text{for } k = 1, 2, \ldots.$$

We can re-express (2.5.20) in terms of the three last definitions as

$$\hat{X}_{k+1,k+1} = \hat{X}_{k,k+1} + a_{k+1}(Y_{k+1} - \hat{Y}_{k,k+1}), \quad \text{for } k = 1, 2, \ldots. \quad (2.5.23)$$

Then, we have

$$\hat{X}_{k,k+1} = \mathsf{E}[X_{k+1} \mid \mathcal{F}_k]$$
$$= \varphi \mathsf{E}[X_k \mid \mathcal{F}_k] + \mathsf{E}[Z_{k+1} \mid \mathcal{F}_k]$$
$$= \varphi \hat{X}_{k,k}, \quad \text{for } k = 1, 2, \ldots. \quad (2.5.24)$$

We can similarly obtain the generalization of (2.5.24) given by

$$\hat{X}_{k,k+j} = \varphi^j \hat{X}_{k,k}, \quad \text{for } k = 1, 2, \ldots \text{ and } j = 0, 1, \ldots.$$

Moreover, we have

$$\begin{aligned}
\hat{Y}_{k,k+1} &= \mathsf{E}[Y_{k+1} \mid \mathcal{F}_k] \\
&= \beta \mathsf{E}[X_{k+1} \mid \mathcal{F}_k] + \mathsf{E}[U_{k+1} \mid \mathcal{F}_k] \\
&= \beta \hat{X}_{k,k+1} \\
&= \beta\varphi \hat{X}_{k,k}, \quad \text{for } k = 1, 2, \ldots,
\end{aligned} \tag{2.5.25}$$

where the last equality is due to (2.5.24). So (2.5.24) and (2.5.25) can be used for the recursive computation of the state predictors (2.5.23).

However we need also the coefficients a_{k+1}, for $k = 1, 2, \ldots$. They can be computed recursively as follows. We have

$$\begin{aligned}
\mathsf{var}(X_{k+1} \mid \mathcal{F}_k) &= \varphi^2 \mathsf{var}(X_k \mid \mathcal{F}_k) + \mathsf{var}(Z_{k+1} \mid \mathcal{F}_k) \\
&= \varphi^2 \mathsf{var}(X_k \mid \mathcal{F}_k) + \sigma_Z^2,
\end{aligned} \tag{2.5.26}$$

$$\begin{aligned}
\mathsf{var}(Y_{k+1} \mid \mathcal{F}_k) &= \beta^2 \mathsf{var}(X_{k+1} \mid \mathcal{F}_k) + \mathsf{var}(U_{k+1} \mid \mathcal{F}_k) \\
&= \beta^2 \mathsf{var}(X_{k+1} \mid \mathcal{F}_k) + \sigma_U^2,
\end{aligned} \tag{2.5.27}$$

and

$$\begin{aligned}
\mathsf{cov}(X_{k+1}, Y_{k+1} \mid \mathcal{F}_k) &= \mathsf{cov}(X_{k+1}, \beta X_{k+1} + U_{k+1} \mid \mathcal{F}_k) \\
&= \beta \mathsf{var}(X_{k+1} \mid \mathcal{F}_k),
\end{aligned} \tag{2.5.28}$$

for $k = 1, 2, \ldots$ With (2.5.27) and (2.5.28) one can re-express (2.5.21) as

$$\mathsf{var}(X_{k+1} \mid \mathcal{F}_{k+1}) = \mathsf{var}(X_{k+1} \mid \mathcal{F}_k) - \frac{\{\beta \mathsf{var}(X_{k+1} \mid \mathcal{F}_k)\}^2}{\beta^2 \mathsf{var}(X_{k+1} \mid \mathcal{F}_k) + \sigma_U^2},$$

for $k = 1, 2, \ldots$. This last formula can be used in (2.5.26) in order to give a formula for $\mathsf{var}(X_{k+1} \mid \mathcal{F}_k)$ in terms of $\mathsf{var}(X_k \mid \mathcal{F}_{k-1})$, for $k = 2, 3, \ldots$. With (2.5.27) and (2.5.28) we can re-express a_{k+1} as

$$a_{k+1} = \frac{1}{\beta}\left(1 + \frac{\sigma_U^2}{\beta^2 \mathsf{var}(X_{k+1} \mid \mathcal{F}_k)}\right)^{-1}, \quad \text{for } k = 1, 2, \ldots,$$

and therefore these coefficients can be obtained recursively.

In order to start the recursion we need $\mathsf{E}[X_1 \mid Y_1]$ and $\mathsf{var}(X_2 \mid Y_1)$. One usually takes their unconditional versions.

2.5.6 *Other stationary time series*

The ARMA time series are certainly very general, but in particular situations other models prove to be more suitable. In this section we describe two alternative models to ARMA. The first model is for integer values and the second one is for nonlinear situations that appear for instance with financial data.

INAR time series

Time series taking nonnegative integer values, namely counting time series, appear in various scientific fields. Various counting time series have been proposed in the literature and these models address various specific features of counting data such as excess of null values or overdispersion. An introduction to this topic is Weiß (2018). The integer-valued AR(1) or INAR(1) time series is based on the binomial thinning operator. The binomial thinning operator $\circ$ with parameter $\tau \in [0, 1]$ is defined by

$$\tau \circ X = \sum_{j=0}^{X} B_j,$$

where $B_0 = 0$, $B_1, B_2, \ldots$ are independent Bernoulli random variables with $\mathsf{P}[B_j = 1] = \tau$ and $\mathsf{P}[B_j = 0] = 1 - \tau$, $\forall j \in \mathbb{N}$, and where X is a $\mathbb{N}$-valued random variable independent of $B_1, B_2, \ldots$. This operator was proposed by Steutel and van Harn (1979).

A practical property of the thinning operator is that

$$\mathsf{E}[\tau \circ X] = \mathsf{E}[\tau X], \ \forall \tau \in [0, 1].$$

Indeed,

$$\mathsf{E}[\tau \circ X] = \mathsf{E}[\mathsf{E}[\tau \circ X | X]] = \mathsf{E}\left[\mathsf{E}\left[\sum_{j=0}^{X} B_j \bigg| X\right]\right] = \mathsf{E}\left[\sum_{j=0}^{X} \mathsf{E}[B_j | X]\right]$$

$$= \mathsf{E}\left[\sum_{j=0}^{X} \mathsf{E}[B_j]\right] = \mathsf{E}\left[\sum_{j=0}^{X} \tau\right] = \mathsf{E}[\tau X] = \tau \mathsf{E}[X].$$

A practical interpretation is as follows. Consider a population of size X at some given time. If, at the next time, each individual disappears with probability $1 - \tau$ and independently of the others, then the number of survivors at the next time is $\tau \circ X$.

Binomial thinning is used for defining integer-valued ARMA time series. For instance, the INAR(1) time series, which was introduced by McKenzie (1985), is given by

$$X_k = \tau \circ X_{k-1} + Z_k, \quad \forall k \in \mathbb{Z}, \tag{2.5.29}$$

where $\tau \in (0, 1)$, $\{Z_k\}_{k \in \mathbb{Z}}$ is a sequence of i.i.d., nonnegative and integer-valued random variables, all thinning operations are performed independently of each other and of $\{Z_k\}_{k \in \mathbb{Z}}$. Moreover, the thinning operations at each time $k \in \mathbb{Z}$ and Z_k are independent of $\{X_j\}_{j \leq k-1}$. It is shown that the INAR(1) time series is stationary. The INAR(1) shares several properties with the AR(1) time series.

Remark 2.5.19 (Galton-Watson process, population dynamics and INAR(1)). *The Galton-Watson process is the stochastic process* $\{X\}_{k\in\mathbb{N}}$ *that is defined by the recurrence equation*

$$X_k = \sum_{j=0}^{X_{k-1}} L_{k-1,j}, \text{ for } k = 1, 2, \ldots,$$

where $X_0 = 1$ *is the initial value,* $L_{k-1,j}$, *for* $j = 1, 2, \ldots, k = 1, 2, \ldots$ *are i.i.d. in* $\mathbb{N}$ *and where* $L_{0,0} = L_{1,0} = \cdots = 0$. *The Galton–Watson process belongs to the class of branching processes. This model is used in many applied sciences, for example in the study of evolutions of population or in genetics. where, at epoch* $k \in \mathbb{N}$, *the number of females descendants from the* j*-th of the* X_{k-1} *females alive at the* $(k-1)$*-th epoch is* $L_{k-1,j}$, *for* $1 \leq j \leq X_{k-1}$, *provided* $X_{k-1} \geq 1$, *and where the lifetime of all individuals is one unit of time.*

Thus a possible interpretation of the INAR(1) *with time domain restricted to* $\mathbb{N}$ *is in terms of the Galton-Watson process with immigration and without emigration. The immigration factor is represented by the additive perturbations term* Z_k, *for* $k = 1, 2, \ldots$, *appearing in (2.5.29). The Bernoulli distribution of the summands of the* INAR(1) *means that each female can have at most one female descendant. However, immigration thwarts rapid extinction of the population.*

A slight variation of the above evolutionary model concerns a population with immigration, without emigration and without births. Consider individuals with unbounded lifetime, where X_{k-1} *are the survivors of the population at time* $k - 1$, $\tau \circ X_{k-1}$ *are the survivors at time* k, Z_k *is the immigration at time* k, *giving* X_k *individuals at time* k, *for* $k = 1, 2, \ldots$.

Nonlinear and GARCH time series

A time series is called linear if it can be represented as the infinite series (2.2.15), in which $\{X_k\}_{k\in\mathbb{Z}}$ represents any WN. The MA(∞) time series of Definition 2.2.18 is a linear and causal time series, in the sense that it is a linear function of present and past values of WN. The casual ARMA(p, q) is the most important instance, cf. Theorem 2.2.12.2. Also, according to (2.4.8), a linear time series results from a time invariant linear filter applied to WN. A time series that does not depend linearly on WN is called nonlinear. A simple example of nonlinear time series is given by

$$X_k = \varphi_1 X_{k-1} + Z_k + \eta X_{k-1} Z_{k-1}, \quad \forall k \in \mathbb{Z},$$

where $\varphi_1, \eta \in \mathbb{R}$ and where $\{Z_k\}_{k \in \mathbb{Z}}$ is WN. Some theory for the analysis of nonlinear time series can be found in Priestley (1981), Section 11.5 and 11.6, for example.

An important class of nonlinear time series the ARCH (autoregressive conditionally heteroscedastic), which was introduced by Engle (1982) in the context of financial markets. Its main motivation is that the conditional variance changes over time. The generalized ARCH (GARCH) time series, which was introduced by Bollerslev (1986), has become popular and successful, mainly because it exhibits the ARMA structure. A general reference on GARCH is Straumann (2005), for example. Let $\{Y_k\}_{k \in \mathbb{Z}}$ represent the values of a financial index or of a individual stock and consider

$$X_k = \log \frac{X_k}{X_{k-1}} = \log\left(1 + \frac{X_k - X_{k-1}}{X_{k-1}}\right) \simeq \frac{Y_k - Y_{k-1}}{Y_{k-1}}, \quad \forall k \in \mathbb{Z}.$$

Thus $\{X_k\}_{k \in \mathbb{Z}}$ represents the approximate relative returns that are obtained with time lag one. Typical observations of $\{X_k\}_{k \in \mathbb{Z}}$ have mean null and display alternation of periods with low and high variability. Another time series of interest is the conditional variance of X_k, given all information known until time $k - 1$, $\forall k \in \mathbb{Z}$. It turns out that the (bivariate) ARMA model is inappropriate in this situation. So a particular nonlinear time series is used for this purpose: it is the GARCH time series defined as follows.

Definition 2.5.20 (GARCH(p, q) time series). *Let $\{Z_k\}_{k \in \mathbb{Z}}$ be WN(1) with independent elements. The bivariate time series $\{(X_k, S_k)\}_{k \in \mathbb{Z}}$ is called:*

- *autoregressive conditionally heteroscedastic of order q, denoted* ARCH(q), *if it solves the equations*

$$X_k = S_k Z_k \quad and$$
$$S_k^2 = \kappa + \theta_1 X_{k-1}^2 + \cdots + \theta_q X_{k-q}^2, \quad \forall k \in \mathbb{Z},$$

  where $q \in \mathbb{N}^*$, $\kappa, \theta_q > 0$ and $\theta_1, \ldots, \theta_{q-1} \geq 0$;
- *generalized autoregressive conditionally heteroscedastic of orders p and q, denoted* GARCH(p, q), *if it solves the equations*

$$X_k = S_k Z_k \quad and$$
$$S_k^2 - \varphi_1 S_{k-1}^2 - \cdots - \varphi_p S_{k-p}^2 = \kappa + \theta_1 X_{k-1}^2 + \cdots + \theta_q X_{k-q}^2, \quad \forall k \in \mathbb{Z},$$

  where $p, q \in \mathbb{N}^*$, $\kappa, \varphi_p, \theta_q > 0$, $\varphi_1, \ldots, \varphi_{p-1}$ and $\theta_1, \ldots, \theta_{q-1} \geq 0$.

Remarks 2.5.21.

(1) We see directly the similitude between $\mathrm{ARCH}(q)$ *and* $\mathrm{MA}(q)$ *and also between* $\mathrm{GARCH}(p,q)$ *and* $\mathrm{ARMA}(p,q)$.

(2) The constraints on the signs of constant and coefficients guarantee the positivity of $\{S_k^2\}_{k \in \mathbb{Z}}$.

(3) The autoregressive structure tells that large (small) values of X_{k-1}^2 *and* S_{k-1}^2 *tend to give large (small) values of* X_k^2 *and* S_k^2, $\forall k \in \mathbb{Z}$.

(4) It is shown that a sufficient condition for (weak) stationarity of the $\mathrm{GARCH}(p,q)$ *time series* $\{(X_k, S_k)\}_{k \in \mathbb{Z}}$ *is given by*

$$\sum_{j=1}^{p} \varphi_j + \sum_{j=1}^{q} \theta_j < 1. \tag{2.5.30}$$

Let $k \in \mathbb{Z}$ and define $\mathcal{F}_k = \sigma(\{Z_j\}_{j \leq k})$, the σ-algebra representing the information available at time k. Then we have

$$\mathsf{var}(X_k \mid \mathcal{F}_{k-1}) = \mathsf{var}(S_k Z_k \mid \mathcal{F}_{k-1}) = S_k^2 \, \mathsf{var}(Z_k \mid \mathcal{F}_{k-1})$$
$$= S_k^2 \, \mathsf{var}(Z_k) = S_k^2.$$

Thus S_k^2 represents the conditional variance of X_k, given the past information. We also find

$$\mathsf{E}[X_k] = \mathsf{E}[\mathsf{E}[S_k Z_k \mid \mathcal{F}_{k-1}]] = \mathsf{E}[S_k \mathsf{E}[Z_k \mid \mathcal{F}_{k-1}]] = \mathsf{E}[S_k \mathsf{E}[Z_k]] = 0.$$

and, for $h = 1, 2, \ldots,$

$$\mathsf{cov}(X_{k+h}, X_k)$$
$$= \mathsf{E}[X_{k+h} X_k] = \mathsf{E}[\mathsf{E}[X_{k+h} X_k \mid \mathcal{F}_{k+h-1}]]$$
$$= \mathsf{E}[\mathsf{E}[S_{k+h} Z_{k+h} X_k \mid \mathcal{F}_{k+h-1}]] = \mathsf{E}[S_{k+h} X_k \mathsf{E}[Z_{k+h} \mid \mathcal{F}_{k+h-1}]]$$
$$= \mathsf{E}[S_{k+h} X_k \mathsf{E}[Z_{k+h}]] = 0.$$

The formula for the variance is longer to show. It is given by

$$\mathsf{var}(X_k) = \frac{\kappa}{1 - \sum_{j=1}^{p} \varphi_j - \sum_{j=1}^{q} \theta_j},$$

which is positive under the stationarity condition (2.5.30). Consequently, $\{X_k\}_{k \in \mathbb{Z}}$ is WN.

Chapter 3

Stationary processes with continuous time

This chapter brings us in the second of the two main parts of this book. The analysis of stationary models is now done with continuous time. A stochastic process can be briefly defined as an ordered set of random variables, $\{X_t\}_{t \in \mathrm{T}}$, with common probability space and with time index t in the set $T \subset \mathbb{R}$, which is usually $T = \mathbb{Z}$, $\mathbb{N}$, $\mathbb{R}$, $\mathbb{R}_+ = [0, \infty)$, $[0, 1]$, etc. The first two cases give processes with discrete time and they have been considered so far. The last three cases give a processes with continuous time. A discrete time process can be viewed as a simplification, in the sense that it can be obtained through discretization of the original continuous time process, for example by setting the original process constant at its level at time k, over the entire time interval $[k, k+1)$, $\forall k \in \mathbb{Z}$. Note that complex-valued stochastic process are also considered here. Their spectral analysis is actually not more complicated. They provide a compact representation of a two-dimensional signal, which is thus expressed as $X_t = R_t \mathrm{e}^{\mathrm{i}\theta_t}$, where the process R_t takes values in $\mathbb{R}$ and the process θ_t in $(-\pi, \pi]$, $\forall t \in \mathbb{R}$. This representation is commonly used in various practical fields such as electrical engineering.

The major topics of this chapter are the following. General concepts and results on the theory of continuous time stochastic processes are presented in Section 3.1. Section 3.2 introduces important stochastic processes: Gaussian processes, Wiener process, selfsimilar processes, fractional Brownian motion, counting processes, compound processes, shot noise processes, point processes and Lévy processes. Section 3.3 is devoted to some regularity properties of stationary processes, namely mean square continuity and mean square differentiability. It thus completes the general presentation of properties of regularity initiated in Section 3.1. Section 3.4 introduces stochastic integration and precisely the mean square integral. At the end, the concept of ergodicity is shortly explained. Section 3.5 presents the spectral distribution. Bochner's theorem and inversion formulae for the spectral distribution are shown. The spectral decomposition of a stationary processes in terms of a stochastic integral is presented in Section 3.6. This section also gives the proof of the spectral theorem, which ensures the existence of the spectral decomposition. The spectral analysis of Gaussian processes is the topic of Section 3.7. Section 3.8 provides the spectral analysis of counting processes, based on Bartlett spectrum. The theory of time invariant linear filters, which is introduced for time series in Section 2.4, is more thoroughly developed for continuous time processes in Section 3.9. Several parts of this chapter are based on Cramér and Leadbetter (1967) and Lindgren (2012).

3.1   Introduction

In this section, general notions and results on the theory of continuous time stochastic processes are studied as follows. Section 3.1.1 gives general definitions and important central theorems. Then Section 3.1.2 introduces stationarity with continuous time. Section 3.1.3 provides results on the sample path regularity of continuous time stochastic processes.

3.1.1   *General definitions and theorems*

Continuous time stochastic processes are introduced in this section. Although there is no restriction to stationarity, most of these processes are directly or indirectly relevant in the context of stationarity. Several important theorems of the theory of stochastic processes are given. Some of them are stated without proof, but their proofs can however be found in more exhaustive books, such as Shiryayev (1984).

Let $n \in \mathbb{N}^*$. The Borel σ-algebra of $\mathbb{R}^n$, denoted $\mathcal{B}(\mathbb{R}^n)$, is the smallest σ-algebra of $\mathbb{R}^n$ containing all sets $B_1 \times \cdots \times B_n$, where $B_1, \ldots, B_n \in \mathcal{B}(\mathbb{R})$, $\mathcal{B}(\mathbb{R})$ denoting the Borel σ-algebra of the real field. These sets are the Borel sets of $\mathbb{R}$. Denote by $T \subset \mathbb{R}$ the time domain of the stochastic process. Then $\mathbb{R}^T$ denotes the space of functions

$$x : T \to \mathbb{R}$$
$$t \mapsto x_t.$$

Let $t_1 < \cdots < t_n \in T$ and $B^{(n)} \in \mathcal{B}(\mathbb{R}^n)$. Then

$$\mathcal{C}_{t_1,\ldots,t_n}(B^{(n)}) = \left\{ x \in \mathbb{R}^T \,\middle|\, (x_{t_1}, \ldots, x_{t_n}) \in B^{(n)} \right\}, \qquad (3.1.1)$$

is a cylinder in $\mathbb{R}^T$ with basis in $\mathbb{R}^n$. It represents the class of functions that is determined through its values on a finite subset of T. These cylinders generate an algebra. However, unless T is finite, they do not generate a σ-algebra. Consider e.g. the set $\left\{ x \in \mathbb{R}^{\mathbb{N}} \mid |\sum_{k=0}^{\infty} x_k| < \infty \right\}$. It depends on the values of the sequence x at a countable set of indices or instants. Therefore it cannot be a cylinder. Nevertheless, it does belong to the cylindrical σ-algebra. The smallest σ-algebra that contains all cylinder sets (3.1.1) is the cylindrical σ-algebra of $\mathbb{R}^T$. It is denoted $\mathcal{B}(\mathbb{R}^T)$.

Examples 3.1.1 (Two $\mathcal{B}(\mathbb{R}^{\mathbb{N}})$-measurable sets).

(1) It holds that

$$\left\{ x \in \mathbb{R}^{\mathbb{N}} \big| x_n \geq 10 \right\}, \, n \geq 0, \, i.o. = \bigcap_{n=0}^{\infty} \bigcup_{m=n}^{\infty} \left\{ x \in \mathbb{R}^{\mathbb{N}} \big| x_m \geq 10 \right\} \in \mathcal{B}(\mathbb{R}^{\mathbb{N}}).$$

(2) It holds that

$$\left\{ x \in \mathbb{R}^{\mathbb{N}} \big| \lim_{n \to \infty} x_n = 0 \right\} = \bigcap_{\varepsilon \in \mathbb{Q}_+^*} \bigcup_{n=0}^{\infty} \bigcap_{m=n}^{\infty} \left\{ x \in \mathbb{R}^{\mathbb{N}} \mid |x_m| < \varepsilon \right\} \in \mathcal{B}(\mathbb{R}^{\mathbb{N}}).$$

One defines a probability measure over $(\mathbb{R}^T, \mathcal{B}(\mathbb{R}^T))$ as follows. Consider the class of finite dimensional distributions (f.d.d.) $\{P_{t_1,\ldots,t_n}\}_{t_1 < \ldots < t_n \in T, n \geq 1}$ over the measure spaces $(\mathbb{R}^n, \mathcal{B}(\mathbb{R}^n))$, for $n = 1, 2, \ldots$, with corresponding d.f. $\{F_{t_1,\ldots,t_n}\}_{t_1 < \ldots < t_n \in T, n \geq 1}$, all together called f.d.d.f. Any one of these two class is called consistent if

$$\lim_{x_k \to \infty} F_{t_1,\ldots,t_k,\ldots,t_n}(x_1, \ldots, x_k, \ldots, x_n) =$$

$$F_{t_1,\ldots,t_{k-1},t_{k+1},\ldots,t_n}(x_1, \ldots, x_{k-1}, x_{x+1}, x_n), \qquad (3.1.2)$$

$\forall t_1 < \cdots < t_k < \cdots < t_n \in T, x_1, \ldots, x_k, \ldots, x_n \in \mathbb{R}, \forall k \leq n$ and $n \in \mathbb{N}^*$.

Theorem 3.1.2 (Kolmogorov extension of measures). *Consider the f.d.d.* $\{P_{t_1,\ldots,t_n}\}_{t_1 < \ldots < t_n \in T}$ *over* $(\mathbb{R}^n, \mathcal{B}(\mathbb{R}^n))$, $\forall n \in \mathbb{N}^*$, *that satisfy the consistency criterium (3.1.2). Then there exists a unique probability measure* P *over* $(\mathbb{R}^T, \mathcal{B}(\mathbb{R}^T))$ *such that*

$$\mathsf{P}\left[\mathcal{C}_{t_1,\ldots,t_n}(B^{(n)}) \right] = \mathsf{P}_{t_1,\ldots,t_n}\left[B^{(n)} \right],$$

$\forall t_1 < \cdots < t_n \in T, B^{(n)} \in \mathcal{B}(\mathbb{R}^n)$ *and* $n \in \mathbb{N}^*$.

Now that we have obtained a probability measure over $(\mathbb{R}^T, \mathcal{B}(\mathbb{R}^T))$, we turn to the precise definition of a stochastic process, as a random element in $(\mathbb{R}^T, \mathcal{B}(\mathbb{R}^T))$.

Definition 3.1.3 (Stochastic process). *A real-valued stochastic process with time domain $T \subset \mathbb{R}$ is given by the measurable map*

$$X : (\Omega, \mathcal{F}) \to \left(\mathbb{R}^T, \mathcal{B}\left(\mathbb{R}^T\right)\right).$$

The following relation between random variable and stochastic process holds. Let $t \in T$ and denote by X_t the projection of the stochastic process X over the t-th component of $\mathbb{R}^T$. Then X_t is a random variable and it is called t-th coordinate of the process. Conversely, each ordered set of random variables

$$X_t : (\Omega, \mathcal{F}) \to (\mathbb{R}, \mathcal{B}(\mathbb{R})), \quad \forall t \in T,$$

determine a stochastic process, namely the process $X = \{X_t\}_{t \in T}$.

In order to show this equivalence, let $n \in \mathbb{N}^*$ and $t_1 < \cdots < t_n \in T$. Then $(X_{t_1}, \ldots, X_{t_n})$ is a random vector, because $\{X_{t_1} \leq x_1, \ldots, X_{t_n} \leq x_n\} \in \mathcal{F}$, $\forall x_1, \ldots, x_n \in \mathbb{R}$, and so $\{(X_{t_1}, \ldots, X_{t_n}) \in B^{(n)}\} \in \mathcal{F}$, $\forall B^{(n)} \in \mathcal{B}(\mathbb{R}^n)$. Then $\{X \in \mathcal{C}_{t_1, \ldots, t_n}(B^{(n)})\} \in \mathcal{F}$, $\forall B^{(n)} \in \mathcal{B}(\mathbb{R}^n)$. By using a basic lemma of measure theory, one passes from all cylinders to the σ-algebra generated by them, i.e. to the cylindrical σ-algebra $\mathcal{B}(\mathbb{R}^T)$. We have thus obtained $\{X \in B\} \in \mathcal{F}$, $\forall B \in \mathcal{B}(\mathbb{R}^T)$. The converse is obvious: let $t > 0$ and $x \in \mathbb{R}$, then $\{X_t \leq x\} = \{X \in \mathcal{C}_t((0, x])\} \in \mathcal{F}$.

Thus Definition 3.1.3 of stochastic process corresponds to the simple interpretation of stochastic process as collection of random variables X_t, with index $t \in T$. For any $\omega \in \Omega$, the function $X_t(\omega)$, $\forall t \in T$, is called a sample path, a trajectory or a realization of the stochastic process $\{X_t\}_{t \in T}$.

We now turn the distribution of a stochastic process.

Definition 3.1.4 (Distribution of a stochastic process). *The distribution of the stochastic process $X = \{X_t\}_{t \in T}$ over $(\Omega, \mathcal{F}, \mathsf{P})$ is given by*

$$\mathsf{P}_X[B] = \mathsf{P}[X \in B], \quad \forall B \in \mathcal{B}\left(\mathbb{R}^T\right).$$

That distribution allows us to define the independence between two stochastic processes.

Definition 3.1.5 (Independent stochastic processes). *The stochastic processes $X^{(1)}, \ldots, X^{(n)}$ over $(\Omega, \mathcal{F}, \mathsf{P})$ are independent if*

$$\mathsf{P}\left[X^{(1)} \in B_1, \ldots, X^{(n)} \in B_n\right] = \mathsf{P}\left[X^{(1)} \in B_1\right] \cdots \mathsf{P}\left[X^{(n)} \in B_n\right],$$

$\forall B_1, \ldots, B_n \in \mathcal{B}\left(\mathbb{R}^T\right).$

Definition 3.1.6 (F.d.d. of a stochastic process). *The f.d.d. of the stochastic process* $\{X_t\}_{t \in T}$ *over* $(\Omega, \mathcal{F}, \mathsf{P})$ *are given by*

$$\mathsf{P}_{X, t_1, \ldots, t_n}\left[B^{(n)}\right] = \mathsf{P}[X \in \mathcal{C}_{t_1, \ldots, t_n}(B^{(n)})] = \mathsf{P}\left[(X_{t_1}, \ldots, X_{t_n}) \in B^{(n)}\right],$$

$\forall t_1 < \cdots < t_n \in T,\ B^{(n)} \in \mathcal{B}(\mathbb{R}^n)$ *and* $n \in \mathbb{N}^*$.

The following important theorem states that a stochastic process can be obtained from a class of f.d.d.

Theorem 3.1.7 (Kolmogorov existence of stochastic process). *Consider the class of probability distributions* $\{\mathsf{P}_{t_1, \ldots, t_n}\}_{t_1 < \ldots < t_n \in T, n \in \mathbb{N}^*}$ *over* $(\mathbb{R}^n, \mathcal{B}(\mathbb{R}^n))$, $\forall n \in \mathbb{N}^*$, *with corresponding d.f.* $\{F_{t_1, \ldots, t_n}\}_{t_1 < \ldots < t_n \in T, n \in \mathbb{N}^*}$ *satisfying the consistency criterium (3.1.2). Then there exist a probability space* $(\Omega, \mathcal{F}, \mathsf{P})$ *and a stochastic process* $\{X_t\}_{t \in T}$ *that is defined over it, such that*

$$\mathsf{P}\left[(X_{t_1}, \ldots, X_{t_n}) \in B^{(n)}\right] = \mathsf{P}[X \in \mathcal{C}_{t_1, \ldots, t_n}(B^{(n)})] = \mathsf{P}_{t_1, \ldots, t_n}\left[B^{(n)}\right],$$

$$(3.1.3)$$

$\forall t_1 < \cdots < t_n \in T,\ B^{(n)} \in \mathcal{B}(\mathbb{R}^n)$ *and* $n \in \mathbb{N}^*$.

Proof. The consistency criterion (3.1.2) allows for the application of Kolmogorov extension theorem 3.1.2, yielding the probability P over $(\mathbb{R}^T, \mathcal{B}(\mathbb{R}^T))$, such that

$$\mathsf{P}[\mathcal{C}_{t_1, \ldots, t_n}(B^{(n)})] = \mathsf{P}_{t_1, \ldots, t_n}[B^{(n)}], \quad \forall B^{(n)} \in \mathcal{B}(\mathbb{R}^n),$$

$\forall t_1 < \cdots < t_n \in T,\ n \in \mathbb{N}^*$. In particular,

$$\mathsf{P}[\{\omega \in \mathbb{R}^T | \omega_{t_1} \le x_1, \ldots, \omega_{t_n} \le x_n\}] = F_{t_1, \ldots, t_n}(x_1, \ldots, x_n),$$

$\forall x_1, \ldots, x_n \in \mathbb{R}$. Choose $\Omega = \mathbb{R}^T$ and $\mathcal{F} = \mathcal{B}(\mathbb{R}^T)$, then the identity $X :$ $(\mathbb{R}^T, \mathcal{B}(\mathbb{R}^T)) \to (\mathbb{R}^T, \mathcal{B}(\mathbb{R}^T))$, i.e. the stochastic process $X_t(\omega) = \omega_t$, $\forall t \in T$, $\omega \in \mathbb{R}^T$, is the desired process. The probability space is $(\mathbb{R}^T, \mathcal{B}(\mathbb{R}^T), \mathsf{P})$. $\square$

Example 3.1.8 (Wiener measure and process). *Consider the function*

$$\phi_t(y|x) = \frac{1}{\sqrt{2\pi t}} \exp\left\{-\frac{1}{2t}(y - x)^2\right\}, \quad \forall x, y \in \mathbb{R} \text{ and } t > 0.$$

Then $\forall x, z \in \mathbb{R}$ and $s, t > 0$ we have

$$\int_{-\infty}^{\infty} \phi_s(y|x)\phi_t(z|y)\mathrm{d}y = \phi_{s+t}(z|x) \tag{3.1.4}$$

$$\Longleftrightarrow \int_{-\infty}^{\infty} \frac{1}{\sqrt{2\pi s}} \exp\left\{-\frac{1}{2s}u^2\right\} \frac{1}{\sqrt{2\pi t}} \exp\left\{-\frac{1}{2t}(z-u-x)^2\right\} \mathrm{d}u$$

$$= \frac{1}{\sqrt{2\pi(s+t)}} \exp\left\{-\frac{1}{2(s+t)}(z-x)^2\right\}$$

$$\Longleftrightarrow \mathcal{N}(0, s) + \mathcal{N}(x, t) \sim \mathcal{N}(x, s+t). \tag{3.1.5}$$

Let F_θ a d.f. with scalar or vector parameter $\theta \in \Theta$. If the convolution identity

$$F_{\theta_1+\theta_2} = F_{\theta_1} * F_{\theta_2}, \quad \forall \theta_1, \theta_2 \in \Theta, \tag{3.1.6}$$

is satisfied, then the class of d.f. $\{F_\theta\}_{\theta \in \Theta}$ is called convolution semigroup. It is the semigroup over the space of d.f. with convolution as inner dyadic operation. If $\theta = (\mu, \sigma^2)$, where $\Theta = \mathbb{R} \times \mathbb{R}_+^*$ and F_θ is the normal d.f. with expectation μ and variance σ^2, then (3.1.6) holds and we have the Gaussian convolution semigroup. It follows from this fact that (3.1.5) holds and thus (3.1.4) holds as well.

The Wiener measure is defined by

$$\mathsf{P}_{t_1,t_2,\ldots,t_n}[I_1 \times I_2 \times \cdots \times I_n]$$

$$= \int_{x_1 \in I_1} \int_{x_2 \in I_2} \cdots \int_{x_n \in I_n} \phi_{t_1}(x_1|0)\phi_{t_2-t_1}(x_2|x_1) \cdots \phi_{t_n-t_{n-1}}(x_n|x_{n-1}),$$

$$\mathrm{d}x_1 \mathrm{d}x_2 \cdots \mathrm{d}x_n,$$

where $I_1, I_2, \ldots, I_n$ are arbitrary intervals of $\mathbb{R}$, $0 \leq t_1 < t_2 < \cdots < t_n$ are arbitrary times and $n \in \mathbb{N}^*$ is arbitrary. Property (3.1.4) can be reformulated as follows,

$$\int_{-\infty}^{\infty} \phi_{t_k-t_{k-1}}(x_k|x_{k-1})\phi_{t_{k+1}-t_k}(x_{k+1}|x_k)\mathrm{d}x_k = \phi_{t_{k+1}-t_{k-1}}(x_{k+1}|x_{k-1}),$$

$\forall x_{k-1}, x_{k+1} \in \mathbb{R}$, from where it follows that

$$\mathsf{P}_{t_1,\ldots,t_k,\ldots,t_n}[I_1 \times \cdots \times I_k \times \cdots \times I_n]$$

$$= \mathsf{P}_{t_1,\ldots,t_{k-1},t_{k+1},\ldots,t_n}[I_1 \times \cdots \times I_{k-1} \times I_{k+1} \times \cdots \times I_n],$$

for $k = 1, \ldots, n$, with suitable conventions at extremities. Therefore the Wiener measure is consistent in the sense of (3.1.2) and Kolmogorov existence theorem 3.1.7 gives us the stochastic process $\{X_t\}_{t \in \mathbb{R}_+}$ that satisfies (3.1.3), which is called Wiener process. It is certainly the most important continuous time stochastic process and its complete presentation is given in Definition 3.2.8.

Two important properties of stochastic processes are the independence and the stationarity of the increments.

Definition 3.1.9 (Stationary increments). *The stochastic process $\{X_t\}_{t\in T}$ has stationary increments if the distribution of $X_t - X_s$ depends on s and t only through $t - s$, $\forall s, t \in T$ such that $s < t$.*

Note the stationarity of the increments should not be confused with the stationarity of the process.

Definition 3.1.10 (Independent increments). *The stochastic process $\{X_t\}_{t\in T}$ has independent increments if, $\forall s_1 < t_1 \leq s_2 < t_2 \leq \cdots \leq s_k < t_k \in T$ and $k \in \mathbb{N}^*$,*

$$X_{t_1} - X_{s_1}, X_{t_2} - X_{s_2}, \ldots, X_{t_k} - X_{s_k},$$

are independent random variables.

The practical meaning of Definition 3.1.10 is the absence of aftereffects or consecutive effects. One can thus exclude that the occurrence of some particular event is systematically followed by the occurrence of another similar event, after a fixed time lag.

Definition 3.1.11 (Markov process). *A Markov or Markovian process $\{X_t\}_{t\in T}$ satisfies the Markov property*

$$\mathsf{P}[X_t \in B | X_s = x_s, X_{r_k} = x_{r_k}, \ldots, X_{r_1} = x_{r_1}] = \mathsf{P}[X_t \in B | X_s = x_s],$$
(3.1.7)

$\forall x_{r_1}, \ldots, x_{r_k}, x_s \in \mathbb{R}$, $r_1 < \cdots < r_k < s < t \in T$, $k \in \mathbb{N}^*$ and $B \in \mathcal{B}(\mathbb{R})$.

A synonym for Markov process with time domain $T = \mathbb{N}$ is Markov chain. A process with independent increments must be a Markov process, but a Markov process must not necessarily possess independent increments. Markov processes are well suited for many practical situations. The Markov property expresses short memory of the process: although past occurrences have some influence on future occurrences, their precise times and values are irrelevant if they do not belong to the immediate past. The class of Markov processes is rather large and it includes many important processes.

Example 3.1.12 (Random walk). *A random walk with discrete time is defined as follows. Let $Y_1, Y_2, \ldots$ be i.i.d. and $\{-1, 1\}$-valued random variables, where $p = \mathsf{P}[Y_1 = 1] \in (0, 1)$ and thus $\mathsf{P}[Y_1 = -1] = 1 - p$, and let $Y_0 = 0$. Then $X_n = \sum_{k=0}^{n} Y_k$, for $n = 0, 1, \ldots$, is a random walk with initial* *value* 0. *The increments are obviously independent and therefore the random walk is a Markov process. Sometimes other distributions of Y_1 than the present two steps distribution are admitted in the definition of the random walk.*

Definition 3.1.13 (Counting process). *A counting process with time domain $T = \mathbb{R}_+$ is the $\mathbb{N}$-valued stochastic process $\{N_t\}_{t\geq 0}$ that satisfies*

$$N_s \leq N_t, \quad \text{a.s.}, \quad \forall 0 \leq s \leq t < \infty.$$

Important stochastic processes are the renewal processes.

Definition 3.1.14 (Renewal process). *Let $D_1, D_2, \ldots$ be positive and i.i.d. random variables, let $T_n = \sum_{i=0}^{n} D_i$, where $D_0 = 0$, and let*

$$N_t = \max\{n \geq 0 | T_n \leq t\}, \quad \forall t \geq 0.$$

Then each one of the three following stochastic process is a renewal process: $\{D_n\}_{n\geq 1}$, $\{T_n\}_{n\geq 1}$ and the counting process $\{N_t\}_{t\geq 0}$.

Consider the renewal processes of Definition 3.1.14 and let G be the d.f. of D_1, then $\forall t \geq 0, n \in \mathbb{N}$, it holds

$$G_n(t) = \mathsf{P}[T_n \leq t] = G^{*n}(t), \tag{3.1.8}$$

where $G^{*0} = \Delta$, which is Dirac's d.f. over $\mathbb{R}$ of Appendix A.8.2,

$$\mathsf{P}[N_t \geq n] = G_n(t), \tag{3.1.9}$$

and

$$\mathsf{P}[N_t = n] = \mathsf{P}[N_t \geq n] - \mathsf{P}[N_t \geq n+1] = G_n(t) - G_{n+1}(t). \tag{3.1.10}$$

In the remaining part of this section we consider exclusively Markov counting processes with time domain $T = \mathbb{R}_+$. As already mentioned, the Markov property is weaker than independence of increments. The distribution of a Markov counting process can be entirely determined through its transition probabilities. These transition probabilities are given by

$$p_{k,k+n}(s,t) = \mathsf{P}[N_t - N_s = n | N_s = k], \forall s \leq t \text{ and for } k, n = 0, 1, \ldots.$$

One can show that these transition probabilities satisfy the following Chapman-Kolmogorov equation,

$$p_{k,k+n}(s,t)$$
$$= \sum_{i=0}^{n} p_{k,k+i}(s,\tau) p_{k+i,k+n}(\tau,t), \forall s \leq \tau \leq t \text{ and for } k, n = 0, 1, \ldots.$$

Through the transition probabilities we can distinguish two types of processes.

Definition 3.1.15 (Homogeneous and inhomogeneous stochastic process). *When $p_{k,k+n}(s,t)$ depends on s and t only through $t - s$, $\forall s \leq t$ and for $k, n = 0, 1, \ldots$, then the Markov counting process is homogeneous, otherwise it is inhomogeneous.*

Remark 3.1.16 (Homogeneity and stationarity of increments). *One should note that homogeneity is a conditional property of the increments, whereas stationarity is the corresponding unconditional property.*

Provided $N_0 = 0$ we define
$$p_n(t) = \mathsf{P}[N_t = n] = p_{0,n}(0,t), \quad \forall t > 0, \ n = 0, 1, \ldots.$$
Let $0 \leq s \leq t$ and $n = 0, 1, \ldots$, then we have
$$\mathsf{P}[N_t - N_s = n] = \sum_{k=0}^{\infty} \mathsf{P}[N_t - N_s = n | N_s = k] \mathsf{P}[N_s = k]$$
$$= \sum_{k=0}^{\infty} p_{k,k+n}(s,t) p_k(s).$$
If the process has stationary increments, then we also have
$$p_n(t - s) = \sum_{k=0}^{\infty} p_{k,k+n}(s,t) p_k(s).$$

3.1.2 *Stationarity*

As with time series, we define the a.c.v.f. of the complex-valued stochastic process $\{X_t\}_{t \in \mathbb{R}}$ by
$$\gamma(t + h, t) = \mathsf{cov}(X_{t+h}, X_t), \quad \forall t, h \in \mathbb{R}.$$

Definition 3.1.17 (Weak and strict stationarity). *The complex-valued stochastic process $\{X_t\}_{t \in \mathbb{R}}$ is (weakly) stationary if, $\forall t, h \in \mathbb{R}$, $\mathsf{E}[|X_t|^2] < \infty$,*

(1) $\mathsf{E}[X_t]$ does not depend on t and
(2) $\gamma(t + h, t) = \gamma(h, 0)$.

In this case we denote by
$$\gamma(h) = \gamma(h, 0), \quad \forall h \in \mathbb{R},$$
the a.c.v.f. of the stationary stochastic process.
  The complex-valued stochastic process $\{X_t\}_{t \in \mathbb{R}}$ is strictly stationary if, $\forall t_1 < \cdots < t_n \in \mathbb{R}$, $h \in \mathbb{R}$ and $n \in \mathbb{N}^*$,
$$(U_{t_1}, V_{t_1}, \ldots, U_{t_n}, V_{t_n}) \sim (U_{t_1+h}, V_{t_1+h}, \ldots, U_{t_n+h}, V_{t_n+h}), \quad (3.1.11)$$
where $X_t = U_t + iV_t$, $\forall t \in \mathbb{R}$.

The equivalence (3.1.11) is indeed the equality of f.d.d. after shifting all times by the amount h. We call the f.d.d. obtained by joining real and complex components of the process by double f.d.d. Note that the particular order in (3.1.11) is not unique. One could have also considered the order of $(U_{t_1}, \ldots, U_{t_n}, V_{t_1}, \ldots, V_{t_n})$.

It is interesting to note that if the real-valued process is stationary and possesses independent increments, then

$$\gamma(h) = \mathsf{cov}(X_{t+h}, X_t) = \mathsf{cov}(X_t + X_{t+h} - X_t, X_t)$$
$$= \mathsf{var}(X_t) + \mathsf{cov}(X_{t+h} - X_t, X_t) = \gamma(0), \quad \forall t, h \in \mathbb{R}.$$

This means that the process has infinitely long memory, according to the terminology that a stochastic process has long memory when its a.c.v.f. decays at slower rate than the exponential.

We now give a simple example of a stationary stochastic process. Its construction uses the homogeneous Poisson process, which is introduced in Section 3.2.4.

Example 3.1.18 (Random telegraph signal). *Consider the stochastic process $\{X_t\}_{t\geq 0}$ such that: it takes values in $\{-1, 1\}$, the number of crossings of the null line is an homogeneous Poisson process with parameter $\lambda > 0$, denoted $\{N_t\}_{t\geq 0}$, and the initial value X_0 takes the values -1 or 1 with same probability $1/2$. This stochastic processes is currently referred to as random telegraph signal. Denote by P_1 the conditional probability $\mathsf{P}_1[\cdot] = \mathsf{P}[\cdot|X_0 = 1]$. Let $t \geq 0$. Then*

$$\mathsf{P}_1[X_t = 1] = \mathsf{P}[N_t \text{ is even}]$$
$$= e^{-\lambda t}\left(1 + \frac{(\lambda t)^2}{2!} + \frac{(\lambda t)^4}{4!} + \cdots\right)$$
$$= e^{-\lambda t} \cosh \lambda t.$$

Similarly,

$$\mathsf{P}_1[X_t = -1] = e^{-\lambda t} \sinh \lambda t.$$

Let S be an independent random variable taking the values -1 or 1 with same probability $1/2$ and let $Y_t = SX_t$, $\forall t \geq 0$. Then the f.d.d. of $\{Y_t\}_{t\geq 0}$ under P_1 are the f.d.d. of $\{X_t\}_{t\geq 0}$ under P. By denoting E_1 the expectation under P_1, we find

$$\mathsf{E}[X_t] = \mathsf{E}_1[Y_t] = \mathsf{E}_1[SX_t] = \mathsf{E}_1[S]\mathsf{E}_1[X_t]$$
$$= 0(e^{-\lambda t} \cosh \lambda t - e^{-\lambda t} \sinh \lambda t) = 0e^{-2\lambda t} = 0.$$

Let $0 \le s < t$. Thus for $x_0 = -1$ and $x_1 = 1$, we have

$$\gamma(s,t) = \mathrm{cov}(X_s, X_t) = \mathsf{E}[X_s X_t] = \mathsf{E}_1[Y_s Y_t] = \mathsf{E}_1[S^2]\mathsf{E}_1[X_s X_t]$$

$$= 1 \sum_{j=0}^{1} \sum_{k=0}^{1} x_j x_k \mathsf{P}_1[X_s = x_j]\mathsf{P}_1[X_t = x_k \mid X_s = x_j].$$

In this last formula, we have

$$\mathsf{P}_1[X_t = 1 \mid X_s = -1] = \mathsf{P}_1[X_t = -1 \mid X_s = 1] = \mathsf{P}_1[X_{t-s} = -1]$$
$$= \mathrm{e}^{-\lambda(t-s)} \sinh \lambda(t - s),$$

and

$$\mathsf{P}_1[X_t = -1 \mid X_s = -1] = \mathsf{P}_1[X_t = 1 \mid X_s = 1] = \mathsf{P}_1[X_{t-s} = 1]$$
$$= \mathrm{e}^{-\lambda(t-s)} \cosh \lambda(t - s),$$

from independence and stationarity of increments of the Poisson process. After simplifications we find $\gamma(s,t) = \mathrm{e}^{-2\lambda(t-s)}$. Therefore, the random telegraph signal is a stationary process with null mean and a.c.v.f.

$$\gamma(h) = \mathrm{e}^{-2\lambda|h|}, \quad \forall h \in \mathbb{R}. \tag{3.1.12}$$

The following theorem gives some important properties of the a.c.v.f. of a (generally complex-valued) stationary process.

Theorem 3.1.19 (Properties of a.c.v.f.). *Let γ be the a.c.v.f. of the generally complex-valued stationary stochastic process $\{X_t\}_{t \in \mathbb{R}}$ with mean zero. Then the following properties hold.*

(1) $\forall t, h \in \mathbb{R}$, $\mathrm{var}(X_{t+h} \pm X_t) = 2\gamma(0) \pm \{\gamma(-h) + \gamma(h)\}$.

(2) $\exists p > 0$ such that $\gamma(p) = \begin{cases} \gamma(0), \\ -\gamma(0) \end{cases} \implies$

$$\{X_t\}_{t \in \mathbb{R}} \text{ is a.s.} \begin{cases} p\text{-periodic}, \\ 2p\text{-periodic} \end{cases} \text{and } \gamma \text{ is} \begin{cases} p\text{-periodic}, \\ 2p\text{-periodic}. \end{cases}$$

(3) If $\gamma(h)$ is continuous at $h = 0$, then it is continuous at any $h \in \mathbb{R}$.

Proof. 1. This result follows immediately from the expansion of the variance.

2. Let $t \in \mathbb{R}$ and $p > 0$. It follows from part 1 that, if $\gamma(p) = -\gamma(0)$, then $X_{t+p} = -X_t + g(t, p)$ a.s., for some deterministic function g. It follows from the fact that the process has mean null that $g(t, p) = 0$ a.s. Thus we have,

$$X_{t+2p} = -X_{t+p} = X_t, \quad \text{a.s.}$$

For the case where $\gamma(p) = \gamma(0)$, p-periodicity can be deduced in a similar way.

3. Let $t, h \in \mathbb{R}$, then it follows from part 1 that

$$\begin{aligned}
\{\gamma(t+h) - \gamma(t)\}^2 &= \{\mathsf{cov}(X_{t+h} - X_t, X_0)\}^2 \\
&\leq \mathsf{var}(X_{t+h} - X_t)\mathsf{var}(X_0) \\
&= \{2\gamma(0) - [\gamma(-h) + \gamma(h)]\}\gamma(0).
\end{aligned}$$

This last expression vanishes as $h \to 0$, whenever γ is continuous at the origin. $\qquad\square$

Remark 3.1.20 (Periodic process and seasonality). *With Theorem 3.1.19.2 we can thus construct a p- or $2p$-periodic stationary process. Let $d = p$ or $2p$ be the period length. We denote this complex-valued and d-periodic process by $\{S_t\}_{t\in\mathbb{R}}$ and we denote by γ_S its a.c.v.f. Assume that there is second complex-valued and stationary process $\{Y_t\}_{t\in\mathbb{R}}$, which is orthogonal to $\{S_t\}_{t\in\mathbb{R}}$, in the sense that $\mathsf{cov}(S_s, Y_t) = 0$, $\forall s, t \in \mathbb{R}$. Then the sum of the processes $X_t = Y_t + S_t$, $\forall t \in \mathbb{R}$, is stationary with a.c.v.f. $\gamma_X = \gamma_Y + \gamma_S$. The d-periodic component $\{S_t\}_{t\in\mathbb{R}}$ is called the seasonality.*

Theorem 2.1.27 characterizes of the a.c.v.f. of a real-valued stationary time series in terms of a n.n.d. function. Theorem 3.1.21 extends this result to continuous time and to complex-valued processes.

Theorem 3.1.21 (Characterization of the a.c.v.f.). *The function $\kappa : \mathbb{R} \to \mathbb{C}$ is the a.c.v.f. of a complex-valued (strictly) stationary stochastic process $\iff \kappa$ is n.n.d.*

Proof. ($\Rightarrow$) We refer to the proof of Theorem 2.1.27, where we replace $k_1 < \cdots < k_n \in \mathbb{Z}$ by $t_1 < \cdots < t_n \in \mathbb{R}$ and $\boldsymbol{c} \in \mathbb{R}^n$ by $\boldsymbol{c} \in \mathbb{C}^n$.

($\Leftarrow$) Let us decompose the n.n.d. function κ at arbitrary $h \in \mathbb{R}$ as $\kappa(h) = \alpha(h) + i\beta(h)$. The Hermitian property of a n.n.d. function ($\kappa(-h) = \overline{\kappa(h)}$, cf. Proposition 2.1.22), gives us

$$\alpha(-h) = \alpha(h) \quad \text{and} \quad \beta(-h) = -\beta(h).$$

Let $n \in \mathbb{N}^*$, $u_j, v_j \in \mathbb{R}$, $c_j = u_j - iv_j$, for $j = 1, \ldots, n$, let $t_1 < \cdots < t_n \in \mathbb{R}$, $\boldsymbol{u} = (u_1, \ldots, u_n)^\top$, $\boldsymbol{v} = (v_1, \ldots, v_n)^\top \in \mathbb{R}^n$ and define

$$q(\boldsymbol{u}, \boldsymbol{v}) = \frac{1}{2}\sum_{j=1}^n \sum_{k=1}^n c_j \overline{c_k}\kappa(t_j - t_k).$$

Then $q(\boldsymbol{u}, \boldsymbol{v}) \geq 0$ implies

$$
\begin{aligned}
q(\boldsymbol{u}, \boldsymbol{v}) &= \frac{1}{2} \sum_{j=1}^{n} \sum_{k=1}^{n} (u_j - iv_j)(u_k + iv_k)\{\alpha(t_j - t_k) + i\beta(t_j - t_k)\} \\
&= \frac{1}{2} \sum_{j=1}^{n} \sum_{k=1}^{n} (u_j u_k + v_j v_k)\alpha(t_j - t_k) - (u_j v_k - v_j u_k)\beta(t_j - t_k).
\end{aligned}
$$
$$(3.1.13)$$

Define the real-valued random vectors

$$
\boldsymbol{U} = (U_{t_1}, \ldots, U_{t_n})^\top \text{ and } \boldsymbol{V} = (V_{t_1}, \ldots, V_{t_n})^\top,
$$

and assume

$$
\begin{pmatrix} \boldsymbol{U} \\ \boldsymbol{V} \end{pmatrix} \sim \mathcal{N}(\boldsymbol{0}, \boldsymbol{\Sigma}), \qquad (3.1.14)
$$

for some n.n.d. matrix $\boldsymbol{\Sigma}$. A possible choice of $\boldsymbol{\Sigma}$ can be obtained by setting

$$
\varphi(\boldsymbol{u}, \boldsymbol{v}) = \mathsf{E}\left[\exp\left\{i\left(\boldsymbol{u}^\top, \boldsymbol{v}^\top\right) \begin{pmatrix} \boldsymbol{U} \\ \boldsymbol{V} \end{pmatrix}\right\}\right] = \exp\left\{-\frac{1}{2}q(\boldsymbol{u}, \boldsymbol{v})\right\}. \qquad (3.1.15)
$$

With this choice and with the formula of the characteristic function (3.2.1), we obtain

$$
\exp\left\{-\frac{1}{2}\left(\boldsymbol{u}^\top, \boldsymbol{v}^\top\right) \boldsymbol{\Sigma} \begin{pmatrix} \boldsymbol{u} \\ \boldsymbol{v} \end{pmatrix}\right\} = \exp\left\{-\frac{1}{2}q(\boldsymbol{u}, \boldsymbol{v})\right\}.
$$

This, (3.1.13) and

$$
\boldsymbol{\Sigma} = \mathsf{E}\left[\begin{pmatrix} \boldsymbol{U}\boldsymbol{U}^\top & \boldsymbol{U}\boldsymbol{V}^\top \\ \boldsymbol{V}\boldsymbol{U}^\top & \boldsymbol{V}\boldsymbol{V}^\top \end{pmatrix}\right],
$$

yield

$$
\mathsf{E}[U_{t_j} U_{t_k}] = \frac{1}{2}\alpha(t_j - t_k), \ \mathsf{E}[V_{t_j} V_{t_k}] = \frac{1}{2}\alpha(t_j - t_k),
$$
$$
\mathsf{E}[U_{t_j} V_{t_k}] = -\frac{1}{2}\beta(t_j - t_k) = \frac{1}{2}\beta(t_k - t_j) \text{ and } \mathsf{E}[V_{t_j} U_{t_k}] = \frac{1}{2}\beta(t_j - t_k),
$$
$$(3.1.16)$$

for $j, k = 1, \ldots, n$. Define

$$
\boldsymbol{X} = (X_{t_1}, \ldots, X_{t_n})^\top,
$$

where $X_{t_j} = U_{t_j} + iV_{t_j}$, for $j = 1, \ldots, n$. Thus we obtain

$$
\begin{aligned}
\mathsf{var}(\boldsymbol{X}) &= \mathsf{E}\left[\boldsymbol{X}\overline{\boldsymbol{X}}^{\top}\right] \\
&= \mathsf{E}\left[(\boldsymbol{U} + i\boldsymbol{V})(\boldsymbol{U} - i\boldsymbol{V})^{\top}\right] \\
&= \mathsf{E}\left[\boldsymbol{U}\boldsymbol{U}^{\top} + \boldsymbol{V}\boldsymbol{V}^{\top} + i\left(\boldsymbol{V}\boldsymbol{U}^{\top} - \boldsymbol{U}\boldsymbol{V}^{\top}\right)\right] \\
&= \frac{1}{2}\left(\alpha(t_j - t_k) + \alpha(t_j - t_k) + i\left[\beta(t_j - t_k) + \beta(t_j - t_k)\right]\right)_{j,k=1,\ldots,n} \\
&= \left(\alpha(t_j - t_k) + i\beta(t_j - t_k)\right)_{j,k=1,\ldots,n} \\
&= \left(\kappa(t_j - t_k)\right)_{j,k=1,\ldots,n}.
\end{aligned}
$$

For the rest, we proceed like in the proof of Theorem 2.1.27, in which the characteristic function is now given by (3.1.15). The consistency condition of Kolmogorov's existence theorem 3.1.7 is satisfied by the double f.d.d. (3.1.14). So there exist two real-valued Gaussian processes $\{U_t\}_{t\in\mathbb{R}}$ and $\{V_t\}_{t\in\mathbb{R}}$ that yield these double f.d.d. We deduce from (3.1.16) that $\{U_t\}_{t\in\mathbb{R}}$ and $\{V_t\}_{t\in\mathbb{R}}$ are jointly strictly stationary, in the sense that they yield double f.d.d. that satisfy (3.1.11) of Definition 3.1.17. Thus $\{X_t\}_{t\in\mathbb{R}}$ is (strictly) stationary. Note that Gaussian processes are introduced in Section 3.2.1. $\square$

3.1.3 Sample path regularity

We first give some results on the sample path regularity of general stochastic processes. Some specific results for stationary processes are presented in the final part of this section. Further specific results for stationary processes, that concern the notions of $\mathcal{L}_2$-continuity and $\mathcal{L}_2$-differentiability, are analyzed in Section 3.3.

We begin by presenting basic definitions and results on the regularity of real functions. Consider $a < x < b$ and the function $g : (a,b) \setminus \{x\} \to \mathbb{R}$, which is discontinuous at $x \in (a,b)$. (The value $g(x)$ may not exist.)

- If $g(x-), g(x+) \in \mathbb{R}$ and $g(x-) = g(x+) \neq g(x)$, then x is a removable discontinuity. This discontinuity has no practical interest because it can be removed by making g continuous at x.
- If $g(x-), g(x+), g(x) \in \mathbb{R}$ and $g(x-) \neq g(x+)$, then x is a discontinuity of the first kind or a jump discontinuity. The value of $g(x)$ has no practical interest.
- In any other situation, precisely when $g(x-) \notin \mathbb{R}$ or $g(x+) \notin \mathbb{R}$, x is a discontinuity of the second kind or an essential discontinuity. (An example is $\sin 1/t$ at $t = 0$.)

- A function is regular if it has no discontinuities of the second kind and if all removable discontinuities have been removed.
- A function is cadlag (which stands for "continue à droite avec limite à gauche") if it is right-continuous and with left limits everywhere. (For example, Dirac's d.f. Δ is cadlag.)
- The jump function of the function g at $t \in (a, b)$ is defined by

$$\Delta g(t) = g(t+) - g(t-). \tag{3.1.17}$$

(The above symbol Δ is always followed by the function of the jump, so that it cannot be confused with the Δ of Dirac's d.f.).

Thus, a cadlag function can only have discontinuities of the first kind. A related result is the following.

Proposition 3.1.22 (Maximal number of discontinuities of a d.f.). *Any d.f. over $\mathbb{R}$ possesses at most countably many discontinuity points.*

Proof. Let F be a d.f. and let U be the set of all discontinuity points of F, i.e. $U = \{x \in \mathbb{R} \mid F(x-) < F(x)\}$. Choose for each $u \in U$ a rational number $q(u)$ such that $F(u-) < q(u) < F(u)$. Thus q maps U to $\mathbb{Q}$. Let us see that q is injective. Let $u, v \in U$, with $u < v$, then

$$F(u-) < q(u) < F(u) \leq F(v-) < q(v) < F(v).$$

Therefore $q(u) < q(v)$: injectivity holds. Because $q(U) \subset \mathbb{Q}$ is countable and q is injective, U must be countable. $\qquad\square$

Thereafter T denotes any real interval (e.g. $[0, 1]$, $\mathbb{R}_+ = [0, \infty)$ or $\mathbb{R} = (-\infty, \infty)$) representing the time domain of a stochastic process. A continuous time stochastic process can be obtained through the f.d.d. $P_{t_1, \ldots, t_n}$, $\forall t_1 < \cdots < t_n \in T$ and $n \in \mathbb{N}^*$, over the spaces $(\mathbb{R}^n, \mathcal{B}(\mathbb{R}^n))$, $\forall n \in \mathbb{N}^*$, whenever they are consistent. The stochastic process obtained is a $\mathcal{B}(\mathbb{R}^T)$-measurable random element. However, some sample path properties (i.e. properties of the function the $\omega \in \mathbb{R}^T$) cannot be re-expressed in terms of sets of $\mathcal{B}(\mathbb{R}^T)$. Such sample paths properties must be added in the definition of the process. We will however see, that some specific constraints on the f.d.d. allow to show that there exists an equivalent process, in a sense defined later, that has the desired sample paths properties. The sample paths properties that we consider are the continuity properties. They are very important for the analysis a hitting or crossing time, which is the first time when the process hits or crosses a given curve of interest (e.g. an horizontal level).

Example 3.1.23 (Nonmeasurable event). *Let $T = [0,1]$, $\Omega = \mathbb{R}^{[0,1]}$ and $X_t(\omega) = \omega_t$, $\forall t \in [0,1], \omega \in \Omega$. Then*

$$A = \{\omega \in \mathbb{R}^{[0,1]} | X_t(\omega) = 0, \forall t \in [0,1]\}$$
$$= \bigcap_{t \in [0,1]} \underbrace{\{\omega \in \mathbb{R}^{[0,1]} | X_t(\omega) = 0\}}_{= \mathcal{C}_t(\{0\}) \in \mathcal{B}(\mathbb{R}^{[0,1]})}$$
$$\notin \mathcal{B}(\mathbb{R}^{[0,1]}),$$

in general. There may indeed be uncountably many intersections of cylinders.

Denote by C the space of continuous functions $T \to \mathbb{R}$. Additional information on the sample paths can be provided by defining the process $\{X_t\}_{0 \leq t \in T}$ over $(C, \mathcal{B}(C))$, where the σ-algebra is the restriction of the initial cylindrical σ-algebra,

$$\mathcal{B}(C) = \mathcal{B}(\mathbb{R}^T) \cap C = \{B \cap C \,|\, B \in \mathcal{B}(\mathbb{R}^T)\}. \tag{3.1.18}$$

Example 3.1.24 (Nonmeasurable event, continuation). *Now we have*

$$A = \bigcap_{t \in [0,1] \cap \mathbb{Q}} \{\omega \in C | X_t(\omega) = 0\} \in \mathcal{B}(C).$$

Remark 3.1.25 (σ-algebra of continuous functions). *Associated with the distance $d(x,y) = \sup_{t \in T} |y(t) - x(t)|$, $\forall x, y \in C$, C becomes a metric space and we can thus define the neighborhoods*

$$N_x(\varepsilon) = \left\{y \in C \,\middle|\, d(x,y) < \varepsilon\right\}, \quad \forall x \in C, \varepsilon > 0.$$

Let us denote the smallest σ-algebra containing all such neighborhoods by $\mathcal{B}^*(C)$. Then one can show that this is the cylindrical σ-algebra (3.1.18), i.e. $\mathcal{B}^*(C) = \mathcal{B}(C)$.

Example 3.1.26 (Processes with same f.d.d.). *Let $T = [0,1]$. Define $X_t = 0$, $\forall t \in [0,1]$, $U \sim \text{Uniform}(0,1)$ and*

$$Y_t = \begin{cases} 0, & \text{if } t \neq U, \\ 1, & \text{if } t = U, \end{cases} \quad \forall t \in [0,1].$$

So $\mathsf{P}[Y_t \neq 0] = \mathsf{P}[U = t] = 0 \Rightarrow \mathsf{P}[Y_t = 0] = 1$, $\forall t \in [0,1]$. Also, $\mathsf{P}[X_t = 0] = 1$, $\forall t \in [0,1]$. So all one-dimensional distributions of $\{X_t\}_{0 \leq t \leq 1}$ and $\{Y_t\}_{0 \leq t \leq 1}$ are the same. Generally, all f.d.d. are the same:

$$\mathsf{P}[Y_{t_1} = 0, \ldots, Y_{t_n} = 0] = \mathsf{P}[U \notin \{t_1, \ldots, t_n\}] = 1,$$

and

$$P[X_{t_1} = 0, \ldots, X_{t_n} = 0] = 1,$$

$\forall t_1 < \cdots < t_n \in [0,1]$ *and for* $n = 1, 2, \ldots$. *So* $\{X_t\}_{0 \le t \le 1}$ *with continuous sample path cannot be distinguished from* $\{Y_t\}_{0 \le t \le 1}$ *with discontinuous sample paths, in terms of f.d.d.*

Definition 3.1.27 (Version of stochastic process). *The stochastic processes* $\{X_t\}_{t \in T}$ *and* $\{Y_t\}_{t \in T}$ *are called (mutual) versions if*

$$P[X_t = Y_t] = 1, \quad \forall t \in T.$$

Sometimes the name modification is used instead of version.

Lemma 3.1.28 (F.d.d. of versions). *If the stochastic processes* $\{X_t\}_{t \in T}$ *and* $\{Y_t\}_{t \in T}$ *are versions, then they possess the same f.d.d.*

Proof. Let $t_1 < \cdots < t_n \in T$ and $n \in \mathbb{N}^*$. It is enough to show that $P\left[\cap_{k=1}^n \{X_{t_k} = Y_{t_k}\}\right] = 1$. Indeed,

$$P\left[\bigcap_{k=1}^n \{X_{t_k} = Y_{t_k}\}\right] = 1 - P\left[\bigcup_{k=1}^n \{X_{t_k} \ne Y_{t_k}\}\right]$$

$$\ge 1 - \sum_{k=1}^n P[X_{t_k} \ne Y_{t_k}] = 1.$$

$\square$

Example 3.1.29 (Stochastic processes with same f.d.d., continuation). *For* $T = [0,1]$, *let* $A_t = \{X_t \ne Y_t\} = \{U = t\}$, *then* $P[A_t] = 0, \forall 0 \le t \le 1$. *So* $\{X_t\}_{0 \le t \le 1}$ *and* $\{Y_t\}_{0 \le t \le 1}$ *are versions. Note however that, for* $A = \cup_{0 \le t \le 1} A_t$, $P[A] = P[\cup_{0 \le t \le 1} \{U = t\}] = 1$. *Consequently,* $P[\cap_{0 \le t \le 1} \{X_t = Y_t\}] = P[\cap_{0 \le t \le 1} A_t^c] = P[A^c] = 0$.

As previously mentioned, it is not always possible to determine the continuity properties of the sample paths of a stochastic process that is defined by its f.d.d. only. However, one can sometimes show the existence of a version that has the desired continuity properties. Then, in the definition of the stochastic process by its f.d.d., one can explicitly refer to that precise version. The existence of the desired version needs some conditions and these conditions must be reported in the definition of the stochastic process.

Stronger than the concept of version is the concept of indistinguishability, of stochastic processes. The outer measure is used in the next definition. The outer measure P^* over $(\Omega, \mathcal{F})$ is defined as follows. For any

$A \subset \Omega$, consider the countable cover $\cup_{n=1}^{\infty} B_n$, where $B_1, B_2, \ldots \in \mathcal{F}$. Then the outer measure of A is $\mathsf{P}^*[A] = \inf \sum_{n=1}^{\infty} \mathsf{P}[B_n]$, where the infimum is taken w.r.t. all countable covers of measurable sets. When A is measurable, then $\mathsf{P}^*[A] = \mathsf{P}[A]$.

Definition 3.1.30 (Indistinguishable stochastic processes). *The stochastic processes $\{X_t\}_{t \in T}$ and $\{Y_t\}_{t \in T}$ defined over the same probability space are called (mutually) indistinguishable if*

$$\mathsf{P}^* \left[\bigcap_{t \in T} \{X_t = Y_t\} \right] = 1,$$

where P^* denotes the outer measure.

The stochastic processes $\{X_t\}_{t \in T}$ and $\{Y_t\}_{t \in T}$ are indistinguishable means that their sample paths are a.s. equal. The stochastic processes $\{X_t\}_{t \in T}$ and $\{Y_t\}_{t \in T}$ are versions means that, at any individual time, they are a.s. equal. Indistinguishable processes are versions.

Example 3.1.31 (Versions with discrete time). *With discrete time, T is countable and with e.g. $T = \mathbb{N}$, any two versions $\{X_n\}_{n \geq 0}$ and $\{Y_n\}_{n \geq 0}$ satisfy*

$$\mathsf{P} \left[\left(\bigcap_{n=0}^{\infty} \{X_n = Y_n\} \right)^c \right] = \mathsf{P} \left[\bigcup_{n=0}^{\infty} \{X_n \neq Y_n\} \right] \leq \sum_{n=0}^{\infty} \underbrace{\mathsf{P}[X_n \neq Y_n]}_{=0} = 0,$$

from Boole's inequality (for countable unions). Thus with discrete time any two versions are indistinguishable.

We now return to the question of existence of a version with some desired continuity properties and provide Kolmogorov's continuity theorem 3.1.32 in this context.

Theorem 3.1.32 (Kolmogorov continuity). *Consider the stochastic process $\{X_t\}_{t \in T}$.*

(1) If $\forall \varepsilon > 0$, $\exists 0 < p < \eta$, $c > 0$, such that for $s < t \in T$,

$$t - s < \varepsilon \implies \mathsf{E}\left[|X_t - X_s|^p\right] \leq c \frac{t - s}{|\log(t - s)|^{1+\eta}},$$

then there exists a version with continuous sample paths.

(2) If $\forall \varepsilon > 0$, $\exists p, \eta, c > 0$, such that for $s \leq t \in T$,

$$t - s < \varepsilon \implies \mathsf{E}[|X_t - X_s|^p] \leq c(t - s)^{1+\eta},$$

then there exists a version with continuous sample paths.

(3) If $\forall \varepsilon > 0$, $\exists p_1, p_2, \eta, c > 0$, such that for $r \leq s \leq t \in T$,

$$t - r < \varepsilon \implies \mathsf{E}[|X_s - X_r|^{p_1}|X_t - X_s|^{p_2}] \leq c(t - r)^{1+\eta},$$

then there exists a version with sample paths without discontinuities of second kind, i.e. right and left limits exist at any interior point of T and one-sided limits exist at each finite boundary point of T.

We can see directly that the condition under Theorem 3.1.32.2 is stronger than the one appearing under Theorem 3.1.32.1 and thus Theorem 3.1.32.2 might appear irrelevant. However, the condition under Theorem 3.1.32.2, which is the Lipschitz continuity between metric spaces, is often more practical to verify and indeed satisfied by various important processes, like the Wiener process. This is shown with details in Example 3.1.33.1.

Examples 3.1.33 (Sample paths of Wiener and Poisson processes).

(1) Wiener process over $\mathbb{R}_+$
The Wiener measure over $(\mathbb{R}^{[0,\infty)}, \mathcal{B}(\mathbb{R}^{[0,\infty)}))$ leads to the stochastic process $\{X_t\}_{t \geq 0}$ over this space, which satisfies in particular $X_t - X_s \sim \mathcal{N}(0, t - s)$, $\forall 0 \leq s < t$. Consequently, with $\mathsf{E}[\{\mathcal{N}(0,1)\}^4] = 3$ one obtains

$$\mathsf{E}\left[|X_t - X_s|^4\right] = 3(t - s)^2.$$

Thus Kolmogorov continuity theorem 3.1.32.2 tells that there exists a version that with continuous sample paths. This particular version is denoted $\{W_t\}_{t \geq 0}$ and called Wiener process or Brownian motion. This version corresponds to Definition 3.2.8.

(2) Poisson process over $\mathbb{R}_+$
Consider the homogeneous Poisson process $\{N_t\}_{t \geq 0}$ with intensity $\lambda > 0$. Then $\forall 0 \leq r \leq s \leq t$,

$$\mathsf{E}[(N_t - N_s)(N_s - N_r)] = \mathsf{E}[N_t - N_s]\mathsf{E}[N_s - N_r]$$
$$= \lambda^2(t - s)(s - r)$$
$$\leq \lambda^2(t - r)^2,$$

so that Kolmogorov continuity theorem 3.1.32.3 tells that there is a version whose only possible sample paths discontinuities are jumps. This property is already known from the definition of the Poisson process as counting process, in the sense of Definition 3.2.4. (The Poisson process is a particular birth process according to Definition 3.2.29.) Nevertheless, we have shown that this continuity property is inherent to the f.d.d. of the Poisson process.

As a general rule and as in Example 3.1.33.1, whenever a stochastic process satisfies the Lipschitz condition of Theorem 3.1.32.2, then it is the continuous version that is retained for the definition. Theorem 3.1.28 tells that the f.d.d. of the two processes are indeed the same.

Definition 3.1.34 (Stochastic continuity). *Let T be any interval of $\mathbb{R}$. The process $\{X_t\}_{t \in T}$ is stochastically continuous at $t \in T$ if*

$$\lim_{h \to 0} \mathsf{P}[|X_{t+h} - X_t| > \varepsilon] = 0, \quad \forall \varepsilon > 0,$$

i.e. $X_{t+h} \xrightarrow{\mathsf{P}} X_t$, as $h \to 0$, with one-sided limit if t is a boundary point of T that belongs to T.

Let us give an example of a discontinuous process.

Example 3.1.35 (Sample paths of Gaussian WN). *Let $\{X_t\}_{t \in \mathbb{R}}$ be a Gaussian process with mean zero and a.c.v.f.*

$$\gamma(h) = \begin{cases} 0, & \text{if } h \neq 0, \\ \sigma^2, & \text{if } h = 0, \end{cases}$$

for some $\sigma > 0$. We refer to Definition 3.2.1 of Gaussian processes. We study the sample paths of $\{X_t\}_{t \in \mathbb{R}}$ and in particular stochastic continuity. Let $\varepsilon, h > 0$ and $t \in \mathbb{R}$. As X_{t+h} and X_t are jointly Gaussian and uncorrelated, we obtain $X_{t+h} - X_t \sim \mathcal{N}(0, 2\sigma^2)$. This implies

$$\mathsf{P}\left[|X_{t+h} - X_t| > \sqrt{2}\sigma\varepsilon\right] = 2\Phi(-\sigma\varepsilon) > 0,$$

not depending on h. Because $\{X_t\}_{t \in \mathbb{R}}$ is nowhere stochastically continuous, its sample paths are nowhere continuous a.s.

As $\sigma^2 \to \infty$, the process $\{X_t\}_{t \in \mathbb{R}}$ tends to continuous time Gaussian WN, which is defined in Section 3.4.2 in terms of a stochastic $\mathcal{L}_2$-integral.

It is direct to see that Kolmogorovs continuity criterion of Theorem 3.1.32.2. implies stochastic continuity. Indeed, assume that the stochastic process $\{X_t\}_{t \in T}$ satisfies the criterion of Theorem 3.1.32.2. Then by Markov inequality we obtain that $\exists p, \eta, c > 0$ such that, for $h > 0$ sufficiently small and for $\delta > 0$,

$$\mathsf{P}[|X_{t+h} - X_t| \geq \delta] \leq \frac{1}{\delta^p} \mathsf{E}[|X_{t+h} - X_t|^p] \leq \frac{c}{\delta^p} h^{1+\eta} \xrightarrow{h \to 0} 0.$$

Thus $\{X_t\}_{t \in T}$ is stochastically continuous.

An elementary example is the following.

Example 3.1.36 (Indicator process). *Consider* $\Omega = [0, 1]$ *with its Borel* σ*-algebra and the Lebesgue measure. Define* $X_t = I_{[0,t]}$, $\forall t \in [0, 1]$. *Then*

$$\mathsf{P}[|X_{t+h} - X_t| > \varepsilon] = \mathsf{P}\left[I_{(t,t+h]}\right] = h \xrightarrow{h \to 0} 0, \quad \forall \varepsilon \in (0, 1], \ t \in [0, 1].$$

The next theorem gives conditions under which a stochastic process admits a cadlag version. It can be used in conjunction with Theorem 3.1.32.3.

Theorem 3.1.37 (Cadlag version of stochastic process). *If the sample paths of a stochastic process have no discontinuities of second kind and if the process is stochastically continuous, then there exists a version with cadlag sample paths.*

Whenever a stochastic process satisfies the conditions of Theorem 3.1.37, then it is the cadlag version that is retained. A large class of stochastic processes are the Lévy processes, introduced in Section 3.2.7. These processes are indeed defined as cadlag processes.

We now provide further results regarding sample path continuity that are specific to stationary stochastic processes. We should note that there exists an alternative notion of continuity, which is w.r.t. the $\mathcal{L}_2$-norm. It is called $\mathcal{L}_2$-continuity or mean square continuity and it is introduced later in Definition 3.3.1. It has an important role in the context of stationary processes. According to Proposition 3.3.3, the stationary process $\{X_t\}_{t \in T}$ is mean square continuous or $\mathcal{L}_2$-continuous iff its a.c.v.f. γ is continuous at the origin. Corollary 3.1.38 below follows directly from Theorem 3.1.32.1, because under stationarity we have

$$\mathsf{E}\left[(X_t - X_s)^2\right] = 2\{\gamma(0) - \gamma(t - s)\}, \quad \forall s \le t \in T.$$

Corollary 3.1.38 (Continuity of stationary process). *Let* $\{X_t\}_{t \in T}$ *be a stationary process with a.c.v.f.* γ. *If* $\exists \eta > 3$ *such that*

$$\gamma(h) = \gamma(0) + \mathrm{O}\left(\frac{h}{|\log h|^\eta}\right), \quad \text{as } h \downarrow 0,$$

then there exists a version with continuous sample paths.

Example 3.1.39. *Consider an a.c.v.f. of the form*

$$\gamma(h) = \gamma(0) + c|h|^\alpha + \mathrm{o}(|h|^\alpha), \quad \text{as } h \to 0, \tag{3.1.19}$$

for some constants $c \in \mathbb{R}^*$ *and* $\alpha \in (1, 2]$. *Let* $\eta > 3$, *then one verifies that*

$$ch^\alpha = \mathrm{O}\left(\frac{h}{|\log h|^\eta}\right), \quad \text{as } h \downarrow 0.$$

So one deduces from Corollary 3.1.38 that any stationary process with
a.c.v.f. of the form (3.1.19) admits a version with continuous sample paths.

Note that $\alpha = 2$ implies that the expansion given in Theorem A.4.6 of
the appendix, in which the characteristic function is replaced by the a.c.r.f.
$\rho = \gamma/\gamma(0)$, holds for some $k \geq 2$. This in turn implies that the second
spectral moment α_2, which is given later in Definition 3.3.7, does exist.
The form (3.1.19) tells that $\alpha_2 = -2c$.

If one would consider the excluded case $\alpha = 4$, for example, then the
form (3.1.19) and Theorem A.4.6 would lead to $\alpha_2 = 0$. This would give
a spectral distribution with total mass at the null frequency. So the corresponding a.c.v.f. could not have the form (3.1.19), meaning that $\alpha = 4$ is
not admissible. More generally, consider the form (3.1.19) with $\alpha = 2 + \delta$,
for some $\delta > 0$. Thus

$$\gamma(h) = \gamma(0) + c|h|^{2+\delta} + o(|h|^{2+\delta}) = \gamma(0) + o(|h|^2), \quad \text{as } h \to 0.$$

Theorem A.4.6 tells that there the second spectral moment is null, i.e. $\alpha_2 =
0$, which is not compatible with (3.1.19).

For an instance of (3.1.19) with the excluded case $\alpha = 1$, we refer to
Example 3.1.43 of the random telegraph signal.

It turns out that the condition given in Corollary 3.1.38 can be weakened
when the stationary process is Gaussian. Although Gaussian processes are
introduced in Section 3.2.1 to come, the next theorem already provides this
weakened condition as well as alternative condition based of the spectral
d.f. This alternative condition is (3.1.20) in Theorem 3.1.40. It requires
the spectral distribution of a stationary process. This type of distributions
is introduced later in Section 3.5, together with a related central result,
namely Bochner's theorem 3.5.1.

Theorem 3.1.40 (Continuity of Gaussian stationary process). *Consider
a Gaussian stationary process with a.c.v.f. γ and spectral d.f. F. If $\exists \eta > 3$
such that*

$$\gamma(h) = \gamma(0) + O\left(\frac{1}{|\log h|^\eta}\right), \quad \text{as } h \downarrow 0,$$

or alternatively that

$$\int_{[0,\infty)} \{\log(1+\alpha)\}^\eta \mathrm{d}F(\alpha) < \infty, \tag{3.1.20}$$

then there exists a version with continuous sample paths.

Let $T \subset \mathbb{R}$ be any interval and let $\{\mathsf{P}_{t_1,\dots,t_n}\}_{t_1 < \cdots < t_n \in T, n \geq 1}$ a class of f.d.d. over $(\mathbb{R}^n, \mathcal{B}(\mathbb{R}^n))$, $\forall n \geq 1$, which is consistent in the sense of (3.1.2). Kolmogorov extension theorem 3.1.2 yields a probability measure over $(\mathbb{R}^T, \mathcal{B}(\mathbb{R}^T))$ with these f.d.d. The existence of a stochastic process $\{X_t\}_{t \in T}$ over a probability space and having the given f.d.d. is guaranteed by Theorem 3.1.7. Assume that these f.d.d. satisfy any of the conditions for continuity given in the last results, precisely in Theorem 3.1.32.1, 3.1.32.2 or in Corollary 3.1.38. Then $\{X_t\}_{t \in T}$ can be replaced by a continuous version of it, which takes values in $(C, \mathcal{B}(C), \mathsf{P})$ instead of $(\mathbb{R}^T, \mathcal{B}(\mathbb{R}^T), \mathsf{P})$.

Another important sample path property is the differentiability. Differentiability w.r.t. the $\mathcal{L}_2$-norm, or simply in $\mathcal{L}_2$ or in mean square, is given in Definition 3.3.5. Sample path and $\mathcal{L}_2$-derivatives are two different stochastic limits of Newton's quotients: according to Lemma A.3.11, if these two limits exist, then they must correspond. The $\mathcal{L}_2$-differentiability of stationary processes is studied in Section 3.3. As with sample path continuity, we are interested in the conditions for the existence of a version with differentiable sample paths. An analogue result to Theorem 3.1.32.2 is the following.

Theorem 3.1.41 (Kolmogorov differentiability). *Consider the stochastic process* $\{X_t\}_{t \in T}$. *If* $\forall \varepsilon > 0$, $\exists\, p, \eta, c > 0$, *such that for any equidistant points* $r < s < t \in T$, *with* $h = s - r = t - s$, *we have*

$$h < \varepsilon \implies \mathsf{E}[|X_t - 2X_s + X_r|^p] \leq ch^{1+p+\eta},$$

then there exists a version with continuously differentiable sample paths.

We now give a specific result for the differentiability of sample paths of stationary Gaussian processes.

Theorem 3.1.42 (Differentiability of Gaussian stationary process). *Consider a Gaussian stationary process with a.c.v.f.* γ *and spectral d.f.* F. *If* $\exists \eta > 3$ *such that*

$$\gamma(h) = \gamma(0) + \frac{\gamma''(0)}{2}h^2 + \mathrm{O}\left(\frac{h^2}{|\log h|^\eta}\right), \quad \text{as } h \downarrow 0,$$

or alternatively that

$$\int_{[0,\infty)} \alpha^2 \{\log(1 + \alpha)\}^\eta \mathrm{d}F(\alpha) < \infty,$$

then there exists a version with continuous sample paths.

Example 3.1.43 (Exponential a.c.v.f.). *Consider the a.c.v.f.*

$$\gamma(h) = e^{-\theta|h|}, \quad \forall h \in \mathbb{R},$$

where $\theta > 0$. This is the a.c.v.f. of the random telegraph signal; cf. (3.1.12). Then

$$\gamma(h) = 1 - \theta h + O(h^2), \text{ as } h \downarrow 0.$$

Because $-\theta h \neq O(h/|\log h|^\eta)$, as $h \downarrow 0$, for some $\eta > 3$, Corollary 3.1.38 cannot guarantee the existence of a version with continuous sample paths. Indeed, the random telegraph signal is discontinuous by construction.

Consider now any Gaussian process, i.e. Gaussian f.d.d., with this a.c.v.f. Then we obtain the Ornstein-Uhlenbeck process of Examples 3.2.21, 3.4.9 and 3.5.11. Because $-\theta h = O(1/|\log h|^\eta)$, as $h \downarrow 0$, for some $\eta > 3$, Theorem 3.1.40 tells that there exists a version with continuous sample paths. Further, because $\gamma''(0)$ does not exist, Theorem 3.1.42 cannot guarantee the existence of a version with differentiable sample paths.

3.2 Important stochastic processes

This section provides an overview of important continuous time stochastic processes that appear in the context of stationarity. A central class of such stochastic processes are the Gaussian ones. They are introduced in Section 3.2.1 in the complex-valued form. An important feature to retain is that, although for real-valued Gaussian processes strict and weak stationarity are equivalent, this is no longer true with complex-valued processes. An important Gaussian process is the Wiener process, which is often called Brownian motion, and it is presented in Section 3.2.2. Its time domain is usually $\mathbb{R}_+$ but it can be extended to $\mathbb{R}$. This leads to a process with orthogonal increments, which is useful for the spectral decomposition of stationary processes. Selfsimilarity is one amongst other appealing properties of the Wiener process. The concept of selfsimilarity is introduced in Section 3.2.3. It roughly means that the shape of the stochastic process possesses shape is invariant w.r.t. any change of the time scale. Nonlinear differentiable functions are not selfsimilar, because they become linear after a sufficiently important magnification. The Lamperti transformations are presented: they convert selfsimilar processes to strict stationary ones and conversely. Section 3.2.3 introduces also the fractional Brownian motion, which is a Gaussian selfsimilar process that generalizes the Wiener process. We then pass to processes taking nonnegative integer values, representing counts. An important class of Markovian counting processes are the birth processes. Their definition as well as a recursive formula for obtaining the transition probabilities are presented in Section 3.2.4. The Poisson process is introduced in Section 3.2.4 as a particular birth process. Compound and shot noise processes are sums of random variables, whose number of summands is a counting processes. These processes are presented in Section 3.2.5. Counting processes can be embedded in the more general class of point processes. They are presented in Section 3.2.6. It is explained later in Section 3.9.5 that Shot noise processes are closely related to filtered point processes. The Wiener and the Poisson processes are two central process that belong to the important class of the Lévy processes. Although this class is not studied here, in the context of stationarity, it is nevertheless briefly presented in Section 3.2.7, for the simple purpose of being complete.

3.2.1 *Gaussian processes*

This section presents the important class of Gaussian processes, provides some of their properties and introduces their spectral decomposition.

Definitions and basic properties

The real- and complex-valued Gaussian or normal random vectors are presented in Section A.8.3. The generalization of these random vectors to stochastic processes leads to the following important and general class of real- and complex-valued stochastic processes.

Definition 3.2.1 (Gaussian process). *If the f.d.d. of a real-valued stochastic process are Gaussian, then the stochastic process is Gaussian.*

If $\{U_t\}_{t \in \mathbb{R}}$ and $\{V_t\}_{t \in \mathbb{R}}$ are two real-valued Gaussian processes, then

$$X_t = U_t + iV_t, \quad \forall t \in \mathbb{R},$$

determines a complex-valued Gaussian process.

Gaussian processes are practical because they inherit the simple and appealing properties of the normal distribution. For instance, if a stochastic process is Gaussian, then the average of the values of the process taken at any times yields a normal random variable. In the practice, the normality assumption is often supported by the data or perhaps after some appropriate transformation of the data. The next proposition provides an equivalent definition of the Gaussian process.

Proposition 3.2.2 (Characterization of Gaussian process). *The stochastic process $\{X_t\}_{t\in\mathbb{R}}$ is a real-valued Gaussian process iff, $\forall t_1 < \cdots < t_n \in \mathbb{R}$, $c_1, \ldots, c_n \in \mathbb{R}$ and $n \in \mathbb{N}^*$,*

$$\sum_{j=1}^{n} c_j X_{t_j},$$

is normally distributed.

Proof. The equivalence with Definition 3.2.1 can be shown as follows. Let $n \in \mathbb{N}^*$, $t_1 < \cdots < t_n \in \mathbb{R}$, $\boldsymbol{c} = (c_1, \ldots, c_n)^\top \in \mathbb{R}^n$, $\boldsymbol{X} = (X_{t_1}, \ldots, X_{t_n})^\top$, $\boldsymbol{\mu} = \mathsf{E}[\boldsymbol{X}]$ and $\boldsymbol{\Sigma} = \mathrm{var}(\boldsymbol{X})$. The normality of $\langle \boldsymbol{c}, \boldsymbol{X}\rangle$ can be re-expressed in terms its characteristic function.

$$\mathsf{E}[\exp\{\mathrm{i}v\langle \boldsymbol{c}, \boldsymbol{X}\rangle\}] = \exp\left\{\mathrm{i}\langle \boldsymbol{c}, \boldsymbol{\mu}\rangle v - \frac{1}{2}\boldsymbol{c}^\top\boldsymbol{\Sigma}\boldsymbol{c}v^2\right\}, \quad \forall v \in \mathbb{R}. \qquad (3.2.1)$$

By defining $\boldsymbol{a} = v\boldsymbol{c}$, we see that the validity of (3.2.1) $\forall \boldsymbol{c} \in \mathbb{R}^n$ and $v \in \mathbb{R}$ is equivalent to its validity $\forall \boldsymbol{a} \in \mathbb{R}^d$. But this last statement means that $\boldsymbol{X}$ follows the normal or the singular normal distribution. $\qquad \square$

Let us now present some other results of Gaussian processes. We remember that in Section 3.1.3 the sample path regularity of stationary Gaussian processes is analyzed: the continuity in Theorem 3.1.40 and the differentiability in Theorem 3.1.42. Regarding the Markov property of Gaussian processes, we have the following result.

Theorem 3.2.3 (Markov property of Gaussian process). *The real-valued Gaussian process $\{X_t\}_{t\in\mathbb{R}}$ is a Markov process $\Longleftrightarrow$*

$$\mathsf{E}[X_t | X_s = x_s, X_{r_k} = x_{r_k}, \ldots, X_{r_1} = x_{r_1}] = \mathsf{E}[X_t | X_s = x_s], \qquad (3.2.2)$$

$\forall x_{r_1}, \ldots, x_{r_k}, x_s \in \mathbb{R}$, $r_1 < \cdots < r_k < s < t \in \mathbb{R}$ and $k \in \mathbb{N}^*$.

Proof. ($\Rightarrow$) This part of the proof is obvious.
($\Leftarrow$) We note that both left and right sides of the Markov property (3.1.7) are Gaussian distributions. Thus (3.1.7) reduces to conditions on mean and variance only. The condition on the mean is exactly (3.2.2). So we only need to show that the condition on the variance is automatically satisfied once (3.2.2) holds. Define

$$R = X_t - \mathsf{E}[X_t | X_s, X_{r_k}, \ldots, X_{r_1}] = X_t - \mathsf{E}[X_t | X_s],$$

where $x_{r_1}, \ldots, x_{r_k}, x_s \in \mathbb{R}$, $r_1 < \cdots < r_k < s < t \in \mathbb{R}$ and $k \in \mathbb{N}^*$. Let $u \in \{r_1, \ldots, r_k, s\}$, then

$$\mathsf{E}[RX_u] = 0.$$

Indeed, R is the residual of the orthogonal projection of the Gaussian random variable X_t onto $\overline{\mathrm{sp}}\{1, X_s, X_{r_k}, \ldots, X_{r_1}\}$, as mentioned in Lemma A.1.9, whereas $\mathsf{E}[RX_u] = \langle R, X_u \rangle$. But R is Gaussian and so it is independent of $X_s, X_{r_k}, \ldots, X_{r_1}$. Consequently,

$$\mathsf{var}(X_t | X_s, X_{r_k}, \ldots, X_{r_1})$$
$$= \mathsf{var}(R | X_s, X_{r_k}, \ldots, X_{r_1}) = \mathsf{var}(R) = \mathsf{var}(R | X_s) = \mathsf{var}(X_t | X_s).$$

$\square$

Stationary Gaussian processes

We remind that Theorem 3.1.21 tells that the a.c.v.f. of a stationary process is characterized in terms of n.n.d. function. According to the proof of Theorem 3.1.21, to each n.n.d. function $\kappa : \mathbb{R} \to \mathbb{C}$, there exists a (strictly) stationary and complex-valued Gaussian process $\{X_t\}_{t \in \mathbb{R}}$ with mean null and a.c.v.f. κ. This result is rather practical, because when the a.c.v.f. of a stationary process is given, we can always assume that the stochastic process is Gaussian. The Gaussian distribution leads to various practical mathematical properties. For instance, it leads to various ways of simulating the process. Simulation algorithms for stationary Gaussian processes are presented in Section 4.7.

It follows immediately from Definitions 3.1.17 and 3.2.1 that for real-valued Gaussian processes, weak and strict stationarity are equivalent. For complex-valued Gaussian processes this equivalence does not hold anymore, but Theorem 3.2.5 holds. Before presenting this theorem and its proof, We recall that for any two complex-valued random variables X and $Y \in \mathcal{L}_2$, we call

$$\mathsf{cov}^*(X, Y) = \mathsf{E}[XY] - \mathsf{E}[X]\mathsf{E}[Y],$$

their pseudo-covariance. Accordingly, we have the following definition.

Definition 3.2.4 (pseudo-a.c.v.f.). *Let $\{X_t\}_{t \in \mathbb{R}}$ be a complex-valued processes in $\mathcal{L}_2$, then the function*

$$\gamma^*(s, t) = \mathsf{cov}^*(X_s, X_t) = \mathsf{E}[X_s X_t] - \mathsf{E}[X_s]\mathsf{E}[X_t], \quad \forall s, t \in \mathbb{R},$$

is called the pseudo-a.c.v.f. of the complex-valued stochastic process.

Theorem 3.2.5 (Strict stationarity of complex-valued Gaussian processes). *Let $\{U_t\}_{t \in \mathbb{R}}$ and $\{V_t\}_{t \in \mathbb{R}}$ be two real-valued Gaussian processes with mean zero. Let $X_t = U_t + iV_t$, $\forall t \in \mathbb{R}$. Then, $\{X_t\}_{t \geq 0}$ is strictly stationary $\iff \gamma(s, t) = \mathsf{E}\left[X_s \overline{X_t}\right]$ and $\gamma^*(s, t) = \mathsf{E}[X_s X_t]$ depend on s and t only through $s - t$, $\forall s, t \in \mathbb{R}$.*

Proof. ($\Leftarrow$) Let $s, t \in \mathbb{R}$. Then

$$\gamma(s,t) = \mathsf{E}[U_s U_t + V_s V_t] + \mathrm{i}\mathsf{E}[U_t V_s - U_s V_t] \quad \text{and}$$
$$\gamma^*(s,t) = \mathsf{E}[U_s U_t - V_s V_t] + \mathrm{i}\mathsf{E}[U_t V_s + U_s V_t],$$

depend on s and t only through $s - t$. By taking the sums and differences of real and imaginary parts, we obtain that

$$\mathsf{E}[U_s U_t], \ \mathsf{E}[V_s V_t], \ \mathsf{E}[U_t V_s] \ \text{and} \ \mathsf{E}[U_s V_t], \tag{3.2.3}$$

do depend on s and t only through $s - t$. This fact can be reformulated as follows. Let $t_1 < \cdots < t_n \in \mathbb{R}$, $h \in \mathbb{R}$ and $n \in \mathbb{N}^*$, then

$$\mathsf{var}((U_{t_1}, V_{t_1}, \ldots, U_{t_n}, V_{t_n})) = \mathsf{var}((U_{t_1+h}, V_{t_1+h}, \ldots, U_{t_n+h}, V_{t_n+h})).$$

Because the processes are Gaussian with mean null, the above covariance matrix determines entirely the distribution of the two random vectors that appear as arguments. Therefore,

$$(U_{t_1}, V_{t_1}, \ldots, U_{t_n}, V_{t_n}) \sim (U_{t_1+h}, V_{t_1+h}, \ldots, U_{t_n+h}, V_{t_n+h}),$$

i.e. the double f.d.d., are invariant under constant time shift.
($\Rightarrow$) This part of the proof is elementary: the expectations in (3.2.3) do depend on s and t only through $s - t$ and so the same holds for $\gamma(s,t)$ and $\gamma^*(s,t)$, $\forall s, t \in \mathbb{R}$. $\qquad\square$

Example 3.2.6 (Nonstationary complex-valued Gaussian process). *Let* $\{R_t\}_{t \in \mathbb{R}}$ *be a real-valued, weakly (and strictly) stationary and Gaussian process. Assume further that its expectation is zero and denote its a.c.v.f. by* γ_R. *Define*

$$X_t = \mathrm{e}^{\mathrm{i}\alpha t} R_t = \cos(\alpha t) R_t + \mathrm{i}\sin(\alpha t) R_t,$$

for some $\alpha \in \mathbb{R}$. *Then,* $\forall s, t \in \mathbb{R}$, *we have*

$$\gamma(s,t) = \mathsf{E}\left[X_s \overline{X_t}\right] = \mathrm{e}^{\mathrm{i}\alpha(s-t)}\gamma_R(s-t) \quad \text{and}$$
$$\gamma^*(s,t) = \mathsf{E}\left[X_s X_t\right] = \mathrm{e}^{\mathrm{i}\alpha(s+t)}\gamma_R(s-t).$$

The dependence on s *and* t *of* $\gamma^*(s,t)$ *cannot be expressed in terms of* $s - t$ *alone, unless* $\alpha = 0$. *So we can deduce from Theorem 3.2.5 that* $\{X_t\}_{t \in \mathbb{R}}$ *is generally not strictly stationary.*

The next example provides the construction of a stationary, Markovian and Gaussian process with time domain $\mathbb{R}$, which gives the Ornstein-Uhlenbeck process. It generalizes a case in Grimmett and Stirzaker (1991), pp. 382–383.

Example 3.2.7 (Ornstein-Uhlenbeck process). *Consider the real-valued stochastic process $\{X_t\}_{t\in\mathbb{R}}$ that is: stationary, Markovian, centered and Gaussian. Denote by γ its a.c.v.f., let $s < t \in \mathbb{R}$ and $h \in \mathbb{R}_+$. It follows from Proposition A.8.4.4 that*

$$\gamma(0)\mathsf{E}[X_{t+h}|X_t] = \gamma(h)X_t.$$

With Markov's property and the above identity we obtain

$$\begin{aligned}
\gamma(0)\mathsf{E}[X_s X_{t+h}] &= \gamma(0)\mathsf{E}[\mathsf{E}[X_s X_{t+h}|X_s, X_t]] \\
&= \gamma(0)\mathsf{E}[X_s \mathsf{E}[X_{t+h}|X_t]] \\
&= \gamma(h)\mathsf{E}[X_s X_t] \\
\Longleftrightarrow \gamma(0)\gamma(t-s+h) &= \gamma(t-s)\gamma(h).
\end{aligned}$$

This is equivalent to the functional equation

$$\gamma(0)\gamma(x+y) = \gamma(x)\gamma(y), \quad \forall x,y \in \mathbb{R}.$$

Given that γ is an a.c.v.f., it must have the form

$$\gamma(h) = \gamma(0)\mathrm{e}^{-\tau|h|}, \quad \forall h \in \mathbb{R},$$

where $\tau > 0$.[1] This is the a.c.v.f. of the Ornstein-Uhlenbeck process, which appears in Examples 3.1.43, 3.2.21, 3.4.9 and 3.5.11.

Spectral decomposition of real-valued Gaussian stationary processes

The spectral decomposition of a real-valued Gaussian stationary process $\{X_t\}_{t\in\mathbb{R}}$ was introduced by Rice (1944, 1945), in the context of noise in radio transmission. It takes the form

$$X_t = \sum_{j=1}^{\infty} C(\alpha_j)\cos(\alpha_j t) + S(\alpha_j)\sin(\alpha_j t), \quad \forall t \in \mathbb{R}, \qquad (3.2.4)$$

where $0 < \alpha_1 < \alpha_2 < \cdots < \infty$ are the frequencies,

$$C(\alpha_j) \sim S(\alpha_j) \sim \mathcal{N}(0, \zeta_j^2),$$

for some $\zeta_j \in \mathbb{R}_+^*$, for $j = 1, 2, \ldots$, and where all these random variables are independent. Consider the random variables $B(\alpha_j)$ and $\theta(\alpha_j)$ such that

$$C(\alpha_j) = B(\alpha_j)\cos-\theta(\alpha_j) \text{ and } S(\alpha_j) = B(\alpha_j)\sin-\theta(\alpha_j), \text{ for } j = 1, 2, \ldots.$$

---

[1]It is known that the only continuous $g : \mathbb{R} \mapsto \mathbb{C}$ that satisfies

$$g(x+y) = g(x)g(y), \quad \forall x,y \in \mathbb{R},$$

is $g(x) = g(0)\mathrm{e}^{cx}$, where $c \in \mathbb{C}$.

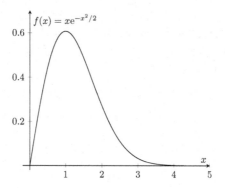

Fig. 3.1:   Rayleigh density.

An equivalent expression of X_t is thus given by

$$X_t = \sum_{j=1}^{\infty} B(\alpha_j) \cos -\theta(\alpha_j) \cos(\alpha_j t) + B(\alpha_j) \sin -\theta(\alpha_j) \sin(\alpha_j t)$$

$$= \sum_{j=1}^{\infty} B(\alpha_j) \cos(\alpha_j t + \theta(\alpha_j)), \quad \forall t \in \mathbb{R}, \tag{3.2.5}$$

where $\theta(\alpha_j)$ is uniformly distributed over $[0, 2\pi)$,

$$B(\alpha_j) \sim \zeta_j R, \tag{3.2.6}$$

for $j = 1, 2, \ldots$, where R is a (standard) Rayleigh distributed random variable, viz. it has density

$$x e^{-\frac{x^2}{2}}, \quad \forall x \geq 0, \tag{3.2.7}$$

and where all these random variables are independent. The graph of the Rayleigh density is given in Figure 3.1. The Rayleigh random variable R is distributed as $\sqrt{2X}$, where X is exponential distributed with mean 1. One finds $\mathsf{E}[R] = \sqrt{\pi/2}$, $\mathsf{E}[R^2] = 2$ and $\mathsf{var}(R) = 2 - \pi/2$. Thus, $\mathsf{E}[B(\alpha_j)] = \sqrt{\pi/2}\zeta_j$ and $\mathsf{var}(B(\alpha_j)) = (2 - \pi/2)\zeta_j^2$, for $j = 1, 2, \ldots$.

The distributions of the random variables in the re-expression (3.2.5) of the process follow from $Z_1 + iZ_2 \sim Re^{i\theta}$, where Z_1, Z_2 are independent standard normal, R is Rayleigh (3.2.7) and θ is independent of R and uniform over $[0, 2\pi)$.[2] Thus $R\cos\theta \sim \mathcal{N}(0, 1)$, which tells that the j-th summand of (3.2.5) is a $\mathcal{N}(0, \zeta_j^2)$ random variable, just like the j-th summand of (3.2.4).

---

[2]This basic result leads to the Box-Müller algorithm for generating pairs of independent standard normal random variables.

3.2.2 Wiener process

This section introduces the Wiener process, which often called Brownian motion in applied fields. Historically, the Wiener process was proposed by R. Brown in 1827 as a mathematical model for the erratic movements of a small particle in a fluid. Around 1900 it was formally analyzed by L. Bachelier, in the context of stock market values, and by A. Einstein in 1905, in the context of movements of molecules in a gas, cf. Einstein (1956). In his model, Einstein studied the movement of a particle whose mass is assumed null, for simplicity. Starting from Einstein's model, P. Langevin developed a stochastic model for the velocity of the particle, under the more realistic assumption that it has positive mass. The velocity at different times is represented by a generalized stochastic process, which is determined through the Langevin equation. This is explained in Example 3.4.9. The mathematical properties of the Brownian motion were investigated by N. Wiener. The Brownian motion or Wiener process is the most important continuous time stochastic process. It has a central role in stochastic integrals and stochastic differential equations. It is used for the representation of Gaussian WN, which is an important process in the context of stationary processes. The Wiener process appears as asymptotic representation of many other stochastic processes and can thus be used as approximation to these processes. In this way it provides a simple method for obtaining asymptotic solutions to many problems that do not have exact solutions. The Wiener process possesses several important and practical properties, such as the selfsimilarity. It is one of the most appealing stochastic model in most scientific fields.

Definition 3.2.8 (Wiener process or Brownian motion over $\mathbb{R}_+$). *The Wiener process, also called Brownian motion, is the real-valued stochastic process $\{W_t\}_{t\geq 0}$, that satisfies the following properties:*

(1) $W_0 = 0$ a.s.;
(2) the sample paths $\{W_t(\omega)\}_{t\geq 0}$, $\forall \omega \in \Omega$, are continuous;
(3) it has independent increments;
(4) $\forall 0 \leq s < t < \infty$, $W_t - W_s \sim \mathcal{N}(0, t-s)$ (implying that it has stationary increments).

Figure 3.2 shows the simulation of 10 sample paths of the Wiener process over the interval $[0, 100]$.

The Wiener process can be intuitively obtained as limit of the following random walk. Let $Y_1, Y_2, \ldots$ be i.i.d. random variables taking values in

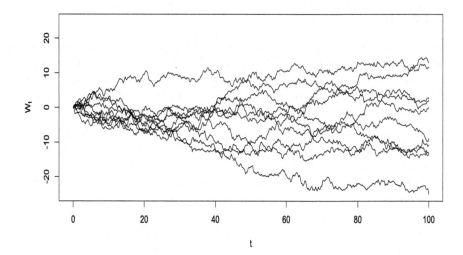

Fig. 3.2:   10 simulated sample paths of Wiener process $\{W_t\}_{t\in[0,100]}$.

$\{-1,1\}$ and with $\mathsf{P}[Y_1 = 1] = \mathsf{P}[Y_1 = -1] = 1/2$. Consider the discrete time stochastic process that starts at the origin and makes jump of magnitude ρY_k at the k-th unit of time, for $k = 1, 2, \ldots$, where $\rho > 0$. If the unit of time is $\tau > 0$, then the position at time $t > 0$ is given by

$$W_{\tau,t} = \rho \sum_{k=0}^{\lfloor \frac{t}{\tau} \rfloor} Y_k,$$

where $Y_0 = 0$. Thus $\mathsf{E}[W_{\tau,t}] = 0$. Consider now ρ as increasing function of τ with $\rho^2 \sim \tau$, as $\tau \to 0$. As $\tau \to 0$, we have

$$\mathsf{var}(W_{\tau,t}) = \rho^2 \left\lfloor \frac{t}{\tau} \right\rfloor \sim \tau \left\lfloor \frac{t}{\tau} \right\rfloor \longrightarrow t,$$

and so the Central limit theorem gives the desired marginal distribution,

$$\frac{W_{\tau,t}}{t} \xrightarrow{\mathrm{d}} \mathcal{N}(0,1).$$

The a.c.v.f. of the Wiener process is given by

$$\mathsf{cov}(W_s, W_t) = s \wedge t, \quad \forall s, t \geq 0.$$

Indeed, $\forall 0 < s \leq t < \infty$,

$$\begin{aligned}
\mathsf{cov}(W_s, W_t) &= \mathsf{cov}(W_s, W_s + W_t - W_s) \\
&= \mathsf{cov}(W_s, W_s) + \mathsf{cov}(W_s, W_t - W_s) \\
&= \mathsf{var}(W_s).
\end{aligned}$$

The mean function and the a.c.v.f. of the Wiener process allow to identify its f.d.d. as normal or singular normal distributions.

Proposition 3.2.9 (F.d.d. of the Wiener process). *Let $n \in \mathbb{N}^*$ and $0 < t_1 < \cdots < t_n$, then*

$$(W_{t_1}, \ldots, W_{t_n}) \sim \mathcal{N}(\mathbf{0}, \boldsymbol{\Sigma}_n),$$

where $\boldsymbol{\Sigma}_n = (t_j \wedge t_k)_{j,k=1,\ldots,n}$ is a n.n.d. matrix.

We show directly that $\boldsymbol{\Sigma}_n$ is n.n.d. as follows. Let $c_1, \ldots, c_n \in \mathbb{R}$, then

$$\sum_{j,k=1}^{n} c_j c_k t_j \wedge t_k = \sum_{j,k=1}^{n} c_j c_k (t_j \wedge t_k - t_1 + t_1)$$

$$= t_1 \left(\sum_{j=1}^{n} c_j \right)^2 + \sum_{j,k=2}^{n} c_j c_k (t_j \wedge t_k - t_1),$$

$$\sum_{j,k=2}^{n} c_j c_k (t_j \wedge t_k - t_2 + t_2 - t_1)$$

$$= (t_2 - t_1) \left(\sum_{j=2}^{n} c_j \right)^2 + \sum_{j,k=3}^{n} c_j c_k (t_j \wedge t_k - t_2),$$

$$\cdots$$

$$\sum_{j,k=n-1}^{n} c_j c_k (t_j \wedge t_k - t_{n-1} + t_{n-1} - t_{n-2})$$

$$= (t_{n-1} - t_{n-2}) \left(\sum_{j=n-1}^{n} c_j \right)^2 + c_n^2 (t_n - t_{n-1}).$$

Definition 3.2.1 of the real-valued Gaussian process and Proposition 3.2.9 for the f.d.d. of the Wiener process lead to the following equivalent definition.

Definition 3.2.10 (Wiener process over $\mathbb{R}_+$, alternative). *The real-valued stochastic process $\{W_t\}_{t \geq 0}$ is the Wiener process if the following properties hold:*

(1) $W_0 = 0$ a.s.;
(2) the paths $\{W_t(\omega)\}_{t \geq 0}$, $\forall \omega \in \Omega$, are continuous;
(3) $\{W_t\}_{t \geq 0}$ is Gaussian;
(4) $\mathsf{E}[W_t] = 0$ and $\mathsf{cov}(W_s, W_t) = s \wedge t$, $\forall s, t \geq 0$.

Let us mention some interesting results of properties. The first one tells that a new Wiener process can be obtained by reversing the time of the original Wiener process.

Theorem 3.2.11 (Time inversion). *Let* $\{W_t\}_{t\geq 0}$ *be a Wiener process with nonnegative time domain, then*

$$V_t = \begin{cases} tW_{\frac{1}{t}}, & \text{if } t > 0, \\ 0, & \text{if } t = 0, \end{cases}$$

$\forall t \geq 0$, *is another Wiener process.*

Proof. Let $s, t \geq 0$. We have $\mathsf{E}[V_t] = 0$ and $\mathrm{cov}(V_s, V_t) = s \wedge t$. We also see directly that $\{V_t\}_{t\geq 0}$ is Gaussian. Consequently, the conditions 1, 3 and 4 of Proposition 3.2.10 hold. It remains to show that condition 2 is satisfied. Any sample path of $\{V_t\}_{t\geq 0}$ is obviously continuous at any point $t > 0$. At $t = 0$, we have

$$
\begin{aligned}
\lim_{t\to 0} V_t &= \lim_{t\to 0} \frac{W_{\frac{1}{t}}}{\frac{1}{t}} \\
&= \lim_{s\to\infty} \frac{W_s}{s} \\
&= \lim_{s\to\infty} \left(\frac{W_{\lfloor s \rfloor}}{\lfloor s \rfloor} + \underbrace{\frac{W_s - W_{\lfloor s \rfloor}}{\lfloor s \rfloor}}_{\to 0} \right) \underbrace{\frac{\lfloor s \rfloor}{s}}_{\to 1} \\
&= \lim_{s\to\infty} \frac{1}{\lfloor s \rfloor} \{ W_1 + (W_2 - W_1) + \cdots + (W_{\lfloor s \rfloor} - W_{\lfloor s \rfloor - 1}) \} \\
&= \mathsf{E}[W_1] \\
&= 0,
\end{aligned}
$$

a.s., from the fact that the Wiener process is bounded over any closed interval (due to continuity) and from the strong law of large numbers. $\square$

Another interesting result, due to Paley and Wiener (1934), is an ingenious method of construction by means of the following random Fourier series,

$$W_t = \sqrt{2} \sum_{j=0}^{\infty} \mathrm{sinc}\left\{ \left(\frac{1}{2} + j \right) \pi t \right\} t Z_j, \quad \forall t \geq 0,$$

where $Z_0, Z_1, \ldots$ are independent standard normal random variables.

We can also mention some limiting results. Let $\{t_n\}_{n\in\mathbb{N}} \in \mathbb{R}_+^{\infty}$ such that $t_n \overset{n\to\infty}{\longrightarrow} \infty$, then

$$\liminf_{n\to\infty} W_{t_n} = -\infty \quad \text{and} \quad \limsup_{n\to\infty} W_{t_n} = \infty, \quad \text{a.s.}$$

The law of iterated logarithm is

$$\limsup_{t\downarrow 0} \frac{W_t}{\sqrt{2t \log \log t^{-1}}} = 1, \quad \text{a.s.}$$

In the context of the analysis of stationary processes, it is useful to extend Definition 3.2.8 of the Wiener process from time domain $\mathbb{R}_+$ to time domain $\mathbb{R}$.

Definition 3.2.12 (Wiener process over $\mathbb{R}$). *The Wiener process with time domain $\mathbb{R}$ is the real-valued stochastic process $\{W_t\}_{t\in\mathbb{R}}$, that satisfies the following properties:*

(1) $W_0 = 0$ a.s.;
(2) the paths $\{W_t(\omega)\}_{t\in\mathbb{R}}$, $\forall \omega \in \Omega$, are continuous;
(3) the increments $W_t - W_s$, $\forall -\infty < s < t < \infty$ are independent;
(4) $\forall -\infty < s < t < \infty$, $W_t - W_s \sim \mathcal{N}(0, t-s)$ (indicating that the increments are stationary).

The Wiener process with time domain $\mathbb{R}$ can be obtained from two independent Wiener processes with time domain $\mathbb{R}_+$ as follows.

Proposition 3.2.13 (Construction of Wiener process over $\mathbb{R}$). *Let $\{W_t^{(1)}\}_{t\in\mathbb{R}_+}$ and $\{W_t^{(2)}\}_{t\in\mathbb{R}_+}$ be two independent Wiener processes. Then*

$$W_t = \begin{cases} W_t^{(1)}, & \text{if } t \geq 0, \\ W_{-t}^{(2)}, & \text{if } t < 0, \end{cases}$$

is a Wiener process with time domain $\mathbb{R}$.

Proof. If $\text{var}(W_t - W_s) = t - s$, $\forall -\infty < s < t < \infty$, would hold, then all properties of Definition 3.2.12 would clearly hold.
    Let $0 \leq s < t$, then $\text{var}(W_t - W_s) = \text{var}(W_t^{(1)} - W_s^{(1)}) = t - s$.
    Let $s < t < 0$, then $\text{var}(W_t - W_s) = \text{var}(W_{-t}^{(2)} - W_{-s}^{(2)}) = t - s$.
    Let $s < 0 \leq t$, then $\text{var}(W_t - W_s) = \text{var}(W_t^{(1)} - W_{-s}^{(2)}) = t - s$. □

This construction allows for the computation of the a.c.v.f.

Proposition 3.2.14 (A.c.v.f. of Wiener process over $\mathbb{R}$). *The a.c.v.f. of the Wiener the process with time domain $\mathbb{R}$ is given by*

$$\gamma(s, t) = \begin{cases} |s| \wedge |t|, & \text{if } \text{sgn}\, s = \text{sgn}\, t, \\ 0, & \text{otherwise.} \end{cases}$$

Proof. Let $0 \leq s < t$, then $\gamma(s,t) = \text{cov}(W_s^{(1)}, W_t^{(1)}) = s$.

Let $s < t < 0$, then $\gamma(s,t) = \text{cov}(W_{-s}^{(2)}, W_{-t}^{(2)}) = \text{cov}(W_{-t}^{(2)} + W_{-s}^{(2)} - W_{-t}^{(2)}, W_{-t}^{(2)}) = \text{cov}(W_{-t}^{(2)}, W_{-t}^{(2)}) = -t$.

Let $s < 0 \leq t$, then $\gamma(s,t) = \text{cov}(W_{-s}^{(2)}, W_t^{(1)}) = 0$. $\qquad\square$

One can easily verify that $\{W_t\}_{t \in \mathbb{R}}$ is stochastic process with orthogonal increments, in the sense given in Definition 3.6.1. Thus, the Wiener process over $\mathbb{R}$ is important for the spectral decomposition of stationary processes.

One can show that the sample paths of the Wiener process are a.s. nowhere differentiable. Because any future increment of the Wiener process is independent of its past values, one may intuitively understand this fact. We can easily show that at any fixed time $t > 0$ the derivative does not exist. Let $h > 0$ and consider

$$\frac{W_{t+h} - W_t}{h} \sim \mathcal{N}(0, h^{-1}).$$

The characteristic function of this Newton's quotient is given by

$$\exp\left\{-\frac{v^2}{2h}\right\} \to \begin{cases} 0, & \text{if } v \neq 0, \\ 1, & \text{if } v = 1, \end{cases} \quad \text{as } h \downarrow 0.$$

The above limit cannot be a characteristic function, because a characteristic function is (uniformly) continuous; cf. Proposition A.4.2.2. Thus this Newton's ratio has no limit in distribution, thus no a.s. limit, i.e. the Wiener process is a.s. not differentiable at time t.

We can however define the stochastic process of derivatives $\{W_t'\}_{t \in \mathbb{R}}$ in terms of the derivative of a generalized stochastic process. This particular derivative process is called Gaussian WN and it is presented in Section 3.4.2.

An important property of the Wiener process is called selfsimilarity. This property is presented in Section 3.2.3 and it is shared by a more general class of process, called fractional Brownian motion.

3.2.3 *Selfsimilar processes and fractional Brownian motion*

This section introduces and analyzes an important property of continuous time stochastic processes called selfsimilarity. A general object is called selfsimilar if its structure remains invariant under changes of scale. Typical examples are fractals. A nonlinear differentiable function is not selfsimilar: by enlarging it over any sufficiently small interval, one finds a line, thus a different structure than the original one. The precise definition of the selfsimilar stochastic process is given. An important selfsimilar stochastic process is the Wiener process and a generalization of the Wiener process that remains selfsimilar is the fractional Brownian motion. It is shown that selfsimilarity is closely related to stationarity and precisely through the Lamperti transformations.

Selfsimilar processes

The selfsimilarity of a stochastic process is defined as follows.

Definition 3.2.15 (Selfsimilarity and h-selfsimilarity). *The stochastic process $\{X_t\}_{t \geq 0}$ is selfsimilar if, $\forall a > 0$, $\exists b > 0$ such that*

$$(X_{at_1}, \ldots, X_{at_n}) \sim (bX_{t_1}, \ldots, bX_{t_n}),$$

$\forall 0 \leq t_1 < \cdots < t_n < \infty$ *and* $n \in \mathbb{N}^*$.

If the stochastic process is selfsimilar with $b(a) = a^h$, for some $h \in \mathbb{R}$, then it is precisely called h-selfsimilar.

Examples 3.2.16 (Selfsimilar processes).

(1) Trivial process. Let $x \in \mathbb{R}$ and $X_t = x$, a.s., $\forall t \geq 0$. Then Definition 3.2.15 holds with $b = 1$, $\forall a > 0$. Any trivial process is 0-selfsimilar.

(2) Wiener process. For the Wiener process $\{W_t\}_{t \geq 0}$, we have

$$(W_{at_1}, \ldots, W_{at_n}) \sim \mathcal{N}\left(\mathbf{0}, ((at_i) \wedge (at_j))_{i,j=1,\ldots,n}\right)$$

$$\sim (a^{\frac{1}{2}} W_{t_1}, \ldots, a^{\frac{1}{2}} W_{t_n}), \tag{3.2.8}$$

$\forall a > 0$, $0 \leq t_1 < \cdots < t_n < \infty$ *and* $n \in \mathbb{N}^*$, *cf. Proposition 3.2.10. The Wiener process is $1/2$-selfsimilar. The equivalence in distribution (3.2.8) is illustrated in Figure 3.3: for $a = 1/10$, 10 sample paths of $\{W_{at}\}_{t \in [0,100]}$ are simulated in (a) and 10 sample paths of $\{a^{1/10} W_t\}_{t \in [0,100]}$ are simulated in (b). The graphs in (a) appear similar to the ones in (b).*

(3) Scaled and drifted Wiener process. Let $\mu \in \mathbb{R}$ and $\sigma > 0$. Then the scaled and drifted stochastic process $\{\mu\sqrt{t} + \sigma W_t\}_{t \geq 0}$ is $1/2$-selfsimilar.

The stochastic continuity of stochastic process, used in the next theorem, is given in Definition 3.1.34.

Theorem 3.2.17. *1. If the stochastic process $\{X_t\}_{t \geq 0}$ is stochastically continuous at $t = 0$ and selfsimilar, then it is h-selfsimilar, for some $h \geq 0$. 2. If, in addition, $\{X_t\}_{t \geq 0}$ is nontrivial with $X_0 = 0$, a.s., then $h > 0$.*

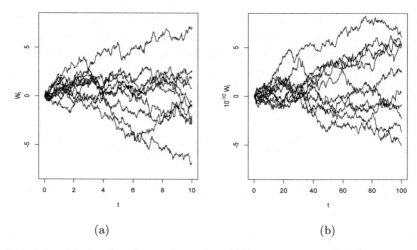

(a)   (b)

Fig. 3.3:   10 simulated sample paths of Wiener process $\{W_{at}\}_{t\in[0,100]}$, in (a), and 10 simulated sample paths of Wiener process $\{a^{1/2}W_t\}_{t\in[0,100]}$, in (b), with $a = 1/10$.

Proof. 1. The claim is obvious if the process is trivial and so we assume it nontrivial. Nontriviality tells that $X_t \neq 0$, a.s., for some $t > 0$. For that $t > 0$ and $\forall a, a' > 0$, we have

$$X_{at} \sim b(a)X_t \text{ and } X_{aa't} \sim b(aa')X_t \implies b(aa') = b(a)b(a') \implies b(a) = a^h,$$

for some $h \in \mathbb{R}$. Now let $a \in (0,1)$, then

$$X_{a^n} \sim b^n(a)X_1, \ \forall n \geq 1, \text{ and } X_{a^n} \xrightarrow{\ P\ } X_0, \text{ as } n \to \infty \implies b(a) \in (0,1], \tag{3.2.9}$$

from stochastic continuity and from the fact that convergence in probability implies convergence in distribution. We have $\forall a \in (0,1), a' > 0$ and for $a'' = aa'$,

$$b(aa') = b(a)b(a') \implies \frac{b(a'')}{b(a)} = b\left(\frac{a''}{a}\right) \implies b(a'') \leq b\left(\frac{a''}{a}\right). \tag{3.2.10}$$

Thus $b(a)$ is a nondecreasing function of a and so $h \geq 0$.
2. We obtain $b(a) \in (0,1)$ from (3.2.9) with $X_0 = 0$ a.s. Thus (3.2.10) gives that $b(a)$ is an increasing function of a and so $h > 0$. $\square$

Remark 3.2.18 (Simulation of a h-selfsimilar process). *There is a simple way of generating a sample path of the h-selfsimilar process $\{X_t\}_{t\geq0}$ over the time interval $[0,s]$, for any possibly large s, once a sample path over $[0,1]$ has been generated. The simulation algorithm is the following:*

(1) Generate $X_0, X_{1/n}, \ldots, X_1$, for some large $n \in \mathbb{N}^*$.
(2) Multiply these values by s^h in order to obtain $(s^h X_0,\ s^h X_{1/n},$
$\ldots, s^h X_1)$.
(3) Redefine the previous vector as $(X_0, X_{s/n}, \ldots, X_s)$.

Selfsimilar processes and strictly stationary processes are related by nonlinear change of time, called Lamperti transformations. We first illustrate this relation through an example, before giving the general result in Theorem 3.2.20.

Example 3.2.19. *Let $\{Y_t\}_{t \in \mathbb{R}}$ be a strictly stationary process and define*

$$X_t = \begin{cases} Y_{\log t}, & \text{if } t > 0, \\ Z, & \text{if } t = 0, \end{cases}$$

for an arbitrarily chosen random variable Z. Thus $\forall a > 0$ and $0 < t_1 < \cdots < t_n$, we have

$$(X_{at_1}, \ldots, X_{at_n}) = (Y_{\log a + \log t_1}, \ldots, Y_{\log a + \log t_n})$$
$$\sim (Y_{\log t_1}, \ldots, Y_{\log t_n})$$
$$\sim (X_{t_1}, \ldots, X_{t_n}).$$

From these equivalences in distribution we deduce directly that $\{X_t\}_{t \geq 0}$ is 0-selfsimilar.

Theorem 3.2.20 (Lamperti transformations). *1. If the stochastic process $\{Y_t\}_{t \in \mathbb{R}}$ is strictly stationary and, for some $h \in \mathbb{R}$,*

$$X_t = \begin{cases} t^h Y_{\log t}, & \text{if } t > 0, \\ 0, & \text{if } t = 0, \end{cases}$$

then $\{X_t\}_{t \geq 0}$ is h-selfsimilar.
2. If the stochastic process $\{X_t\}_{t \geq 0}$ is h-selfsimilar and

$$Y_t = e^{-ht} X_{e^t}, \quad \forall t \in \mathbb{R},$$

for some $h \in \mathbb{R}$, then $\{Y_t\}_{t \in \mathbb{R}}$ is strictly stationary.

Proof. 1. Let $0 < t_1 < \cdots < t_n < \infty$, $n \in \mathbb{N}^*$ and $a > 0$. Then,

$$(X_{at_1}, \ldots, X_{at_n}) = (a^h t_1^h Y_{\log a + \log t_1}, \ldots, a^h t_n^h Y_{\log a + \log t_n})$$
$$\sim (a^h t_1^h Y_{\log t_1}, \ldots, a^h t_n^h Y_{\log t_n})$$
$$= (a^h X_{t_1}, \ldots, a^h X_{t_n}).$$

2. Let $-\infty < t_1 < \cdots < t_n < \infty$, $n \in \mathbb{N}^*$ and $a \in \mathbb{R}$, then

$$(Y_{t_1+a}, \ldots, Y_{t_n+a}) = (\mathrm{e}^{-ha}\mathrm{e}^{-ht_1}X_{\mathrm{e}^a\mathrm{e}^{t_1}}, \ldots, \mathrm{e}^{-ha}\mathrm{e}^{-ht_n}X_{\mathrm{e}^a\mathrm{e}^{t_n}})$$
$$\sim (\mathrm{e}^{-ht_1}X_{\mathrm{e}^{t_1}}, \ldots, \mathrm{e}^{-ht_n}X_{\mathrm{e}^{t_n}})$$
$$\sim (Y_{t_1}, \ldots, Y_{t_n}).$$

$\square$

We can thus create selfsimilar processes from stationary processes and conversely. The next example gives the construction of the stationary Ornstein-Uhlenbeck process from the 1/2-selfsimilar Wiener process.

Example 3.2.21 (Ornstein-Uhlenbeck process). *Consider the 1/2-selfsimilar Wiener process* $\{W_t\}_{t\geq 0}$. *It follows from Theorem 3.2.20.2 that*

$$Y_t = \mathrm{e}^{-\frac{t}{2}}W_{\mathrm{e}^t}, \quad \forall t \in \mathbb{R},$$

is strictly stationary. Clearly, $\{Y_t\}_{t\in\mathbb{R}}$ *is Gaussian with mean zero and its a.c.v.f. is given by*

$$\gamma(h) = \mathsf{E}[Y_{t+h}Y_t] = \mathrm{e}^{-\left(t+\frac{h}{2}\right)}\mathsf{E}[W_{\mathrm{e}^{t+h}}W_{\mathrm{e}^t}] = \mathrm{e}^{-\frac{|h|}{2}}, \quad \forall h \in \mathbb{R}.$$

The stochastic process $\{\sigma Y_{\tau t/2}\}_{t\in\mathbb{R}}$, *for some* $\sigma, \tau > 0$, *has a.c.v.f. (3.5.7) and we have thus obtained the Ornstein-Uhlenbeck process, which appears in other examples, e.g. from the Langevin equation in Example 3.4.9 or in Example 3.2.7.*

Example 3.2.22 (Selfsimilar cosine process). *Let* $Y_t = b\cos(\alpha t + \theta)$, $\forall t \in \mathbb{R}$, *where* $b \in \mathbb{R}$, $\alpha \geq 0$ *and where* θ *is a random angle that is uniformly distributed over* $[-\pi, \pi)$. *Because the addition of any constant to* θ *does not alter its circular uniformity,* $\{Y_t\}_{t\in\mathbb{R}}$ *is strictly stationary. Let* $h \in \mathbb{R}$ *and*

$$X_t = \begin{cases} bt^h\cos(\alpha\log t + \theta), & \text{if } t > 0, \\ 0, & \text{if } t = 0. \end{cases}$$

It follows from the Lamperti transformation of Theorem 3.2.20.1 that $\{X_t\}_{t\geq 0}$ *is h-selfsimilar.*

Fractional Brownian motion

An important h-selfsimilar stochastic process that generalizes the Wiener process in various ways is the fractional Brownian motion, which is given in Definition 3.2.24. It was introduced by Mandelbrot and van Ness (1968) and it has been widely applied in statistical studies that exhibit selfsimilarity and long range dependence, for example in telecommunication. We start the presentation with the following related result.

Proposition 3.2.23. *Let $\{X_t\}_{t\geq 0}$ be a h-selfsimilar stochastic process, for some $h > 0$, with stationary increments and satisfying $\mathsf{E}[X_1^2] < \infty$. Then*

$$\mathsf{E}[X_s X_t] = \frac{\mathsf{E}[X_1^2]}{2}\left(s^{2h} + t^{2h} - |s - t|^{2h}\right), \quad \forall s, t \geq 0.$$

Proof. Let $s, t \geq 0$. It follows from $h > 0$ that $X_0 = 0$, a.s. Thus $\mathsf{E}[(X_s - X_t)^2] = \mathsf{E}[X_{|s-t|}^2]$. Consequently,

$$\mathsf{E}[X_s X_t] = \frac{1}{2}\left(\mathsf{E}[X_s^2] + \mathsf{E}[X_t^2] - \mathsf{E}[(X_s - X_t)^2]\right)$$

$$= \frac{1}{2}\left(\mathsf{E}[X_s^2] + \mathsf{E}[X_t^2] - \mathsf{E}\left[X_{|s-t|}^2\right]\right).$$

It follows from selfsimilarity that this is equal to the desired result. $\square$

Definition 3.2.24 (Fractional Brownian motion). *The real-valued stochastic process $\{B_t\}_{t\geq 0}$ is a fractional Brownian motion if:*

(1) it has continuous sample paths;
(2) it is Gaussian;
(3) for some index $h \in (0, 1]$,

$$\mathsf{E}[B_t] = 0 \quad and \quad \mathsf{E}[B_s B_t] = \frac{\mathsf{E}[B_1^2]}{2}\left(s^{2h} + t^{2h} - |s - t|^{2h}\right), \quad \forall s, t \geq 0.$$

Remarks 3.2.25. *(1) We have in particular $\mathsf{E}[B_0^2] = 0$, so that $B_0 = 0$ a.s.*

(2) We can easily control that the fractional Brownian motion is h-selfsimilar.

(3) If we consider $h = 1/2$ in Definition 3.2.24, then we find a rescaled Wiener process. Indeed, let $s, t \geq 0$, then

$$\mathrm{cov}(B_s, B_t) = \mathsf{E}[B_s B_t] = \mathsf{E}[B_1^2](s \wedge t).$$

(4) The variance of the fractional Brownian motion, which is given by

$$\mathrm{var}(B_t) = \mathsf{E}[B_1^2]t^{2h}$$

increases w.r.t. t, $\forall t \geq 0$, at $\begin{cases} slower\ than \\ exactly \\ faster\ than \end{cases}$ linear rate iff

$$h \begin{cases} \in (0, 1/2), \\ = 1/2, \\ \in (1/2, 1). \end{cases}$$

Proposition 3.2.26 (Stationarity of increments of fractional Brownian motion). *1. The fractional Brownian motion has stationary increments.*
2. The fractional Brownian motion has independent increments iff it has selfsimilarity index $h = 1/2$.

Proof. 1. Denote by $\{B_t\}_{t \geq 0}$ the fractional Brownian motion with selfsimilarity index $h \in (0, 1]$. Let $s, t \geq 0$, then $\mathrm{var}(B_{t+s} - B_t) = s^{2h} \mathsf{E}[B_1^2]$. Therefore, given that the process is Gaussian, the distribution of the increment $B_{t+s} - B_t$ depends only on the length of the time interval, namely on s.

2. As noted above, if $h = 1/2$, then $\{B_t\}_{t \geq 0}$ is a rescaled Wiener process and thus it has independent increments. Conversely, if $\{B_t\}_{t \geq 0}$ has independent increments, then, $\forall 0 \leq s \leq t$,

$$\begin{aligned}
0 &= \mathsf{E}[B_s(B_t - B_s)] \\
&= \mathsf{E}[B_s B_t] - \mathsf{E}\left[B_s^2\right] \\
&= \frac{1}{2}\mathsf{E}\left[B_1^2\right]\{s^{2h} + t^{2h} - (t-s)^{2h} - 2s^{2h}\} \\
&= \frac{1}{2}\mathsf{E}\left[B_1^2\right]\{t^{2h} - s^{2h} - (t-s)^{2h}\}.
\end{aligned}$$

This implies $h = 1/2$. In order to show this, we note that the function $f(x) = x^{2h}$ is strictly $\begin{cases} \text{concave, if } h \in (0, 1/2), \\ \text{convex, if } h \in (1/2, 1], \end{cases}$ $\forall x \geq 0$. Thus, $\forall 0 \leq s < t$,

$$f(t) - f(s) \begin{cases} < \\ > \end{cases} f(t-s), \text{ if } h \in \begin{cases} (0, 1/2), \\ (1/2, 1]. \end{cases} \qquad \square$$

Proposition 3.2.27 (Correlation of increments of fractional Brownian motion). *The fractional Brownian motion with index h has* $\begin{cases} negatively \\ positively \end{cases}$ *correlated increments if* $\begin{cases} h \in (0, 1/2), \\ h \in (1/2, 1]. \end{cases}$

Proof. Denote by $\{B_t\}_{t \geq 0}$ the fractional Brownian motion with index h, consider $n \in \mathbb{N}^*$ and $0 \leq s_1 \leq t_1 < s_2 \leq t_2 < \cdots < s_n \leq t_n$. Then, $\forall j, k \in \{1, \dots, n\}$ with $j < k$,

$$\begin{aligned}
\mathsf{E}[(B_{t_j} &- B_{s_j})(B_{t_k} - B_{s_k})] \\
&= \left(\mathsf{E}[B_{t_j} B_{t_k}] - \mathsf{E}[B_{t_j} B_{s_k}] - \mathsf{E}[B_{s_j} B_{t_k}] + \mathsf{E}[B_{s_j} B_{s_k}]\right) \\
&= \frac{\mathsf{E}\left[B_1^2\right]}{2}\{f(t_k - s_j) - f(t_k - t_j) - f(s_k - s_j) + f(s_k - t_j)\},
\end{aligned}$$

where $f(x) = x^{2h}$, $\forall x \geq 0$. Thus

$$E[(B_{t_j} - B_{s_j})(B_{t_k} - B_{s_k})] \begin{cases} < \\ > \end{cases} 0 \Longleftrightarrow$$

$$f(t_k - s_j) - f(s_k - s_j) \begin{cases} < \\ > \end{cases} f(t_k - t_j) - f(s_k - t_j) \Longleftarrow$$

$$f \text{ is strictly} \begin{cases} \text{concave,} \\ \text{convex.} \end{cases}$$

This last analysis together with the fact that f is strictly $\begin{cases} \text{concave, if } h \in (0, 1/2), \\ \text{convex, if } h \in (1/2, 1], \end{cases}$ give the desired result. □

Remark 3.2.28 (Long range dependence of fractional Brownian motion). *Thus the fractional Brownian motion is a practical stochastic model for certain phenomena with long range dependence or long memory when* $h \in (1/2, 1]$. *Clearly, the closer is h to one, the more convex becomes the function f, in the proof of Proposition 3.2.27. Thus the covariance between the increments increases as the index h approaches one. Consequently the process inherits longer range dependence. Stationary processes with long range dependence are analyzed at the end of Section 3.3, see in particular Table 3.1, and in Section 4.3.*

3.2.4 *Counting processes*

This section presents the following classes of counting processes: the inhomogeneous birth processes, the subclass of the Poisson processes, the processes with stationary stream and the regular processes.

Inhomogeneous birth processes

We introduced the counting process in Definition 3.1.13 and the Markov process in Definition 3.1.11. Birth processes represent an important class of Markovian counting processes to which the notorious Poisson process belongs. Their definition is the following.

Definition 3.2.29 (Birth processes). *The Markovian counting process* $\{N_t\}_{t \geq 0}$ *is a inhomogeneous birth process, if the transition probabilities take* *the following form,*

$$p_{k,k+n}(t, t+h) = \mathsf{P}[N_{t+h} - N_t = n | N_t = k]$$

$$= \begin{cases} 1 - \lambda_k(t)h + o(h), & \text{if } n = 0, \\ \lambda_k(t)h + o(h), & \text{if } n = 1, \\ o(h), & \text{if } n > 1, \end{cases}$$

$\forall t > 0$, *for* $k = 0, 1, \ldots$ *and as* $h \to 0$, *where*

$$\lambda_k : \mathbb{R}_+ \to \mathbb{R}_+$$

$$t \mapsto \lambda_k(t), \tag{3.2.11}$$

is nonnegative and continuous, for $k = 0, 1, \ldots$. *The functions (3.2.11) are called transition intensity functions.*

An important result from pp. 59–60 of Grandell (1997) is the following. Each class of nonnegative and continuous functions (3.2.11) that satisfy the condition

$$\sum_{k=n}^{\infty} \left(\max_{0 \le s < t} \lambda_k(s) \right)^{-1} = \infty, \quad \forall t > 0, \ n \in \mathbb{N}, \tag{3.2.12}$$

is indeed the class of transition intensity functions of some counting process, because (3.2.12) guarantees that $\sum_{n=0}^{\infty} p_{k,k+n}(s, t) = 1$, $\forall 0 \le s \le t < \infty$ and $k = 0, 1, \ldots$. Because the sum in (3.2.12) decreases with increasing values of t or of n, we understand that the divergence in (3.2.12) is required for arbitrarily large values of t and n. Consequently, condition (3.2.12) controls the growth of the transition intensity functions.

Note that we have

$$\lambda_k(t) = \lim_{h \to 0} \frac{p_{k,k+1}(t, t+h)}{h}, \quad \text{for } k = 0, 1, \ldots \text{ and } \forall t \ge 0.$$

When $\lambda_0(t), \lambda_1(t), \ldots$ are independent of t, $\forall t \ge 0$, we obtain a homogeneous birth process. When $\lambda_0(t) = \lambda_1(t) = \ldots, \forall t \ge 0$, then we have a process with independent increments.

Theorem 3.2.30 allows for the computation of the transition probabilities from the transition intensity functions.

Theorem 3.2.30 (Recursive formulae for the transition probabilities). *The transition probabilities* $p_{k,k+n}(s, t)$, $\forall 0 \le s \le t$ *and for* $k, n = 0, 1, \ldots$, *of an inhomogeneous birth process with initial values*

$$p_{k,k}(s, s) = 1 \text{ and } p_{k,k+n}(s, s) = 0, \ \forall s \ge 0, \text{ for } k = 0, 1, \ldots, \ n = 1, 2, \ldots,$$

can be obtained by the following recursive formulae,

$$p_{k,k}(s,t) = \exp\left\{ - \int_s^t \lambda_k(x)\mathrm{d}x \right\},$$

$$p_{k,k+n}(s,t) = \int_s^t \lambda_{k+n-1}(y)p_{k,k+n-1}(s,y)\exp\left\{ - \int_y^t \lambda_{k+n}(x)\mathrm{d}x \right\}\mathrm{d}y,$$

$\forall 0 \le s \le t$ *and for* $k = 0, 1, \dots$ *and* $n = 1, 2, \dots$.

Poisson processes

The Poisson process is the simplest and also the most important counting process. It represents the WN of counting processes. It is a particular birth process and so its transition probabilities can be obtained by Theorem 3.2.30.

Let $\lambda > 0$ be a constant. The homogeneous Poisson process is the birth process obtained by setting $\lambda_0(t) = \lambda_1(t) = \cdots = \lambda$, $\forall t \ge 0$. The constant λ is called intensity of rate of (occurrences) of the process.

Theorem 3.2.31 (Transition probabilities of homogeneous Poisson process). *The transition probabilities of the homogeneous Poisson process with intensity* $\lambda > 0$ *are given by*

$$p_{k,k+n}(s,t) = \mathrm{e}^{-\lambda(t-s)}\frac{\{\lambda(t-s)\}^n}{n!},$$

$\forall 0 \le s \le t$, $k, n = 0, 1, \dots$. *In particular,*

$$p_n(t) = \mathrm{e}^{-\lambda t}\frac{(\lambda t)^n}{n!}, \quad \forall t > 0 \text{ and for } n = 0, 1, \dots.$$

Proof. We first show the formula for transition probabilities with the help of Theorem 3.2.30. It is trivial for $n = 0$ and, assuming it hold for $n - 1$, $n \ge 1$, then, $\forall 0 \le s \le t$, we have

$$\begin{aligned}
p_{k,k+n}(s,t) &= \int_s^t \lambda \mathrm{e}^{-\lambda(y-s)}\frac{\{\lambda(y-s)\}^{n-1}}{(n-1)!}\exp\left\{ - \int_y^t \lambda \mathrm{d}x \right\}\mathrm{d}y \\
&= \frac{\lambda^n \mathrm{e}^{-\lambda(t-s)}}{(n-1)!}\int_s^y (y-s)^{n-1}\mathrm{d}y \\
&= \frac{\lambda^n \mathrm{e}^{-\lambda(t-s)}}{(n-1)!}\frac{(t-s)^n}{n}.
\end{aligned}$$

Then we have

$$p_{k,k+n}(s,t) = p_{0,n}(s,t) = p_{0,n}(0, t-s) = p_n(t-s),$$

$\forall 0 \le s \le t$, $k, n = 0, 1, \dots$. $\qquad\square$

We define the times of occurrences of the homogeneous Poisson process $\{N_t\}_{t \geq 0}$ by

$$T_n = \inf\left\{t \geq 0 \mid N_t \geqslant n\right\}, \quad \text{for } n = 0, 1, \ldots.$$

We can conversely obtain the process $\{T_n\}_{n \in \mathbb{N}}$ from the counting process as follows,

$$N_t = \max\left\{n \geq 0 \mid T_n \leqslant t\right\}, \quad \forall t \geq 0.$$

We define the times between consecutive occurrences by

$$D_n = T_n - T_{n-1}, \quad \text{for } n = 1, 2, \ldots.$$

The three stochastic processes $\{N_t\}_{t \geqslant 0}$, $\{T_n\}_{n \in \mathbb{N}}$ and $\{D_n\}_{n \in \mathbb{N}^*}$ are three representations of the same Poisson process. It follows from Proposition 3.2.32 below that we have a renewal process in the sense of Definition 3.1.14.

Proposition 3.2.32 (Interoccurrence times of homogeneous Poisson process). *Let $\{N_t\}_{t \geqslant 0}$ be a homogeneous Poisson process with intensity $\lambda > 0$ and inter-occurrence times $\{D_n\}_{n \in \mathbb{N}}$. Then $D_1, D_2, \ldots$ are independent and exponentially distributed random variables, with same parameter λ.*

Proof. Let $t, t_1 > 0$, then

$$\mathsf{P}[D_1 > t] = \mathsf{P}[N_t = 0] = \mathrm{e}^{-\lambda t} \qquad \text{and}$$

$$\begin{aligned}
\mathsf{P}[D_2 > t \mid D_1 = t_1] &= \mathsf{P}[N_{t_1+t} - N_{t_1} = 0 \mid N_{t_1} = 1] \\
&= \mathsf{P}[N_{t_1+t} - N_{t_1} = 0] \\
&= \mathsf{P}[N_t = 0] = \mathrm{e}^{-\lambda t}.
\end{aligned}$$

Thus D_1 and D_2 are i.i.d. Similarly, for $n \geqslant 1$ and $t, d_1, \ldots, d_n > 0$,

$$\begin{aligned}
\mathsf{P}[D_{n+1} > t \mid D_1 = d_1, \ldots, D_n = d_n] &= \mathsf{P}[N_{t_n+t} - N_{t_n} = 0 \mid N_{t_n} = n] \\
&= \mathsf{P}[N_{t_n+t} - N_{t_n} = 0] \\
&= \mathsf{P}[N_t = 0] = \mathrm{e}^{-\lambda t},
\end{aligned}$$

where $t_n = d_1 + \cdots + d_n$. □

The inter-occurrence times entirely determine the Poisson process and so this proposition is indeed a characterization of the homogeneous Poisson process. Proposition 3.2.32 can be generalized to the inhomogeneous Poisson process; see e.g. Gatto (2020), pp. 80–82, where it is shown that the interoccurrence times are i.i.d. iff the Poisson process is homogeneous.

The homogeneous Poisson process can used for constructing various stationary processes. A famous illustration is the random telegraph signal that is presented in Example 3.1.18. Two further stationary processes obtained from the Poisson process are the Poisson process over the circle, cf. Example 3.2.33, and the process of the increments of the Poisson process, cf. Example 3.2.34.

Example 3.2.33 (Wrapped Poisson process). *Let $\{N_t\}_{t \in \mathbb{R}_+}$ be an homogeneous Poisson process with parameter $\lambda > 0$ and let the random variable J be independent and uniformly distributed over $\{1, \ldots, m\}$, for some integer $m \geq 2$. Denote by $\theta_m = \exp\{i2\pi/m\}$ the m-th root of 1. The following process makes jumps around the unit circle $\mathcal{A}_1$: it starts at a random angle, from which it makes counter-clockwise rotations of angle $2\pi/m$, at random times following the Poisson process. It is thus given by*

$$X_t = \theta_m^{J+N_t}, \quad \forall t \geq 0.$$

Let $t, h \in \mathbb{R}_+$. It is direct to verify that $\mathsf{E}\left[\theta_m^J\right] = 0$. Consequently, $\mathsf{E}[X_t] = 0$. Further, one finds

$$\mathsf{cov}(X_{t+h}, X_t) = \mathsf{E}\left[X_{t+h}\overline{X_t}\right] = \mathsf{E}\left[\theta_m^{N_{t+h}}\theta_m^{-N_t}\right]$$
$$= \mathsf{E}\left[\theta_m^{N_h}\right] = \exp\left\{\lambda h(\theta_m - 1)\right\}.$$

Consequently, $\{X_t\}_{t \in \mathbb{R}_+}$ is stationary with a.c.v.f. given by

$$\gamma(h) = \exp\left\{\lambda|h|(\theta_m - 1)\right\}, \quad \forall h \in \mathbb{R}.$$

Such a stochastic process is useful for modelling the planar trajectory of a particle that moves forward and changes direction at random times. Another important application would be for modelling directions of winds.

Planar directions can be represented by angles instead of unit vectors. The angular representation of $\{X_t\}_{t \in \mathbb{R}_+}$ is given by

$$\alpha_t = \frac{2\pi}{m}(J + N_t) \bmod 2\pi, \quad \forall t \geq 0.$$

We call any one of the two representations $\{X_t\}_{t \in \mathbb{R}_+}$ and $\{\alpha_t\}_{t \in \mathbb{R}_+}$ wrapped Poisson process, because they arise from wrapping the Poisson process around the unit circle.

Example 3.2.34 (Increments of Poisson process). *Let $\{N_t\}_{t \in \mathbb{R}_+}$ be an homogeneous Poisson process with parameter $\lambda > 0$. Define the process of increments of length $l > 0$ as*

$$X_t = N_{t+l} - N_t, \quad \forall t \in \mathbb{R}_+.$$

Let $t, h \in \mathbb{R}_+$. It is simple to verify that $\mathsf{E}[X_t] = \lambda l$ and that

$$\operatorname{cov}(X_{t+h}, X_t) = \begin{cases} \lambda(l - h), & \text{if } h < l, \\ 0, & \text{if } h \geq l. \end{cases}$$

From this last result we deduce that $\{X_t\}_{t \in \mathbb{R}_+}$ is stationary with a.c.v.f.

$$\gamma(h) = \lambda(l - |h|)\mathsf{I}\{|h| < l\}, \quad \forall h \in \mathbb{R}.$$

We can extend the homogeneous Poisson process from time domain $\mathbb{R}_+$ to time domain $\mathbb{R}$ by following the same principle used for the Wiener process in Proposition 3.2.13.

Definition 3.2.35 (Poisson process with time domain $\mathbb{R}$). *Let $\{N_t^{(1)}\}_{t \in \mathbb{R}_+}$ and $\{N_t^{(2)}\}_{t \in \mathbb{R}_+}$ be two independent homogeneous Poisson processes with same intensity $\lambda > 0$. Then*

$$N_t = \begin{cases} N_t^{(1)}, & \text{if } t \geq 0, \\ -N_{(-t)-}^{(2)}, & \text{if } t < 0, \end{cases}$$

determines a homogeneous Poisson process with time domain $\mathbb{R}$ and intensity λ.

Denote by $\{T_n\}_{n \geq 1}$ the occurrence times of $\{N_t^{(1)}\}_{t \in \mathbb{R}_+}$ and by $\{-T_{-n}\}_{n \geq 0}$ the occurrence times of $\{N_t^{(2)}\}_{t \in \mathbb{R}_+}$. Then $\{T_n\}_{n \in \mathbb{Z}}$ are the occurrence times of $\{N_t\}_{t \in \mathbb{R}}$.

Remarks 3.2.36. *(1) The Poisson process over $\mathbb{R}$ has $T_0 < 0$ a.s.*
(2) The particular transform of $\{N_t^{(2)}\}_{t \in \mathbb{R}_+}$ in Definition 3.2.35 makes any sample path of $\{N_t\}_{t \in \mathbb{R}}$ a cadlag function.
(3) Let $t > 0$, then $N_t^{(1)}$ is the number of the occurrences in $[0, t]$ and $N_t^{(2)}$ is the number of occurrences in $[-t, 0]$. Let $s < t$, then $N_t - N_s$ is the number of occurrences in $(s, t]$. If $s > 0$ and $t \geq 0$, then $N_t - N_{-s} = N_t^{(1)} + N_{s-}^{(2)}$. If $0 < s < t$, then $N_{-s} - N_{-t} = N_{t-}^{(2)} - N_{s-}^{(2)}$.
(4) The particular definition of the occurrence times implies that all but one of the interoccurrence times are exponentially distributed with parameter λ, the exception being $T_1 - T_0 \sim \text{Gamma}(2, \lambda)$. Indeed, T_1 and $-T_0$ are independent and Exponential(λ) distributed.
(5) Definition 3.1.13 of the general counting process can be extended to $T = \mathbb{R}$ by allowing the process to be $\mathbb{Z}$-valued, instead of $\mathbb{N}$-valued.

The case $\lambda_0(t) = \lambda_1(t) = \cdots = \lambda(t)$, $\forall t \geq 0$, yields the inhomogeneous Poisson process. The function $\lambda : \mathbb{R}_+ \to \mathbb{R}_+$ is called the intensity function of the inhomogeneous Poisson process. The increments are no longer stationary, but are still independent. Theorem 3.2.37 provides the transition probabilities of the inhomogeneous Poisson process. It follows from the recursive formulae of Theorem 3.2.30.

Theorem 3.2.37 (Transition probabilities of inhomogeneous Poisson process). *The inhomogeneous Poisson process with intensity function* $\lambda : \mathbb{R}_+ \to \mathbb{R}_+$ *has the transition probabilities*

$$p_{0,n}(s,t) = p_{k,k+n}(s,t) = \exp\left\{-\int_s^t \lambda(x)\mathrm{d}x\right\} \frac{\left\{\int_s^t \lambda(x)\mathrm{d}x\right\}^n}{n!}, \quad (3.2.13)$$

$\forall 0 \le s \le t$ *and for* $k, n = 0, 1, \ldots$.

Processes with stationary stream and regular processes

We now consider counting processes obtained from a generalization of the homogeneous Poisson process over $\mathbb{R}$. In particular, these processes can have dependent increments. This generalization includes for example renewal processes. It requires the following alternative version of the stationarity of increments, which was proposed by Khintchine (1960); see also p. 53 of Cramér and Leadbetter (1967).

Definition 3.2.38 (Stationary stream and regular process). *Let* $\{N_t\}_{t \in \mathbb{R}}$ *be a counting process and denote* $N((s,t]) = N_t - N_s$, $\forall s < t \in \mathbb{R}$.

(1) *The counting process* $\{N_t\}_{t \in \mathbb{R}}$ *possesses a stationary stream of events or occurrences if,* $\forall k \in \mathbb{N}^*$, $\forall s_j < t_j \in \mathbb{R}$, *for* $j = 1, \ldots, k$, *and* $\forall h > 0$,

$$(N((s_1, t_1]), \ldots, N((s_k, t_k]))$$
$$\sim (N((s_1 + h, t_1 + h]), \ldots, N((s_k + h, t_k + h])).$$

(2) *When* $\{N_t\}_{t \in \mathbb{R}}$ *possesses a stationary stream of events, it is regular (or orderly) if,* $\forall t \in \mathbb{R}$,

$$\mathsf{P}[N((0, h]) \ge 2] = \mathrm{o}(h), \quad \text{as } h \to 0.$$

Definition 3.2.38.1 with $k = 1$ corresponds indeed to stationarity of increments of Definition 3.1.9. In fact, if $\{N_t\}_{t \in \mathbb{R}}$ would have independent increments, then Definition 3.2.38.1 could be simplified to $k = 1$.

Whenever the counting process $\{N_t\}_{t \in \mathbb{R}}$ possesses stationary increments, it holds that

$$\mathsf{P}[N((0, h]) \ge 1] = \lambda h + \mathrm{o}(h), \quad \text{as } h \to 0, \quad (3.2.14)$$

for some $\lambda \in [0, \infty]$.

In order to show (3.2.14), we use the following result from Khintchine (1960), Section 7. Let $a > 0$ and $g : (0, a] \to \mathbb{R}_+$ be nondecreasing and such that, whenever $x, y, x + y \in (0, a)$, the inequality $g(x + y) \leq g(x) + g(y)$ holds. Then, $g(x)/x \xrightarrow{x \to 0} \lambda$, for some $\lambda \in [0, \infty]$, with $\lambda = 0$ iff $g(a) = 0$.

The function $g(t) = \mathsf{P}[N((0, t]) \geq 1]$, $\forall t \geq 0$, is nondecreasing and $g(t_1 + t_2) \leq g(t_1) + g(t_2)$, $\forall t_1, t_2 \geq 0$. Indeed, an occurrence over $(0, t_1 + t_2]$ implies an occurrence over $(0, t_1]$ or an occurrence over $(t_1, t_1 + t_2]$, the increments being stationary. Thus $g(t)/t \xrightarrow{t \to 0} \lambda$, for some $\lambda \in [0, \infty]$. Note also that $\lambda = 0$ iff $\mathsf{P}[N((0, t]) \geq 1] = 0$, $\forall t \geq 0$, i.e. no occurrences a.s.

Assume that the counting process $\{N_t\}_{t \in \mathbb{R}}$ has stationary stream and that it is regular. Then (3.2.14) holds and $\lambda \in [0, \infty]$ is called intensity. For $n = 2, 3, \ldots$, we have $\mathsf{P}[N((0, h]) \geq n] \leq \mathsf{P}[N((0, h]) \geq 2] = \mathrm{o}(h)$, as $h \to 0$. This result and (3.2.14) give

$$\mathsf{E}[N((0, h])] = \sum_{n=1}^{\infty} \mathsf{P}[N((0, h]) \geq n]$$

$$= \lambda h + \mathrm{o}(h) + \underbrace{\sum_{n=2}^{\infty} \mathsf{P}[N((0, h]) \geq n]}_{\substack{=\mathrm{o}(h), \\ \text{uniformly w.r.t. } n}}$$

$$= \lambda h + \mathrm{o}(h), \quad \text{as } h \to 0. \tag{3.2.15}$$

So we have, for small values of h, $\lambda \simeq \mathsf{E}[N((0, h])]/h$.

To any counting process with stationary stream of events, we define its covariance intensity function as follows.

Definition 3.2.39 (Covariance intensity function). *Let $\{N_t\}_{t \in \mathbb{R}}$ be a counting process with stationary stream of events. The function*

$$w(s) = \lim_{h \downarrow 0} \frac{\mathrm{cov}(N((t, t + h]), N((t + s, t + s + h]))}{h^2}$$

$$= \lim_{h \downarrow 0} \frac{\mathrm{cov}(N_{t+h} - N_t, N_{t+s+h} - N_{t+s})}{h^2}, \quad \forall s \neq 0, t \in \mathbb{R},$$

is called covariance intensity function, whenever the limit exists.

Note that the covariance intensity function is symmetric, i.e. $w(-s) = w(s)$, $\forall s \neq 0$.

For a counting process $\{N_t\}_{t \in \mathbb{R}}$ with stationary stream and that is regular, the following symbolic notation is suggested by (3.2.15): $\forall s \neq 0, t \in \mathbb{R}$,

$$\lambda = \frac{\mathsf{E}[\mathrm{d}N_t]}{\mathrm{d}t} \quad \text{and} \quad w(s) = \frac{\mathsf{E}[\mathrm{d}N_t \mathrm{d}N_{t+s}]}{(\mathrm{d}t)^2} - \lambda^2. \tag{3.2.16}$$

The computation (3.2.15) can be adapted in order to show

$$\mathsf{E}[\mathrm{d}N_t] = \mathsf{E}\left[(\mathrm{d}N_t)^2\right], \quad \forall t \in \mathbb{R}.$$

Thus, if one would define $w(0)$ by setting $s = 0$ in (3.2.16), then $w(0)$ would have higher order than any $w(s)$ with $s \neq 0$. Following this argument, one defines the complete covariance intensity function by

$$\tilde{w}(s) = \lambda\delta(s) + w(s), \quad \forall s \in \mathbb{R}, \tag{3.2.17}$$

in which $w(0) = \lim_{h \to 0} w(h)$ and where δ is Dirac's δ function; cf. Section A.8.2.

The homogeneous Poisson process has stationary stream of events, it is regular and it satisfies $w(s) = 0$, $\forall s \neq 0$.

3.2.5 *Compound and shot noise processes*

Counting processes can be used for the construction of compound processes and of shot noise processes. These processes are introduced in this section.

Compound processes

A compound process is a stochastic process that takes the form of a random sum, precisely of a sum where the number of summands is a counting process. Consider the counting process $\{N_t\}_{t \geq 0}$ and the i.i.d. random variables $X_1, X_2, \ldots$, then

$$Z_t = \sum_{k=0}^{N_t} X_k, \quad \forall t \geq 0, \tag{3.2.18}$$

where $X_0 = 0$, is a compound process. Just like sample paths of counting processes, the sample paths of the compound process possess only jumps, but the magnitudes of the jumps are now stochastic. The sample paths are cadlag functions and the discontinuity points are given by the occurrence times of the counting process. In what follows, we assume counting process and summands independent, although this is generally not required. Let $t \geq 0$. Expectation and variance of Z_t are given by

$$\mathsf{E}[Z_t] = \mathsf{E}[\mathsf{E}[Z_t|N_t]] = \mathsf{E}[N_t\mathsf{E}[X_1]] = \mathsf{E}[N_t]\mathsf{E}[X_1],$$

and

$$\begin{aligned}
\mathrm{var}(Z_t) &= \mathsf{E}[\mathrm{var}(Z_t|N_t)] + \mathrm{var}(\mathsf{E}[Z_t|N_t]) \\
&= \mathsf{E}[N_t\mathrm{var}(X_1)] + \mathrm{var}(N_t\mathsf{E}[X_1]) \\
&= \mathsf{E}[N_t]\mathrm{var}(X_1) + \mathrm{var}(N_t)\mathsf{E}^2[X_1].
\end{aligned}$$

The moment generating function of Z_t is

$$M_{Z_t}(v) = \mathsf{E}\left[e^{vZ_t}\right] = \mathsf{E}\left[\mathsf{E}[e^{vZ_t}|N_t]\right] = \mathsf{E}\left[\left(\mathsf{E}\left[e^{vX_1}\right]\right)^{N_t}\right]$$

$$= \mathsf{E}\left[\exp\left\{\log \mathsf{E}\left[e^{vX_1}\right] N_t\right\}\right] = M_{N_t}(\log M_X(v)), \qquad (3.2.19)$$

where $M_X(v) = \mathsf{E}[e^{vX_1}]$ and $M_{N_t}(v) = \mathsf{E}[e^{vN_t}]$ are the moment generating functions of X_1 and N_t and where $v \in \mathbb{R}$ belongs to the domains of definition of these moment generating functions.

We end with a stationary compound Poisson process over the circle.

Example 3.2.40 (Wrapped compound Poisson process). *We extend the wrapped Poisson process over the circle of Example 3.2.33 to the compound Poisson. Let $\{N_t\}_{t\in\mathbb{R}_+}$ be an homogeneous Poisson process with parameter $\lambda > 0$ and consider the compound Poisson process (3.2.18), where the sizes of the jumps $X_1, X_2, \ldots$ are independent of $\{N_t\}_{t\in\mathbb{R}_+}$. Let the random variable J be independent of all previous random elements and uniformly distributed over $\{1, \ldots, m\}$, for some integer $m \geq 2$. Denote the m-th root of 1 by $\theta_m = \exp\{\mathrm{i}2\pi/m\}$. We consider the wrapped compound Poisson process*

$$Y_t = \theta_m^{J+Z_t}, \quad \forall t \geq 0.$$

Let $0 < T_1 < T_2 < \cdots$ be the occurrence times of the Poisson process. The following process makes jumps around the unit circle $\mathcal{A}_1$: it starts at random angle $2\pi J/m$, from which it makes a counter-clockwise rotation of angle $2\pi X_j/m \bmod 2\pi$ at time T_j, for $j = 1, 2, \ldots$.

Let $t, h \in \mathbb{R}_+$. One verifies directly that $\mathsf{E}[Y_t] = 0$. Then, one computes

$$\mathrm{cov}(Y_{t+h}, Y_t) = \mathsf{E}\left[Y_{t+h}\overline{Y_t}\right] = \mathsf{E}\left[\theta_m^{Z_{t+h}}\theta_m^{-Z_t}\right]$$

$$= \mathsf{E}\left[\theta_m^{Z_h}\right] = \mathsf{E}\left[\exp\left\{\log\theta_m Z_t\right\}\right]$$

$$= M_{Z_t}(\log\theta_m) = M_{N_t}(\log M_X(\log\theta_m))$$

$$= \exp\left\{\lambda h\left[M_X(\log\theta_m) - 1\right]\right\}$$

$$= \exp\left\{\lambda h\left[M_X\left(\mathrm{i}\frac{2\pi}{m}\right) - 1\right]\right\},$$

by using the notation of (3.2.19). We have thus obtained that $\{Y_t\}_{t\in\mathbb{R}_+}$ is stationary with a.c.v.f.

$$\gamma(h) = \exp\left\{\lambda|h|\left[\varphi_X\left(\frac{2\pi}{m}\right) - 1\right]\right\}, \quad \forall h \in \mathbb{R},$$

where φ_X denotes the characteristic function of X_1. This process can represent the planar trajectory of a particle moving forward and changing direction at Poisson times.

The angular representation of $\{Y_t\}_{t\in\mathbb{R}_+}$ is given by

$$\alpha_t = \frac{2\pi}{m}(J + Z_t) \bmod 2\pi, \quad \forall t \geq 0.$$

The processes $\{Y_t\}_{t\in\mathbb{R}_+}$ and $\{\alpha_t\}_{t\in\mathbb{R}_+}$ are called wrapped compound Poisson process.

Shot noise process

The shot noise process was originally proposed for the study of random fluctuations of current in vacuum tubes. It is commonly used in electronics, optics and telecommunication, for example. It is a particular compound process and we consider the case where the counting process is Poisson.

Definition 3.2.41 (Shot noise process). *Let $\{N_t\}_{t\geq 0}$ be a Poisson process with occurrence times $0 < T_1 < T_2 < \ldots$, a.s. For a given measurable function $\psi : \mathbb{R} \to \mathbb{R}$ that is nonincreasing and null at negative arguments, the shot noise Poisson process is given by*

$$Z_t = \sum_{j=1}^{N_t} \psi(t - T_j), \text{ over } \{N_t \geqslant 1\}, \forall t > 0. \tag{3.2.20}$$

One can illustrate this process with electrical current inside a wire. The current is a stream of moving charge carriers that are electrons or another type of particle, depending on the conductor. Let us suppose that they are electrons and that at a given extremity of the wire they deliver instantaneously current at times $T_1, T_2, \ldots$. As soon a delivered, the current intensity decays during time following some function ψ, which is determined from physical properties of the conductor.

Note that this shot noise process can be re-expressed in terms of the following Riemann-Stieltjes stochastic integral,

$$Z_t = \sum_{j=1}^{\infty} \psi(t - T_j) = \int_{(0,\infty)} \psi(t - s) \mathrm{d}N_s, \quad \forall t > 0.$$

This basic shot noise model can be generalized to allow for jumps of random amplitude as follows,

$$Z_t = \sum_{j=1}^{N_t} X_j \, \psi(t - T_j), \text{ over } \{N_t \geqslant 1\}, \forall t > 0,$$

where $X_1, X_2, \ldots$ are i.i.d. and independent of the epochs $T_1, T_2, \ldots$.

Examples 3.2.42 (Insurers total claim amounts). *Let $t > 0$ and assume that $X_1, X_2, \ldots > 0$ represent individual claim amounts received by an insurance.*

(1) *For $\psi(s) = 1\{s \geqslant 0\}$, $\forall s \geq 0$, one simply obtains the total claim amount at time t, i.e. $Z_t = \sum_{j=1}^{N_t} X_j$, over $\{N_t \geq 1\}$ and $\forall t > 0$.*

(2) *A nondecreasing function ψ with $\psi(s) \overset{s \to \infty}{\longrightarrow} 1$ represents delayed settlement of the individual claim amounts from the insurer.*

(3) *If, starting from time T_j, the individual claim amount X_j is subject to the interest rate $r \in \mathbb{R}$, for $j = 1, 2 \ldots$, then*

$$Z_t = \sum_{j=1}^{N_t} e^{r(t-T_j)} X_j, \quad \text{over } \{N_t \geq 1\}, \tag{3.2.21}$$

is the total claim amount at time $t > 0$ compound with the interest rate r.

Although the distribution of the shot noise process at a given time can generally not be obtained analytically, the saddlepoint approximation provides a very accurate approximation to it: we refer to Gatto (2010, 2012). The saddlepoint approximation is a large deviations approximation. This topic in introduced in Section 4.8.1.

One can extend the time domain of the counting process from $\mathbb{R}_+$ to $\mathbb{R}$ according to Definition 3.2.35 and extend the shot noise process (3.2.20) to

$$Z_t = \sum_{j=-\infty}^{\infty} \psi(t - T_j) = \int_{\mathbb{R}} \psi(t - s) \mathrm{d}N_s, \quad \forall t \in \mathbb{R}. \tag{3.2.22}$$

3.2.6 Point processes

Counting processes can be embedded in the general class of point processes, which is briefly introduced in this section. The theory presented is not central to the analysis of stationary processes, but it allows to derive useful properties and generalizations of counting processes. A more complete reference is Embrechts et al. (2013), pp. 220–232.

Definition 3.2.43 (Counting measure, point measure and point process). *Let E be a subset of $\bar{\mathbb{R}}^d$, for some $d \geq 1$, and let $\mathcal{E} = \mathcal{B}(E)$.*

(1) Let $x_1, x_2, \ldots \in E$, then a counting measure is given by

$$m : \mathcal{E} \to \mathbb{N} \cup \{\infty\}$$

$$A \mapsto \sum_{k=1}^{\infty} \Delta_{x_k}(A),$$

where Δ_x is the Dirac distribution defined in Section A.8.2.

(2) If $m(A) < \infty$, $\forall A \subset E$ compact, then m is a point measure.

(3) Denote by $M_p(E)$ is the space of all point measures over $(E, \mathcal{E})$, with a σ-algebra $\mathcal{M}_p(E)$. Then a point process over E is the mapping

$$N : (\Omega, \mathcal{F}) \to (M_p(E), \mathcal{M}_p(E)).$$

Thus N is a random element having point measures in $M_p(E)$ as realizations. It can be represented as $\{N(A)\}_{A \in \mathcal{E}}$, where $\forall A \in \mathcal{E}$, $N(A)$ is an (extended) random variable with values in $\mathbb{N} \cup \{\infty\}$. Consider the random variables $X_1, X_2, \ldots$ taking values in E. Then, according to Definition 3.2.43.1,

$$N(A) = \sum_{k=1}^{\infty} \Delta_{X_k}(A), \quad \forall A \in \mathcal{E},$$

is a random counting measure and regularity conditions we obtain the point process.

Some practical examples are the following.

Example 3.2.44 (Point process of exceedances). *A point process that is widely used, for example in insurance in order to model individual claim amounts above a certain threshold, is the following. Let $X_1, \ldots, X_n$ be random variables, let $E = (0, 1]$ and let $u \in \mathbb{R}$. Then*

$$N_n(A) = \sum_{k=1}^{n} \Delta_{\frac{k}{n}}(A) \mathsf{I}\{X_k > u\}, \quad \forall A \in \mathcal{B}((0, 1]),$$

is the point process of exceedances. We have $N_n((0, 1]) = 0 \Leftrightarrow X_{(n)} \leq u$ and, for $k = 1, \ldots, n$,

$$N_n((0, 1]) < k \Longleftrightarrow \operatorname{card}\{1 \leq j \leq n | X_j > u\} < k \Longleftrightarrow X_{(n-k+1)} \leq u,$$

where $X_{(1)} \leq \ldots \leq X_{(n)}$ are the ordered $X_1, \ldots, X_n$.

Alternatively, for $E = (0, 1] \times (u, \infty)$, one can define the point process of exceedances as

$$N_n(A) = \sum_{k=1}^{n} \Delta_{\left(\frac{k}{n}, X_k\right)}(A), \quad \forall A \in \mathcal{B}((0, 1] \times (u, \infty)).$$

For the application in insurance, $X_1, \ldots, X_n$ are individual claim amounts at times $1/n, \ldots, n/n$ and u is any positive threshold.

Example 3.2.45 (Renewal counting process). *Let $Y_1, Y_2, \ldots$ be positive and i.i.d. random variables, let $T_n = Y_1 + \cdots + Y_n$ and let $N_t = \operatorname{card}\{k \in \mathbb{N}^* | T_k \leq t\}, \forall n \in \mathbb{N}^*, t \geq 0$. Then*

$$N = \sum_{k=1}^{\infty} \Delta_{T_k},$$

is a point process over $E = [0, \infty)$ and $N([0,t]) = N_t, \forall t \geq 0$, defines a renewal counting process.

Example 3.2.46. *(Point process of exceedances at renewal times) Let $X_1, X_2, \ldots$ be i.i.d. random variables and independent of the renewal process $\{N_t\}_{t \geq 0}$ of Example 3.2.45. Consider the point process over $E = \mathbb{R}_+ \times \mathbb{R}$ given by*

$$\tilde{N} = \sum_{k=1}^{\infty} \Delta_{(T_k, X_k)}.$$

Then $\tilde{N}([0,t] \times \mathbb{R}) = N_t$ and $\tilde{N}((s,t] \times (u, \infty)) = \operatorname{card}\{j \in \mathbb{N}^* \mid s < T_j \leq t, X_j > u\}$ gives the number of exceedances of u during $(s,t], \forall 0 \leq s < t$.

The distribution of the point process N is $\mathsf{P}_N[A] = \mathsf{P}[N \in A], \forall A \in \mathcal{M}_p(E)$. It is entirely determined by the f.d.d.
$\mathsf{P}[N(A_1) = k_1, \ldots, N(A_l) = k_l], \forall k_1, \ldots, k_l \in \mathbb{N}, A_1, \ldots, A_l \in \mathcal{E}$ and $l \in \mathbb{N}^*$. It is also entirely determined by its Laplace functional, which is defined as follows.

Definition 3.2.47 (Laplace functional). *The Laplace functional of the point process N at $g : (E, \mathcal{E}) \to (\mathbb{R}_+, \mathcal{B}(\mathbb{R}_+))$ is given by*

$$\Psi_N(g) = \mathsf{E}\left[\exp\left\{-\int_E g \, dN\right\}\right] = \int_{\mathcal{M}_p(E)} \exp\left\{-\int_E g \, dm\right\} d\mathsf{P}_N(m),$$

where $\mathsf{P}_N[A] = \mathsf{P}[N \in A], \forall A \in \mathcal{M}_p(E)$.

Examples 3.2.48 (Two Laplace functionals).

(1) If $N = \sum_{j=1}^{\infty} \Delta_{X_j}$, for some random variables $X_1, X_2, \ldots$, then

$$\Psi_N(g) = \mathsf{E}\left[\exp\left\{-\sum_{j=1}^{\infty} g(X_j)\right\}\right].$$

(2) If $g = \sum_{j=1}^{l} v_j \mathbb{1}_{A_j}$, for $v_j \geq 0$, $A_j \in \mathcal{E}$ and $j = 1, \ldots, l$, then

$$\Psi_N(g) = \mathsf{E}\left[\exp\left\{-v_1 N(A_1) - \cdots - v_l N(A_l)\right\}\right], \forall l \in \mathbb{N}^*,$$

which gives the Laplace transforms of the f.d.d.

Definition 3.2.49 (Poisson random measure). *Let μ be a measure over $(E, \mathcal{E})$ such that $\mu(A) < \infty$, $\forall A \subset E$ compact. The point process N over $E \in \bar{\mathbb{R}}^d$ is a Poisson random measure (P.r.m.) with mean measure μ, denoted PRM(μ), if $\forall k \in \mathbb{N}$, $A \in \mathcal{E}$,*

$$P[N(A) = k] = \begin{cases} e^{-\mu(A)} \frac{\mu^k(A)}{k!}, & \text{if } \mu(A) < \infty, \\ 0, & \text{if } \mu(A) = \infty, \end{cases}$$

and if, $\forall l \in \mathbb{N}^*$ and $A_1, \ldots, A_l \in \mathcal{E}$ disjoint, $N(A_1), \ldots, N(A_l)$ are independent random variables.

The P.r.m. appears as limit of many point processes. If $\mu = \lambda L$, for some $\lambda > 0$, with L denoting the Lebesgue measure over E, then the P.r.m. is called homogeneous with rate λ. If μ is absolutely continuous w.r.t. L, then its density exists and it is called the intensity of the inhomogeneous P.r.m.

It can be shown that the Laplace functional of PRM(μ) over E is given by

$$\Psi_N(g) = \exp\left\{ -\int_E \left(1 - e^{-g(x)}\right) d\mu(x) \right\}, \ \forall g : (E, \mathcal{E}) \to (\mathbb{R}_+, \mathcal{B}(\mathbb{R}_+)).$$

An important property of p.r.m. is the following.

Proposition 3.2.50 (Invariance of P.r.m.). *Let N be a PRM(μ) over $E \subset \bar{\mathbb{R}}^d$ and $T : E \to \tilde{E}$, for $\tilde{E} \subset \bar{\mathbb{R}}^l$, be a measurable transform of the points of N. Then the transformed point process $\tilde{N}$ is a PRM($\mu \circ T^{-1}$) over $\tilde{E}$.*

Proof. Consider the representation $\tilde{N} = \sum_{j=1}^{\infty} \Delta_{T(X_j)}$. Then the Laplace functional of $\tilde{N}$ is given by

$$\Psi_{\tilde{N}}(g) = \mathsf{E}\left[\exp\left\{ -\int_{\tilde{E}} g \, d\tilde{N} \right\} \right] = \mathsf{E}\left[\exp\left\{ -\sum_{j=1}^{\infty} g(T(X_j)) \right\} \right]$$

$$= \mathsf{E}\left[\exp\left\{ -\int_E g(T) dN \right\} \right] = \exp\left\{ -\int_E \left(1 - e^{-g(T(x))}\right) d\mu(x) \right\}$$

$$= \exp\left\{ -\int_{\tilde{E}} \left(1 - e^{-g(x)}\right) d(\mu \circ T^{-1})(x) \right\},$$

$\forall g : (E, \mathcal{E}) \to (\mathbb{R}_+, \mathcal{B}(\mathbb{R}_+))$. $\qquad\square$

Example 3.2.51 (Scale invariant P.r.m.). *$X_1, X_2, \ldots$ be the points of N, which is a PRM(λL) over $E = \bar{\mathbb{R}}_+$, for some $\lambda > 0$. Let T be the exponential function. Then $\tilde{N} = \sum_{j=1}^{\infty} \Delta_{\exp\{X_j\}}$ is PRM($\tilde{\mu}$) over $\tilde{E} = [1, \infty)$, where,*

$\forall 1 \leq a < b$, $\tilde{\mu}((a,b]) = \lambda(\log b - \log a) = \lambda \log(b/a)$. *Thus this P.r.m. has the scale invariance property*

$$\tilde{N}((a,b]) \sim \tilde{N}(s(a,b]), \quad \forall s \geq a^{-1}.$$

3.2.7 *Lévy processes*

This section has no essential role in this introduction to stationary processes and its scope is merely to provide a deeper understanding of the central processes presented in the past sections, namely of the Wiener, the Poisson and the compound processes. These processes have some essential common structure and they belong to the class of Lévy processes. We remember that in Example 3.1.8, the Wiener process is obtained from the Wiener measure and from and Kolmogorov's existence theorem 3.1.7. The Wiener process possesses the structure of a convolution semigroup, which we remind here.

A semigroup is a set together with an internal associative binary operation. If $\{Q_t\}_{t \in \mathbb{R}}$ are probability distributions (over a common measure space) that satisfy

$$Q_{s+t} = Q_s * Q_t, \quad \forall s, t \in \mathbb{R}, \tag{3.2.23}$$

then $(\{Q_t\}_{t \in T}, *)$ has the structure of a convolution semigroup: the semigroup of probability distributions with the convolution as internal binary operation.

When Q_t denotes the distribution of the Wiener process at time t, i.e. W_t, $\forall t \geq 0$, we obtain the Gaussian convolution semigroup. Indeed,

$$\underbrace{W_{s+t}}_{\sim \mathcal{N}(0,s+t)} \sim \underbrace{W_s}_{\sim \mathcal{N}(0,s)} + \underbrace{W_{s+t} - W_s}_{\sim \mathcal{N}(0,t)}, \quad \forall s, t \geq 0,$$

where $\mathcal{N}(0, s)$ and $\mathcal{N}(0, t)$ are independent normal random variables. The convolution semigroup can in fact be obtained with any stochastic process with independent and stationary increments that starts at the origin: this is mainly the definition of the Lévy process.

Definition 3.2.52 (Lévy process). *The stochastic process $\{X_t\}_{t \geq 0}$ is called Lévy process, if it satisfies the three following conditions:*

(1) $X_0 = 0$ a.s.;
(2) it possesses stationary and independent increments;
(3) it is everywhere stochastically continuous, i.e., $\forall t \geq 0$,

$$\lim_{h \to 0} \mathsf{P}[|X_{t+h} - X_t| > \varepsilon] = 0, \quad \forall \varepsilon > 0;$$

(4) it possesses cadlag sample paths.

Stochastic continuity, which appears already in Definition 3.1.34, is a rather weak condition that does not imply the continuity of the sample paths. For simplicity, we defined the Lévy process with cadlag sample facts. This property is usually not part of the definition. However one can show that any Lévy process has a cadlag version and therefore we are a simply considering that version in the definition. These sample paths can have infinitely many jumps per unit of time.

Some basic Lévy processes are the following: the Wiener process, the homogeneous Poisson process and the homogeneous compound Poisson process. The sum of these basic processes are again Lévy processes. In particular, a linear drift plus a homogeneous compound Poisson process plus a Wiener process is used to model the capital of an insurance company. This process is called risk process and it was introduced by Lundberg (1903).

An important related concept is infinite divisibility.

Definition 3.2.53 (Infinite divisibility). *A random variable is called infinitely divisible if it is distributed as the sum of n i.i.d. random variables, for $n = 1, 2, \ldots$. In this case, its distribution is also called infinitely divisible.*

Some infinitely divisible distributions are the Poisson, the negative binomial, the gamma and the Cauchy. Infinitely divisible distributions are related to Lévy processes.

Theorem 3.2.54 (Infinite divisibility of Lévy process). *If $\{X_t\}_{t \geq 0}$ is a Lévy-Process, then X_t is infinitely divisible, $\forall t > 0$.*

Proof. Let $t > 0$ and $n = 1, 2, \ldots$. Then

$$X_t = \underbrace{X_0 + X_{\frac{t}{n}}}_{=R_{1,n}} \underbrace{-X_{\frac{t}{n}} + X_{\frac{2t}{n}}}_{=R_{2,n}} - \cdots \underbrace{-X_{\frac{(n-1)t}{n}} + X_t}_{=R_{n,n}}.$$

It follows from Definition 3.2.52.1 and 3.2.52.2 that $R_{1,n}, \ldots, R_{n,n}$ are i.i.d. $\qquad \square$

The following Lévy-Khintchine representation gives the characteristic function of any infinitely divisible distribution. The measure ν over $(\mathbb{R}^*, \mathcal{B}(\mathbb{R}^*))$ is called a Lévy measure if it satisfies

$$\int_{\mathbb{R}^*} \left(1 \wedge x^2\right) \mathrm{d}\nu(x) < \infty.$$

Theorem 3.2.55 (Lévy-Khintchine representation). *The probability distribution* Q *over* $(\mathbb{R}, \mathcal{B}(\mathbb{R}))$ *is infinitely divisible iff* $\exists b \in \mathbb{R}, a \geq 0$ *and* ν *a Lévy measure over* $(\mathbb{R}^*, \mathcal{B}(\mathbb{R}^*))$, *such that,* $\forall v \in \mathbb{R}$, $\int_{\mathbb{R}} e^{ivx} dQ(x) = e^{\eta(v)}$, *where*

$$\eta(v) = ibv - \frac{1}{2}av^2 + \int_{\mathbb{R}^*} \left(e^{ivx} - 1 - ivx\, l\{x \in (-1,1)\}\right) d\nu(x).$$

The function η is called characteristic exponent.

Examples 3.2.56 (Infinitely divisible distributions).

(1) Let Q *be the Gaussian distribution with mean* b *and variance* a. *The characteristic exponent* η *is obtained with* $\nu = 0$.

(2) Let Q *be the Poisson distribution with parameter* λ. *The characteristic exponent* η *is obtained with* $a = b = 0$ *and* $\nu = \lambda \Delta_1$.

(3) Let Q *be the compound Poisson distribution with Poisson parameter* λ *and distribution of summands* ρ, *over* $\mathbb{R}^*$. *Then* $a = 0$, $b = \lambda \int_{(-1,1)\setminus\{0\}} x d\rho(x)$, *and* $\nu = \lambda \rho$.

It follows immediately from Theorem 3.2.54 and from the Lévy-Khintchine representation of Theorem 3.2.55 that,

Proposition 3.2.57 (Characteristic function of Lévy process). *For any Lévy process* $\{X_t\}_{t \geq 0}$, *there exists a characteristic exponent* η *such that*

$$\mathsf{E}\left[e^{ivX_t}\right] = e^{t\eta(v)}, \quad \forall v \in \mathbb{R}, \ t \geq 0. \tag{3.2.24}$$

Proof. This result follows from Theorem 3.2.54 and from the Lévy-Khintchine representation of Theorem 3.2.55. Define $g_v(t) = \mathsf{E}[e^{ivX_t}]$, $\forall t \geq 0, v \in \mathbb{R}$. It follows from Definition 3.2.52.1 and 2 that $g_v(t+h) = \mathsf{E}[e^{iv(X_{t+h}-X_t)}]\mathsf{E}[e^{ivX_t}] = g_v(h)g_v(t)$, where $v \in \mathbb{R}$, $t \geq 0$ and $h \in \mathbb{R}$. We deduce from Definition 3.2.52.3, Lemma A.4.7 and the remark of p. 181 that $g_v(t) = e^{t\xi(v)}$, for some $\xi : \mathbb{R} \to \mathbb{C}$. Infinite divisibility tells $g_v(1) = e^{\xi(v)} = e^{\eta(v)}$, meaning that $\xi = \eta$. $\square$

Examples 3.2.56 combined with Proposition 3.2.57 yield Gaussian, Poisson and compound Poisson processes. We note that the only Gaussian Lévy process is the Wiener process, with possible drift and rescaling.

The Lévy measure provides all information about the existence of moments.

Theorem 3.2.58 (Moments of Lévy process). *Let ν denote the Lévy measure of the Lévy process $\{X_t\}_{t\geq 0}$. We then have, $\forall r > 0$,*

$$\mathsf{E}[|X_t|^r] < \infty, \; \forall t > 0 \Longleftrightarrow \int_{|x|\geq 1} |x|^r \mathrm{d}\nu(x) < \infty.$$

Thus, the case $r = 2$ is important in the context of stationarity.

The Lévy measure provides the information about the discontinuous part of the sample paths of the Lévy process. The value $\nu(B)$ gives the expected number of jumps of size in $B \in \mathcal{B}(\mathbb{R}^*)$ occurring during any chosen time interval of length one, i.e. per unit of time. If ν is a finite measure, viz. $\lambda = \nu(\mathbb{R}^*) < \infty$, then $\nu(B)/\nu(\mathbb{R}^*)$, $\forall B \in \mathcal{B}(\mathbb{R}^*)$, id the probability distribution of the jumps according to their sizes. In this case, λ is the expected number of jumps per unit of time. If ν is an infinite measure, viz. $\nu(\mathbb{R}^*) = \infty$, then infinitely many jumps of small sizes per unit of time are expected. That is, $\nu(B) = \infty$, for some $B \in \mathcal{B}(\mathbb{R}^*)$ such that $0 \in \overline{B}$. This explanation is summarized as follows.

Theorem 3.2.59 (Number of jumps of Lévy process). *Let ν denote the Lévy measure of a Lévy process. If $\nu(\mathbb{R}^*) < \infty$, then the sample paths of the Lévy process have a.s. finitely many jumps per unit of time.[3] If $\nu(\mathbb{R}^*) = \infty$, then the sample paths have a.s. infinitely many jumps per unit of time.[4]*

We now give two examples relating Lévy processes with stationarity.

Example 3.2.60 (Increments of Lévy process). *Let $\{X_t\}_{t\in\mathbb{R}_+}$ be a Lévy process with characteristic exponent η. Define the process of increments of length $l > 0$ as*

$$Y_t = X_{t+l} - X_t, \quad \forall t \in \mathbb{R}_+.$$

Let $t \in \mathbb{R}_+$. We obtain from $Y_t \sim X_l$ that

$$\varphi(v) = \mathsf{E}\left[e^{ivY_t}\right] = \mathsf{E}\left[e^{ivX_l}\right] = e^{l\eta(v)}, \quad \forall v \in \mathbb{R}.$$

Consequently, $\mathsf{E}[Y_t] = -i\varphi'(0) = -il\eta'(0)$. Let $h \in \mathbb{R}_+$. One shows directly that

$$\mathsf{cov}(Y_{t+h}, Y_t) = \begin{cases} \mathsf{var}(X_{t+l}) - \mathsf{var}(X_{t+h}), & \text{if } h < l, \\ 0, & \text{if } h \geq l. \end{cases}$$

---

[3]This referred to as the finite activity of the Lévy process.
[4]This referred to as the infinite activity of the Lévy process.

However,

$$\mathrm{var}(X_t) = \mathsf{E}[X_t^2] - \mathsf{E}^2[X_t]$$

$$= -\left(\frac{\mathrm{d}}{\mathrm{d}v}\right)^2 e^{t\eta(v)}\bigg|_{v=0} + \left(\frac{\mathrm{d}}{\mathrm{d}v} e^{t\eta(v)}\bigg|_{v=0}\right)^2$$

$$= -t\left\{\eta''(0) + t[\eta'(0)]^2\right\} + \{t\eta'(0)\}^2$$

$$= -t\eta''(0).$$

This leads to

$$\mathrm{cov}(Y_{t+h}, Y_t) = \begin{cases} -\eta''(0)(l-h), & \text{if } h < l, \\ 0, & \text{if } h \geq l. \end{cases}$$

From this last result we deduce that $\{Y_t\}_{t\in\mathbb{R}_+}$ *is stationary with a.c.v.f.*

$$\gamma(h) = -\eta''(0)(l - |h|)\mathsf{I}\{|h| < l\}, \quad \forall h \in \mathbb{R}.$$

The particular case where $\{X_t\}_{t\in\mathbb{R}_+}$ *is an homogeneous Poisson process corresponds to Example 3.2.34.*

Example 3.2.61 (Generalized Ornstein-Uhlenbeck process). *A generalized Ornstein-Uhlenbeck process is one that uses a Lévy process, instead of Gaussian WN as in Example 3.4.9, as driving process in the Langevin equation (3.4.16). This process was proposed by Barndorff-Nielsen and Shephard (2001) and it is thus the solution of*

$$\mathrm{d}X_t + \tau X_t \mathrm{d}t = \sqrt{2\tau}\sigma \mathrm{d}Z_t, \quad \forall t \geq 0,$$

where $\{Z_t\}_{t\geq 0}$ *is the driving Lévy process. In the above equation, we use the parameterization of Example 3.7.4. When the driving Lévy process is a compound process, then the solution* $\{X_t\}_{t\in\mathbb{R}}$ *is the shot noise process (3.2.21).*

3.3 Mean square properties of stationary processes

The last part of Section 1.2.2 provides a short study of the relationship between smoothness or regularity and decay in the tails, within a Fourier pair: high regularity in the signal, i.e. in the original function, is related to fast decay in the tails of the Fourier transform. This section analyzes this relationship when the original function is a stationary stochastic process with continuous time. Thus, this section begins by introducing some infinitesimal mean square properties of stationary stochastic processes: mean square or $\mathcal{L}_2$-continuity and mean square or $\mathcal{L}_2$-differentiability. Then, these properties are related with the regularity of the a.c.v.f. and with the decay of the spectral distribution. We note that various definitions and results regarding sample path regularity of stochastic processes are already presented in Section 3.1.3. In particular, specific results for stationary processes can be found at the end of Section 3.1.3. This section is thus a continuation of this study and it specifically oriented towards $\mathcal{L}_2$-continuity and $\mathcal{L}_2$-differentiability.

Let us begin by defining $\mathcal{L}_2$-continuity of a stochastic process.

Definition 3.3.1 ($\mathcal{L}_2$-continuity). *The complex-valued stochastic process* $\{X_t\}_{t \in \mathbb{R}}$ *is* $\mathcal{L}_2$-*continuous or continuous in mean square at* $t \in \mathbb{R}$, *if*

$$\mathsf{E}\left[|X_{t+h} - X_t|^2\right] \overset{h \to 0}{\longrightarrow} 0.$$

Examples 3.3.2 (Two $\mathcal{L}_2$-continuous processes).

(1) The Wiener process is $\mathcal{L}_2$-*continuous, because* $\forall t \in \mathbb{R}_+$,

$$\mathsf{E}\left[|W_{t+h} - W_t|^2\right] = h \overset{h \to 0}{\longrightarrow} 0.$$

(2) The inhomogeneous Poisson process is $\mathcal{L}_2$-*continuous, because* $\forall t \in \mathbb{R}_+$,

$$\mathsf{E}\left[|N_{t+h} - N_t|^2\right] = \int_t^{t+h} \lambda(s)\mathrm{d}s \overset{h \to 0}{\longrightarrow} 0.$$

Note that the sample paths are a.s. discontinuous, because jumps with magnitude one do appear a.s.

Proposition 3.3.3 ($\mathcal{L}_2$-continuity). *The complex-valued stationary stochastic process* $\{X_t\}_{t \in \mathbb{R}}$ *with a.c.v.f.* γ *is* $\mathcal{L}_2$-*continuous iff its a.c.v.f. is continuous at the origin, i.e.* $\gamma(h) \overset{h \to 0}{\longrightarrow} \gamma(0)$.

Proof. Let $t \in \mathbb{R}$, then

$$\mathsf{E}\left[|X_{t+h} - X_t|^2\right] = \mathsf{var}(X_{t+h} - X_t) = 2\gamma(0) - \{\gamma(h) + \gamma(-h)\} \overset{h \to 0}{\longrightarrow} 0,$$

iff $\gamma(h) \overset{h \to 0}{\longrightarrow} \gamma(0)$. □

Example 3.3.4 (Random telegraph signal). *The random telegraph signal is introduced in Example 3.1.18. It is stationary and* $\mathcal{L}_2$-*continuous, because its a.c.v.f.* γ *satisfies* $\gamma(h) = \mathrm{e}^{-2\lambda|h|} \overset{h \to 0}{\longrightarrow} 1 = \gamma(0)$.

Definition 3.3.5 ($\mathcal{L}_2$-differentiability). *The complex-valued stochastic process* $\{X_t\}_{t \in \mathbb{R}}$ *is* $\mathcal{L}_2$-*differentiable or mean square differentiable at* $t \in \mathbb{R}$ *if there exists a random variable* $X_t' \in \mathcal{L}_2$, *such that*

$$\frac{X_{t+h} - X_t}{h} \overset{\mathcal{L}_2}{\longrightarrow} X_t'.$$

Note that we use the same notation for the $\mathcal{L}_2$-derivative and for the sample path derivative of a stochastic process. This should however not lead to ambiguities, because the precise meaning of the derivative should be easily deduced from its current context.

Examples 3.3.6 (Two non-$\mathcal{L}_2$-differentiable processes). *Two basic processes that are not $\mathcal{L}_2$-differentiable are the following.*

(1) The Wiener process is nowhere $\mathcal{L}_2$-differentiable, because $\forall t \in \mathbb{R}_+$,

$$\mathsf{E}\left[\left|\frac{W_{t+h} - W_t}{h}\right|^2\right] = \frac{1}{h} \xrightarrow{h\downarrow 0} \infty.$$

(2) The inhomogeneous Poisson process is nowhere $\mathcal{L}_2$-differentiable, because $\forall t \in \mathbb{R}_+$,

$$\mathsf{E}\left[\left|\frac{N_{t+h} - N_t}{h}\right|^2\right] = \frac{1}{h}\frac{\int_t^{t+h}\lambda(s)\mathrm{d}s}{h} = \frac{1}{h}\{\lambda(t) + \mathrm{o}(1)\} \xrightarrow{h\downarrow 0} \infty.$$

We now define the moments of the spectral distribution of a stationary process, which is however introduced later, in Section 3.5, where Bochner's theorem 3.5.1 is the central result.

Definition 3.3.7 (Spectral moment). *We denote by*

$$\alpha_k = \int_{\mathbb{R}} \alpha^k \mathrm{d}F(\alpha),$$

the absolute moment of order k of the spectral d.f. F, density or distribution, and call it k-th spectral moment, for $k = 1, 2, \ldots$. We say that the k-th spectral moment exists whenever $\int_{\mathbb{R}} |\alpha|^k \mathrm{d}F(\alpha) < \infty$, for $k = 1, 2, \ldots$.

The following lemma is useful for the proof of Proposition 3.3.10.

Lemma 3.3.8 (Second spectral moment and a.c.v.f.). *Let γ and F be a pair of a.c.v.f. and spectral d.f. of some real-valued and stationary process. Let $\alpha_2 = \int_{\mathbb{R}} \alpha^2 \mathrm{d}F(\alpha)$ be the second spectral moment. Then the following statements hold.*

(1) $\alpha_2 = 2 \lim\limits_{h \to 0} \dfrac{\gamma(0) - \gamma(h)}{h^2}$.

(2) If α_2 exists, viz. $\alpha_2 < \infty$, then γ'' exists, it is continuous and $\gamma''(0) = -\alpha_2$.

(3) If $\gamma''(0)$ exists, then α_2 exists.

Proof. 1. Assume $\alpha_2 < \infty$. Then

$$2\frac{\gamma(0) - \gamma(h)}{h^2} = 2\int_{\mathbb{R}} \frac{1 - \cos h\alpha}{(h\alpha)^2}\alpha^2 dF(\alpha) = \int_{\mathbb{R}} \{1 + O((h\alpha)^2\}\alpha^2 dF(\alpha),$$

as $h\alpha \to 0$, inside the integral. The limit as $h \to 0$ can be taken and can be inserted inside the integrals, because $0 \le (1 - \cos h\alpha)(h\alpha)^{-2} \le 1/2$ allows for the application of Dominated convergence theorem A.5.3. Assume $\alpha_2 = \infty$. Then $(1 - \cos h\alpha)(h\alpha)^{-2} \overset{h \to 0}{\longrightarrow} 1/2$ together with Fatou's lemma A.5.5 give the desired result.

2. The assumption $\alpha_2 < \infty$ together with Corollary A.5.4 allow for the following double interchange between differentiation and integration,

$$\begin{aligned}
\gamma''(h) &= \frac{d^2}{dh^2}\int_{\mathbb{R}} \cos h\alpha \, dF(\alpha) \\
&= \int_{\mathbb{R}} \frac{d^2}{dh^2} \cos h\alpha \, dF(\alpha) \\
&= -\int_{\mathbb{R}} \alpha^2 \cos h\alpha \, dF(\alpha), \quad \forall h \in \mathbb{R}.
\end{aligned} \tag{3.3.1}$$

3. Assume that $\gamma''(0)$ exists. Let $h, l \in \mathbb{R}$ and $g(h, l) = \gamma(h) - \gamma(h - l)$. Then

$$\frac{g(h, h) - g(0, h)}{h^2} = \frac{\gamma(h) - \gamma(0) - \gamma(0) + \gamma(-h)}{h^2} = 2\frac{\gamma(h) - \gamma(0)}{h^2}.$$

Thus

$$2\frac{\gamma(h) - \gamma(0)}{h^2} = \frac{g_1'(uh, h)}{h},$$

for some $u \in (0, 1)$, from the Mean value theorem. The partial derivative $g_1'(uh, h)$ does exist for h over a neighborhood of zero, because $\gamma''(0)$ exists. By evaluating $g_1'(uh, h)$ we find, as $h \to 0$,

$$\begin{aligned}
2\frac{\gamma(h) - \gamma(0)}{h^2} &= \frac{\gamma'(uh) - \gamma'(uh - h)}{h} \\
&= u\gamma''(0) + o(1) - (u - 1)\gamma''(0) + o(1) \\
&= \gamma''(0) + o(1).
\end{aligned}$$

Part 1 of the lemma yields $\gamma''(0) = \alpha_2$. $\qquad\square$

Remarks 3.3.9. *Lemma 3.3.8.2 and 3.3.8.3 yield the following facts.*

(1) $\gamma''(0)$ exists iff α_2 exists.
(2) If $\gamma''(0)$ exists, then γ'' exists and it is continuous.

The proof of Lemma 3.3.8.2, precisely (3.3.1), yields the following fact.

(3) If α_2 exists, then γ'' exists and it is given by

$$\gamma''(h) = -\int_{\mathbb{R}} \alpha^2 \cos h\alpha \, dF(\alpha), \quad \forall h \in \mathbb{R}.$$

Proposition 3.3.10 ($\mathcal{L}_2$-differentiability). *Let $\{X_t\}_{t\in\mathbb{R}}$ be a real-valued and stationary stochastic process with a.c.v.f. γ and spectral d.f. F. Then the following statements hold.*

(1) The process $\{X_t\}_{t\in\mathbb{R}}$ is $\mathcal{L}_2$-differentiable $\iff \gamma''$ exists and it is continuous

(2) The process of $\mathcal{L}_2$-derivatives $\{X_t'\}_{t\in\mathbb{R}}$ is stationary with mean zero. Its a.c.v.f., denoted $\gamma_{X'}(h) = \mathrm{cov}(X_{t+h}', X_t')$, is given by

$$\gamma_{X'}(h) = -\gamma''(h) = \int_{\mathbb{R}} \alpha^2 \cos \alpha h \, dF(\alpha), \quad \forall t, h \in \mathbb{R}.$$

Proof. 1. ($\Leftarrow$) Let $t, h, l \in \mathbb{R}$ and $g(h,l) = \gamma(h) - \gamma(h-l)$. Then for some $u, v \in (0,1)$,

$$
\begin{aligned}
\left\langle \frac{X_{t+h} - X_t}{h}, \frac{X_{t+l} - X_t}{l} \right\rangle &= \frac{\gamma(h-l) - \gamma(h) - \gamma(-l) + \gamma(0)}{hl} \\
&= -\frac{g(h,l) - g(0,l)}{hl} \\
&= -\frac{g_1'(uh,l)}{l} \\
&= -\frac{g_1'(uh,0) + lg_{12}''(uh,vl)}{l} \\
&= -g_{12}''(uh,vl) \\
&= -\gamma''(uh - vl),
\end{aligned}
$$

from the Mean value theorem. The continuity of γ'' over a neighborhood of zero implies $\lim_{h,l\to 0} \gamma''(uh - vl) = \gamma''(0)$. This allows for the application of Loève's criterion of Lemma A.1.6 in order to deduce

$$\frac{X_{t+h} - X_t}{h} \xrightarrow{\mathcal{L}_2} X_t',$$

for some $X_t' \in \mathcal{L}_2$.

($\Rightarrow$) The $\mathcal{L}_2$-differentiability of $\{X_t\}_{t\in\mathbb{R}}$ means that for arbitrary $t \in \mathbb{R}$, we have

$$\infty > \mathsf{E}[|X_t'|^2] = \lim_{h\to 0} \mathsf{E}\left[\left|\frac{X_{t+h} - X_t}{h}\right|^2\right] = 2\lim_{h\to 0} \frac{\gamma(0) - \gamma(h)}{h^2} = \alpha_2,$$

from Lemma 3.3.8.1. Lemma 3.3.8.2 tells that γ'' exists and it is continuous
2. There are two equalities to show and the second one is exactly

Remark 3.3.9.3. Regarding the first equality, let $s, t, h, l \in \mathbb{R}$ and, $g(h, l) = \gamma(h) - \gamma(h - l)$. Then for some $u, v \in (0, 1)$,

$$
\left\langle \frac{X_{t+s+h} - X_{t+s}}{h}, \frac{X_{t+l} - X_t}{l} \right\rangle = \frac{\gamma(s + h - l) - \gamma(s + h) - \gamma(s - l) + \gamma(s)}{hl}
$$

$$
= -\frac{g(s + h, l) - g(s, l)}{hl}
$$

$$
= -\frac{g_1'(s + uh, l)}{l}
$$

$$
= -\frac{g_1'(s + uh, 0) + l g_{12}''(s + uh, vl)}{l}
$$

$$
= -g_{12}''(s + uh, vl)
$$

$$
= -\gamma''(s + uh - vl),
$$

from the Mean value theorem. The continuity of γ'' implies $\lim_{h,l \to 0} \gamma''(s + uh - vl) = -\gamma''(s)$. $\square$

Remarks 3.3.11. *(1) Proposition 3.3.10.1 can be rewritten as follows:*

$$\{X_t\}_{t \in \mathbb{R}} \text{ is } \mathcal{L}_2\text{-differentiable} \iff \gamma''(0) \text{ exists} \iff \alpha_2 \text{ exists}.$$

The first equivalence is due to Remark 3.3.9.2 and the second equivalence is due to Remark 3.3.9.1. The existence of α_2 or, equivalently, of $\gamma''(0)$ enforces smoothness on $\{X_t\}_{t \in \mathbb{R}}$.

(2) One can iteratively apply the previous results in order to obtain results for higher order $\mathcal{L}_2$-derivatives of a stationary process. Let $\{X_t\}_{t \in \mathbb{R}}$ be a real-valued stationary process with a.c.v.f. γ and spectral d.f. F. For $k = 1, 2, \ldots$, if the derivative of order $2k$ of γ at zero, i.e. $\gamma^{(2k)}(0)$, exists, or if spectral moment of order $2k$, i.e. α_{2k}, exists, then the process of $\mathcal{L}_2$-derivatives of order k denoted $\{X_t^{(k)}\}_{t \in \mathbb{R}}$[5] exists, has mean zero and it is stationary. Denote by $\gamma_{X^{(k)}}$ its a.c.v.f. The following formulae hold for $k = 0, 1, \ldots$:

$$\gamma^{(2k)}(h) = (-1)^k \int_{\mathbb{R}} \alpha^k \cos h\alpha \, dF(\alpha),$$

and

$$\gamma_{X^{(k)}}(h) = (-1)^k \gamma^{(2k)}(h), \quad \forall h \in \mathbb{R}.$$

Moreover, for $j, k = 0, 1, \ldots$,

$$\mathsf{cov}\left(X_{t+h}^{(j)}, X_t^{(k)}\right) = (-1)^k \gamma^{(j+k)}(h), \quad \forall t, h \in \mathbb{R}. \tag{3.3.2}$$

---

[5]As already mentioned, we use the same notation for the $\mathcal{L}_2$-derivative and for the sample path derivative of a stochastic process.

Table 3.1: Relations between differentiability of a.c.v.f., shape of spectral distribution, existence of spectral moments, smoothness of stationary process and memory of stationary process.

A.c.v.f. differentiability	Spectral distribution shape	Spectral moments existence	Process smoothness	Process memory
high order	light-tailed	high order	considerable	long
low order	heavy-tailed	low order	poor	short

One deduces for instance that X_t and X_t' are uncorrelated, because

$$\mathsf{cov}(X_t, X_t') = -\gamma'(0) = -\frac{\mathrm{d}}{\mathrm{d}h} \int_{\mathbb{R}} \cos h\alpha \mathrm{d}F(\alpha)|_{h=0} = 0, \quad \forall t \in \mathbb{R}.$$

One can conclude this section by summarizing the implications between existence of spectral moments and regularity, in terms of smoothness and memory, of the stationary stochastic process.

- Consider the low frequency case, where the process possesses relatively few high frequencies. The spectral distribution is typically light-tailed and, as a consequence, spectral moments of high order do exist. Thus the process admits high order $\mathcal{L}_2$-derivatives and it is smooth in this sense. This situation can lead to long range dependence, i.e. long memory.
- Consider the high frequency case, where the process possesses relatively many high frequencies. Because the spectral distribution is typically heavy-tailed, spectral moments of high order do not exist. Thus the stochastic process admits low order $\mathcal{L}_2$-derivatives only, telling that it is irregular and possibly noisy. In this case, the process does typically exhibit independence at large time lags, i.e. short memory.

Table 3.1 provides a summary of the different characteristics of spectral distributions.

3.4 Stochastic integrals

This section introduces stochastic integration together with some important related results. Section 3.4.1 presents two useful stochastic integrals, that are central to the study of stationary processes. They are mean square or $\mathcal{L}_2$-integrals, in the sense that they are defined $\mathcal{L}_2$-limits of either Riemann or Riemann-Stieltjes sums. From the given definition of

$\mathcal{L}_2$-integral, Section 3.4.2 introduces continuous time Gaussian WN, which generalizes the discrete time WN used with time series. The $\mathcal{L}_2$-integral allows also to introduce the concept of ergodicity of a stochastic process, which is the topic of Section 3.4.3. Then, in Section 3.4.4, generalized and purely random stochastic processes are presented. It is thanks to these concepts that we can give a precise meaning to continuous time WN.

3.4.1 *Two mean square integrals*

This section presents two basic stochastic integrals that involve a complex-valued function and a complex-valued stochastic process. These integral are defined as $\mathcal{L}_2$-limits of Riemann or Riemann-Stieltjes sums. They have a central role in the context of stationary processes. Indeed, the stationary process at any time can be re-expressed in terms of one of these integrals, giving the spectral decomposition of the process.

Theorem 3.4.1 (Two $\mathcal{L}_2$-integrals). *Let* $-\infty < a < b < \infty$ *and let* $\{X_t\}_{t\in\mathbb{R}}$ *be a complex-valued stochastic process in* $\mathcal{L}_2$ *with mean null and a.c.v.f.* $\gamma(s,t) = \mathsf{E}[X_s\overline{X_t}]$, $\forall s,t \in \mathbb{R}$.

(1) For any function $g : \mathbb{R} \to \mathbb{C}$ *such that*

$$q_1 = \int_a^b \int_a^b g(s)\overline{g(t)}\gamma(s,t)\mathrm{dt}\mathrm{ds},$$

is a finite complex number, the stochastic integral

$$I_1 = \int_a^b g(t)X_t\mathrm{dt},$$

exists as $\mathcal{L}_2$-limit of the analogue Riemann sum; cf. (3.4.1). Moreover,

$$\mathsf{E}[I_1] = 0 \text{ and } \mathsf{E}\left[|I_1|^2\right] = q_1.$$

(2) For any function $g : \mathbb{R} \to \mathbb{C}$ *such that*

$$q_2 = \int_{[a,b]^2} g(s)\overline{g(t)}\mathrm{d}\gamma(s,t),$$

is finite, the stochastic integral

$$I_2 = \int_{[a,b]} g(t)\mathrm{d}X_t,$$

exists as $\mathcal{L}_2$-limit of the analogue Riemann-Stieltjes sum; cf. (3.4.3). Moreover,

$$\mathsf{E}[I_2] = 0 \text{ and } \mathsf{E}[|I_2|^2] = q_2.$$

Proof. 1. Define

$$S_n = \sum_{j=1}^{n} g(\tau_{n,j}) X_{\tau_{n,j}} (t_{n,j} - t_{n,j-1}), \tag{3.4.1}$$

where

$$a = t_{n,0} < t_{n,1} < \cdots < t_{n,n-1} < t_{n,n} = b \tag{3.4.2}$$

and $t_{n,j-1} \leq \tau_{n,j} \leq t_{n,j}$, for $j = 1, \ldots, n$ and $\forall n \in \mathbb{N}^*$. Then, $\forall m, n \in \mathbb{N}^*$, we have

$$\mathsf{E}\left[S_m \overline{S_n}\right] = \sum_{j=1}^{m} \sum_{k=1}^{n} g(\tau_{m,j}) \overline{g(\tau_{n,k})} \gamma(\tau_{m,j}, \tau_{n,k})(t_{m,j} - t_{m,j-1})(t_{n,k} - t_{n,k-1}).$$

The finiteness of the integral q_1 means that the above double sum, i.e. $\mathsf{E}[S_m \overline{S_n}]$, converges precisely to q_1. The convergence takes place as $m, n \to \infty$ with vanishing mesh of the partition (3.4.2), i.e. with $\max_{j=1,\ldots,n} t_{n,j} - t_{n,j-1} \overset{n\to\infty}{\longrightarrow} 0$. It then follows from Loève's criterion, given in Lemma A.1.6, that $S_n \overset{\mathcal{L}_2}{\longrightarrow} I_1$, from some $I_1 \in \mathcal{L}_2$.

The continuity of the scalar product implies that $\mathsf{E}[I_1] = 0$. The continuity of the scalar product implies also $\mathsf{E}\left[S_m \overline{S_n}\right] \overset{m,n\to\infty}{\longrightarrow} \mathsf{E}\left[I_1 \overline{I_1}\right] = \mathsf{E}\left[|I_1|^2\right]$. The uniqueness of the limit implies $\mathsf{E}\left[|I_1|^2\right] = q_1$.

2. Define

$$S_n = \sum_{j=1}^{n} g(\tau_{n,j}) \left(X_{t_{n,j}} - X_{t_{n,j-1}} \right). \tag{3.4.3}$$

Then

$$\mathsf{E}\left[S_m \overline{S_n}\right] = \sum_{j=1}^{m} \sum_{k=1}^{n} g(\tau_{m,j}) \overline{g(\tau_{n,k})} \mathsf{E}\left[(X_{t_{m,j}} - X_{t_{m,j-1}}) \overline{(X_{t_{n,k}} - X_{t_{n,k-1}})}\right]$$

$$= \sum_{j=1}^{m} \sum_{k=1}^{n} g(\tau_{m,j}) \overline{g(\tau_{n,k})} \{\gamma(t_{m,j}, t_{n,k}) - \gamma(t_{m,j}, t_{n,k-1})$$

$$-\gamma(t_{m,j-1}, t_{n,k}) + \gamma(t_{m,j-1}, t_{n,k-1})\}.$$

This last expression is a Riemann-Stieltjes sum, where $\gamma(s,t)$ is first differentiated w.r.t. s and then each one of the two terms of this differentiation is differentiated w.r.t. t. The finiteness of the integral q_2 means that the double sum above, i.e. $\mathsf{E}[S_m \overline{S_n}]$, converges to q_2. With Loève's criterion we obtain that $S_n \overset{\mathcal{L}_2}{\longrightarrow} I_2$, from some $I_2 \in \mathcal{L}_2$.

The continuity of the scalar product implies that $\mathsf{E}[I_2] = 0$. The continuity of the scalar product implies also $\mathsf{E}\left[S_m \overline{S_n}\right] \overset{m,n\to\infty}{\longrightarrow} \mathsf{E}\left[I_2 \overline{I_2}\right] = \mathsf{E}\left[|I_2|^2\right]$. The uniqueness of the limit implies $\mathsf{E}[|I_2|^2] = q_2$. $\qquad\square$

Remarks 3.4.2.

(1) *If the Riemann integral q_1 of Theorem 3.4.1 converges, as $a \to -\infty$, as $b \to \infty$ or with both limits simultaneously, then the corresponding stochastic integral I_1 converges in $\mathcal{L}_2$, under the corresponding asymptotics of a and b. The same holds with the Riemann-Stieltjes integral q_2 of Theorem 3.4.1 and the corresponding stochastic integral I_2.*

(2) *It follows from the above given proof that the stochastic integrals I_1 and I_2 inherit the basic properties of deterministic integrals. Let $u, v \in \mathbb{C}$, $-\infty < a < b < \infty$ and $g, h : \mathbb{R} \to \mathbb{C}$. Then we have, for example,*

$$\int_{[a,b]} \{ug(t) + vh(t)\} \mathrm{d}X_t = u \int_{[a,b]} g(t) \mathrm{d}X_t + v \int_{[a,b]} h(t) \mathrm{d}X_t.$$

Similarly, the usual partial integration formula holds,

$$\int_{[a,b]} g(t) \mathrm{d}X_t = g(b)X_b - g(a)X_a - \int_{[a,b]} X_t \mathrm{d}g(t). \qquad (3.4.4)$$

By selecting the intermediate points of the partition in the proof of Theorem 3.4.1 as $\tau_{n,j} = t_{n,j}$, for $j = 1, \ldots, n$, we obtain

$$\sum_{j=1}^{n} g(t_{n,j})(X_{t_{n,j}} - X_{t_{n,j-1}})$$

$$= g(t_{n,n})X_{t_{n,n}} - g(t_{n,1})X_{t_{n,0}} - \sum_{j=2}^{n} X_{t_{n,j-1}}\{g(t_{n,j}) - g(t_{n,j-1})\}.$$

The $\mathcal{L}_2$-limit of this equation, as the mesh of the partition vanishes, is given by (3.4.4).

The a.c.v.f. of a stochastic process can be generalized to the covariances between two different complex-valued stochastic processes.

Definition 3.4.3 (Crosscovariance function). *Let $\{X_t\}_{t \in \mathbb{R}}$ and $\{Y_t\}_{t \in \mathbb{R}}$ be two complex-valued stochastic processes, then the crosscovariance function between the stochastic processes $\{X_t\}_{t \in \mathbb{R}}$ and $\{Y_t\}_{t \in \mathbb{R}}$ is given by*

$$\gamma_{XY}(s, t) = \mathsf{cov}(X_s, Y_t), \quad \forall s, t \in \mathbb{R}.$$

Theorem 3.4.4 provides generalizations of the equalities $\mathsf{E}\left[|I_1|^2\right] = q_1$ and $\mathsf{E}\left[|I_2|^2\right] = q_2$ that appear in Theorem 3.4.1.

Theorem 3.4.4. *Let $\{X_t\}_{t \in \mathbb{R}}$ and $\{Y_t\}_{t \in \mathbb{R}}$ be two complex-valued stochastic processes in $\mathcal{L}_2$ with mean null and crosscovariance function γ_{XY}. Then, $\forall g : \mathbb{R} \to \mathbb{C}$ and $h : \mathbb{R} \to \mathbb{C}$ such that the two integrals appearing on the* *right sides of (3.4.5) and (3.4.6) are finite complex numbers and* $\forall - \infty < a < b < \infty$ *and* $-\infty < c < d < \infty$, *we have*

$$\mathsf{E}\left[\int_a^b g(s)X_s \mathrm{d}s \int_c^d \overline{h(t)Y_t}\mathrm{d}t\right] = \int_a^b \int_c^d g(s)\overline{h(t)}\gamma_{XY}(s,t)\mathrm{d}s\mathrm{d}t, \quad (3.4.5)$$

and

$$\mathsf{E}\left[\int_{[a,b]} g(s)\mathrm{d}X_s \int_{[c,d]} \overline{h(t)\mathrm{d}Y_t}\right] = \int_{[a,b]\times[c,d]} g(s)\overline{h(t)}\mathrm{d}\gamma_{XY}(s,t). \quad (3.4.6)$$

It would be natural to ask whether these stochastic integrals, that are $\mathcal{L}_2$-limits, would be a.s. equal to the analogue a.s. limits of Riemann or Riemann-Stieltjes sums, whenever these two types of limits would exist. Such an a.s. equality would certainly be desired. The answer to this question is positive and it can be justified with the help of Lemma A.3.11.

3.4.2 Gaussian white noise

In the study of stationary processes, the most important $\mathcal{L}_2$-integrals presented Section 3.4.1 are based on the Wiener process: the process $\{X_t\}_{t\in\mathbb{R}}$ that appears in definitions and results is mainly the Wiener process over $\mathbb{R}$. The $\mathcal{L}_2$-integral with the Wiener process leads to the central concept of Gaussian WN, which is not a stochastic process in the exact sense but rather a so-called generalized stochastic process. This generalized stochastic process is introduced in Section 3.4.4.

Thus, let $\{X_t\}_{t\in\mathbb{R}}$ be equal to the Wiener process $\{W_t\}_{t\in\mathbb{R}}$. Its a.c.v.f. γ is given by Proposition 3.2.14. Thus the integration measure of q_2 in Theorem 3.4.1.2 assigns total mass to the diagonal $s = t$. Precisely, $\mathrm{d}\gamma(s,t)$ is equal to: the Lebesgue measure over the diagonal $s = t$, $\forall s, t \geq 0$; minus the Lebesgue measure over the diagonal $s = t$, $\forall s, t < 0$; zero, $\forall s, t \in \mathbb{R}$ such that sgn$s \neq$ sgn$t$. Thus, $\forall 0 \leq a < b < \infty$, we find $q_2 = \int_a^b |g(t)|^2\mathrm{d}t$. Otherwise, $\forall -\infty < a < 0 < b < \infty$, we find $q_2 = \int_0^b |g(t)|^2\mathrm{d}t - \int_a^0 |g(t)|^2\mathrm{d}t$. Theorem 3.4.1.2 tells that whenever q_2 is finite, the stochastic integral

$$I_2 = \int_{[a,b]} g(t)\mathrm{d}W_t,$$

exists as a $\mathcal{L}_2$-limit. It also follows from (3.4.6) in Theorem 3.4.4 that, for any real-valued functions g and h and $-\infty < a < c < b < d < \infty$,

$$\mathsf{E}\left[\int_{[a,b]} g(s)\mathrm{d}W_s \int_{[c,d]} h(t)\mathrm{d}W_t\right] = \int_c^b g(t)h(t)\mathrm{d}t, \quad (3.4.7)$$

whenever the integral on the right side exists.

We have already mentioned that the Wiener process takes a central role in the theory of stationary processes, also because it provides Gaussian WN. From a formal perspective, if we redenote the symbol dW_t, which appears in the stochastic integral defined as a $\mathcal{L}_2$-limit, as $W_t'dt$, then we call $\{W_t'\}_{t \in \mathbb{R}}$ Gaussian WN. With this formal notation we can re-express (3.4.7) as

$$\text{cov}\left(\int_a^b g(s)W_s'ds, \int_c^d h(t)W_t'dt \right) = \int_c^b g(t)h(t)dt.$$

This equation characterizes Gaussian WN, which is thus not a proper stochastic process but a generalized stochastic process. The notation W_t' is clearly not one of the two known derivatives: the sample paths of the Wiener process are nowhere differentiable and the Wiener process is not $\mathcal{L}_2$-differentiable, cf. Examples 3.3.6. This particular derivative, which is denoted by W_t' just as other derivatives, is defined through the notion of generalized stochastic process introduced in Section 3.4.4.

3.4.3 *Ergodicity*

This short section presents an ergodicity property of a $\mathcal{L}_2$-integrable stochastic process. The concept of ergodicity originates from statistical mechanics. When a characteristic of a stochastic process at any fixed time (like the mean value of the process at some fixed time) can be re-established from the analogue characteristic observed over the infinite time horizon (like the mean of all values taken by the process during the infinite time horizon), then the process is called ergodic (relative to the property). Thus, an ergodic process will eventually take all of its possible values. Classical ergodic theorems refer to the $\mathcal{L}_2$-integrals

$$\frac{1}{\tau} \int_0^\tau X_t dt \text{ and } \frac{1}{2\tau} \int_{-\tau}^\tau X_t dt, \tag{3.4.8}$$

for large values of τ. The next result considers the first of these two integrals.

Corollary 3.4.5 (Ergodic mean). *Consider the complex-valued and square-integrable stochastic process* $\{X_t\}_{t \geq 0}$, *with* $\mathsf{E}[X_t] = 0$, $\forall t \geq 0$, *and with a.c.v.f.* γ.
1. If the a.c.v.f. γ *satisfies*

$$\lim_{\tau \to \infty} \frac{1}{\tau^2} \int_0^\tau \int_0^\tau \gamma(s,t)ds dt = 0, \tag{3.4.9}$$

then

$$\lim_{\tau \to \infty} \frac{1}{\tau} \int_0^\tau X_t dt \xrightarrow{\mathcal{L}_2} 0, \quad \text{as } \tau \to \infty. \tag{3.4.10}$$

2. If $\{X_t\}_{t \geq 0}$ *is stationary with a.c.v.f.* γ *that satisfies*

$$\lim_{\tau \to \infty} \frac{1}{\tau} \int_0^\tau \gamma(s) ds = 0,$$

then (3.4.10) holds.

Proof. 1. It follows from Theorem 3.4.1.1 that, under the condition (3.4.9) the first of the stochastic integrals of (3.4.8) exists as a $\mathcal{L}_2$-limit for τ sufficiently large. Moreover,

$$\mathsf{E}\left[\left| \frac{1}{\tau} \int_0^\tau X_t dt \right|^2 \right] = \frac{1}{\tau^2} \int_0^\tau \int_0^\tau \gamma(s,t) ds dt \xrightarrow{\tau \to \infty} 0.$$

2. This part can be seen as follows,

$$\int_0^\tau \int_0^\tau \gamma(s,t) ds dt = \int_0^\tau \int_0^\tau \gamma(s-t) ds dt = \int_0^\tau \int_{-t}^{\tau-t} \gamma(u) du dt$$

$$= \int_{-\tau}^\tau \int_0^\tau \mathrm{I}\{-t \leq u \leq \tau - t\} dt \gamma(u) du = \int_{-\tau}^\tau (\tau - |u|)\gamma(u) du. \tag{3.4.11}$$

$\square$

Note that if one replaces (3.4.9) by a slightly stronger condition, then one can show that (3.4.10) holds with a.s. convergence instead of $\mathcal{L}_2$-convergence. The proof of this fact is given at p. 94 of Cramér and Leadbetter (1967).

3.4.4 *Generalized and purely random stochastic processes*

This section introduces the notion of generalized stochastic processes and it mainly follows Yaglom (1962), pp. 207–213. In the context of stationary time series, WN is the time series with a.c.v.f. taking value zero at any (integer) non-zero time lag and taking a finite positive value at lag zero. Thus, the values of the time series are uncorrelated at non-null time lags. Regarding the frequency domain, the WN time series is the one with constant spectral density over $[-\pi, \pi]$: all frequencies are equally represented in the time series (cf. Example 2.4.16). This characteristic of WN, expressed in time domain and re-expressed in frequency domain, justify the name of purely random time series or discrete time process.

By extending this notion of purely random process from discrete to continuous time, we obtain a stationary stochastic process whose values are uncorrelated at any two arbitrarily close instants:

$$\gamma(h) = 0, \quad \forall h \neq 0.$$

As with the time series, the spectral density is a constant function:

$$f(\alpha) = f_0, \quad \forall \alpha \in \mathbb{R},$$

for some $f_0 \in (0, \infty)$. Thus, this purely random process or continuous time WN has nonintegrable spectral density, because the integration domain (that was $(-\pi, \pi]$ for time series) has now become $\mathbb{R}$. Referring to end of Section 3.3, because no spectral moment exists, continuous time WN has no $\mathcal{L}_2$-derivatives and it is irregular or noisy in this sense. WN is also completely memoryless, in the sense that it exhibits uncorrelation at any nonvanishing time lag. Although the discontinuity of γ at zero does not allow for the application of Bochner's theorem 3.5.1, which follows, the formal application of Fourier inversion formula leads to

$$\gamma(h) = \int_{-\infty}^{\infty} e^{i\alpha h} f(\alpha) d\alpha = f_0 \int_{-\infty}^{\infty} e^{i\alpha h} d\alpha = \begin{cases} \infty, & \text{if } h = 0, \\ 0, & \text{if } h \neq 0. \end{cases} \quad (3.4.12)$$

If one would take as a.c.v.f. γ equal to Dirac's density δ (cf. Section A.8.2) times σ^2, for some $\sigma > 0$, then Fourier inversion formula leads to

$$f(\alpha) = \frac{1}{2\pi} \int_{-\infty}^{\infty} e^{-i\alpha s} \gamma(s) ds = \frac{\sigma^2}{2\pi}, \quad \forall \alpha \in \mathbb{R}.$$

These considerations alone motivate the need of a more general definition of a stochastic process with continuous time, which would have WN with continuous time as example. This particular definition shares similarities with the linear filter, which is introduced later in Section 3.9.1. We merely mention that the impulse response function of the linear filter, $\psi : \mathbb{R} \to \mathbb{R}$, leads to the output of the linear filter of $\{X_t\}_{t \in \mathbb{R}}$. This output at time $t \in \mathbb{R}$ takes the following form,

$$\int_{-\infty}^{\infty} \psi(s) X_{t-s} ds = \int_{-\infty}^{\infty} \psi(t-s) X_s ds. \quad (3.4.13)$$

An output in the form of a single random variable, instead of a stochastic process, can be obtained by the replacing $\psi(t-s)$ by $w(s)$ in (3.4.13), for some function w. This gives Definition 3.4.6 of the generalized stochastic processes, as functional of the so-called weighting function w.

Definition 3.4.6 (Generalized stochastic process). *Let $w : \mathbb{R} \to \mathbb{R}$ be a weighting function and let $\{X_t\}_{t\in\mathbb{R}}$ be a complex-valued stochastic process of $\mathcal{L}_2$. Then the mean square integral*

$$X(w) = \int_{-\infty}^{\infty} w(t)X_t \mathrm{d}t, \qquad (3.4.14)$$

is the generalized stochastic process of the ordinary stochastic process $\{X_t\}_{t\in\mathbb{R}}$ at weighting function w. This generalized stochastic process has point values X_t, $\forall t \in \mathbb{R}$.

The stochastic integral (3.4.14) is of the type of Theorem 3.4.1.1.

Assume that $\mathcal{W}$ is any suitable set of weighting functions such that, $\forall w \in \mathcal{W}$, $X(w)$ exists. The generalized stochastic process $\{X(w)\}_{w\in\mathcal{W}}$ may be understood from the fact that any measure of $\{X_t\}_{t\in\mathbb{R}}$ depends on a physical device. Any physical device is subject to some inertia which leads to operation of averaging the point values X_t, $\forall t \in \mathbb{R}$. The pure or exact measure of the point values of the generalized process is only an ideal concept which is not available in the practice.

Remarks 3.4.7 (Characteristics of generalized process).

(1) The generalized stochastic process inherits the linearity of the integral, that is

$$X(a_1 w_1 + a_2 w_2) = a_1 X(w_1) + a_2 X(w_2), \ \forall a_1, a_2 \in \mathbb{R} \text{ and } w_1, w_2 \in \mathcal{W}. \qquad (3.4.15)$$

(2) Any ordinary process $\{X_t\}_{t\in\mathbb{R}}$ leads to a generalized stochastic process $\{X(w)\}_{w\in\mathcal{W}}$ through (3.4.14). However, a generalized process $\{X(w)\}_{w\in\mathcal{W}}$ is not necessarily defined from $\{X_t\}_{t\in\mathbb{R}}$ through (3.4.14).

(3) A generalized stochastic process can be specified by an adequate adaptation of the concept of the f.d.d. and of Kolmogorov existence theorem 3.1.7, in which any finite number of times is replaced by a finite number of weight functions.

(4) One can deduce from the linearity property (3.4.15) that the set $\mathcal{W}$ should be taken as a space, in the sense that the generalized stochastic process at linear combinations of the weight functions should exist. Moreover, the larger the space $\mathcal{W}$, the narrower the corresponding generalized stochastic process, in the following sense. Let $\mathcal{V} \subset \mathcal{W}$, then

$$\{X(w)\}_{w\in\mathcal{W}} \text{ exists} \implies \{X(w)\}_{w\in\mathcal{V}} \text{ exists}.$$

However,

$$\{X(w)\}_{w \in \mathcal{V}} \ \text{exists} \ \nRightarrow \ \{X(w)\}_{w \in \mathcal{W}} \ \text{exists}.$$

Thus $\mathcal{W}$ should be taken as any not too wide functional space.
An important and practical choice for $\mathcal{W}$ is the space of functions that are $n \geq 1$ times differentiable and vanishing outside a finite interval: we denote this space by $\mathcal{W}_n$.

Assume the weighting function $w \in \mathcal{W}_1$ differentiable, with derivative denoted w', and the stochastic process $\{X_t\}_{t \in \mathbb{R}}$ mean square differentiable, with mean square derivative at t denoted X'_t. Partial integration gives $-\int_{-\infty}^{\infty} w'(t) X_t \mathrm{d}t = \int_{-\infty}^{\infty} w(t) X'_t \mathrm{d}t$. Thus we can always define the derivative of the ordinary process $\{X_t\}_{t \in \mathbb{R}}$ in terms of a generalized process and precisely as follows,

$$X'(w) = -\int_{-\infty}^{\infty} w'(t) X_t \mathrm{d}t, \quad \forall w \in \mathcal{W}_1.$$

The following definition is motivated by the fact that the above integral does exist even without the differentiability assumption on X_t.

Definition 3.4.8 (Derivative of generalized stochastic process). *The derivative of the generalized stochastic process of $\{X_t\}_{t \in \mathbb{R}}$ at the weighting function $w \in \mathcal{W}_1$ is given by*

$$X'(w) = -X(w').$$

According to Definition 3.4.8, an ordinary stochastic process can always be regarded as differentiable, except that the derivative may be a generalized stochastic process instead of an ordinary one. The central example is the Gaussian WN, introduced in Section 3.4.2. The sample paths of the Wiener process $\{W_t\}_{t \in \mathbb{R}}$ are not differentiable, as explained at the end of Section 3.2.2, and the Wiener process is not $\mathcal{L}_2$-differentiable, see Examples 3.3.6. However the generalized stochastic process $\{W'_t\}_{t \in \mathbb{R}}$ can be defined according to Definition 3.4.8. It has (nonintegrable) constant spectral density over $\mathbb{R}$ and a.c.v.f. (3.4.12).

Example 3.4.9 (Langevin equation). *A notorious illustration of Gaussian WN is provided by the Langevin equation. Consider a particle suspended in some homogeneous fluid. Denote by m the mass of the particle, by V_t its velocity and by V'_t its acceleration, both at time $t \in \mathbb{R}$. The coefficient of friction f depends on the viscosity of the fluid and on the shape and the size of the particle. Thus, the particle is subject to the resistance from* *friction given by* fV_t, *which is a force acting in the opposite direction of its velocity. Newton's third law, sometimes referred to as the action-reaction law, is given by*

$$fV_t = -mV_t'.$$

However, this system is perturbed by an independent Gaussian WN term W_t' generated from the sum of many very small collisions of the surrounding molecules of the fluid. This additive description of the perturbation justifies its Gaussian nature. The perturbed model gives the Langevin equation

$$m\mathrm{d}V_t + fV_t\mathrm{d}t = \sigma\mathrm{d}W_t \ \text{ or } \ V_t' + \frac{f}{m}V_t = \frac{\sigma}{m}W_t', \ \ \forall t \in \mathbb{R}, \qquad (3.4.16)$$

for some scaling parameter $\sigma > 0$. The solution of this stochastic differential equation is the Ornstein-Uhlenbeck process $\{V_t\}_{t\in\mathbb{R}}$ introduced by Uhlenbeck and Ornstein (1930, 1954). It can be given in the form of a stochastic integral, as shown in Example 3.7.4. The Ornstein-Uhlenbeck process is obtained as the stationary, Markovian and Gaussian stochastic process, in Example 3.2.7, and through the Lamperti transformation, in Example 3.2.21.

3.5   Spectral distribution and autocovariance function

We now analyze two central and related topics: the spectral distribution and the a.c.v.f. Section 3.5.1 presents and shows Bochner's theorem, which essentially states that a.c.v.f. and spectral distribution form a Fourier pair. Then, Section 3.5.2 provides some explicit inversion formulae for spectral density, d.f. and a.c.v.f. When the stationary process is real-valued, the spectral distribution is symmetric and thus the one-sided spectral distribution, which defined over $\mathbb{R}_+$ only, is often considered. This is the topic of Section 3.5.3. Sampling a continuous time stationary process usually means collecting its values at equidistant epochs. It turns out that the spectral distribution of the sampled process is obtained by wrapping the initial spectral distribution around the circle. This is explained in Section 3.5.4.

3.5.1   *Spectral distribution and Bochner's theorem*

A central theorem in the analysis of time series is Herglotz's theorem 2.4.10: in view of the characterization of the a.c.v.f. in terms of n.n.d. function $\mathbb{Z} \to \mathbb{C}$ given by Theorem 2.4.6, it states that every a.c.v.f. of a stationary time series is the Fourier transform of its spectral distribution. The analogue version for stationary continuous time stochastic processes is Bochner's theorem 3.5.1 given below: in view of the characterization of the a.c.v.f. in terms of n.n.d. function $\mathbb{R} \to \mathbb{C}$ given by Theorem 3.1.21, it states that every continuous a.c.v.f. of a stationary continuous time stochastic process is the Fourier transform of its spectral distribution. Bochner's theorem provides a remarkable analogy of the pair, a.c.v.f. and spectral distribution, with the pair, characteristic function and probability distribution. The precise inversion formula for the spectral distribution, which is analogue to the inversion formula for the probability distribution, cf. Theorem A.4.3, will be presented later in Theorem 3.5.4.

Theorem 3.5.1 (Bochner). *The continuous function $\kappa : \mathbb{R} \to \mathbb{C}$ is n.n.d.*
$\Longleftrightarrow$

$$\kappa(h) = \int_{\mathbb{R}} e^{ih\alpha} dF(\alpha), \quad \forall h \in \mathbb{R},$$

for some d.f. F over $\mathbb{R}$, with finite mass and with $F(-\infty) = 0$.

Proof. ($\Leftarrow$) Given $\kappa(h) = \int_{\mathbb{R}} e^{ih\alpha} dF(\alpha)$, $\forall h \in \mathbb{R}$, we have $\forall -\infty < t_1 < \cdots < t_n < \infty$, $a_1, \ldots, a_n \in \mathbb{C}$ and $n \in \mathbb{N}^*$,

$$\sum_{j=1}^{n} \sum_{k=1}^{n} a_j \overline{a_k} \kappa(t_j - t_k) = \int_{\mathbb{R}} \sum_{j=1}^{n} a_j e^{it_j \alpha} \sum_{k=1}^{n} \overline{a_k} e^{-it_k \alpha} dF(\alpha)$$

$$= \int_{\mathbb{R}} \left| \sum_{j=1}^{n} a_j e^{it_j \alpha} \right|^2 dF(\alpha)$$

$$\geq 0.$$

($\Rightarrow$) Let $\tau \in \mathbb{R}_+^*$ and consider the arbitrary partition $0 = t_0 < t_1 < \cdots < t_{n-1} < t_n = \tau$. Let $\alpha \in \mathbb{R}$. Because κ is assumed n.n.d., we find

$$f_\tau(\alpha) = \frac{1}{2\pi\tau} \int_0^\tau \int_0^\tau \kappa(s - t) e^{-i\alpha(s-t)} ds dt$$

$$= \lim_{n \to \infty} \sum_{j=1}^{n} \sum_{k=1}^{n} \kappa(t_j - t_k) e^{-i\alpha t_j} (t_j - t_{j-1}) \overline{e^{-i\alpha t_k} (t_k - t_{k-1})}$$

$$\geq 0,$$

where the mesh of the partition vanishes as $n \to \infty$.

We need to show that f_τ is an integrable function over $\mathbb{R}$. The assumption that κ is n.n.d. implies that $\kappa(\cdot)e^{-i\alpha\cdot}$ is an Hermitian function over $\mathbb{R}$, for fixed α. We can then follow the computation in (3.4.11), in order to obtain

$$f_\tau(\alpha) = \frac{1}{2\pi} \int_{-\tau}^{\tau} \left(1 - \frac{|s|}{\tau}\right) \kappa(s) e^{-i\alpha s} ds = \frac{1}{2\pi} \int_{-\infty}^{\infty} p\left(\frac{s}{\tau}\right) \kappa(s) e^{-i\alpha s} ds,$$

(3.5.1)

where p denotes here the triangular probability density given by (2.4.4). Let $\delta \in \mathbb{R}_+^*$. We have

$$
\begin{aligned}
u(\delta, \tau) &= \int_{-\infty}^{\infty} p\left(\frac{\alpha}{2\delta}\right) g(\alpha, \tau) d\alpha \\
&= \frac{1}{2\pi} \int_{-\infty}^{\infty} p\left(\frac{\alpha}{2\delta}\right) \int_{-\infty}^{\infty} p\left(\frac{s}{\tau}\right) \kappa(s) e^{-i\alpha s} ds \, d\alpha \\
&= \frac{1}{2\pi} \int_{-\tau}^{\tau} p\left(\frac{s}{\tau}\right) \kappa(s) \int_{-2\delta}^{2\delta} p\left(\frac{\alpha}{2\delta}\right) e^{-i\alpha s} d\alpha \, ds,
\end{aligned}
$$

because the fact that integrand and integration domain are bounded allows for the application of Fubini's theorem. The inner integral can be evaluated as follows,

$$
\begin{aligned}
& \int_{-2\delta}^{2\delta} \left(1 - \frac{|\alpha|}{2\delta}\right) e^{-i\alpha s} d\alpha \\
&= 2 \int_0^{2\delta} \left(1 - \frac{\alpha}{2\delta}\right) \cos \alpha s \, d\alpha \\
&= 2 \left\{ -\int_0^{2\delta} \left(-\frac{1}{2\delta}\right) \frac{\sin \alpha s}{s} d\alpha + \left[\left(1 - \frac{\alpha}{2\delta}\right) \frac{\sin \alpha s}{s}\right]_0^{2\delta} \right\} \\
&= \frac{1}{\delta} \int_0^{2\delta} \frac{\sin \alpha s}{s} d\alpha \\
&= \frac{1 - \cos 2\delta s}{\delta s^2} \\
&= 2\delta \operatorname{sinc}^2 \delta s, \quad \forall s \in [-\tau, \tau].
\end{aligned}
$$

Thus we obtain

$$
\begin{aligned}
u(\delta, \tau) &= \frac{\delta}{\pi} \int_{-\tau}^{\tau} p\left(\frac{s}{\tau}\right) \kappa(s) \operatorname{sinc}^2 \delta s \, ds \\
&\le \frac{1}{\pi} \int_{-\delta\tau}^{\delta\tau} \left| p\left(\frac{s}{\delta\tau}\right) \kappa\left(\frac{s}{\delta}\right) \operatorname{sinc}^2 s \right| ds \\
&\le \frac{2}{\pi} \kappa(0) \int_0^{\delta\tau} \operatorname{sinc}^2 s \, ds \\
&\le \kappa(0),
\end{aligned}
$$

because $\int_0^\infty \mathrm{sinc}^2 s\,ds = \pi/2$. Because for $\alpha \in \mathbb{R}$ fixed the function $p(\alpha/(2\delta))f_\tau(\alpha) \geq 0$ is increasing w.r.t. δ, Monotone convergence theorem A.5.1 yields

$$\int_{-\infty}^\infty f_\tau(\alpha)d\alpha = \lim_{\delta\to\infty} u(\delta, \tau) \leq \kappa(0). \qquad (3.5.2)$$

Thus $f_\tau(\alpha) \geq 0$, $\forall \alpha \in \mathbb{R}$, is integrable and, obviously, $p(\cdot/\tau)\kappa(\cdot)$ is integrable too. Therefore, (3.5.1) shows that these two integrable functions form a Fourier pair. Consequently, we obtain

$$p\left(\frac{s}{\tau}\right)\kappa(s) = \int_{-\infty}^\infty e^{i\alpha s} f_\tau(\alpha)d\alpha, \quad \forall s \in \mathbb{R}. \qquad (3.5.3)$$

This last equation at $s = 0$ allows to refine (3.5.2) with

$$\kappa(0) = \int_{-\infty}^\infty f_\tau(\alpha)d\alpha.$$

This equation tells that the function $h_\tau = f_\tau/\kappa(0)$ is a probability density. We can indeed exclude the trivial case $\kappa(0) = 0$. Then, it follows from (3.5.3) that

$$\varphi_\tau(s) = p\left(\frac{s}{\tau}\right)\frac{\kappa(s)}{\kappa(0)}, \quad \forall s \in \mathbb{R},$$

is the characteristic function of the density h_τ. By assumption we have that

$$\lim_{\tau\to\infty} \varphi_\tau(s) = \frac{\kappa(s)}{\kappa(0)}, \quad \forall s \in \mathbb{R},$$

is continuous and in particular at $s = 0$. Thus from Lévy continuity theorem A.3.17 we obtain that $\kappa(s)/\kappa(0)$ is the characteristic function of some probability d.f. H. We have thus obtained that

$$\frac{\kappa(s)}{\kappa(0)} = \int_{-\infty}^\infty e^{is\alpha}dH(\alpha), \quad \forall s \in \mathbb{R},$$

and the desired result is obtained by selecting the d.f. $F = \kappa(0)H$. $\qquad\square$

Remark 3.5.2 (Proofs of Herglotz and Bochner's theorems). *It is instructive to compare the proofs of Herglotz theorem 2.4.10 and Bochner's theorem 3.5.1. In both proofs parts ($\Leftarrow$) are similar. Parts ($\Rightarrow$) can be compared as follows. In the proof of Herglotz theorem, the function f_n, where $n \in \mathbb{N}^*$, is clearly integrable over $[-\pi, \pi]$. In the proof of Bochner's theorem, a quite long justification of the integrability over $\mathbb{R}$ of the function f_τ, where $\tau \in \mathbb{R}_+^*$, is required; cf. second paragraph of ($\Rightarrow$). With Herglotz theorem (2.4.6) gives the Fourier coefficient of f_n, whereas with Bochner theorem*

(3.5.3) gives the Fourier transform of f_τ. The last arguments of Herglotz theorem use Helly's theorem, whereas Bochner's theorem uses Lévy continuity theorem. The use of Helly's theorem is possible because the d.f. of Herglotz theorem have bounded support, $[-\pi, \pi]$, whereas they have unbounded support, $\mathbb{R}$, in Bochner's theorem. We refer to Remark A.3.20 about Helly's theorem when d.f. have bounded or unbounded support.

3.5.2 *Inversion formulae*

This section concerns inversion formulae for spectral d.f. and density of a stationary process. The main results are given in Theorem 3.5.4. Before presenting this theorem, some comments on the analogy with inversion formulae for probability d.f. and density are given under Remarks 3.5.3.

Remarks 3.5.3 (Analogy between a.c.v.f. and characteristic function).

(1) *Many useful properties of the a.c.v.f. can be directly deduced from Bochner theorem 3.5.1. For instance, the product of two continuous a.c.v.f. has the spectral distribution given by the convolution of the two original spectral distributions. Consequently, the product of two continuous a.c.v.f. is another a.c.v.f. Moreover, given the characterization in terms of n.n.d. function of Theorem 3.1.21, we know that every characteristic function is the a.c.r.f. of a continuous time (real- or complex-valued) stationary process.*

(2) *Another consequence of Theorem 3.5.1 is that every continuous a.c.v.f. γ that is normalized as $\varphi(h) = \gamma(h)/\gamma(0)$, $\forall h \in \mathbb{R}$, i.e. every continuous a.c.r.f., is a characteristic function. Note that any characteristic function is continuous; cf. Proposition A.4.2.2. Given this fact, Theorem A.4.3 for the inversion of characteristic functions can be adapted to the case where the characteristic function φ is replaced by the a.c.v.f. γ. In this case, the probability d.f. of Theorem A.4.3 becomes a spectral d.f. with mass $\gamma(0) < \infty$. The same reasoning holds for Corollary A.4.4, which expresses the probability density as the Fourier inversion of the characteristic function. Corollary A.4.5 can be adapted so to tell that a continuous a.c.v.f. determines the spectral d.f. uniquely. For completeness or clarity, we state in Theorem 3.5.4 below the adaptation to the a.c.v.f. of Theorem A.4.3 and its Corollary A.4.4 that are given for the characteristic function.*

Theorem 3.5.4 (Inversion formulae for spectral d.f. and density). *Let γ be a continuous a.c.v.f. with spectral d.f. F.*

(1) If $\int_{-\infty}^{\infty} |\gamma(s)| \mathrm{d}s < \infty$, then F is absolutely continuous with bounded density f and it holds that, $\forall \alpha \in \mathbb{R}$,

$$f(\alpha) = \frac{1}{2\pi} \int_{-\infty}^{\infty} \exp\{-i\alpha s\} \gamma(s) \mathrm{d}s. \tag{3.5.4}$$

(2) Let $F^*(\alpha) = \{F(\alpha) + F(\alpha-)\}/2$, $\forall \alpha \in \mathbb{R}$. Then, $\forall \alpha < \beta \in \mathbb{R}$,

$$F^*(\beta) - F^*(\alpha) = \frac{1}{2\pi} \lim_{u \to \infty} \int_{-u}^{u} \frac{\exp\{-i\beta s\} - \exp\{-i\alpha s\}}{-is} \gamma(s) \mathrm{d}s. \tag{3.5.5}$$

Remarks 3.5.5.

(1) The limit in (3.5.5) is Cauchy's principal value of the integral (from $-\infty$ to ∞).

(2) The integrability condition of the a.c.v.f. that appears in Theorem 3.5.4.1 is a sufficient but not a necessary condition. In fact (3.5.4) holds at any point $\alpha \in \mathbb{R}$ where the spectral density f is continuous, left- and right-differentiable, without further condition. One may also consider the limit of (3.5.5) given by

$$f^*(\alpha) = \lim_{\varepsilon \to 0} \frac{F^*(\alpha + \varepsilon) - F^*(\alpha - \varepsilon)}{2\varepsilon}, \quad \forall \alpha \in \mathbb{R}.$$

If f is continuous at α, then we obtain the desired result $f^*(\alpha) = f(\alpha)$. The next example is an instance of a nonintegrable a.c.v.f. for which this last equation does not hold at discontinuity points of f.

Example 3.5.6 (A.c.v.f. sinc). *We want to find the spectral density of the a.c.v.f.*

$$\gamma(h) = \operatorname{sinc} h, \quad \forall h \in \mathbb{R},$$

cf. Definition A.8.1. As shown in (A.8.1), the sinc *function is not integrable. This function is however continuous and (3.5.5) gives, $\forall \varepsilon \in (0,1)$,*

$$\frac{F^*(\alpha + \varepsilon) - F^*(\alpha - \varepsilon)}{2\varepsilon}$$

$$= \frac{1}{2\pi} \lim_{u \to \infty} \int_{-u}^{u} \frac{\exp\{-i(\alpha + \varepsilon)s\} - \exp\{-i(\alpha - \varepsilon)s\}}{-is2\varepsilon} \text{sinc } s \, ds$$

$$= \frac{1}{2\pi} \lim_{u \to \infty} \int_{-u}^{u} e^{-i\alpha s} \text{sinc } s \text{ sinc } \varepsilon s \, ds$$

$$= \begin{cases} \frac{1}{2}, & \text{if } |\alpha| < 1 - \varepsilon, \\ \frac{1}{4}\left(1 + \frac{1-|\alpha|}{\varepsilon}\right), & \text{if } 1 - \varepsilon \leq |\alpha| \leq 1 + \varepsilon, \\ 0, & \text{if } |\alpha| > 1 + \varepsilon. \end{cases} \tag{3.5.6}$$

The last equality of (3.5.6) can be shown as follows. We know that for a given $a > 0$, the function $\text{sinc } ah$, $\forall h \in \mathbb{R}$, is the a.c.v.f. of the spectral density corresponding to the uniform probability density over $(-a, a)$, cf. (A.8.2). Consequently, the convolution of two uniform probability densities, one over $(-1, 1)$ and the other one over $(-\varepsilon, \varepsilon)$, gives the last equality in (3.5.6); cf. Remark 3.5.3.1. (The computation of this convolution is an entertaining exercise.) Thus,

$$f^*(\alpha) = \lim_{\varepsilon \to 0} \frac{F^*(\alpha + \varepsilon) - F^*(\alpha - \varepsilon)}{2\varepsilon} = \begin{cases} \frac{1}{2}, & \text{if } |\alpha| < 1, \\ \frac{1}{4}, & \text{if } |\alpha| = 1, \\ 0, & \text{if } |\alpha| > 1. \end{cases}$$

But f^* is the uniform density over $(-1, 1)$ with special values at the boundary points ± 1. Up to these two boundary values, we have the uniform density given in (A.8.2). We have thus illustrated Remark 3.5.5.2.

An absolutely continuous spectral distribution usually leads to noisier time signal, than a discrete spectral distribution does. The spectral distribution can also be a mixture of an absolutely continuous distribution and a discrete distribution. In such situations it is typical that the continuous part carries the noise and the discrete part carries the signal.

3.5.3 *One-sided spectral distribution*

Whenever a stationary stochastic process is real-valued, its spectral distribution is symmetric around the origin. In this situation, it is current practice to define the one-sided spectral d.f. over $\mathbb{R}_+$ solely, by cumulating the mass of the spectral frequencies from the origin towards positive and negative directions simultaneously. Precisely, let F be the spectral d.f. of a real-valued and stationary process. Then the one-sided spectral d.f. is given by

$$F_1(\alpha) = \begin{cases} 0, & \text{if } \alpha < 0, \\ F(\alpha) - F((-\alpha)-), & \text{if } \alpha \geq 0. \end{cases}$$

Let γ be the a.c.v.f. of the spectral d.f. F. Then symmetry means that

$$F((-\alpha)-) = \gamma(0) - F(\alpha), \quad \forall \alpha \in \mathbb{R}.$$

The construction of the one-sided spectral d.f. is equivalent to reflect the mass of negative frequencies towards their positive counterpart, so to double the mass of the positive frequencies, and then to correct by subtracting the total mass, so to leave the total mass unchanged. This procedure leads to the following definition.

Definition 3.5.7 (One-sided spectral d.f.). *Let F be the spectral d.f. of a real-valued stationary process. Its one-sided spectral d.f. is given by*

$$F_1(\alpha) = \begin{cases} 0, & \text{if } \alpha < 0, \\ 2F(\alpha) - \gamma(0), & \text{if } \alpha \geq 0. \end{cases}$$

Remarks 3.5.8 (Properties of the one-sided spectral d.f.). *Let F be the spectral d.f. of a real-valued stationary process and let F_1 its one-sided spectral d.f. Then the following properties hold.*

(1) $F_1(0-) = 0$ and $F_1(\infty) = F(\infty) = \gamma(0)$.

(2) $F(\alpha)$ has a jump of size $m > 0$ at $\alpha = 0 \Longrightarrow F_1(0) = F(0) - F(0-) = m$.

(3) $F(\alpha)$ has jump of size $m > 0$ at $\alpha > 0$, i.e. $F(\alpha)$ has jump of size $m > 0$ at $-\alpha \Longrightarrow F_1(\alpha)$ has jump of size $2m$ at α.

(4) F has density f over $\mathbb{R} \Longrightarrow F_1$ has density $f_1 = 2f$ over $\mathbb{R}_+$.

Two specific inversion formulae for one-sided spectral distribution follow immediately from Theorem 3.5.4.

Corollary 3.5.9 (Inversion formulae for one-sided spectral density and d.f.). *Let γ be a continuous a.c.v.f. with one-sided spectral d.f. F_1.*

(1) If $\int_{-\infty}^{\infty} |\gamma(s)| \mathrm{d}s < \infty$, then F_1 is absolutely continuous with bounded density f_1 and it holds that, $\forall \alpha \in \mathbb{R}_+^*$,

$$f_1(\alpha) = \frac{1}{\pi} \int_0^{\infty} \cos \alpha s \, \gamma(s) \mathrm{d}s.$$

(2) *Let* $\alpha \in \mathbb{R}_+^*$ *be any continuity point of* F_1, *then*

$$F_1(\alpha) = \frac{2}{\pi} \int_0^\infty \frac{\sin \alpha s}{s} \gamma(s) \mathrm{d}s.$$

3.5.4 *Sampling and wrapping*

Consider the complex-valued and stationary stochastic process $\{X_t\}_{t \in \mathbb{R}}$ with spectral d.f. F and a.c.v.f. γ. Suppose that $\{X_t\}_{t \in \mathbb{R}}$ can be sampled at any time $k \in \mathbb{Z}$. Then, the a.c.v.f. of the sampled process $\{X_k\}_{k \in \mathbb{Z}}$ is simply obtained by restricting the a.c.v.f. of $\{X_t\}_{t \in \mathbb{R}}$ to all integer arguments. Consequently, the spectral d.f. of the sampled process $\{X_k\}_{k \in \mathbb{Z}}$ is obtained by wrapping F, the spectral d.f. of $\{X_t\}_{t \in \mathbb{R}}$, around the unit circle. This is shown in Lemma 3.5.10. The rule to retain is the following: sampling at integer times along the time domain corresponds to wrapping around the unit circle in frequency domain.

Lemma 3.5.10 (Wrapped spectral distribution). *Let* $\{X_t\}_{t \in \mathbb{R}}$ *be a stationary process with spectral d.f.* F.

(1) *Then the sampled process* $\{X_k\}_{k \in \mathbb{Z}}$ *has the wrapped spectral d.f. at* $\alpha \in (-\pi, \pi]$ *given by*

$$F^\circ(\alpha) = \sum_{k=-\infty}^\infty F(\alpha + 2k\pi) - F((2k-1)\pi).$$

(2) *If* F *is absolutely continuous with density* f, *then* F° *is also absolutely continuous and it possesses the wrapped density at* $\alpha \in (-\pi, \pi]$ *given by*

$$f^\circ(\alpha) = \sum_{k=-\infty}^\infty f(\alpha + 2k\pi).$$

(3) *Wrapped and non-wrapped distributions possess the same same total mass, i.e.*

$$F^\circ(\pi) = F(\infty) = \gamma(0).$$

Proof. The a.c.v.f. of the sampled process $\{X_k\}_{k\in\mathbb{Z}}$ is the a.c.v.f. at integer values of the complete process $\{X_t\}_{t\in\mathbb{R}}$. We have indeed, $\forall k \in \mathbb{Z}$,

$$
\begin{aligned}
\gamma(k) &= \int_{\mathbb{R}} e^{i\alpha k} dF(\alpha) \\
&= \sum_{j=-\infty}^{\infty} \int_{(-\pi+2j\pi,\pi+2j\pi]} e^{i\alpha k} dF(\alpha) \\
&= \int_{(-\pi,\pi]} \sum_{j=-\infty}^{\infty} e^{i(\alpha+2j\pi)k} dF(\alpha + 2j\pi) \\
&= \int_{(-\pi,\pi]} e^{i\alpha k} d\left\{ \sum_{j=-\infty}^{\infty} F(\alpha + 2j\pi) \right\} \\
&= \int_{(-\pi,\pi]} e^{i\alpha k} d\left\{ \sum_{j=-\infty}^{\infty} F(\alpha + 2j\pi) - F((2j-1)\pi) \right\}.
\end{aligned}
$$

It follows from Herglotz's theorem 2.4.10 that

$$
F^\circ(\alpha) = \sum_{k=-\infty}^{\infty} F(\alpha + 2k\pi) - F((2k-1)\pi), \quad \forall \alpha \in [-\pi, \pi],
$$

is the unique d.f. with finite mass, with $F^\circ(-\pi) = 0$ and with Fourier coefficients $\gamma(h)$, $\forall h \in \mathbb{Z}$. $\qquad \square$

Example 3.5.11 (Wrapped Cauchy spectral distribution). *We can find in Remark 2.4.33 that the Cauchy probability distribution has density (2.4.11), characteristic function (2.4.14) and that it leads to the wrapped Cauchy circular density (2.4.12). This density can be multiplied by σ^2, with $\sigma > 0$, and become the spectral density of the* AR(1) *time series with WN variance equal to σ^2, as explained in Example 2.4.32.*

Consider now a continuous time stationary process with a.c.v.f. given by the constant σ^2 times the characteristic function of the Cauchy distribution, precisely by

$$
\gamma(h) = \sigma^2 e^{-\tau|h|}, \quad \forall h \in \mathbb{R}, \tag{3.5.7}
$$

for some $\sigma, \tau > 0$. This stochastic process sampled at integer times has spectral density equal to σ^2 times the wrapped Cauchy density (2.4.11) with parameters $\mu = 0$ and τ. It is thus the AR(1) *time series. The original or nonsampled process can be the Ornstein-Uhlenbeck process, which was introduced in Example 3.2.7 as the stationary, Markovian and Gaussian* *stochastic process, in Example 3.4.9 as solution of the Langevin stochastic differential equation and in Example 3.2.21 as the Lamperti transformation of the 1/2-selfsimilar Wiener process. But while the Ornstein-Uhlenbeck process is Gaussian, the AR(1) time series is not necessarily Gaussian.*

Remark 3.5.12 (Aliasing). *The wrapping procedure is a many-to-one map. Precisely, infinitely many spectral densities over $\mathbb{R}$ with same total mass can be wrapped to a single spectral density, over $(-\pi, \pi]$, of a sampled process. In order to see this, let f° be the spectral density with total mass $\sigma^2 < \infty$, of a sampled process, and let $\{p_k\}_{k \in \mathbb{Z}}$ be any probability function. Then*

$$f(\alpha) = \sum_{k=-\infty}^{\infty} p_k f^\circ(\alpha) \mathbb{I}\left\{\alpha \in ((2k-1)\pi, (2k+1)\pi]\right\}, \quad \forall \alpha \in \mathbb{R},$$

is a spectral density over $\mathbb{R}$ with same total mass σ^2. It is also the particular unwrap of f° associated to the chosen probability distribution $\{p_k\}_{k \in \mathbb{Z}}$. This means that the original or the nonsampled spectral distribution, and thus stationary process, cannot be retrieved after sampling. This impossibility of identification is often called aliasing.

Remark 3.5.13 (Sampling over general time lattice and Nyquist frequency). *Consider the more general situation where the sampling times belong to the lattice $\delta \mathbb{Z} = \{\ldots, -\delta, 0, \delta, \ldots\}$, for some $\delta \in \mathbb{R}^*_+$. Then Lemma 3.5.10 can be quite easily generalized, so to tell that the initial spectral d.f. F over $\mathbb{R}$ is wrapped around the circle of circumference $2\pi/\delta$. After wrapping, all relevant frequencies will be in $(-\pi/\delta, \pi/\delta]$. Thus, the sampled process $\{X_k\}_{k \in \delta \mathbb{Z}}$ has wrapped spectral d.f. at $\alpha \in (-\pi/\delta, \pi/\delta]$ given by*

$$F^\circ(\alpha) = \sum_{k=-\infty}^{\infty} F\left(\alpha + \frac{2k\pi}{\delta}\right) - F\left(\frac{(2k-1)\pi}{\delta}\right). \tag{3.5.8}$$

If F is absolutely continuous with density f, then F° is also so and the wrapped density at $\alpha \in (-\pi/\delta, \pi/\delta]$ is given by

$$f^\circ(\alpha) = \sum_{k=-\infty}^{\infty} f\left(\alpha + \frac{2k\pi}{\delta}\right). \tag{3.5.9}$$

One can readily verify these two formulae by following the proof of Lemma 3.5.10, essentially after noting that the a.c.v.f. of the sampled process is given by $\gamma(\delta k)$, $\forall k \in \mathbb{Z}$, and that

$$\exp\left\{i\alpha\delta k\right\} = \exp\left\{i\left(\alpha + \frac{2j\pi}{\delta}\right)\delta k\right\}, \quad \forall j, k \in \mathbb{Z} \text{ and } \alpha \in \mathbb{R}.$$

When the non-wrapped spectral distribution is null at all frequencies α such that $|\alpha| > \pi/\delta$, then the wrapping operation is without effect on the spectral distribution. In other terms, sampling such a continuous time process over the time lattice $\delta\mathbb{Z}$ does not lead to loss of information, because the all frequencies until highest frequency have been captured. This highest frequency is π/δ and it is called Nyquist frequency. Thus, the Nyquist frequency provides the minimal rate at which a signal needs to be sampled in order to retain all of the information. Sampling this process at time lags larger than δ is often called undersampling.

3.6 Spectral decomposition of stationary processes and the spectral theorem

This part is devoted to the spectral decomposition of a stationary process and it is divided in the three following parts. Section 3.6.1 presents the spectral analysis of real-valued stationary processes that have at most countably many frequencies. In Section 3.6.2 the process is complex-valued and the frequencies are not necessarily countable. Then the Spectral decomposition theorem 3.6.4 of stationary process is shown. This theorem justifies the existence of the spectral process. Section 3.6.3 provides the spectral analysis of complex-valued stationary processes with at most countably many frequencies. It is the adaptation of Section 3.6.1 to the complex case.

3.6.1 *Discrete spectral process and distribution of a real process*

This section considers the spectral decomposition of a real-valued stationary process with at most countably many frequencies over $\mathbb{R}_+$. This spectral decomposition is already introduced in (3.2.5). It is the analogue of the spectral decomposition of complex-valued time series given in Section 2.4.1, in which finitely many frequencies restricted to $(-\pi, \pi]$ are considered.

We thus consider the real-valued and stationary stochastic process $\{X_t\}_{t\in\mathbb{R}}$ given by

$$X_t = B(\alpha_0) + 2\sum_{j=1}^{\infty} B(\alpha_j)\cos(\alpha_j t + \theta(\alpha_j)), \quad \forall t \in \mathbb{R}, \qquad (3.6.1)$$

where $0 = \alpha_0 < \alpha_1 < \cdots < \infty$ are the frequencies, $\theta(\alpha_j)$, for $j = 1, 2, \ldots$, uniformly distributed over $[0, 2\pi)$, are the phases, $B(\alpha_j)$, for $j = 0, 1, \ldots$, real-valued and square-integrable, are the amplitudes and where all these random variables are independent. Note that the distributions of $B(\alpha_j)$, for $j = 0, 1, \ldots,$ need not be specified in order to have stationarity. When these distributions are the Rayleigh ones given in (3.2.6), then $\{X_t\}_{t \in \mathbb{R}}$ is a stationary Gaussian process. The number of summands in (3.6.1) may of course also be finite. Define

$$\alpha_{-j} = -\alpha_j, \quad \text{for} \ \ j = 1, 2, \ldots,$$

and assume

$$B(-\alpha) = B(\alpha), \ \theta(-\alpha) = -\theta(\alpha), \quad \forall \alpha \in \mathbb{R}. \tag{3.6.2}$$

Then we can re-express (3.6.1) as

$$
\begin{aligned}
X_t &= B(\alpha_0) + 2 \sum_{j=1}^{\infty} B(\alpha_j) \frac{e^{i\{\alpha_j t + \theta(\alpha_j)\}} + e^{-i\{\alpha_j t + \theta(\alpha_j)\}}}{2} \\
&= B(\alpha_0) + \sum_{j=-\infty, j \neq 0}^{\infty} e^{i\alpha_j t} B(\alpha_j) e^{i\theta(\alpha_j)} \\
&= \sum_{j=-\infty}^{\infty} e^{i\alpha_j t} A(\alpha_j), \tag{3.6.3}
\end{aligned}
$$

where

$$A(\alpha_j) = B(\alpha_j) e^{i\theta(\alpha_j)}, \quad \forall j \in \mathbb{Z}. \tag{3.6.4}$$

The complex coefficients in (3.6.4) modulate the harmonics in (3.6.3). They are uncorrelated and satisfy the Hermitian property

$$A(\alpha_{-j}) = \overline{A(\alpha_j)}, \quad \forall j \in \mathbb{Z}.$$

We clearly have $\mathsf{E}[X_t] = \mathsf{E}[B(0)]$, $\forall t \in \mathbb{R}$, and we assume $\mathsf{E}[B(0)] = 0$.

We now derive the spectral distribution. We have, $\forall t, h \in \mathbb{R}$ and $j \in \mathbb{N}$,

$$
\begin{aligned}
&\mathsf{E}[B(\alpha_j) \cos(\alpha_j(t + h) + \theta(\alpha_j)) B(\alpha_j) \cos(\alpha_j t + \theta(\alpha_j))] \\
&= \mathsf{E}\left[B^2(\alpha_j)\right] \mathsf{E}\left[\cos^2(\alpha_j t + \theta(\alpha_j))\right] \cos \alpha_j h \\
&= \frac{1}{2} \mathsf{E}\left[B^2(\alpha_j)\right] \left(\mathsf{E}\left[\cos 2\{\alpha_j t + \theta(\alpha_j)\}\right] + 1\right) \cos \alpha_j h \\
&= \frac{1}{2} \mathsf{E}\left[B^2(\alpha_j)\right] \cos \alpha_j h.
\end{aligned}
$$

Define

$$\sigma_j^2 = \mathsf{E}\left[|A(\alpha_j)|^2\right] = \mathsf{E}\left[B^2(\alpha_j)\right], \quad \text{where} \ \sigma_j > 0, \quad \forall j \in \mathbb{Z}. \tag{3.6.5}$$

We deduce from the above computation that the a.c.v.f. of $\{X_t\}_{t \in \mathbb{R}}$ is given by

$$\gamma(h) = \sigma_0^2 + 2 \sum_{j=1}^{\infty} \sigma_j^2 \cos \alpha_j h = \sum_{j=-\infty}^{\infty} \sigma_j^2 e^{i\alpha_j h}, \quad \forall h \in \mathbb{R}, \tag{3.6.6}$$

because $\sigma_{-j} = \sigma_j$, for $j = 1, 2, \ldots$. The total spectral mass is thus

$$\gamma(0) = \sigma_0^2 + 2 \sum_{j=1}^{\infty} \sigma_j^2,$$

which is finite because of the square-integrability of the process. Thus, the one-sided spectral d.f. is given by

$$F_1(\alpha) = \sigma_0^2 + 2 \sum_{j=1}^{\infty} \sigma_j^2 \, 1\{\alpha_j \leq \alpha\}, \quad \forall \alpha \geq 0, \tag{3.6.7}$$

and the usual spectral d.f. is given by

$$F(\alpha) = \sum_{j=-\infty}^{\infty} \sigma_j^2 \, 1\{\alpha_j \leq \alpha\}, \quad \forall \alpha \in \mathbb{R}. \tag{3.6.8}$$

The stochastic process $\{A(\alpha_j)\}_{j \in \mathbb{Z}}$ with the properties given above, in particular with the orthogonality property (3.6.9), is the spectral process. It is a discrete process as there are countably many frequencies. The next section presents the spectral process with frequencies in the continuum, which will be denoted by $\{Z_\alpha\}_{\alpha \in \mathbb{R}}$.

3.6.2 *Spectral decomposition*

The stationary processes analyzed in Section 3.6.1 are real-valued and have at most countably many frequencies. In this section we consider the most general complex-valued stationary process $\{X_t\}_{t \in \mathbb{R}}$ for which the set of all possible frequencies is not necessarily countable. In the countable case, the complex-valued random coefficients (3.6.4) have the following properties: $\forall j, k \in \mathbb{Z}$, $\mathsf{E}[A(\alpha_j)] = 0$ and

$$\mathsf{E}\left[A(\alpha_j)\overline{A(\alpha_k)}\right] = \begin{cases} \sigma_j^2, & \text{if } |j| = |k|, \\ 0, & \text{if } |j| \neq |k|. \end{cases} \tag{3.6.9}$$

For any stationary process $\{X_t\}_{t \in \mathbb{R}}$, we now define the continuous process that modulates the frequencies, called spectral process and denoted by $\{Z_\alpha\}_{\alpha \in \mathbb{R}}$. The increments of the spectral process inherit the properties of $\{A(\alpha_j)\}_{j \in \mathbb{Z}}$ and in particular the orthogonality property (3.6.9).

Definition 3.6.1 (Spectral process). *Let $\{X_t\}_{t \in \mathbb{R}}$ be a complex-valued and stationary process. Its spectral process $\{Z_\alpha\}_{\alpha \in \mathbb{R}}$, whenever it exists, is the complex-valued stochastic process defined through the mean square stochastic integral*

$$X_t = \int_{\mathbb{R}} e^{i\alpha t} dZ_\alpha, \quad \text{a.s.}, \forall t \in \mathbb{R}, \tag{3.6.10}$$

together with the additional conditions:

$$\mathsf{E}[Z_\alpha] = 0, \quad \forall \alpha \in \mathbb{R},$$

$$\mathsf{E}\left[(Z_{\alpha_2} - Z_{\alpha_1})\overline{(Z_{\alpha_4} - Z_{\alpha_3})}\right] = 0, \ \forall -\infty < \alpha_1 < \alpha_2 < \alpha_3 < \alpha_4 < \infty,$$

(3.6.11)

i.e. it has orthogonal increments, and

$$\mathsf{E}\left[|Z_{\alpha_2} - Z_{\alpha_1}|^2\right] = F(\alpha_2) - F(\alpha_1), \ \forall -\infty < \alpha_1 < \alpha_2 < \infty, \quad (3.6.12)$$

where F is the spectral d.f. of $\{X_t\}_{t\in\mathbb{R}}$.

We use the following symbolic representation of (3.6.11) and (3.6.12): $\forall \alpha, \beta \in \mathbb{R}$,

$$\mathsf{d}\,\mathsf{cov}(Z_\alpha, Z_\beta) = \mathsf{dE}\left[Z_\alpha \overline{Z_\beta}\right] = \mathsf{E}\left[\mathsf{d}Z_\alpha \overline{\mathsf{d}Z_\beta}\right] = \begin{cases} \mathsf{d}F(\alpha), & \text{if } \alpha = \beta, \\ 0, & \text{if } \alpha \neq \beta. \end{cases}$$

(3.6.13)

The intuition behind the spectral decomposition (3.6.10) is that, for given $t \in \mathbb{R}$, X_t is the sum of infinitesimal small orthogonal harmonics

$$e^{it\alpha}\mathsf{d}Z_\alpha,$$

with oscillations regulated by $e^{it\alpha}$ and with random amplitude and phase regulated by $\mathsf{d}Z_\alpha$, $\forall \alpha \in \mathbb{R}$. The magnitude of $\mathsf{d}Z_\alpha$ depends on $\mathsf{d}F(\alpha)$.

Remarks 3.6.2.

(1) *In Definition 3.6.1, the spectral process appears always through its increments, with exception of the null expectation requirement. Consequently, one can always assume $Z_0 = 0$ a.s.*

(2) *A consequence of (3.6.12) is that the spectral process has a.s. null increment over any region with spectral measure null.*

(3) *Another consequence of (3.6.12) is that, if the spectral d.f. F is continuous at $\alpha \in \mathbb{R}$, then the spectral process $\{Z_\alpha\}_{\alpha\in\mathbb{R}}$ is $\mathcal{L}_2$-continuous at α.*

Proposition 3.6.3 (Properties of spectral process). *Consider the spectral process of Definition 3.6.1. Then the stochastic process $X_t = \int_{\mathbb{R}} e^{it\alpha}\mathsf{d}Z_\alpha$, $\forall t \in \mathbb{R}$, as given in (3.6.10), has the following properties.*

(1) $\mathsf{d}Z_{-\alpha} = \overline{\mathsf{d}Z_\alpha}$, $\forall \alpha \in \mathbb{R} \iff \{X_t\}_{t\in\mathbb{R}}$ *is real-valued.*

$\mathsf{d}Z_{-\alpha} = \overline{\mathsf{d}Z_\alpha} \iff \arg \mathsf{d}Z_{-\alpha} = -\arg \mathsf{d}Z_\alpha$ *and* $|\mathsf{d}Z_{-\alpha}| = |\mathsf{d}Z_\alpha|$.

(2) $\mathsf{E}[X_t] = 0$, $\forall t \in \mathbb{R}$.

(3) The a.c.v.f. of $\{X_t\}_{t \in \mathbb{R}}$ is given by

$$\gamma(h) = \int_{\mathbb{R}} e^{i\alpha h} dF(\alpha), \quad \forall h \in \mathbb{R},$$

for some d.f. F, and so the process is stationary.

Proof. 1. We have, $\forall t \in \mathbb{R}$,

$$\overline{X_t} = \overline{\int_{\mathbb{R}} e^{i\alpha t} dZ_\alpha} = \int_{\mathbb{R}} e^{-i\alpha t} \overline{dZ_\alpha} = \int_{\mathbb{R}} e^{-i\alpha t} dZ_{-\alpha} = \int_{\mathbb{R}} e^{i\alpha t} dZ_\alpha = X_t, \quad \text{a.s.}$$

We refer to Definition A.6.1 for the penultimate equality above.

2. We have that, $\forall t \in \mathbb{R}$, X_t is the mean square limit of a sequence of Riemann-Stieltjes partial sums with mean zero. Thus, the continuity of the scalar product gives the desired result.

3. We have, $\forall s, t \in \mathbb{R}$,

$$\text{cov}(X_s, X_t) = \mathsf{E}\left[\int_{\mathbb{R}} e^{i\alpha s} dZ_\alpha \overline{\int_{\mathbb{R}} e^{i\beta t} dZ_\beta}\right]$$

$$= \int_{\mathbb{R}} \int_{\mathbb{R}} e^{i(\alpha s - \beta t)} \mathsf{E}\left[dZ_\alpha \overline{dZ_\beta}\right]$$

$$= \int_{\mathbb{R}} e^{i\alpha(s-t)} dF(\alpha),$$

where the second equality is due to (3.4.6) and the third one is due to (3.6.13). $\qquad \square$

We now present and prove the Spectral decomposition theorem 3.6.4. It states that every complex-valued, stationary and mean square continuous stochastic process $\{X_t\}_{t \in \mathbb{R}}$ admits the spectral decomposition given in Definition 3.6.1.

For this purpose we need some definitions. We define the two following Hilbert spaces:

$$\mathcal{H}_X = \overline{\text{sp}}\{X_t | t \in \mathbb{R}\} \quad \text{and} \quad \mathcal{H}_F = \left\{g : \mathbb{R} \to \mathbb{C} \,\middle|\, \int_{\mathbb{R}} |g|^2 dF < \infty\right\}, \quad (3.6.14)$$

where $\overline{\text{sp}}$ denotes the closed span of Definition A.1.7. In $\mathcal{H}_X$, we define the following scalar product and norm:

$$\langle U, V \rangle_{\mathcal{H}_X} = \mathsf{E}\left[U\overline{V}\right] \quad \text{and} \quad \|U\|_{\mathcal{H}_X} = \{\langle U, U \rangle_{\mathcal{H}_X}\}^{\frac{1}{2}}.$$

Note that $Y \in \mathcal{H}_X$ can be complex-valued even when $\{X_t\}_{t \in \mathbb{R}}$ is real-valued, because the coefficients of the linear combinations can be complex. In $\mathcal{H}_F$, we define the following scalar product and norm:

$$\langle u, v \rangle_{\mathcal{H}_F} = \int_{\mathbb{R}} u\overline{v} dF \quad \text{and} \quad \|u\|_{\mathcal{H}_F} = \{\langle u, u \rangle_{\mathcal{H}_F}\}^{\frac{1}{2}}.$$

Theorem 3.6.4 (Spectral decomposition of stationary process). *If the complex-valued, square-integrable and stationary stochastic process $\{X_t\}_{t \in \mathbb{R}}$ has mean zero and spectral d.f. F over $\mathbb{R}$, then there exists a stochastic process $\{Z_\alpha\}_{\alpha \in \mathbb{R}}$, the spectral process of Definition 3.6.1, that satisfies all properties of that definition and the additional property $Z_\alpha \in \mathcal{H}_X$, $\forall \alpha \in \mathbb{R}$.*

Proof. Consider the Hilbert spaces $\mathcal{H}_X$ and $\mathcal{H}_F$ defined in (3.6.14). Let $t \in \mathbb{R}$, then

$$\|X_t\|_{\mathcal{H}_X} = \mathsf{E}\left[|X_t|^2\right] = \gamma(0) = \int_{\mathbb{R}} |e^{i\alpha t}|^2 dF(\alpha) = \|e^{i \cdot t}\|_{\mathcal{H}_F}.$$

More generally, let $s, t \in \mathbb{R}$, then

$$\langle X_s, X_t \rangle_{\mathcal{H}_X} = \mathsf{E}[X_s \overline{X_t}] = \gamma(s - t) = \int_{\mathbb{R}} e^{i\alpha(s-t)} dF(\alpha) = \langle e^{i \cdot s}, e^{i \cdot t} \rangle_{\mathcal{H}_F}.$$

The first part of this proof consists in extending this last equality between scalar products to all elements of $\mathcal{H}_X$ and of $\mathcal{H}_F$. We thus establish an isometry between the Hilbert spaces $\mathcal{H}_X$ and $\mathcal{H}_F$, whose respective spanning elements are X_t, $\forall t \in \mathbb{R}$, and $e^{i \cdot t}$, $\forall t \in \mathbb{R}$. In the second part of the proof, we want to find $\{Z_\alpha\}_{\alpha \in \mathbb{R}}$, the spectral process of $\{X_t\}_{t \in \mathbb{R}}$. Our isometry allows for the shift of the search of $Z_\alpha \in \mathcal{H}_X$ to the search of the corresponding element $g_\alpha \in \mathcal{H}_F$. By taking this last function equal to a single step function centered at α, we show that the corresponding element $Z_\alpha \in \mathcal{H}_X$ is indeed the one appearing in the spectral decomposition (3.6.10).

Let $n \in \mathbb{N}^*$, $-\infty < t_1 < \cdots < t_n < \infty$, $-\infty < t_1' < \cdots < t_n' < \infty$, $z_1, \ldots, z_n \in \mathbb{C}$ and $z_1', \ldots, z_n' \in \mathbb{C}$. Define $Y_n = z_1 X_{t_1} + \cdots + z_n X_{t_n}$ and $Y_n' = z_1' X_{t_1'} + \cdots + z_n' X_{t_n'} \in \mathcal{H}_X$. Define the corresponding elements in $\mathcal{H}_F$ given by $g_n = z_1 e^{i \cdot t_1} + \cdots + z_n e^{i \cdot t_n}$ and $g_n' = z_1' e^{i \cdot t_1'} + \cdots + z_n' e^{i \cdot t_n'}$. Then one can easily verify that

$$\langle Y_n, Y_n' \rangle_{\mathcal{H}_X} = \langle g_n, g_n' \rangle_{\mathcal{H}_F},$$

and thus that

$$\|Y_n - Y_n'\|_{\mathcal{H}_X} = \|g_n - g_n'\|_{\mathcal{H}_F}. \tag{3.6.15}$$

From (3.6.15) we deduce that $Y_n = Y_n'$ P-a.s. iff $g_n = g_n'$ F-a.e.

From (3.6.15) we also deduce that to any Cauchy sequence $\{Y_n\}_{n \geq 1}$ in $\mathcal{H}_X$, of the form of finite linear combinations of elements of $\{X_t\}_{t \in \mathbb{R}}$, there exists a corresponding Cauchy sequence $\{g_n\}_{n \geq 1}$ in $\mathcal{H}_F$, of the form of finite linear combinations of elements of $\{e^{i \cdot t}\}_{t \in \mathbb{R}}$. The converse holds too. The completeness property of the Hilbert space tells that this equivalence holds for the limiting values of the Cauchy sequences of the given forms: to each Cauchy sequence limit $Y \in \mathcal{H}_X$, there exists a corresponding Cauchy sequence limit $g \in \mathcal{H}_F$ and conversely.

After obtaining this correspondence for limits of Cauchy sequences of particular forms, we extend it to the entire spaces $\mathcal{H}_X$ and $\mathcal{H}_F$. The space $\mathcal{H}_X$ is already defined as closed span, i.e. limits of linear combinations, so that no additional arguments are required. The extension from the limits of Cauchy sequences to the space $\mathcal{H}_F$ can be obtained with the Stone-Weierstrass theorem. A consequence of that theorem is indeed that the set of all possible linear combinations of elements of $\{e^{i \cdot t}\}_{t \in \mathbb{R}}$ is dense in the Hilbert space $\mathcal{H}_F$. We have thus an isometry between $\mathcal{H}_X$ and $\mathcal{H}_F$: the correspondence between the elements of these spaces can be formalized through the linear and bijective operator $A : \mathcal{H}_X \to \mathcal{H}_F$ such that, $\forall u, v \in \mathcal{H}_X$, $\langle Au, Av \rangle_{\mathcal{H}_F} = \langle u, v \rangle_{\mathcal{H}_X}$; cf. Definition A.1.3.

Define the descending single step function centered at $\alpha_0 \in \mathbb{R}$, by $s_{\alpha_0}(\alpha) = I\{\alpha \leq \alpha_0\}$, $\forall \alpha \in \mathbb{R}$. Clearly, $s_{\alpha_0} \in \mathcal{H}_F$ and

$$\|s_{\alpha_0}\|_{\mathcal{H}_F}^2 = \int_{\mathbb{R}} |s_{\alpha_0}|^2 \mathrm{d}F = \int_{(-\infty, \alpha_0]} \mathrm{d}F = F(\alpha_0) - F(-\infty) = F(\alpha_0).$$

Let $\alpha \in \mathbb{R}$ and let $Z_\alpha \in \mathcal{H}_X$ and $s_\alpha \in \mathcal{H}_F$ be corresponding elements. As mentioned above, this means that Z_α is equal to some linear operator applied to s_α. It follows from linearity that, for $-\infty < \alpha_1 < \alpha_2 < \alpha_3 < \alpha_4 < \infty$, $Z_{\alpha_2} - Z_{\alpha_1}$ and $s_{\alpha_2} - s_{\alpha_1}$ are corresponding elements and so are $Z_{\alpha_4} - Z_{\alpha_3}$ and $s_{\alpha_4} - s_{\alpha_3}$. Thus,

$$\begin{aligned} \langle Z_{\alpha_2} - Z_{\alpha_1}, Z_{\alpha_4} - Z_{\alpha_3} \rangle_{\mathcal{H}_X} &= \langle s_{\alpha_2} - s_{\alpha_1}, s_{\alpha_4} - s_{\alpha_3} \rangle_{\mathcal{H}_F} \\ &= \int_{\mathbb{R}} (s_{\alpha_2} - s_{\alpha_1})(s_{\alpha_4} - s_{\alpha_3}) \mathrm{d}F \\ &= \int_{(\alpha_1, \alpha_2] \cap (\alpha_3, \alpha_4]} \mathrm{d}F \\ &= 0. \end{aligned}$$

So (3.6.11) holds, i.e. the increments are orthogonal. Also,

$$\|Z_{\alpha_2} - Z_{\alpha_1}\|_{\mathcal{H}_X}^2 = \|s_{\alpha_2} - s_{\alpha_1}\|_{\mathcal{H}_F}^2 = \int_{(\alpha_1, \alpha_2]} \mathrm{d}F = F(\alpha_2) - F(\alpha_1).$$

So (3.6.12) holds.

It remains to show that $\{Z_\alpha\}_{\alpha \in \mathbb{R}}$ is the spectral process of the stationary process $\{X_t\}_{t \in \mathbb{R}}$, i.e. the stochastic integral equation (3.6.10) holds. Let $t \in \mathbb{R}$ and consider the partition of $\mathbb{R}$ given by $-\infty < \alpha_{n,0} < \cdots < \alpha_{n,n} < \infty$,

$\forall n \in \mathbb{N}^*$. We assume that, as $n \to \infty$, the mesh of the partition vanishes, $\alpha_{n,0} \to -\infty$ and $\alpha_{n,n} \to \infty$. We need to show that $S_{t,n} \xrightarrow{\mathcal{L}_2} X_t$, where for given $n \in \mathbb{N}^*$,

$$S_{t,n} = \sum_{j=1}^{n} e^{i\nu_{n,j}t} \left(Z_{\alpha_{n,j}} - Z_{\alpha_{n,j-1}} \right),$$

with $\alpha_{n,j-1} \leq \nu_{n,j} \leq \alpha_{n,j}$, for $j = 1, \ldots, n$. The element in $\mathcal{H}_F$ corresponding to $S_{n,t} \in \mathcal{H}_X$ is $g_{n,t}$, given by

$$g_{n,t}(\alpha) = \sum_{j=1}^{n} e^{i\nu_{n,j}t} \left\{ s_{\alpha_{n,j}}(\alpha) - s_{\alpha_{n,j-1}}(\alpha) \right\}$$

$$= \sum_{j=1}^{n} e^{i\nu_{n,j}t} \mathbb{I} \left\{ \alpha_{n,j-1} < \alpha \leq \alpha_{n,j} \right\}, \quad \forall \alpha \in \mathbb{R}.$$

We have that $g_{n,t} \to e^{i \cdot t}$, in $\mathcal{H}_F$-norm. Because the correspondence of elements in $\mathcal{H}_F$ with elements is $\mathcal{H}_X$ is preserved at the limit, we deduce that $e^{i \cdot t} \in \mathcal{H}_F$ and $\int_{\mathbb{R}} e^{i\alpha t} dZ_\alpha \in \mathcal{H}_X$ are corresponding elements. But we also know that $e^{i \cdot t} \in \mathcal{H}_F$ and $X_t \in \mathcal{H}_X$ are corresponding elements. These two last statements imply that $X_t = \int_{\mathbb{R}} e^{i\alpha t} dZ_\alpha$, a.s., i.e. (3.6.10) holds. $\qquad \square$

We now provide a generalization of the spectral decomposition (3.6.10) to an arbitrary random variable in $\mathcal{H}_X$.

Corollary 3.6.5 (Spectral decomposition of random variable). *Consider the complex-valued stationary process $\{X_t\}_{t \in \mathbb{R}}$ having spectral process $\{Z_\alpha\}_{\alpha \in \mathbb{R}}$ and spectral d.f. F. Then, $\forall Y \in \mathcal{H}_X$, $\exists g \in \mathcal{H}_F$ such that*

$$Y = \int_{\mathbb{R}} g(\alpha) dZ_\alpha, \quad \text{a.s.,}$$

and

$$\mathsf{E}[|Y|^2] = \int_{\mathbb{R}} |g(\alpha)|^2 dF(\alpha) \quad \text{i.e.} \quad \|Y\|_{\mathcal{H}_X} = \|g\|_{\mathcal{H}_F}.$$

Proof. By definition, any $Y \in \mathcal{H}_X$ is a $\mathcal{L}_2$-limit of linear combinations of random variables of $\{X_t\}_{t \in \mathbb{R}}$. Therefore,

$$Y_n = \sum_{j=1}^{n} z_{n,j} X_{t_n,j} \xrightarrow{\mathcal{L}_2} Y,$$

for some $z_{n,j} \in \mathbb{C}$ and $t_{n,j} \in \mathbb{R}$, for $j = 1, \ldots, n$, for given $n \in \mathbb{N}^*$. It follows from the Spectral decomposition theorem 3.6.4 that there exists a stochastic process $\{Z_\alpha\}_{\alpha \in \mathbb{R}}$, such that

$$X_{t_{n,j}} = \int_{\mathbb{R}} e^{i\alpha t_{n,j}} dZ_\alpha, \quad \text{a.s., for } j = 1, \ldots, n.$$

Thus we have, a.s.,

$$\int_{\mathbb{R}} \sum_{j=1}^{n} z_{n,j} e^{i\alpha t_{n,j}} dZ_\alpha = Y_n \xrightarrow{\mathcal{L}_2} Y.$$

But

$$g_n = \sum_{j=1}^{n} z_{n,j} e^{i \cdot t_{n,j}} \longrightarrow g, \quad \text{in } \mathcal{H}_F,$$

for some $g \in \mathcal{H}_F$, implies the first desired result $Y = \int_{\mathbb{R}} g(\alpha) dZ_\alpha$, a.s. Indeed, we know from the proof of Theorem 3.6.4 that, in the isometry between $\mathcal{H}_X$ and $\mathcal{H}_F$, $\int_{\mathbb{R}} e^{i\alpha t} dZ_\alpha \in \mathcal{H}_X$ and $e^{i \cdot t} \in \mathcal{H}_F$, $\forall t \in \mathbb{R}$, and thus $Y_n \in \mathcal{H}_X$ and $g_n \in \mathcal{H}_F$ are corresponding functions, $\forall n \in \mathbb{N}^*$. In the limit, $Y \in \mathcal{H}_X$ and $g \in \mathcal{H}_F$ are corresponding functions. Consequently we have

$$\left\| \int_{\mathbb{R}} g_n(\alpha) dZ_\alpha - \int_{\mathbb{R}} g(\alpha) dZ_\alpha \right\|_{\mathcal{H}_X} = \left\| \int_{\mathbb{R}} \{g_n(\alpha) - g(\alpha)\} dZ_\alpha \right\|_{\mathcal{H}_X}$$
$$= \|g_n - g\|_{\mathcal{H}_F} \xrightarrow{n \to \infty} 0.$$

Moreover we have $\|Y\|_{\mathcal{H}_X} = \|g\|_{\mathcal{H}_F}$. $\qquad \square$

3.6.3 *Discrete spectral process and distribution*

Real-valued stationary processes with at most countably many frequencies are presented in Section 3.6.1. We now consider complex-valued stationary processes for which the spectral process and the spectral d.f. are step functions with jumps at the frequencies

$$-\infty < \cdots < \alpha_{-1} < \alpha_0 = 0 < \alpha_1 < \cdots < \infty.$$

We use the notation of the jump of a function introduced in (3.1.17) and define accordingly $\Delta Z_{\alpha_j} = A(\alpha_j)$, $\forall j \in \mathbb{Z}$, that are defined in (3.6.4), but now with the more general above given frequencies and without the symmetry restrictions that are given in (3.6.2). Thus, $\forall t \in \mathbb{R}$,

$$
\begin{aligned}
X_t &= \sum_{j=-\infty}^{-\infty} e^{i\alpha_j t} \Delta Z_{\alpha_j} \\
&= \sum_{j=-\infty}^{-\infty} e^{i\alpha_j t} A(\alpha_j) \\
&= B(0) + \sum_{j=-\infty, j \neq 0}^{-\infty} B(\alpha_j) e^{i\{\alpha_j t + \theta(\alpha_j)\}} \\
&= B(0) + \sum_{j=-\infty, j \neq 0}^{-\infty} B(\alpha_j) \cos(\alpha_j t + \theta(\alpha_j)) + i\, B(\alpha_j) \sin(\alpha_j t + \theta(\alpha_j)).
\end{aligned}
$$

$$(3.6.16)$$

In the real-valued case we consider $\alpha_{-j} = -\alpha_j$ and impose the restrictions (3.6.2), that are equivalent to $\Delta Z_{-\alpha_j} = \overline{\Delta Z_{\alpha_j}}$, $\forall j \in \mathbb{Z}$. In that case the sine part of (3.6.16) vanishes and we retrieve (3.6.1). We define $\sigma_j^2 = \mathsf{E}[|\Delta Z_{\alpha_j}|^2]$, $\forall j \in \mathbb{Z}$, in correspondence with (3.6.5) for the real-valued process. The formula for the spectral d.f. of the real-valued process (3.6.8) generalizes without modifications to complex-valued processes. Note that (3.6.13) gives $\Delta F(\alpha_j) = \mathsf{E}[|\Delta Z_{\alpha_j}|^2] = \sigma_j^2$, $\forall j \in \mathbb{Z}$. Thus

$$
\gamma(h) = \sum_{j=-\infty}^{\infty} e^{i\alpha_j h} \Delta F(\alpha_j), \quad \forall h \in \mathbb{R},
$$

which corresponds to (3.9.16) of the real-valued process.

The following theorem provides some inversion formulae for a stationary stochastic process with discrete spectrum. It is given without proof.

Theorem 3.6.6 (Inversion formulae for discrete spectrum). *Let $\{X_t\}_{t \in \mathbb{R}}$ be a complex-valued and stationary process with spectral process $\{Z_\alpha\}_{\alpha \in \mathbb{R}}$ and spectral d.f. F, both of them having only jumps that occur at the frequencies $-\infty < \cdots < \alpha_{-1} < \alpha_0 = 0 < \alpha_1 < \cdots < \infty$. Then the following formulae hold,*

$$
\frac{1}{t} \int_0^t e^{-i\alpha_j s} \gamma(s) ds \xrightarrow{t \to \infty} \Delta F(\alpha_j) \text{ and}
$$

$$
\frac{1}{t} \int_0^t e^{-i\alpha_j s} X_s ds \xrightarrow{\mathcal{L}_2} \Delta Z_{\alpha_k}, \quad \forall j \in \mathbb{Z}.
$$

3.7 Spectral analysis of Gaussian processes

We now analyze the spectral analysis of Gaussian processes. Section 3.7.1 presents the spectral decomposition of real-valued Gaussian processes. It also provides a discretization scheme in frequency domain that can be used for the simulation of the sample paths. Section 3.7.2 provides two methods of construction of approximate Gaussian WN: one method yields a process with truncated frequencies and another method is obtained from the Ornstein-Uhlenbeck process.

3.7.1 *Spectral decomposition of real Gaussian processes*

Let $\{X_t\}_{t\in\mathbb{R}}$ be a real-valued, stationary and Gaussian stochastic process and let $\{Z_\alpha\}_{\alpha\in\mathbb{R}}$ be its spectral process. It follows from $Z_\alpha \in \mathcal{H}_X$, $\forall \alpha \in \mathbb{R}$, and from the characterization of the Gaussian of Proposition 3.2.2, that $\{Z_\alpha\}_{\alpha\in\mathbb{R}}$ is complex-valued Gaussian, in the sense of Definition 3.2.1. A real-valued Gaussian stationary processes $\{X_t\}_{t\in\mathbb{R}}$ with mean zero is entirely determined by its a.c.v.f γ or, alternatively, by its spectral d.f. F. Thus, if the spectral d.f. F is given, then the spectral process $\{Z_\alpha\}_{\alpha\in\mathbb{R}}$, whose existence is guaranteed by Theorem 3.6.4, is a determined Gaussian process. We should note that the same spectral d.f. F, or same a.c.v.f. γ, may however also corresponds to some other non-Gaussian stationary process with mean zero.

We now give a decomposition of the real-valued and stationary process $\{X_t\}_{t\in\mathbb{R}}$, which is however not restricted to Gaussian processes. We have, $\forall t \in \mathbb{R}$,

$$
\begin{aligned}
X_t &= \Delta Z_0 + \int_{\mathbb{R}^*} e^{i\alpha t} dZ_\alpha \\
&= \Delta Z_0 + \int_{(0,\infty)} e^{i\alpha t} dZ_\alpha + \int_{(-\infty,0)} e^{i\alpha t} dZ_\alpha \\
&= \Delta Z_0 + \int_{(0,\infty)} e^{i\alpha t} dZ_\alpha + \int_{(0,\infty)} e^{-i\alpha t} dZ_{-\alpha} \\
&= \Delta Z_0 + \int_{(0,\infty)} \cos \alpha t (dZ_\alpha + dZ_{-\alpha}) + \int_{(0,\infty)} \sin \alpha t\, i(dZ_\alpha - dZ_{-\alpha}) \\
&= \Delta Z_0 + \int_{(0,\infty)} \cos \alpha t\, dU_\alpha + \int_{(0,\infty)} \sin \alpha t\, dV_\alpha, \quad \text{a.s.,} \quad (3.7.1)
\end{aligned}
$$

where

$$
dU_\alpha = dZ_\alpha + dZ_{-\alpha} \quad \text{and} \quad dV_\alpha = i(dZ_\alpha - dZ_{-\alpha}), \quad \forall \alpha > 0. \quad (3.7.2)
$$

Note that dU_α and dV_α are real-valued iff

$$dZ_{-\alpha} = \overline{dZ_\alpha}, \quad \forall \alpha > 0. \tag{3.7.3}$$

Obviously, (3.7.2) and (3.7.3) imply

$$dU_\alpha = 2\operatorname{Re} dZ_\alpha \quad \text{and} \quad dV_\alpha = -2\operatorname{Im} dZ_\alpha, \quad \forall \alpha > 0. \tag{3.7.4}$$

Moreover, if (3.7.3) holds and ΔZ_0 is real-valued, then $\{X_t\}_{t\in\mathbb{R}}$ is real-valued. In fact the converse is also true, as already mentioned in Proposition 3.6.3.1. Define $\Delta U_0 = \Delta Z_0$ and $\Delta V_0 = 0$. Assume $\{X_t\}_{t\in\mathbb{R}}$ real-valued. Let $\alpha > 0$. By recalling (3.6.13) and Definition 3.5.7, we find

$$
\begin{aligned}
\mathsf{E}[(dU_\alpha)^2] &= \mathsf{E}[(dZ_\alpha + dZ_{-\alpha})^2] \\
&= \mathsf{E}[(dZ_\alpha)^2] + \mathsf{E}[(dZ_{-\alpha})^2] + 2\mathsf{E}[dZ_\alpha dZ_{-\alpha}] \\
&= \mathsf{E}[dZ_\alpha \overline{dZ_{-\alpha}}] + \mathsf{E}[dZ_{-\alpha}\overline{dZ_\alpha}] + 2\mathsf{E}[dZ_\alpha \overline{dZ_\alpha}] \\
&= 2dF(\alpha) = dF_1(\alpha)
\end{aligned}
$$

and

$$
\begin{aligned}
\mathsf{E}[(dV_\alpha)^2] &= \mathsf{E}[\{i(dZ_\alpha - dZ_{-\alpha})\}^2] \\
&= -\mathsf{E}[(dZ_\alpha)^2] - \mathsf{E}[(dZ_{-\alpha})^2] + 2\mathsf{E}[dZ_\alpha dZ_{-\alpha}] \\
&= -\mathsf{E}[dZ_\alpha \overline{dZ_{-\alpha}}] - \mathsf{E}[dZ_{-\alpha}\overline{dZ_\alpha}] + 2\mathsf{E}[dZ_\alpha \overline{dZ_\alpha}] \\
&= 2dF(\alpha) = dF_1(\alpha).
\end{aligned}
$$

Let $\alpha_1, \alpha_2 > 0$ with $\alpha_1 \neq \alpha_2$. Then

$$
\begin{aligned}
\mathsf{E}[dU_{\alpha_1} & dV_{\alpha_2}] \\
&= i(\mathsf{E}[dZ_{\alpha_1} dZ_{\alpha_2}] - \mathsf{E}[dZ_{\alpha_1} dZ_{-\alpha_2}] + \mathsf{E}[dZ_{-\alpha_1} dZ_{\alpha_2}] - \mathsf{E}[dZ_{-\alpha_1} dZ_{-\alpha_2}]) \\
&= i(\mathsf{E}[dZ_{\alpha_1} \overline{dZ_{-\alpha_2}}] - \mathsf{E}[dZ_{\alpha_1} \overline{dZ_{\alpha_2}}] + \mathsf{E}[dZ_{-\alpha_1} \overline{dZ_{-\alpha_2}}] - \mathsf{E}[dZ_{-\alpha_1} \overline{dZ_{\alpha_2}}]) \\
&= 0.
\end{aligned}
$$

The decomposition (3.7.1) is now applied to the real-valued process $\{X_t\}_{t\in\mathbb{R}}$ that admits a spectral density over $\mathbb{R}$, denoted by f. Define the stochastic processes

$$W_{1,\alpha} = \int_{\{v \in (0,\alpha]\,|\,f(v)>0\}} \frac{dU_v}{\sqrt{2f(v)}} \quad \text{and} \quad W_{2,\alpha} = \int_{\{v \in (0,\alpha]\,|\,f(v)>0\}} \frac{dV_v}{\sqrt{2f(v)}}, \tag{3.7.5}$$

$\forall \alpha > 0$. Then we can re-express (3.7.1) a.s. as

$$X_t = \Delta U_0 + \int_{(0,\infty)} \cos \alpha t \, \sqrt{2f(\alpha)}\, dW_{1,\alpha} + \int_{(0,\infty)} \sin \alpha t \, \sqrt{2f(\alpha)} dW_{2,\alpha}, \tag{3.7.6}$$

$\forall t \in \mathbb{R}$. Indeed, we know from (3.6.13) that $\|dZ_\alpha\|^2 = f(\alpha)d\alpha$, so that dZ_α vanishes a.s. whenever α belongs to any subset of $\mathbb{R}$ of F-measure null. Then, it follows from (3.7.4) that $\|dZ_\alpha\| = 0$ implies $dU_\alpha = dV_\alpha = 0$ a.s., $\forall \alpha > 0$.

Assume now that $\{X_t\}_{t \in \mathbb{R}}$ is Gaussian with spectral density f over $\mathbb{R}$. Thus, $\forall \alpha \in \mathbb{R}$, $\|dZ_\alpha\|^2 = f(\alpha)d\alpha$, implying that $\Delta U_0 = \Delta Z_0$ appearing in (3.7.6) vanishes a.s.: it can be removed from the equation.

Moreover, we have

$$\mathsf{E}\left[(dW_{1,\alpha})^2\right] = \mathsf{E}\left[(dW_{2,\alpha})^2\right] = d\alpha, \quad \forall \alpha > 0. \tag{3.7.7}$$

Therefore $\{dW_{1,\alpha}\}_{t>0}$ and $\{dW_{2,\alpha}\}_{t>0}$ are in fact the increments of the two independent Wiener processes $\{W_{1,\alpha}\}_{t\geq 0}$ and $\{W_{2,\alpha}\}_{t\geq 0}$, respectively.

Remark 3.7.1 (Discretization in frequency domain of real-valued Gaussian process). *The representation (3.7.6) of a stationary and real-valued Gaussian process can be discretized in the frequency domain as follows. For some large integer n, let $C_1, \ldots, C_n$ and $S_1, \ldots, S_n$ be independent standard normal random variables and consider $\delta > 0$ small. An approximation to $\{X_t\}_{t \in \mathbb{R}}$, is given by*

$$X_{n,t} = \sum_{j=1}^{n} \sqrt{2[F(j\delta) - F((j-1)\delta)]} C_j \cos j\delta t$$

$$+ \sum_{j=1}^{n} \sqrt{2[F(j\delta) - F((j-1)\delta)]} S_j \sin j\delta t,$$

$\forall t \in \mathbb{R}$. *This approximation is a discretization of (3.7.6) and it is in fact a special case of the spectral decomposition (3.2.4). We can thus rewrite it as (3.2.5), giving*

$$X_{n,t} = \sum_{j=1}^{n} \sqrt{2[F(j\delta) - F((j-1)\delta)]} B_j \cos(j\delta t + \theta_j), \ \forall t \in \mathbb{R}, \tag{3.7.8}$$

where

$$B_j = \left(C_j^2 + S_j^2\right)^{\frac{1}{2}} \quad \text{and} \quad \theta_j = -\arg\{C_j + iS_j\}, \quad \text{for } j = 1, \ldots, n.$$

As already mentioned just after (3.2.5) in Section 3.2.1, B_j follows the Rayleigh distribution (3.2.7), θ_j is uniformly distributed over $[0, 2\pi)$ and B_j and θ_j are independent, for $j = 1, \ldots, n$. One can relate (3.7.8) with (3.2.5) simply by setting

$$\zeta_j = \begin{cases} \sqrt{2[F(j\delta) - F((j-1)\delta)]}, & \text{for } j = 1, \ldots, n, \\ 0, & \text{for } j = n+1, n+2, \ldots. \end{cases}$$

The discretization scheme (3.7.8) can be used for the simulation of the sample paths of $\{X_t\}_{t\in\mathbb{R}}$. An efficient algorithm that uses the Fast Fourier transform (FFT) is presented in Section 4.7.3. For the FFT, refer to Section A.10.

Remark 3.7.2. *Let us re-express the integral representation (3.7.6) in terms of the complex-valued Wiener process*

$$\widetilde{W}_\alpha = \begin{cases} 2^{-\frac{1}{2}}(W_{1,-\alpha} - \mathrm{i}\,W_{2,-\alpha}), & \text{if } \alpha < 0, \\ 2^{-\frac{1}{2}}(W_{1,\alpha} + \mathrm{i}\,W_{2,\alpha}), & \text{if } \alpha > 0. \end{cases}$$

The identities (3.7.7) imply $\mathsf{E}[|\mathrm{d}\widetilde{W}_\alpha|^2] = \mathrm{d}\alpha$, at any $\alpha \neq 0$. We can directly control that, $\forall t \in \mathbb{R}$,

$$\int_{(0,\infty)} \cos\alpha t \sqrt{2f(\alpha)}\,\mathrm{d}W_{1,\alpha} + \int_{(0,\infty)} \sin\alpha t \sqrt{2f(\alpha)}\,\mathrm{d}W_{2,\alpha}$$

$$= \int_{\mathbb{R}} \mathrm{e}^{\mathrm{i}\alpha t}\sqrt{f(\alpha)}\mathrm{d}\widetilde{W}_\alpha,$$

leading to the compact formulation

$$X_t = \Delta U_0 + = \int_{\mathbb{R}^*} \mathrm{e}^{\mathrm{i}\alpha t}\sqrt{f(\alpha)}\mathrm{d}\widetilde{W}_\alpha.$$

3.7.2   Gaussian white noise and Ornstein-Uhlenbeck process

In Section 3.4.4, the notion of WN is introduced under the name of purely random stochastic processes as a generalized stochastic processes. The WN processes considered in this section are Gaussian and they are obtained from the processes $\{W'_{1,\alpha}\}_{\alpha\in\mathbb{R}}$ and $\{W'_{2,\alpha}\}_{\alpha\in\mathbb{R}}$ that are defined in (3.7.5). We use these two stochastic processes for constructing two new Gaussian WN processes. The first one has a truncation for the high frequencies and the remaining low frequencies are equally represented. The second one derives from the stationary Ornstein-Uhlenbeck process, with large values of its parameters.

Example 3.7.3 (Gaussian WN with truncated frequencies). *The spectral d.f. of the a.c.v.f. $\operatorname{sinc} h = h^{-1}\sin h$ is given by (A.8.2). Therefore the a.c.v.f. $\gamma(h) = \sigma^2\operatorname{sinc} lh$, $\forall h \in \mathbb{R}$, has spectral density*

$$f(\alpha) = \frac{\sigma^2}{2l}\mathsf{I}\{|\alpha| < l\}, \quad \forall \alpha \in \mathbb{R},$$

where $l, \sigma > 0$. If $\{X_t\}_{t \in \mathbb{R}}$ is a Gaussian process with mean zero and with a.c.v.f. γ, i.e. with spectral density f, then the representation (3.7.6) implies that this process can be expressed as

$$X_t = \frac{\sigma}{\sqrt{l}} \left(\int_{(0,l)} \cos \alpha t \, \mathrm{d}W_{1,\alpha} + \int_{(0,l)} \sin \alpha t \, \mathrm{d}W_{2,\alpha} \right), \quad \text{a.s.,} \quad \forall t \in \mathbb{R},$$

where $\{W_{1,\alpha}\}_{\alpha \geq 0}$ and $\{W_{2,\alpha}\}_{\alpha \geq 0}$ are two independent Wiener processes. We call $\{X_t\}_{t \in \mathbb{R}}$ Gaussian WN with truncated frequencies.

Example 3.7.4 (Gaussian WN from Ornstein-Uhlenbeck process). *The Ornstein-Uhlenbeck process is introduced in Example 3.4.9 as the solution of the Langevin stochastic differential equation driven by Gaussian WN. The process at time $t \in \mathbb{R}$ represents the velocity of a particle suspended in some homogeneous fluid. It appears in Example 3.5.11 as a stochastic process that, upon sampling at integer times, has spectral density proportional to the wrapped Cauchy probability density and corresponds thus to the AR(1) time series; cf. Remark 2.4.33. It is also the Lamperti transformation of the 1/2-selfsimilar Wiener process, cf. Example 3.2.21. The Ornstein-Uhlenbeck process is the stationary Gaussian process whose a.c.v.f. is given by a constant times the characteristic function of the Cauchy distribution, precisely by (3.5.7). It follows from Theorem 3.5.4.1 that the Ornstein-Uhlenbeck process admits a spectral density. It is given by*

$$f(\alpha) = \frac{\sigma^2}{\pi} \frac{\tau}{\tau^2 + \alpha^2}, \quad \forall \alpha \in \mathbb{R}, \tag{3.7.9}$$

for some $\sigma, \tau > 0$. It follows from Proposition 3.3.3 that the Ornstein-Uhlenbeck process is $\mathcal{L}_2$-continuous, whereas it follows from Proposition 3.3.10.1 or from Remark 3.3.11.1 that it is not $\mathcal{L}_2$-differentiable. According to Table 3.1, the Ornstein-Uhlenbeck process is not smooth and has short memory only.

By treating the Langevin equation (3.4.16) as a deterministic first order linear differential equation, we can apply the method of integrating factors; cf. Section A.9. We multiply it by $\mathrm{e}^{\tau t}$ and obtain $\mathrm{e}^{\tau t}(X_t' + \tau X_t) = \mathrm{e}^{\tau t} \sqrt{2\tau} \sigma W_t'$, i.e. $(\mathrm{e}^{\tau t} X_t)' = \mathrm{e}^{\tau t} \sqrt{2\tau} \sigma W_t'$, where the velocity at time t is now denoted X_t. By integration we find $X_t = \sigma \sqrt{2\tau} \int \mathrm{e}^{-\tau(t-s)} W_s' \mathrm{d}s$. This result is also obtained in Section 3.9.4, where it is explained that the Ornstein-Uhlenbeck process $\{X_t\}_{t \in \mathbb{R}}$ admits the stationary $\mathcal{L}_2$-integral representation

$$X_t = \sigma \sqrt{2\tau} \int_{(-\infty,t]} \mathrm{e}^{-\tau(t-s)} \mathrm{d}W_s, \quad \text{a.s.,} \quad \forall t \in \mathbb{R}. \tag{3.7.10}$$

We now examine the Ornstein-Uhlenbeck process asymptotically, for large values of σ and τ, with fixed ratio $\sigma^2/\tau = c/2$, for some $c > 0$. In this case, we see that the integrand in (3.7.10) consists essentially of increments $\mathrm{d}W_s$, for s close to t, implying that there is short term approximate independence. From the side of the a.c.v.f. (3.5.7), we have

$$\gamma(h) = \frac{c\tau}{2}\mathrm{e}^{-\tau|h|} \to c\,\delta(h), \quad \forall h \in \mathbb{R}.$$

The memory of the process vanishes asymptotically at exponential rate. The asymptotic decay of the memory can also be seen with the spectral density (3.7.9). It converges to the constant $c/(2\pi)$, meaning that all frequencies are equally represented. Consequently, the Ornstein-Uhlenbeck process behaves asymptotically like Gaussian WN.

3.8 Spectral analysis of counting processes

This section introduces a specific spectral analysis for counting processes. The Bartlett spectral distribution of a point process is defined and obtained for the homogeneous Poisson process and for the Cox process.

Let $\{N_t\}_{t\in\mathbb{R}}$ be a counting process with stationary stream of events and regular, in the sense of Definition 3.2.38, with covariance intensity function w, as given in Definition 3.2.39, and with complete covariance intensity function $\tilde{w}$, as defined in (3.2.17). If w is integrable, then in analogy with Theorem 3.5.4 we compute the inverse Fourier transform

$$f(\alpha) = \frac{1}{2\pi} \int_{-\infty}^{\infty} \exp\{-\mathrm{i}\alpha s\} w(s)\mathrm{d}s, \quad \forall \alpha \in \mathbb{R},$$

which is a well-defined and bounded function. In this case we define the Bartlett spectral density of the counting process $\{N_t\}_{t\in\mathbb{R}}$ as

$$\tilde{f}(\alpha) = \frac{1}{2\pi} \int_{-\infty}^{\infty} \exp\{-\mathrm{i}\alpha s\} \tilde{w}(s)\mathrm{d}s = \frac{\lambda}{2\pi} + f(\alpha), \quad \forall \alpha \in \mathbb{R},$$

where $\lambda \in (0, \infty)$ is given by (3.2.16). Although Bartlett spectral density is not precisely the spectral density of the process, it does nevertheless provide information about the frequencies or the periodicities of the process. In particular, we note that the first term of $\tilde{f}$, which is given by

$$\frac{\lambda}{2\pi}, \tag{3.8.1}$$

is Bartlett spectral density of the homogeneous Poisson process with rate λ. Indeed, the covariance intensity function w of Definition 3.2.39 is null for the homogeneous Poisson process. Because this first term is also the spectral density of WN, according to Section 3.4.4, we have obtained an analogy between the homogeneous Poisson process and WN.

There is an alternative definition of Bartlett spectrum that can be generally applied to the point process of Definition 3.1.13.3. Let N be a point process over $(\mathbb{R}, \mathcal{B}(\mathbb{R}))$ with stationary stream of events and regular, in the sense of Definition 3.2.38. Define the operator

$$\Theta_N(g) = \int_{\mathbb{R}} g(t) \mathrm{d}N(t), \qquad (3.8.2)$$

where $g : \mathbb{R} \to \mathbb{C}$ is any integrable and square-integrable function, i.e. $g \in \mathcal{L}_1(\mathbb{R}) \cap \mathcal{L}_2(\mathbb{R})$. The Bartlett spectral d.f. of the point process N is defined as follows.

Definition 3.8.1 (Bartlett spectral d.f. of a point process). *Let N be a point process over $(\mathbb{R}, \mathcal{B}(\mathbb{R}))$ with stationary stream of events, regular and bounded. Consider the function $g_j : \mathbb{R} \to \mathbb{C}$ in $\mathcal{L}_1(\mathbb{R}) \cap \mathcal{L}_2(\mathbb{R})$ with integral null and denote by $\hat{g}_j$ its Fourier transform*

$$\hat{g}_j(\alpha) = \int_{-\infty}^{\infty} \mathrm{e}^{-\mathrm{i}\alpha t} g_j(t) \mathrm{d}t, \quad \text{for } j = 1, 2. \qquad (3.8.3)$$

The d.f. F that satisfies

$$\mathsf{cov}(\Theta_N(g_1), \Theta_N(g_2)) = \int_{\mathbb{R}} \hat{g}_1(\alpha) \overline{\hat{g}_2(\alpha)} \mathrm{d}F(\alpha),$$

is Bartlett spectral d.f. of the point process N.

Bartlett spectral d.f. is symmetric around the origin and its total mass may be infinite.

Example 3.8.2 (Homogeneous Poisson process). *Let N be the point process obtained from the homogeneous Poisson process over $\mathbb{R}$, with occurrence times $\{T_n\}_{n \in \mathbb{Z}}$ and with rate $\lambda > 0$. Consider the partition of $\mathbb{R}$ obtained with the points $t_k = kh$, for some $h > 0$ small. Let $g \in \mathcal{L}_1(\mathbb{R}) \cap \mathcal{L}_2(\mathbb{R})$ with integral null, then we have, as $h \to 0$,*

$$\Theta_N(g) = \sum_{n=-\infty}^{\infty} g(T_n) \sim \sum_{k=-\infty}^{\infty} g(t_k) \sum_{n=-\infty}^{\infty} \mathrm{I}\{t_k < T_n \le t_{k+1}\},$$

because when h is small, there is at most one occurrence in each interval $(t_k, t_{k+1}]$ of the partition. Also, the number of events in each interval of the partition are independent and Poisson distributed with parameter λh.

Thus the indicators appearing in the above formula are independent random variables with variance λh. Consequently, as $h \to 0$, we have

$$\text{var}(\Theta_N(g)) \sim \lambda \sum_{k=-\infty}^{\infty} |g(t_k)|^2 h \sim \lambda \int_{-\infty}^{\infty} |g(t)|^2 \mathrm{d}t = \frac{\lambda}{2\pi} \int_{-\infty}^{\infty} |\hat{g}(\alpha)|^2 \mathrm{d}\alpha,$$

(3.8.4)

where the last equality is Parseval-Plancherel formula (1.2.12). The comparison of this last expression with Definition 3.8.1 implies that Bartlett spectral d.f. F of the point process.

$$\mathrm{d}F(\alpha) = \frac{\lambda}{2\pi} \mathrm{d}\alpha, \quad \forall \alpha \in \mathbb{R}.$$

Therefore, the alternative Bartlett spectral d.f. corresponds to Bartlett spectral d.f. of the homogeneous Poisson process, whose density is given in (3.8.1).

Example 3.8.3 (Cox process). *The inhomogeneous Poisson process with time domain $\mathbb{R}_+$ is obtained from the birth process of Definition 3.2.29 with transition intensity functions $\lambda_0(t) = \lambda_1(t) = \cdots = \lambda(t)$, $\forall t \geq 0$. The time domain of the process can then be extended from $\mathbb{R}_+$ to $\mathbb{R}$, with the construction of Definition 3.2.35. If the function $\{\lambda(t)\}_{t \in \mathbb{R}}$ is considered as a nonnegative stochastic process, which we redenote by $\{L_t\}_{t \in \mathbb{R}}$ in this case, then we obtain the doubly stochastic Poisson or Cox process, introduced by Cox (1955). The Cox process can also be viewed as the inhomogeneous version of the mixed Poisson process, which is a Poisson process with stochastic rate. A practical property of the mixed Poisson process is overdispersion, namely that the variance of the process is larger than or equal to the expectation. The mixed Poisson process can also be viewed from the perspective of Bayesian statistical modelling, where fixed parametric quantities of the model are considered random.*

In this example $\{L_t\}_{t \in \mathbb{R}}$ is a stationary stochastic process with spectral density f_L and we define $\lambda_0 = \mathsf{E}[L_t]$, $\forall t \in \mathbb{R}$. Let N be the point process associated to this Cox process and let $g \in \mathcal{L}_1(\mathbb{R}) \cap \mathcal{L}_2(\mathbb{R})$ with integral null.

For given $t \in \mathbb{R}$, denote $\mathcal{F}_t = \sigma(\{L_s\}_{s \leq t})$, i.e. the σ-algebra generated by $\{L_s\}_{s \leq t}$. Thus $\{\mathcal{F}_t\}_{t \in \mathbb{R}}$ is a filtration. By following the same arguments that lead to (3.8.4), we can obtain

$$\mathsf{E}\left[\Theta_N(g)|\{\mathcal{F}_t\}_{t \in \mathbb{R}}\right] = \int_{-\infty}^{\infty} g(t) L_t \mathrm{d}t,$$

and

$$\text{var}\left(\Theta_N(g)|\{\mathcal{F}_t\}_{t \in \mathbb{R}}\right) = \int_{-\infty}^{\infty} |g(t)|^2 L_t \mathrm{d}t.$$

We have

$$\mathsf{E}\left[\int_{-\infty}^{\infty}|g(t)|^2 L_t \mathrm{d}t\right] = \lambda_0 \int_{-\infty}^{\infty}|g(t)|^2 \mathrm{d}t = \frac{\lambda_0}{2\pi}\int_{-\infty}^{\infty}|\hat{g}(\alpha)|^2 \mathrm{d}\alpha,$$

as a consequence of Parseval-Plancherel formula (1.2.12). By applying Theorem 3.4.5 and precisely (3.4.4), we easily show that

$$\mathsf{E}\left[\left(\int_{-\infty}^{\infty}g(t)L_t\mathrm{d}t\right)^2\right] = \int_{-\infty}^{\infty}|\hat{g}(\alpha)|^2 f_L(\alpha)\mathrm{d}\alpha.$$

Thus whenever g has null integral, we obtain

$$\mathsf{var}\left(\Theta_N(g)\right) = \mathsf{E}\left[\int_{-\infty}^{\infty}|g(t)|^2 L_t\mathrm{d}t\right] + \mathsf{var}\left(\int_{-\infty}^{\infty}g(t)L_t\mathrm{d}t\right)$$

$$= \int_{-\infty}^{\infty}|\hat{g}(\alpha)|^2\left(\frac{\lambda_0}{2\pi}+f_L(\alpha)\right)\mathrm{d}\alpha.$$

Consequently, Bartlett spectral d.f. F of the doubly stochastic Poisson process or Cox process is given by

$$\mathrm{d}F(\alpha) = \left(\frac{\lambda_0}{2\pi}+f_L(\alpha)\right)\mathrm{d}\alpha, \quad \forall\alpha \in \mathbb{R}. \tag{3.8.5}$$

3.9 Time invariant linear filters

Linear filters have a central role because most practical operators of stationary stochastic processes like weighted sums, integrals and derivatives are linear filters. Section 3.9.1 provides definitions that are related to these filters. Section 3.9.2 presents two classical linear filters together with their differential equations: the exponential smoothing filter of resistance-capacitance and the harmonic oscillator filter. In Section 3.9.3, the solution of a linear differential equation driven by a stationary process is obtained. The short Section 3.9.4 studies the case where the driving process is WN. Section 3.9.5 recasts the shot noise process in the framework of filtered point process.

3.9.1 *Definitions*

Most practical operators of stationary stochastic processes like weighted sums, integrals and derivatives can be represented in terms of linear filters. These filters have important practical roles such as prediction, completion of missing data and noise reduction. The analysis of these filters is generally more convenient in the frequency domain of the stochastic process and in fact filters provide an important rationale for the spectral analysis of stochastic processes. This section is the continuous time analogue of Section 2.4.3 of time series analysis. The type of filter considered here takes the form of the convolution acting between the stochastic process and the so-called impulse response function.

Definition 3.9.1 (Time invariant and causal linear filter via impulse response function). *Consider the stationary stochastic process* $X = \{X_t\}_{t \in \mathbb{R}}$ *and the function* $\psi : \mathbb{R} \to \mathbb{R}$, *called impulse response function. Then the filtered process is given by*

$$Y_t = \int_{-\infty}^{\infty} \psi(s) X_{t-s} \mathrm{d}s, \quad \forall t \in \mathbb{R}. \tag{3.9.1}$$

The linear filter is called time invariant whenever the impulse response function ψ in (3.9.1) does not depend on the time t. In this case (3.9.1) is a convolution.

If $\psi(s) = 0$, $\forall s < 0$, then the filter is called causal.

Remarks 3.9.2.

(1) *We call the operation in (3.9.1) linear filtering because it is the mathematical representation of a physical filter with input $\{X_t\}_{t \in \mathbb{R}}$ and output $\{Y_t\}_{t \in \mathbb{R}}$.*

(2) *With a causal filter, Y_t does not depend on X_s, $\forall s > t$. Thus the filtered process does not depend on future values of the original process. Causality is a practical condition and its importance is emphasized by its alternative name: a causal filter is also called physically realizable, cf. e.g. Cramér and Leadbetter (1967) p. 141.*

(3) *When the input is an impulse, i.e. $X_t = \delta(t)$ (the Dirac function), then the output or the response of the system is $Y_t = \psi(t)$, $\forall t \in \mathbb{R}$. For this reason, ψ is called impulse response function of the filter.*

Let us re-express the time invariant linear filter of (3.9.1) as the linear operator Ψ satisfying $Y = \Psi(X)$, with $Y = \{Y_t\}_{t \in \mathbb{R}}$ in $\mathcal{L}_2$. For any stationary stochastic processes $X^{(1)}$, $X^{(2)}$, X and $\forall a_1, a_2, a \in \mathbb{R}$, the operator Ψ satisfies

$$\Psi \left(a_1 X^{(1)} + a_2 X^{(2)} \right) = a_1 \Psi \left(X^{(1)} \right) + a_2 \Psi \left(X^{(2)} \right),$$

and

$$\Psi \left(X_{(a)} \right) = \Psi_a \left(X \right),$$

where $X_{(a)} = \{X_{a+t}\}_{t \in \mathbb{R}}$ and Ψ_a is the operator identified by the impulse response function $\psi(a + s)$, $\forall s \in \mathbb{R}$. In particular, we have $X_{(0)} = X$ and

$\Psi_0 = \Psi$. These two properties characterize the operator Ψ and thus the time invariant linear filter.

By inserting the spectral decomposition of the process that is given in Definition 3.6.1 in (3.9.1), or rather in

$$Y_t = \int_{-\infty}^{\infty} \psi(t - s) X_s \mathrm{d}s, \quad \forall t \in \mathbb{R}, \tag{3.9.2}$$

one obtains the alternative representation

$$Y_t = \int_{\mathbb{R}} \xi(\alpha) e^{i\alpha t} \mathrm{d}Z_\alpha, \quad \text{a.s.}, \forall t \in \mathbb{R}, \tag{3.9.3}$$

where

$$\xi(\alpha) = \int_{-\infty}^{\infty} e^{-i\alpha v} \psi(v) \mathrm{d}v, \quad \forall \alpha \in \mathbb{R}. \tag{3.9.4}$$

By assuming ψ integrable, i.e. $\int_{-\infty}^{\infty} |\psi(v)| \mathrm{d}v < \infty$, the integral (3.9.4) exists. We note that (3.9.4) is the continuous time analogue of the transfer function of time series (2.4.9). It is also called transfer function. By inserting it in (3.9.3), we obtain the initial convolution formulae of the filter (3.9.1) and (3.9.2), a.s. Any one of these convolution formulae together with Corollary 3.6.5 yield $\mathsf{E}\left[|Y_t|^2\right] = \int_{\mathbb{R}} |\xi(\alpha)|^2 \mathrm{d}F(\alpha)$, $\forall t \in \mathbb{R}$, so that

$$\int_{\mathbb{R}} |\xi(\alpha)|^2 \mathrm{d}F(\alpha) < \infty, \tag{3.9.5}$$

where F is the spectral d.f. of $\{X_t\}_{t \in \mathbb{R}}$. Thus (3.9.5) is the condition to assign to ξ if the filter were defined directly through (3.9.3). Note that the impulse response function can be obtained by

$$\psi(v) = \frac{1}{2\pi} \int_{-\infty}^{\infty} e^{i\alpha v} \xi(\alpha) \mathrm{d}\alpha, \quad \forall v \in \mathbb{R}.$$

These developments lead to the following alternative definition of the time invariant linear filter, based on the spectrum of the process.

Definition 3.9.3 (Time invariant filter via transfer function). *A time invariant linear filter is the linear operator that transforms the stationary stochastic process $X = \{X_t\}_{t \in \mathbb{R}}$, having spectral process $\{Z_\alpha\}_{\alpha \in \mathbb{R}}$ and spectral d.f. F, to the stationary stochastic process (3.9.3). The function ξ is called transfer function and it satisfies (3.9.5). The spectral process and the spectral d.f. are given in Definition 3.6.1.*

Some precise illustrations are the following.

Examples 3.9.4 (Time invariant linear filters). *Some important examples of time invariant linear filters with their impulse and transfer functions are the following.*

(1) *Best low pass filter*
Consider the transfer function $\xi(\alpha) = \mathbb{1}\{|\alpha| \leq a\}$, *for some* $a > 0$. *The effect of this function is the elimination of all frequencies* α *such that* $|\alpha| > a$. *The best low pass filter for time series is given in Example 2.4.28. We can indeed see in (3.9.3) that the part* $\{Z_\alpha\}_{|\alpha|>a}$ *is eliminated from* $\{Z_\alpha\}_{\alpha \in \mathbb{R}}$, *in the filtered process* $\{Y_t\}_{t \in \mathbb{R}}$. *The impulse response function is given by*

$$\psi(v) = \frac{1}{2\pi} \int_{-\infty}^{\infty} e^{i\alpha v}\xi(\alpha)\mathrm{d}\alpha = \frac{a}{\pi}\mathrm{sinc}\,av, \quad \forall v \in \mathbb{R}.$$

As shown in Section A.8.1, the function ψ *is not integrable.*

(2) *Derivation filter*
Assume that the stationary process $\{X_t\}_{t \in \mathbb{R}}$ is differentiable with derivative $\{X_t'\}_{t \in \mathbb{R}}$. Let $t \in \mathbb{R}$ and consider the Newton ratio

$$\frac{X_{t+\varepsilon} - X_t}{\varepsilon} = -\int_{-\infty}^{\infty} \frac{\delta(s-t-\varepsilon) - \delta(s-t)}{-\varepsilon}X_s\mathrm{d}s$$

$$\xrightarrow{\varepsilon \to 0} -\int_{-\infty}^{\infty} \delta'(s-t)X_s\mathrm{d}s,$$

where the generalized function δ' is defined through the partial integration rule

$$-\int_{-\infty}^{\infty} \delta'(s-t)p(s)\mathrm{d}s = \int_{-\infty}^{\infty} \delta(s-t)p'(s)\mathrm{d}s = p'(t),$$

for any differentiable function p. Thus the impulse response function is $\psi(v) = -\delta'(v)$, $\forall v \in \mathbb{R}$, whereas the transfer function is

$$\xi(\alpha) = \int_{-\infty}^{\infty} e^{-i\alpha v}\{-\delta'(v)\}\mathrm{d}v = i\alpha \int_{-\infty}^{\infty} e^{-i\alpha v}\delta(v)\mathrm{d}v = i\alpha, \quad \forall \alpha \in \mathbb{R}.$$

An alternative derivation of this transfer function is presented in Example 3.9.7.

(3) *Time shift filter*
Let $a \in \mathbb{R}$ and consider the time shifted process $Y_t = X_{t-a}$, $\forall t \in \mathbb{R}$. The impulse response function is $\psi(v) = \delta(v - a)$, $\forall v \in \mathbb{R}$. The transfer function is $\xi(\alpha) = e^{-ia\alpha}$, $\forall \alpha \in \mathbb{R}$.

(4) Exponential smoothing filter
    Let a > 0 and consider the exponentially smoothed process

$$Y_t = \int_{-\infty}^{t} e^{-a(t-s)} X_s ds, \quad \forall t \in \mathbb{R}.$$

The impulse response function is given by $\psi(v) = e^{-av}\mathrm{I}\{v \geq 0\}$, $\forall v \in \mathbb{R}$, and the transfer function is given by $\xi(\alpha) = 1/(a + i\alpha)$, $\forall \alpha \in \mathbb{R}$.

Let us apply the two transfer functions ξ and ξ^* to the stationary process $\{X_t\}_{t\in\mathbb{R}}$ with spectral process $\{Z_\alpha\}_{\alpha\in\mathbb{R}}$. We obtain the filtered processes

$$Y_t = \int_{\mathbb{R}} \xi(\alpha) e^{i\alpha t} dZ_\alpha \text{ and } Y_t^* = \int_{\mathbb{R}} \xi^*(\alpha) e^{i\alpha t} dZ_\alpha, \quad \forall t \in \mathbb{R}.$$

We remember that $\{Z_\alpha\}_{\alpha\in\mathbb{R}}$ has mean zero and that it possesses uncorrelated increments satisfying (3.6.13). Consequently the crosscovariance function (cf. Definition 3.4.3) is given by

$$\gamma_{YY^*}(t+h, t) = \mathsf{cov}(Y_{t+h}, Y_t^*) = \int_{\mathbb{R}} \xi(\alpha)\overline{\xi^*(\alpha)} e^{ih\alpha} dF(\alpha), \quad \forall h, t \in \mathbb{R}.$$
(3.9.6)

Example 3.9.5. *Consider the transfer function of the best low pass filter $\xi(\alpha) = \mathrm{I}\{|\alpha| \leq a\}$ and the one of the time shift $\xi^*(\alpha) = e^{-ib\alpha}$, $\forall \alpha \in \mathbb{R}$, for some $a > 0$ and $b \in \mathbb{R}$. Then*

$$\gamma_{YY^*}(t+h, t) = \int_{[-a,a]} e^{i(b+h)\alpha} dF(\alpha), \quad \forall h, t \in \mathbb{R}.$$

By choosing $\xi = \xi^*$ in (3.9.6) we obtain that the filtered process $\{Y_t\}_{t\in\mathbb{R}}$ is also stationary and that its a.c.v.f. is given by

$$\gamma_Y(h) = \int_{\mathbb{R}} e^{ih\alpha} |\xi(\alpha)|^2 dF(\alpha), \quad \forall h \in \mathbb{R}. \qquad (3.9.7)$$

From this expression we deduce directly that the spectral d.f. of the filtered process $\{Y_t\}_{t\in\mathbb{R}}$ is given by

$$dF_Y(\alpha) = |\xi(\alpha)|^2 dF(\alpha), \quad \forall \alpha \in \mathbb{R}. \qquad (3.9.8)$$

As in Definition 2.4.23.2 with time series, the function $|\xi(\alpha)|^2$, $\forall \alpha \in \mathbb{R}$, is called the power transfer function. The analogue formula in time domain can be obtained by (3.4.5) in Theorem 3.4.4. It is given by

$$\gamma_Y(u) = \int_{-\infty}^{\infty} \int_{-\infty}^{\infty} \psi(r)\overline{\psi(s)} \gamma(u+s-r) ds dr, \quad \forall u \in \mathbb{R}. \qquad (3.9.9)$$

This formula is more complicated than the version in frequency domain (3.9.7). Let $t \in \mathbb{R}$. Regarding the mean of the filtered process we have

$$\mathsf{E}[Y_t] = \mathsf{E}[X_t] \int_{-\infty}^{\infty} \psi(s)\mathrm{d}s = \mathsf{E}[X_t]\xi(0) = 0, \qquad (3.9.10)$$

from the assumption $\mathsf{E}[X_t] = 0$, $\forall t \in \mathbb{R}$. The quantity $\xi(0)$ is called gain of the filter.

Let $t \in \mathbb{R}$, $n \in \mathbb{N}^*$ and define

$$Y_t = \int_{\mathbb{R}} \xi(\alpha)\mathrm{e}^{i\alpha t}\mathrm{d}Z_\alpha \text{ and } Y_{n,t} = \int_{\mathbb{R}} \xi_n(\alpha)\mathrm{e}^{i\alpha t}\mathrm{d}Z_\alpha,$$

where $\xi, \xi_n \in \mathcal{H}_F$, defined in (3.6.14). We know from Theorem 3.6.4 that $Y_{n,t} - Y_t \in \mathcal{H}_X$, defined in (3.6.14). Thus we can use Corollary 3.6.5 in order to justify that

$$\|Y_{n,t} - Y_t\|_{\mathcal{H}_X}^2 = \|\xi_n \mathrm{e}^{i \cdot t} - \xi \mathrm{e}^{i \cdot t}\|_{\mathcal{H}_F}^2 = \|\xi_n - \xi\|_{\mathcal{H}_F}^2.$$

Consequently,

$$Y_{n,t} \xrightarrow{\mathcal{L}_2} Y_t \text{ iff } \xi_n \to \xi, \text{ in } \mathcal{H}_F. \qquad (3.9.11)$$

Example 3.9.6 (Filtering Gaussian WN). *Continuous time Gaussian WN is introduced as a generalized stochastic process in Section 3.4.2. We have already mentioned that WN is a random process where all frequencies are equally represented, just like white objects reflect all visible wave frequencies of light. The spectral density of WN is nonintegrable and it can expressed as*

$$f(\alpha) = \frac{\sigma^2}{2\pi}, \qquad \forall \alpha \in \mathbb{R}, \qquad (3.9.12)$$

for some $\sigma > 0$. Let $\{W_t\}_{t \in \mathbb{R}}$ be a Wiener process. Gaussian WN can be obtained from the generalized derivative process $\{W_t'\}_{t \in \mathbb{R}}$ and by filtering it with the impulse response function ψ, one obtains

$$Y_t = \int_{\mathbb{R}} \psi(t - s)W_s'\mathrm{d}s, \qquad \forall t \in \mathbb{R}.$$

The a.c.v.f. of the filtered process $\{Y_t\}_{t \in \mathbb{R}}$ can be directly obtained from (3.4.7) and it is given by

$$\begin{aligned}
\gamma_Y(u) &= \mathsf{E}\left[\int_{-\infty}^{\infty} \psi(t + u - v)W_v'\mathrm{d}v \int_{-\infty}^{\infty} \psi(t - v)W_v'\mathrm{d}v\right] \\
&= \int_{-\infty}^{\infty} \psi(u + v)\psi(v)\mathrm{d}v, \qquad \forall t, u \in \mathbb{R}.
\end{aligned}$$

The transfer function ξ is given by (3.9.4) and the spectral density is obtained from (3.9.8) as

$$f_Y(\alpha) = \frac{(\sigma|\xi(\alpha)|)^2}{2\pi}, \quad \forall \alpha \in \mathbb{R}.$$

Because we want the condition (3.9.5) to hold, we set the condition $\int_{-\infty}^{\infty} |\psi(v)|^2 dv < \infty$ to the impulse response function. Indeed, by Parseval-Plancherel formula (1.2.12), it is equivalent to $\int_{-\infty}^{\infty} |\xi(\alpha)|^2 d\alpha < \infty$.

Example 3.9.7 (Derivation filter, continuation). *We provide a justification of the formula of the derivation filter obtained in Example 3.9.4.2, now without the use of the δ function. Assume that the square-integrable and stationary stochastic process $\{X_t\}_{t\in\mathbb{R}}$ is $\mathcal{L}_2$-differentiable with derivative $\{X_t'\}_{t\in\mathbb{R}}$, in the sense of Definition 3.3.5. A sufficient and necessary condition for the existence of $\{X_t'\}_{t\in\mathbb{R}}$ is $\int_{\mathbb{R}} \alpha^2 dF(\alpha) < \infty$; cf. Remarks 3.3.11.1. Let $t, \varepsilon \in \mathbb{R}$, then*

$$\frac{X_{t+\varepsilon} - X_t}{\varepsilon} = \int_{\mathbb{R}} \frac{e^{i\varepsilon\alpha} - 1}{\varepsilon} e^{it\alpha} dZ_\alpha, \quad \text{a.s.}$$

It follows from (3.9.11) that, for given $\alpha \in \mathbb{R}$ and as $\varepsilon \to 0$,

$$\frac{X_{t+\varepsilon} - X_t}{\varepsilon} \xrightarrow{\mathcal{L}_2} X_t', \quad \text{for some} \quad X_t' \in \mathcal{H}_X,$$

iff $\dfrac{e^{i\varepsilon\alpha} - 1}{\varepsilon} \to \xi$, in $\mathcal{H}_F$, for some $\xi \in \mathcal{H}_F$.

When the limit in $\mathcal{H}_F$ given by ξ exists, it must coincide with $d/d\varepsilon$ $e^{i\varepsilon\alpha}|_{\varepsilon=0} = i\alpha$. Therefore we obtain

$$X_t' = i \int_{\mathbb{R}} \alpha e^{it\alpha} dZ_\alpha, \quad \text{a.s.}$$

The transfer function of derivation is thus given by

$$\xi(\alpha) = i\alpha = |\alpha| e^{i\frac{\pi}{2} \operatorname{sgn}\alpha}, \qquad (3.9.13)$$

and the spectral d.f. of $\{X_t'\}_{t\in\mathbb{R}}$ is given by

$$dF_{X'}(\alpha) = |\xi(\alpha)|^2 dF(\alpha) = \alpha^2 dF(\alpha), \quad \forall \alpha \in \mathbb{R},$$

with F denoting the spectral d.f. of $\{X_t\}_{t\in\mathbb{R}}$.

The polar form of the transfer function leads to the following interpretation:

$$Y_t = \int_{\mathbb{R}} \underbrace{|\xi(\alpha)|}_{\substack{\text{amplification of the} \\ \text{harmonic of frequency } \alpha}} \exp\{i[t\alpha + \underbrace{\arg\xi(\alpha)}_{\substack{\text{phasing of the} \\ \text{harmonic of frequency } \alpha}}]\}dZ_\alpha, \quad \text{a.s.},$$

$\forall t \in \mathbb{R}$. The polar form of the transfer function of the derivative is given in (3.9.13): the harmonic of frequency α is amplified by $|\alpha|$ and phased by $\pi/2 \operatorname{sgn}\alpha$.

3.9.2   *Some linear filters and their differential equations*

This section shows how linear filters can be re-expressed as linear differential equations. Some applications in engineering are presented.

Exponential smoothing and resistance-capacitance

An electronic circuit made of one resistor and one capacitor is called resistance-capacitance filter. At any time $t \in \mathbb{R}$, X_t designates the potential difference at the input of the circuit and Y_t designates the potential difference at the output of the circuit. The circuit reduces high variations of potential difference. The relation between input X_t and output Y_t is given by

$$aY_t' + Y_t = X_t. \tag{3.9.14}$$

This differential equation can be solved with the method of integrating factors, cf. Section A.9, and one finds the particular solution

$$Y_t = \frac{1}{a} \int_{-\infty}^{t} e^{-\frac{t-s}{a}} X_s ds,$$

if $e^{t/a} Y_t \to 0$, as $t \to -\infty$. We have the exponential smoothing of Example 3.9.4.4: the impulse response function is $\psi(v) = a^{-1}e^{-v/a} I\{v \geq 0\}$, $\forall v \in \mathbb{R}$, and the transfer function is $\xi(\alpha) = (1 + ia\alpha)^{-1}$, $\forall \alpha \in \mathbb{R}$. Consequently, the spectral density of $\{Y_t\}_{t \in \mathbb{R}}$ is given by

$$f_Y(\alpha) = \left| \frac{1}{1 + ia\alpha} \right|^2 f(\alpha) = \frac{1}{1 + (a\alpha)^2} f(\alpha), \quad \forall \alpha \in \mathbb{R}, \tag{3.9.15}$$

f being the spectral density of the input $\{X_t\}_{t \in \mathbb{R}}$. Thus the resistance-capacitance filter mitigates high frequencies.

   Note that if the spectral density of the input would be constant, then the spectral density of the output would be the one of the Ornstein-Uhlenbeck process given in (3.7.9).

   Note also that the impulse response function ψ solves the homogeneous correspondent of the differential equation (3.9.14), over $\mathbb{R}_+$ and with initial condition $\psi(0) = 1/a$.

   An alternative computation of the spectral density f_Y is the following. Denote by γ, γ_Y and $\gamma_{Y'}$ the a.c.v.f. of $\{X_t\}_{t \in \mathbb{R}}$, $\{Y_t\}_{t \in \mathbb{R}}$ and $\{Y_t'\}_{t \in \mathbb{R}}$ respectively. By applying (3.3.2), one obtains that

$$\gamma(h) = \text{cov}(aY_t' + Y_t, aY_{t+h}' + Y_{t+h}) = a^2 \gamma_{Y'}(h) + \gamma_Y(h), \quad \forall t, h \in \mathbb{R}.$$

This is equivalent to

$$\int_{\infty}^{\infty} e^{ih\alpha} f(\alpha) d\alpha = a^2 \int_{\infty}^{\infty} e^{ih\alpha} \alpha^2 f_Y(\alpha) d\alpha + \int_{\infty}^{\infty} e^{ih\alpha} f_Y(\alpha) d\alpha, \quad \forall h \in \mathbb{R},$$

which leads directly to (3.9.15).

Harmonic oscillator

The linear oscillator is a basic model in mechanical and electrical engineering. A schematic description is provided by a cart with mass $m > 0$ whose left side is attached to a wall by a spring. At horizontal position $y = 0$, the spring exerts no force on the cart. When the cart is displaced by the distance $y > 0$ to the right, the spring exerts the elastic force $F_s = -ky$ on it, for some spring constant $k > 0$. We consider the three following situations.

(1) *Simple harmonic oscillator*: the cart experiences neither friction nor external force.
From $my'' = F_s$ we obtain

$$y'' + \frac{k}{m}y = 0.$$

One verifies directly that the general solution is $y(t) = a_1 \cos \sqrt{k/m}t + a_2 \sin \sqrt{k/m}t$. With the initial conditions $y(0) = y_0 > 0$ and $y'(0) = 0$ (meaning that the cart is pulled of y_0 to the right and released) we obtain $a_1 = 0$ and $a_2 = y_0$. The solution to the initial value problem is $y(t) = y_0 \cos \sqrt{k/m}t$, $\forall t \geq 0$.

(2) *Damped harmonic oscillator*: the cart experiences friction but no external force.
In this more realistic model, friction slows down oscillation. The friction generates the dumping force $F_d = -cy'$, for some constant $c > 0$, which is a characteristic of the medium (e.g. the fluid viscosity). From $my'' = F_s + F_d$ we obtain

$$y'' + \frac{c}{m}y' + \frac{k}{m}y = 0. \tag{3.9.16}$$

Theorem A.9.1 gives the general solution, $y(t) = a_1 e^{r_1 t} + a_2 e^{r_2 t}$, where

$$r_1 = -b + i\beta, \ r_2 = -b - i\beta, \ b = \frac{c}{2m} \text{ and } \beta = \sqrt{\frac{k}{m} - \left(\frac{c}{2m}\right)^2}. \tag{3.9.17}$$

Assume $\beta > 0$, i.e.

$$c < 2\sqrt{km}. \tag{3.9.18}$$

Note that the roots r_1 and r_2 lie on the left half of the complex plane and thus the differential equation is stable in the sense of Definition 3.9.8. This is the case of small damping, precisely where the damping force is smaller than the spring force. The general solution can now be rewritten as

$$y(t) = e^{-bt}(a_1 \cos \beta t + a_2 \sin \beta t). \tag{3.9.19}$$

With the initial conditions of the non-damped oscillator we obtain $a_1 = y_0$ and $a_2 = bx_0/\beta$, giving

$$y(t) = y_0 e^{-bt}\left(\cos \beta t + \frac{b}{\beta}\sin \beta t\right), \quad \forall t \geq 0. \tag{3.9.20}$$

We can note that if the characteristic of the medium (e.g. the viscosity) c increases, then the frequency of the oscillator β decreases.

(3) *Stochastically driven damped harmonic oscillator:* the cart experiences friction and external force.

The cart is subject to friction and to the random external force F_e (magnetic field, wind, etc.) that takes the form of a stationary process $\{X_t\}_{t\in\mathbb{R}}$. From $my'' = F_s + F_d + F_e$ we obtain the second order linear differential equation of the displacement of the cart

$$Y_t'' + \frac{c}{m}Y_t' + \frac{k}{m}Y_t = \frac{X_t}{m}. \tag{3.9.21}$$

Consider the simple instance of the spectral decomposition (3.6.1) for which $X_t = B\cos(\lambda t + \theta)$, $\forall t \in \mathbb{R}$, for some square-integrable random variable B, for some independent and uniformly distributed over $[0, 2\pi)$ random variable θ and for some fixed frequency $\lambda > 0$. We search for a particular solution by the method of undetermined coefficients. We insert the guess $Y_t = A_1 \cos \lambda t + A_2 \sin \lambda t$ into the differential equation and obtain $(k - m\lambda^2)A_1 + c\lambda A_2 = B$ and $-c\lambda A_1 + (k - m\lambda^2)A_2 = 0$. By solving these equations we obtain

$$A_1 = \frac{(k - m\lambda^2)B}{(k - m\lambda^2)^2 + (c\lambda)^2} \quad \text{and} \quad A_2 = \frac{c\lambda B}{(k - m\lambda^2)^2 + (c\lambda)^2}.$$

The particular solution is thus

$$\frac{B}{(k - m\lambda^2)^2 + (c\lambda)^2}\{(k - m\lambda^2)\cos(\lambda t + \theta) + c\lambda \sin(\lambda t + \theta)\}$$

$$= \frac{B}{\{(k - m\lambda^2)^2 + (c\lambda)^2\}^{\frac{1}{2}}}\cos(\lambda t + \theta - \xi),$$

where $\xi = \arg\{k - m\lambda^2 + ic\lambda\}$. Note that $\tau = \theta - \xi$ is also uniformly distributed over the unit circle. By adding this particular solution to (4.9.4), i.e. to the solution of the homogeneous equation under the small damping assumption (3.9.18), we obtain the general solution of (3.9.21) given by

$$Y_t = e^{-bt}(a_1 \cos \beta t + a_2 \sin \beta t) + \frac{B}{\{(k - m\lambda^2)^2 + (c\lambda)^2\}^{\frac{1}{2}}} \cos(\lambda t + \tau),$$

$\forall t \geq 0$.

When the differential equation is driven by an arbitrary stationary process $\{X_t\}_{t \in \mathbb{R}}$, we can provide a stationary solution for the small damping case in the form of a linear filter (3.9.1). According to Theorem 3.9.9 of Section 3.9.3 to come, a particular solution is given by

$$Y_t = \int_{-\infty}^t \psi(t - s) X_s \mathrm{d}s, \quad \forall t \in \mathbb{R}, \qquad (3.9.22)$$

where

$$\psi(u) = \frac{1}{m\beta} e^{-bu} \sin \beta u, \quad \forall u \in \mathbb{R}.$$

Indeed, we have already found the general solution of the associated differential equation $m\psi'' + c\psi' + k\psi = 0$. This is the equation of the damped harmonic oscillator (3.9.16), whose general solution is given by (3.9.19). With the initial conditions of Theorem 3.9.9, viz. $\psi(0) = 0$ and $\psi'(0) = 1/m$, we obtain $a_1 = 0$ and $a_2 = k/(m\beta)$. Note that (3.9.22) is only one of the possible solutions (that are obtained by adding to it the solutions to the homogeneous equation).

The analysis in the frequency domain appears simpler. Denote by F the spectral d.f. of the stationary driving force $\{X_t\}_{t \in \mathbb{R}}$ and by F_Y the spectral d.f. of the stationary displacement of the cart $\{Y_t\}_{t \in \mathbb{R}}$. Then $\mathrm{d}F(\alpha) = |\xi(\alpha)|^2 \mathrm{d}F_Y(\alpha)$, $\forall \alpha \in \mathbb{R}$, where ξ is the transfer function. It is obtained by repeated application of the single derivation filter (3.9.13), yielding

$$\xi(\alpha) = -m\alpha^2 + ic\alpha + k,$$

and thus

$$\mathrm{d}F_Y(\alpha) = \{(k - m\alpha^2)^2 + (c\alpha)^2\}^{-1} \mathrm{d}F(\alpha), \quad \forall \alpha \in \mathbb{R}.$$

The frequency $\alpha_0 = \sqrt{k/m}$ solves $k - m\alpha^2 = 0$ and it is called resonance frequency, because frequencies α close to α_0 are substantially amplified. Precisely, small frequency variations of the input force around α_0 lead to large movements of the cart.

3.9.3   *Solutions of linear differential equations driven by stationary processes*

We consider the linear stochastic differential equation

$$a_0 Y_t^{(n)} + a_1 Y_t^{(n-1)} + \cdots + a_n Y_t = X_t, \quad \forall t \in \mathbb{R}, \tag{3.9.23}$$

where $a_0, \ldots, a_n \in \mathbb{R}$ and where the driving stochastic process $\{X_t\}_{t \in \mathbb{R}}$ is stationary and sufficiently differentiable. Define the characteristic equation

$$z^n p(z) = a_0 z^n + \cdots + a_{n-1} z + a_n = 0. \tag{3.9.24}$$

Definition 3.9.8 (Stable and unstable linear differential equation). *The linear differential equation (3.9.23) is called stable if all roots of the characteristic equation (3.9.24) possess a real part that is negative, i.e. if all roots belong to the left complex half plane. Otherwise, the linear differential equation is called unstable.*

As a result, any solution of a stable homogeneous differential equation vanishes as $t \to \infty$.

Theorem 3.9.9 (Linear differential equation driven by stationary process). *If the linear differential equation (3.9.23) is stable and if the driving process $\{X_t\}_{t \in \mathbb{R}}$ is stationary, then there exists a stationary process $\{Y_t\}_{t \in \mathbb{R}}$ that solves the differential equation. This solution can be expressed under the form of the time invariant linear filter*

$$Y_t = \int_{-\infty}^{t} \psi(t-s) X_s \mathrm{d}s, \quad \forall t \in \mathbb{R},$$

where the impulse response function ψ is a solution of the linear differential equation

$$a_0 \psi^{(n)}(u) + a_1 \psi^{(n-1)}(u) + \cdots + a_n \psi(u) = 0, \quad \forall u \geq 0,$$

under the initial conditions

$$\psi(0) = \frac{a_1}{a_0}, \qquad\qquad\qquad\qquad \text{if } n = 1,$$

$$\psi(0) = \cdots = \psi^{(n-2)}(0) = 0 \text{ and } \psi^{(n-1)}(0) = \frac{1}{a_0}, \qquad \text{if } n = 2, 3, \ldots.$$

Moreover, the impulse response function ψ is integrable.

The stability assumption guarantees that the fluctuations of the driving process $\{X_t\}_{t \in \mathbb{R}}$ are attenuated as time passes. In some cases, the attenuation occurs at exponential rate. A minor generalization of the differential equation (3.9.23) is the subject of Proposition 3.9.10.

Proposition 3.9.10. *Consider the linear differential equation*

$$a_0 Y_t^{(n)} + a_1 Y_t^{(n-1)} + \cdots + a_n Y_t = b_0 X_t^{(m)} + b_1 X_t^{(m-1)} + \cdots + b_m X_t,$$

$\forall t \in \mathbb{R}$, *where* $a_0, \ldots, a_n, b_0, \ldots, b_m \in \mathbb{R}$ *and where the processes* $\{X_t\}_{t \in \mathbb{R}}$ *and* $\{Y_t\}_{t \in \mathbb{R}}$ *are stationary and sufficiently differentiable. Then*

$$\left| \sum_{j=1}^{n} a_j (i\alpha)^{n-j} \right|^2 dF_Y(\alpha) = \left| \sum_{j=1}^{m} b_j (i\alpha)^{m-j} \right|^2 dF_X(\alpha), \quad \forall \alpha \in \mathbb{R},$$

where F_X *and* F_Y *are the spectral d.f. of* $\{X_t\}_{t \in \mathbb{R}}$ *and* $\{Y_t\}_{t \in \mathbb{R}}$, *respectively.*

Proof. By using the transfer function of the j-th derivative $\xi(\alpha) = (i\alpha)^j$, we obtain from (3.9.3) that

$$U_t = \sum_{j=1}^{n} a_j Y_t^{(n-j)} = \int_{\mathbb{R}} e^{i\alpha t} \sum_{j=1}^{n} a_j (i\alpha)^{n-j} dZ_\alpha, \quad \forall t \in \mathbb{R},$$

for some spectral process $\{Z_\alpha\}_{\alpha \in \mathbb{R}}$. According to (3.6.13), the spectral d.f. of $\{U_t\}_{t \in \mathbb{R}}$ is given by

$$dF_U(\alpha) = \mathsf{E}\left[\sum_{j=1}^{n} a_j (i\alpha)^{n-j} dZ_\alpha \overline{\sum_{j=1}^{n} a_j (i\alpha)^{n-j} dZ_\alpha} \right]$$

$$= \left| \sum_{j=1}^{n} a_j (i\alpha)^{n-j} \right|^2 dF_Y(\alpha), \quad \forall \alpha \in \mathbb{R}. \qquad (3.9.25)$$

The same developments can be done with the process $\{X_t\}_{t \in \mathbb{R}}$ and yield

$$dF_U(\alpha) = \left| \sum_{j=1}^{m} b_j (i\alpha)^{m-j} \right|^2 dF_X(\alpha), \quad \forall \alpha \in \mathbb{R}.$$

$\square$

3.9.4 *Solutions of linear differential equations driven by white noise*

The Langevin differential equation is introduced in Example 3.4.9 as a model for the velocity of a particle suspended in homogeneous fluid. This differential equation, which is precisely given in (3.4.16), can be rewritten in the form of (3.9.23) with $n = 1$ and with Gaussian WN driving process

$\{X_t\}_{t\in\mathbb{R}} = \sigma\{W_t'\}_{t\in\mathbb{R}}$, where $\{W_t\}_{t\in\mathbb{R}}$ is the Wiener process and $\sigma > 0$. So the velocity process $\{Y_t\}_{t\in\mathbb{R}}$ is given by

$$a_0 Y_t' + a_1 Y_t = \sigma W_t', \quad \forall t \in \mathbb{R},$$

for some $a_0, a_1 > 0$. Its solution is the Ornstein-Uhlenbeck process, introduced in Example 3.7.4, and an integral representation is given in (3.7.10). That integral can also be obtained with Theorem 3.9.9 and the result is

$$Y_t = \sigma \int_{(-\infty,t]} \psi(t-s)\mathrm{d}W_s, \quad \forall t \in \mathbb{R},$$

where $\psi(u) = 1/a_0 \exp\{-a_1/a_0 u\}$. Note that partial integration formula (3.4.4) gives

$$Y_t = \frac{\sigma}{a_0}\left(W_t - a_1 \int_{-\infty}^t \psi(t-s)W_s\,\mathrm{d}s\right), \quad \forall t \in \mathbb{R}.$$

3.9.5   Shot noise and filtered point process

The shot noise process is introduced in Section 3.2.5. In this section we consider the shot noise process with time domain $\mathbb{R}$ defined in (3.2.22). We can note the similarity between the integral formula of the shot noise process in (3.2.22) and the integral formula of the time invariant linear filter (3.9.2). We show that it is closely related to the theory of filters and this relation allows us to obtain its spectral distribution, precisely its Bartlett spectral d.f., together with its a.c.v.f. The main result is the following.

Proposition 3.9.11 (Spectral d.f. and a.c.v.f. of shot noise process). *Consider the counting process $\{N_t\}_{t\in\mathbb{R}}$ that possesses a stationary stream of events and that is regular, in the sense of Definition 3.2.38. For $\psi \in \mathcal{L}_1(\mathbb{R}) \cap \mathcal{L}_2(\mathbb{R})$, define the shot noise process*

$$X_t = \sum_{j=-\infty}^{\infty} \psi(t - T_j) = \int_{\mathbb{R}} \psi(t-s)\mathrm{d}N_s, \quad \forall t \in \mathbb{R},$$

where $\{T_n\}_{n\in\mathbb{Z}}$ is the process of occurrence times. Then the shot noise process $\{X_t\}_{t\in\mathbb{R}}$ is stationary with a.c.v.f. given by

$$\gamma_X(u) = \int_{\mathbb{R}} \mathrm{e}^{iu\alpha}|\xi(\alpha)|^2 \mathrm{d}F(\alpha), \quad \forall u \in \mathbb{R}, \qquad (3.9.26)$$

where F is Bartlett spectral d.f. of the associated point process, as given in Definition 3.8.1, and where ξ is the transfer function of the time invariant linear filter with impulse response function ψ, as given in (3.9.4).

An alternative formula for the a.c.v.f. is given by

$$\gamma_X(u) = \int_{\mathbb{R}} \widehat{\psi_u}(\alpha)\overline{\widehat{\psi_0}(\alpha)}\mathrm{d}F(\alpha), \quad \forall u \in \mathbb{R},$$

where $\psi_t(s) = \psi(t - s)$, $\forall s, t \in \mathbb{R}$, and where, for given $t \in \mathbb{R}$,

$$\widehat{\psi_t}(\alpha) = \int_{-\infty}^{\infty} \mathrm{e}^{-\mathrm{i}\alpha s}\psi_t(s)\mathrm{d}s, \quad \forall \alpha \in \mathbb{R}.$$

The spectral d.f. of $\{X_t\}_{t \in \mathbb{R}}$ is given by

$$\mathrm{d}F_X(\alpha) = |\xi(\alpha)|^2\mathrm{d}F(\alpha), \quad \forall \alpha \in \mathbb{R}.$$

Note the similarity of the relation (3.9.26), between the a.c.v.f. of the shot noise and Bartlett's spectral d.f. of the counting process, with the relation (3.9.7), between the a.c.v.f. of the filtered process and the spectral d.f. of the input process.

Proof. Let us re-express the shot noise in terms of the point process N over $(\mathbb{R}, \mathcal{B}(\mathbb{R}))$, with stationary stream of events and regular, in the sense of Definition 3.2.38. Thus

$$X_t = \int_{\mathbb{R}} \psi(t - s)\mathrm{d}N(s), \quad \forall t \in \mathbb{R},$$

can be re-expressed by means of the operator Θ_N given in (3.8.2). Let $t, u \in \mathbb{R}$, then

$$X_t = \int_{\mathbb{R}} \psi_t(s)\mathrm{d}N(s) = \Theta_N(\psi_t) \text{ and } X_{t+u} = \int_{\mathbb{R}} \psi_{t+u}(s)\mathrm{d}N(s) = \Theta_N(\psi_{t+u}).$$

Because $\psi \in \mathcal{L}_1(\mathbb{R}) \cap \mathcal{L}_2(\mathbb{R})$, we can obtain Bartlett's spectral d.f. of N that is given in Definition 3.8.1. Let us denote this d.f. by F, then it is given by

$$\mathsf{cov}(X_{t+u}, X_t) = \mathsf{cov}(\Theta_N(\psi_{t+u}), \Theta_N(\psi_t)) = \int_{\mathbb{R}} \widehat{\psi_{t+u}}(\alpha)\overline{\widehat{\psi_t}(\alpha)}\mathrm{d}F(\alpha),$$

where, according to (3.8.3),

$$\widehat{\psi_t}(\alpha) = \int_{-\infty}^{\infty} \mathrm{e}^{-\mathrm{i}\alpha s}\psi_t(s)\mathrm{d}s = \mathrm{e}^{-\mathrm{i}t\alpha} \int_{-\infty}^{\infty} \mathrm{e}^{\mathrm{i}\alpha v}\psi(v)\mathrm{d}v = \mathrm{e}^{-\mathrm{i}t\alpha}\xi(-\alpha), \,\forall \alpha \in \mathbb{R},$$

where ξ is the transfer function of the time invariant linear filter with impulse response function ψ, cf. (3.9.4). Therefore we obtain

$$\mathsf{cov}(X_{t+u}, X_t) = \int_{\mathbb{R}} \mathrm{e}^{-\mathrm{i}(t+u)\alpha}\xi(-\alpha)\mathrm{e}^{\mathrm{i}t\alpha}\overline{\xi(-\alpha)}\mathrm{d}F(\alpha)$$

$$= \int_{\mathbb{R}} \mathrm{e}^{-\mathrm{i}u\alpha}|\xi(-\alpha)|^2\mathrm{d}F(\alpha)$$

$$= \int_{\mathbb{R}} \mathrm{e}^{\mathrm{i}u\alpha}|\xi(\alpha)|^2\mathrm{d}F(-\alpha)$$

$$= \int_{\mathbb{R}} \mathrm{e}^{\mathrm{i}u\alpha}|\xi(\alpha)|^2\mathrm{d}F(\alpha),$$

where the last equality is due to the symmetry of F. One can refer to Appendix A.6 for details on the two last equalities. $\qquad\square$

Example 3.9.12 (Shot noise process driven by an homogeneous Poisson process). *Let $\{N_t\}_{t\in\mathbb{R}}$ be a homogeneous Poisson process with occurrence rate $\lambda > 0$. Bartlett's spectral density of the homogeneous Poisson process is given by (3.8.1): it is $f(\alpha) = \lambda/(2\pi)$, $\forall\alpha \in \mathbb{R}$. We can directly apply Proposition 3.9.11 and obtain the spectral density of the Poisson shot noise process, respectively by*

$$f_X(\alpha) = \frac{\lambda}{2\pi}|\xi(\alpha)|^2, \quad \forall\alpha \in \mathbb{R},$$

where ξ is the transfer function of ψ, cf. (3.9.4). The a.c.v.f. can be obtained either by

$$\gamma_X(u) = \frac{\lambda}{2\pi}\int_{-\infty}^{\infty}\mathrm{e}^{iu\alpha}|\xi(\alpha)|^2\mathrm{d}\alpha, \quad \forall u \in \mathbb{R},$$

or by

$$\gamma_X(u) = \frac{\lambda}{2\pi}\int_{-\infty}^{\infty}\widehat{\psi_u}(\alpha)\overline{\widehat{\psi_0}(\alpha)}\mathrm{d}\alpha = \lambda\int_{-\infty}^{\infty}\psi_u(s)\psi_0(s)\mathrm{d}s$$

$$= \lambda\int_{-\infty}^{\infty}\psi(s-u)\psi(s)\mathrm{d}s, \quad \forall u \in \mathbb{R},$$

according to Parseval-Plancherel formula (1.2.12).

Example 3.9.13 (Shot noise process driven by a Cox process). *Let $\{N_t\}_{t\in\mathbb{R}}$ be the doubly stochastic or Cox process with random intensity given by the nonnegative stochastic process $\{L_t\}_{t\in\mathbb{R}}$. This process is assumed stationary: its expectation is $\lambda_0 = \mathsf{E}[L_t]$, $\forall t \in \mathbb{R}$, and its spectral density is f_L. This process is introduced in Example 3.8.3 and we refer to its section for more explanations. Bartlett's spectral density of the Cox process is given by (3.8.5): it is $f(\alpha) = \lambda_0/(2\pi) + f_L(\alpha)$, $\forall\alpha \in \mathbb{R}$. We can directly apply Proposition 3.9.11 and obtain the spectral density of the Cox shot noise process as*

$$f_X(\alpha) = |\xi(\alpha)|^2\left(\frac{\lambda_0}{2\pi} + f_L(\alpha)\right), \quad \forall\alpha \in \mathbb{R},$$

where ξ is the transfer function of ψ, cf. (3.9.4).

Chapter 4

Selected topics on stationary models

This chapter is a selection of special topics on the analysis of stationary models. Some of these topics are only briefly introduced. Section 4.1 introduces stationary random fields, that generalize stationary processes in the sense that the one-dimensional time domain is generalized to a multidimensional domain, this in order to include planar or spatial coordinates. Section 4.2 presents time series of planar directions, called circular time series. These directions are angles and the definition of stationarity requires a specific measure of correlation, called circular correlation. Section 4.3 introduces the long range dependence or long memory, which is described by the importance of low frequencies and the scarcity of high frequencies. The third topic is given in Section 4.4 and it concerns the spectral density that is unbounded at the origin and in this way nonintegrable, giving rise to the concept of intrinsic stationarity. An intrinsic stationary process is nonstationary but it can be made stationary through simple linear filtering. The next topic concerns the unstable systems and it is presented in Section 4.5. It concerns processes defined through an unstable differential equation. Then, in Section 4.6, the Hilbert transform and the related process called envelope are presented. The Hilbert transform is a stationary process obtained by applying a particular filter. An important computational topic is the stochastic simulation of stationary Gaussian processes. Section 4.7 presents four different simulation algorithms that use: Choleski factorization, circulant embedding, spectral distribution and ARMA approximation. A brief survey of large deviations theory followed by an application to the AR(1) time series is given in Section 4.8. The last topic of this chapter, given in Section 4.9, concerns the application of information theory to time series and to their spectral distributions.

4.1 Stationary random fields

If we extend the univariate time domain of a stationary process to a multivariate one, like $\mathbb{R}$ to $\mathbb{R}^d$, for some $d \geq 2$, then we obtain random element called random field. Thus, we denote by $\{X_t\}_{t \in \mathbb{R}^d}$ a random field whose random variables take values in $\mathbb{R}$ or in $\mathbb{C}$. Some common random fields have planar and temporal coordinates, thus $d = 2 + 1 = 3$, and spatial and temporal coordinates, thus $d = 3 + 1 = 4$. An example with $d = 4$ arises from the theory of turbulence, where $X_{(s_1, s_2, s_3, t)}$ represents the velocity at location $(s_1, s_2, s_3) \in \mathbb{R}^3$ within a turbulent fluid and at time $t \in \mathbb{R}$.

Whenever $\mathsf{E}[|X_t|^2] < \infty$, $\forall t \in \mathbb{R}^d$, we can define the a.c.v.f. of the random field by

$$\gamma(t + h, t) = \mathsf{cov}(X_{t+h}, X_t), \quad \forall t, h \in \mathbb{R}^d.$$

Definition 4.1.1 (Stationary random field and its a.c.v.f.). *The random field $\{X_t\}_{t \in \mathbb{R}^d}$ is called stationary if $\mathsf{E}\left[|X_t|^2\right] < \infty$ and if $\mathsf{E}[X_t]$ and $\gamma(t + h, t)$ do not depend on t, $\forall t, h \in \mathbb{R}^d$. In this case we define the a.c.v.f. of the stationary random field by*

$$\gamma(h) = \gamma(h, 0), \quad \forall h \in \mathbb{R}^d.$$

Many fundamental properties and results of the spectral analysis of stationary stochastic processes (where $d = 1$) have their generalizations to stationary random fields (where $d \geq 2$). Thus, a central result is that a stationary random field with a.c.v.f. γ possesses a spectral d.f. F over $\mathbb{R}^d$, such that

$$\gamma(h) = \int_{\mathbb{R}^d} \exp\{i\langle h, \alpha \rangle\} dF(\alpha), \quad \forall h \in \mathbb{R}^d.$$

Further, there exists a (complex-valued) spectral random field $\{Z_\alpha\}_{\alpha \in \mathbb{R}^d}$ defined through the mean square stochastic integral

$$X_t = \int_{\mathbb{R}^d} \exp\{i\langle t, \alpha \rangle\} dZ_\alpha, \quad \text{a.s.}, \forall t \in \mathbb{R}^d,$$

where

$$\mathsf{E}[Z_\alpha] = 0, \quad \forall \alpha \in \mathbb{R}^d,$$

and where, $\forall \alpha, \beta \in \mathbb{R}^d$,

$$\mathsf{E}\left[dZ_\alpha \overline{dZ_\beta}\right] = \begin{cases} dF(\alpha), & \text{if } \alpha = \beta, \\ 0, & \text{if } \alpha \neq \beta. \end{cases}$$

In general, $\gamma(\boldsymbol{h})$ depends on $\boldsymbol{h}$ through $\|\boldsymbol{h}\|$ and $\arg \boldsymbol{h}$, unless $\boldsymbol{h} = 0$ (in which case no direction would exist). But sometimes $\arg \boldsymbol{h}$ is irrelevant and in this case we have an isotropic random field: if the a.c.v.f. of a stationary random field does not depend on the direction of its argument, then the random field is called isotropic. In this case, we can for example denote $\lambda(\|\boldsymbol{h}\|) = \gamma(\boldsymbol{h})$ and it can be shown that

$$\lambda(l) = \int_{\mathbb{R}_+} (lx)^{1-\frac{d}{2}} J_{\frac{d}{2}-1}(lx) \mathrm{d}G(x), \quad \forall l \in \mathbb{R}_+,$$

where G is a d.f. with finite mass over $\mathbb{R}_+$ and where

$$J_\nu(z) = \frac{1}{\pi} \int_0^\pi \cos(z \sin \alpha - \nu \alpha) \mathrm{d}\alpha - \frac{\sin(\nu \pi)}{\pi} \int_0^\infty \exp\{-z \sinh t - \nu t\} \mathrm{d}t,$$

$\forall z \in \mathbb{C}$ such that $|\arg z| < \pi/2$, is the Bessel function J of order $\nu \in \mathbb{C}$, cf. e.g. 9.1.22 at p. 360 of Abramowitz and Stegum (1972).

Two complete introductions to the theory stationary random fields can be found in Sections 15-17 of Yaglom (1962) and in Chapter 7 of Lindgren (2012).

Some important applications of the theory of stationary random fields appear in oceanography. In this case the real-valued random variable $X_{(s_1,s_2,t)}$ represents the surface of the sea, where $(s_1, s_2) \in \mathbb{R}^2$ is the coordinate vector of a location at sea and $t \in \mathbb{R}$ is the time. Stationary random fields for sea waves have an important role in naval engineering. In this context, St. Denis and Pierson (1953) extended the spectral decomposition (3.2.5) of the Gaussian stationary process to

$$X_{(s_1,s_2,t)}$$

$$= \sum_{j=1}^\infty B(\alpha_j, \kappa_{1,j}, \kappa_{2,j}) \cos(\alpha_j t - \kappa_{1,j} s_1 - \kappa_{2,j} s_2 + \theta(\alpha_j, \kappa_{1,j}, \kappa_{2,j})),$$

where: $t \in \mathbb{R}$ is the time, $(s_1, s_2) \in \mathbb{R}^2$ is the coordinate of a location over the surface of the ocean, $0 < \alpha_1 < \alpha_2 < \cdots < \infty$ are the waves frequencies at any fixed location (s_1, s_2), $0 < \kappa_{1,1} < \kappa_{1,2} < \cdots < \infty$ and $0 < \kappa_{2,1} < \kappa_{2,2} < \cdots < \infty$ determine the shapes of the waves at any fixed time t, $\theta(\alpha_j, \kappa_{1,j}, \kappa_{2,j})$, for $j = 1, 2, \ldots$ are uniformly distributed over $[0, 2\pi)$, $B(\alpha_j, \kappa_{1,j}, \kappa_{2,j})$, for $j = 1, 2, \ldots$ are Rayleigh distributed with different scales and where all these random variables are independent. At any fixed time t, the shape of the waves is represented by the sum of cosine functions and the planar null level is determined by all coordinate values (s_1, s_2) that are along the lines

$$\{\alpha_j t - \kappa_{1,j} s_1 - \kappa_{2,j} s_2 + \theta(\alpha_j, \kappa_{1,j}, \kappa_{2,j})\} \mathrm{mod} \pi = \frac{\pi}{2}.$$

4.2 Circular time series

Two continuous time stationary stochastic process over the unit circle are the Poisson and the compound Poisson, that are presented in Examples 3.2.33 and 3.2.40, respectively. This section concerns discrete time only and it presents general methods for constructing time series over the circle, namely for planar directions or circular data. Stochastic models for circular data are important in many disciplines, such as as statistical mechanics, crystallography, astronomy, meteorology and earth sciences, both as experimental and observational data. Two main references on this topic are Mardia and Jupp (2000) and Jammalamadaka and SenGupta (2001). One can refer to Gatto and Jammalamadaka (2015) for a short introduction and to Pewsey and Garcia-Portugués (2021) for a recent review. An important application of these models is for directions taken by winds and it is presented in Part I of Breckling (1989). We should note that time series of directions with associated magnitudes (such as velocities of winds or of sea currents) can be directly modeled by complex-valued time series. So the special methods of this section may not be necessary for these situations; see also Gonella (1972).

One of the earliest study on correlation between values of time series of directions is due to Watson and Beran (1967), who propose the following measure of correlation. Assume that $X_1, \ldots, X_n$ is a sample from a time series of unit vectors, representing planar directions. Then the empirical correlation coefficient at time lag one is given by

$$\frac{1}{n} \sum_{k=1}^{n-1} X_{k+1} X_k.$$

In this section we model directions by angles instead of unit vectors. We thus present time series for angles, namely for data defined modulo 2π. These are called circular time series and they are (weakly) stationary, according to a measure of correlation for circular random variables introduced in Section 4.2.1. After presenting this circular correlation, five types of circular time series are introduced: radially projected or offset, in Section 4.2.2, wrapped AR (WAR) of order p (WAR(p)), in Section 4.2.3, linked ARMA (LARMA) of orders p, q (LARMA(p, q)), in Section 4.2.4, circular AR (CAR) of order p (CAR(p)), in Section 4.2.5, and von Mises (vM) of order p (vM(p)) time series, in Section 4.2.6. A synthesis with comments on model selection is given in Section 4.2.7. This short introduction to circular time series is mainly based on Fisher and Lee (1994). Other references are

Section 7.2 of Fisher (1993), Part II of Breckling (1989) and Section 11.5 of Mardia and Jupp (2000).

4.2.1 *Circular correlation*

A measure of correlation between two circular random variables viz. random angles α and β is due to Jammalamadaka and Sarma (1988) and it takes the following form,

$$\begin{aligned}
\mathsf{ccorr}'(\alpha, \beta) &= \mathsf{corr}(\sin(\alpha - \mu), \sin(\beta - \nu)) \\
&= \frac{\mathsf{E}[\sin(\alpha - \mu)\sin(\beta - \nu)]}{\{\mathsf{E}[\sin^2(\alpha - \mu)]\mathsf{E}[\sin^2(\beta - \nu)]\}^{\frac{1}{2}}},
\end{aligned} \tag{4.2.1}$$

where $\mu = \arg\mathsf{E}[e^{i\alpha}]$ is the mean direction of α and $\nu = \arg\mathsf{E}[e^{i\beta}]$ the one of β. We can note that $\mathsf{E}[\sin(\alpha - \mu)] = \mathsf{E}[\sin(\beta - \nu)] = 0$. Thus, like in the linear case, $\sin(\alpha - \mu)$ and $\sin(\beta - \nu)$ are deviations from an indicator of main tendency. This definition requires unequivocally defined mean directions μ and ν. This is not the case under the isotropic or circular uniform distribution.

The correlation coefficient between the linear random variables X and Y can be expressed as

$$\mathsf{corr}(X, Y) = \frac{\mathsf{E}[(X_1 - X_2)(Y_1 - Y_2)]}{\{\mathsf{E}[(X_1 - X_2)^2]\mathsf{E}[(Y_1 - Y_2)^2]\}^{\frac{1}{2}}},$$

where the random vectors (X_1, Y_1) and (X_2, Y_2) are independent and distributed as (X, Y). This formula allows Fisher and Lee (1983) to define the circular correlation between the circular random variable α and β as

$$\mathsf{ccorr}(\alpha, \beta) = \frac{\mathsf{E}\left[\sin(\alpha_1 - \alpha_2)\sin(\beta_1 - \beta_2)\right]}{\left\{\mathsf{E}[\sin^2(\alpha_1 - \alpha_2)]\mathsf{E}[\sin^2(\beta_1 - \beta_2)]\right\}^{\frac{1}{2}}},$$

where (α_1, β_1) and (α_2, β_2) are independent and distributed as (α, β). However this definition differs algebraically from (4.2.1) and gives indeed different values, in general.

The two circular correlation coefficients above inherit many properties from the linear one. The most important are the following. Let α and β denote two circular random variables. Then the following properties hold and the hold with ccorr replaced by ccorr' as well.

- $\mathsf{ccorr}(\alpha, \beta) \in [-1, 1]$.
- If α and β are independent, then $\mathsf{ccorr}(\alpha, \beta) = 0$. The converse doe not hold, in general.

- ccorr is invariant w.r.t. the choice of the null directions that are selected for either α or β.
- $\text{ccorr}(\alpha, \beta) = \pm 1$ iff $\alpha = (\pm\beta + \xi_0) \bmod 2\pi$, for some fixed $\xi_0 \in [0, 2\pi)$.

We note that invariance of statistical procedures w.r.t. the choice of the null direction is very important, because the choice of the null direction is essentially arbitrary. These properties can be directly controlled. We refer to Section 8.2 of Jammalamadaka and SenGupta (2001) for further properties of ccorr' and to Fisher and Lee (1983) for further properties of ccorr.

4.2.2   *Radially projected time series*

A time series taking values over the unit circle can be constructed by radial projection of a planar time series or, equivalently, of a complex-valued time series. Precisely, let $\{X_k\}_{k\in\mathbb{Z}}$ be a complex-valued time series and consider

$$\beta_k = \arg X_k, \quad \forall k \in \mathbb{Z}.$$

Then $\{\beta_k\}_{k\in\mathbb{Z}}$ is the radially projected or offset circular time series. An offset distribution is the marginal distribution of the directional component of a multivariate distribution. Thus $\{\beta_k\}_{k\in\mathbb{Z}}$ has marginal offset distributions.

An important situation is when $\{\text{Re}\,X_k\}_{k\in\mathbb{Z}}$ and $\{\text{Im}\,X_k\}_{k\in\mathbb{Z}}$ are two independent Gaussian stationary time series with mean zero and variance one. Then, as noted at p. 182, the circular time series $\{\beta_k\}_{k\in\mathbb{Z}}$ possesses isotropic, i.e. uniform, marginal distributions. Application of Theorem 3.2.5 tells that $\{X_k\}_{k\in\mathbb{Z}}$ is strictly stationary. Thus $\{\beta_k\}_{k\in\mathbb{Z}}$ is strictly stationary as well. We have so obtained a simple stationary circular time series.

Consider independent random vectors satisfying

$$\begin{pmatrix} U_1 \\ U_2 \end{pmatrix} \sim \begin{pmatrix} V_1 \\ V_2 \end{pmatrix} \sim \mathcal{N}\left(\begin{pmatrix} 0 \\ 0 \end{pmatrix}, \sigma^2 \begin{pmatrix} 1 & \rho \\ \rho & 1 \end{pmatrix} \right),$$

for some $\rho \in [-1, 1]$ and $\sigma > 0$. Denote $\beta_j = \arg(U_j, V_j)$, for $j = 1, 2$. It is shown in Fisher and Lee (1994) that

$$\text{ccorr}(\beta_1, \beta_2) = \left\{ \frac{\pi}{4}\rho(1 - \rho^2)\,{}_2F_1\left(\frac{3}{2}, \frac{3}{2}; 2; \rho^2 \right) \right\}^2, \tag{4.2.2}$$

where ${}_2F_1$ is the hypergeometric function defined by

$$\,{}_2F_1(a, b; c; z) = \sum_{n=0}^{\infty} \frac{(a)_n (b)_n}{(c)_n} \frac{z^n}{n!}, \quad \forall z \in \mathcal{D}_0^{\circ}, \tag{4.2.3}$$

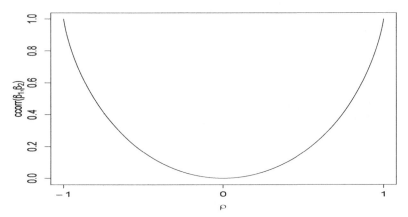

Fig. 4.1: ccorr(β_1, β_2) for radially projected Gaussian random vectors given by (4.2.2).

where $a, b \in \mathbb{C}$, $c \in \mathbb{C}\backslash\{0, -1, \ldots\}$; cf. 15.1.1 at p. 556 of Abramowitz and Stegum (1972).[1] Figure 4.1 shows ccorr(β_1, β_2) for radially projected Gaussian random vectors, according to (4.2.2).

We obtain directly from (4.2.2) that, when the above defined Gaussian time series Re $\{X_k\}_{k \in \mathbb{Z}}$ and Im $\{X_k\}_{k \in \mathbb{Z}}$ possess common a.c.r.f. ρ, the circular a.c.r.f. (c.a.c.r.f.) of $\{\beta_k\}_{k \in \mathbb{Z}}$ computed from the measure ccorr is given by

$$\rho^\circ(h) = \left\{ \frac{\pi}{4}\rho(h)\{1 - \rho^2(h)\} \, {}_2F_1\left(\frac{3}{2}, \frac{3}{2}; 2; \rho^2(h)\right) \right\}^2, \quad \forall h \in \mathbb{Z}. \quad (4.2.4)$$

Thus, according to this measure of circular correlation, $\{\beta_k\}_{k \in \mathbb{Z}}$ is stationary. Figure 4.2 shows the c.a.c.r.f. (4.2.4), for radially projected Gaussian

---

[1]The hypergeometric function can in principle be computed by a suitable truncation of the series. Some hints are the following. Denote the coefficients of the series (4.2.3) by $h_n = (a)_n (b)_n / [(c)_n n!]$, for $n = 0, 1, \ldots$. Thus $h_0 = 1$ and we compute the following coefficients recursively with

$$h_{n+1} = h_n \frac{(a+n)(b+n)}{(c+n)} \frac{1}{n+1}, \quad \text{for } n = 0, 1, \ldots.$$

Define $s_0 = 1$. We then obtain partial sum approximations to (4.2.3) at $z \in \mathcal{D}_0^\circ$ as follows,

$$s_{n+1} = s_n + h_{n+1} z^{n+1}, \quad \text{for } n = 0, 1, \ldots.$$

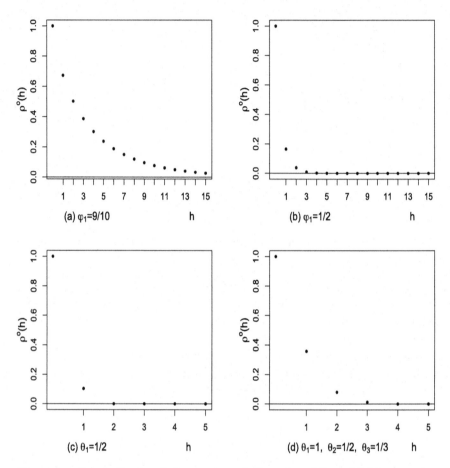

Fig. 4.2:   C.a.c.r.f. of radially projected Gaussian AR and MA given by
(4.2.4) for AR(1) with $\varphi_1 = 9/10$, in (a), AR(1) with $\varphi_1 = 1/2$, in (b),
MA(1) with $\theta_1 = 1/2$, in (c), and MA(2) with $\theta_1 = 1$, $\theta_2 = 1/2$, $\theta_3 = 1/3$,
in (d).

AR and MA time series: precisely for AR(1) with $\varphi_1 = 9/10$, in (a), AR(1)
with $\varphi_1 = 1/2$, in (b), MA(1) with $\theta_1 = 1/2$, in (c), and MA(2) with $\theta_1 = 1$,
$\theta_2 = 1/2$, $\theta_3 = 1/3$, in (d). These c.a.c.r.f. can be compared with the a.c.r.f.
of the AR(1) with $\varphi_1 = 1/2$ given in Figure 2.10 and with the a.c.r.f. of the
MA(1) with $\theta_1 = 1/2$ given in Figure 2.9. These c.a.c.r.f. are quite similar
to their corresponding a.c.r.f. Because this time series possesses isotropic
marginal distributions, it is thus useful whenever the data exhibit isotropy.

Remarks 4.2.1 (Related models). *Let us give some comments on the strong assumption made on the distribution of $\{X_k\}_{k\in\mathbb{Z}}$.*

(1) If $\{X_k\}_{k\in\mathbb{Z}}$ would have nonzero mean, then the marginal distribution of the directional time series $\{\beta_k\}_{k\in\mathbb{Z}}$ would be the offset normal. Its density, which can be found e.g. at p. 43 of Jammalamadaka and SenGupta (2001), appears rather intractable.

(2) Consider as before $\{X_k\}_{k\in\mathbb{Z}}$ with zero mean and with $\{\operatorname{Re} X_k\}_{k\in\mathbb{Z}}$ and $\{\operatorname{Im} X_k\}_{k\in\mathbb{Z}}$ independent. However, assume now that the real part has variance σ_1^2 and that the imaginary part has variance σ_2^2. Then the circular density of the radial projection β_k, has $\forall k \in \mathbb{Z}$ the tractable form

$$g(\alpha) = \frac{\sqrt{1-r^2}}{2\pi(1 - r\cos 2\alpha)}, \quad \forall \alpha \in (-\pi, \pi],$$

where

$$r = \frac{\sigma_1^2 - \sigma_2^2}{\sigma_1^2 + \sigma_2^2}.$$

Isotropy is indeed attained with $\sigma_1^2 = \sigma_2^2$.

(3) Another tractable offset normal offset normal distribution is obtained when $\{\operatorname{Re} X_k\}_{k\in\mathbb{Z}}$ and $\{\operatorname{Im} X_k\}_{k\in\mathbb{Z}}$ have zero means, unit variances and correlation $\rho = \operatorname{corr}(\operatorname{Re} X_k, \operatorname{Im} X_k)$, $\forall k \in \mathbb{Z}$. Then the circular density of the radial projection is given by

$$g(\alpha) = \frac{\sqrt{1-\rho^2}}{2\pi(1 - \rho\sin 2\alpha)}, \quad \forall \alpha \in (-\pi, \pi].$$

Isotropy corresponds to $\rho = 0$.

(4) In this context it is interesting to note that Gaussian conditional offset distributions, that are obtained by conditioning bivariate Gaussian vectors so to have unit length, possess very appealing forms. These are the von Mises (vM) and the generalized von Mises (GvM) distributions, whose densities are given in (4.2.12) and (4.9.1), respectively. Proof of this is given in Gatto and Jammalamadaka (2007).

4.2.3 *Wrapped AR time series*

A time series taking values over the unit circle can be constructed by wrapping a real-valued time series around the circle. Precisely, let $\{X_k\}_{k\in\mathbb{Z}}$ be a real-valued time series, then wrapped the time series is given by

$$\beta_k = X_k \bmod 2\pi, \quad \forall k \in \mathbb{Z}.$$

An important case is obtained when $\{X_k\}_{k \in \mathbb{Z}}$ is the causal AR(p): $\{\beta_k\}_{k \in \mathbb{Z}}$ is called WAR(p). Assume that $\varphi(z) = 1 - \varphi_1 z - \cdots - \varphi_p z^p \neq 0$, $\forall z \in \mathcal{D}_0$. Then $\{X_k\}_{k \in \mathbb{Z}}$ is causal and we have $X_k = \sum_{j=0}^{\infty} \psi_j Z_{k-j}$, $\forall k \in \mathbb{Z}$, with $\sum_{j=0}^{\infty} |\psi_j| < \infty$; cf. Theorem 2.2.12. In this situation the WAR(p) admits the explicit representation

$$\beta_k = \left(\sum_{j=0}^{\infty} \psi_j Z_{k-j} \right) \bmod 2\pi, \quad \forall k \in \mathbb{Z}.$$

The coefficients $\{\psi_j\}_{j \in \mathbb{N}}$ can be obtained recursively, according to Example 2.3.2.

Although only the circular time series $\{\beta_k\}_{k \in \mathbb{Z}}$ is observed, fitting the WAR(p) to a sample requires estimation of the AR(p) coefficients $\varphi_1, \ldots, \varphi_p$ and of the WN variance σ^2, that describe the unobserved time series $\{X_k\}_{k \in \mathbb{Z}}$. We have precisely, $\forall k \in \mathbb{Z}$, $X_k = \beta_k + 2\pi j_k$, where j_k is an unobserved integer. We have thus an estimation problem with missing data.

It is shown in Fisher and Lee (1983) that if

$$\begin{pmatrix} X_1 \\ X_2 \end{pmatrix} \sim \mathcal{N} \left(\begin{pmatrix} 0 \\ 0 \end{pmatrix}, \begin{pmatrix} \sigma_1^2 & \rho \sigma_1 \sigma_2 \\ \rho \sigma_1 \sigma_2 & \sigma_2^2 \end{pmatrix} \right), \qquad (4.2.5)$$

for some $\rho \in [-1, 1]$ and $\sigma_1, \sigma_2 > 0$, then

$$\mathsf{ccorr}(\beta_1, \beta_2) = \frac{\sinh(2\rho \sigma_1 \sigma_2)}{\{\sinh(2\sigma_1^2) \sinh(2\sigma_2^2)\}^{\frac{1}{2}}}, \qquad (4.2.6)$$

where $\beta_1 = X_1 \bmod 2\pi$ and $\beta_2 = X_2 \bmod 2\pi$. Figure 4.3 shows the graphs of the circular correlation (4.2.6) of two wrapped normal random variables, with the Gaussian distribution (4.2.5) with $\sigma_1 = \sigma_2 = 1/2$, for the solid line, $\sigma_1 = \sigma_2 = 1$, for the dashed line, and $\sigma_1 = \sigma_2 = 2$, for the dotted line. We note that the wrapped normal density with mean μ and variance σ^2 is given by

$$\phi^\circ(\alpha | \mu, \sigma^2) = \sum_{k=-\infty}^{\infty} \frac{1}{\sigma} \phi \left(\frac{\alpha + 2k\pi - \mu}{\sigma} \right)$$

$$= \frac{1}{\sigma \sqrt{2\pi}} \sum_{k=-\infty}^{\infty} e^{\frac{-(\alpha - \mu - 2k\pi)^2}{2\sigma^2}}, \quad \forall \alpha \in (-\pi, \pi], \qquad (4.2.7)$$

Denote as usual by γ and ρ the a.c.v.f. and a.c.r.f. of $\{X_k\}_{k \in \mathbb{Z}}$. One of the Yule-Walker equations (2.3.4) leads to

$$\gamma(0) - \varphi_1 \gamma(1) - \cdots - \varphi_p \gamma(p) = \sigma^2,$$

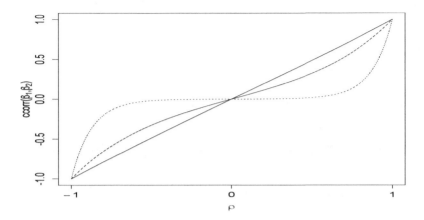

Fig. 4.3: $\mathsf{ccorr}(\beta_1, \beta_2)$ for wrapped normal random variables given by (4.2.6) with $\sigma_1 = \sigma_2 = 1/2$ (solid line), $\sigma_1 = \sigma_2 = 1$ (dashed line) and $\sigma_1 = \sigma_2 = 2$ (dotted line).

which can re-expressed as

$$\mathsf{var}(X_k) = \frac{\sigma^2}{1 - \varphi_1\rho(1) - \cdots - \varphi_p\rho(p)}, \quad \forall k \in \mathbb{Z}. \qquad (4.2.8)$$

Thus, with (4.2.6) and (4.2.8) we obtain the c.a.c.r.f. of the Gaussian WAR(p) time series as

$$\rho^\circ(h) = \frac{\sinh(2\rho(h)\sigma^2\{1 - \varphi_1\rho(1) - \cdots - \varphi_p\rho(p)\}^{-1})}{\sinh(2\sigma^2\{1 - \varphi_1\rho(1) - \cdots - \varphi_p\rho(p)\}^{-1})}, \quad \forall h \in \mathbb{Z}. \qquad (4.2.9)$$

Figure 4.4 shows the c.a.c.r.f. of the Gaussian WAR(1) time series given by (4.2.9) for $\varphi_1 = 1/2$, $\sigma^2 = 1$, in (a), $\varphi_1 = -1/2$, $\sigma^2 = 1$, in (b), $\varphi_1 = 1/2$, $\sigma^2 = 2$, in (c), and $\varphi_1 = -1/2$, $\sigma^2 = 2$, in (d). These c.a.c.r.f. can be compared with the a.c.r.f. of the AR(1) with $\varphi_1 = -1/2$ and $\varphi_1 = 1/2$ that are given in Figure 2.10. The c.a.c.r.f. behaves like the a.c.r.f. When the variance of the unwrapped time series is large, the wrapped time series becomes rather uniformly spread around the circle. The WAR(p) with large variance is thus convenient when data indicate isotropy.

4.2.4 *Linked ARMA time series*

A time series taking values over the unit circle can be obtained by applying a link function to a real-valued time series. We call link function any

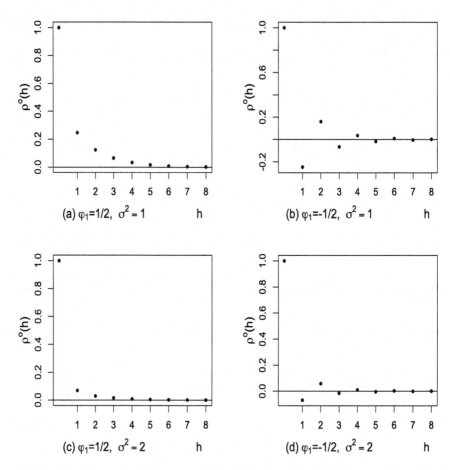

Fig. 4.4: C.a.c.r.f. of Gaussian WAR(1) given by (4.2.9) for $\varphi_1 = 1/2$, $\sigma^2 = 1$, in (a), $\varphi_1 = -1/2$, $\sigma^2 = 1$, in (b), $\varphi_1 = 1/2$, $\sigma^2 = 2$, in (c), and $\varphi_1 = -1/2$, $\sigma^2 = 2$, in (d).

increasing function $g : \mathbb{R} \to (-\pi, \pi)$ such that $g(0) = 0$. Some link functions are

$$g(x) = 2 \arctan x, \ g(x) = \pi \frac{e^x - 1}{e^x + 1} \ \text{ and } \ g(x) = 2\pi \left\{ F(x) - \frac{1}{2} \right\}, \quad (4.2.10)$$

where F is an increasing probability d.f. The second function above is the logit link and the probit link function is obtained by the third function with $F = \Phi$, the standard normal d.f. Let X be a real-valued random variable with density f_X and define the linked circular random variable $\beta = g(X)$. If g is differentiable and if f_β denotes the circular density of β, then we have

$f_X(x) = f_\beta(g(x))g'(x)$, $\forall x \in \mathbb{R}$. Consider $g(x) = 2\arctan x$, then $g'(x) = 2/(1+x^2)$ and thus, unless f_β is very concentrated, f_X behaves in the tails like x^{-2}. This heavy-tailed behavior should however be avoided whenever X is desired square-integrable. The probit link is thus more appropriate this situation and it should be preferred for the following application to time series.

Let $\{X_k\}_{k \in \mathbb{Z}}$ be a real-valued time series and define the linked circular time series by

$$\beta_k = \{\mu + g(X_k)\} \bmod 2\pi, \quad \forall k \in \mathbb{Z},$$

for some angle $\mu \in (-\pi, \pi]$. When $\{X_k\}_{k \in \mathbb{Z}}$ is ARMA(p,q), the linked time series $\{\beta_k\}_{k \in \mathbb{Z}}$ is called LARMA(p,q). Conversely, if $\{\beta_k\}_{k \in \mathbb{Z}}$ is LARMA(p,q), then $X_k = g^{-1}(\beta_k - \mu)$, $\forall k \in \mathbb{Z}$, is ARMA(p,q). The LAR or LAR(p) time series is obtained with $q = 0$ whereas the LMA or LMA(q) corresponds to $p = 0$.

The LARMA(p,q) is stationary and its c.a.c.r.f. is given by

$$\rho^\circ(h) = \mathsf{ccorr}(g(X_{k+h}), g(X_k)), \quad \forall k, h \in \mathbb{Z}. \tag{4.2.11}$$

4.2.5 *Circular AR time series*

The circular normal or von Mises (vM) distribution represents within circular statistics what the normal distribution represents in linear statistics. Its density is given by

$$f(\alpha \mid \mu, \kappa) = \frac{1}{2\pi I_0(\kappa)} \exp\{\kappa \cos(\alpha - \mu)\}, \quad \forall \alpha \in (-\pi, \pi], \tag{4.2.12}$$

where $\mu \in (-\pi, \pi]$ is the mean direction, $\kappa \geq 0$ is the concentration and where $I_n(z) = (2\pi)^{-1} \int_0^{2\pi} \cos n\alpha \, \exp\{z \cos \alpha\} d\alpha$, $\forall z \in \mathbb{C}$, is the modified Bessel function I of integer order n; see e.g. Abramowitz and Stegum (1972), p. 376. We denote any circular random variable following this distribution by vM(μ, κ). Figure 4.5 shows the graphs of three vM densities with common parameter $\mu = 0$. The density with solid line has $\kappa = 1$, the density with dashed line has $\kappa = 3$ and $\kappa = 10$ is drawn with dotted line.

The CAR(p), for some $p \in \mathbb{N}^*$, is the time series $\{\beta_k\}_{k \in \mathbb{Z}}$ such that, $\forall k \in \mathbb{Z}$, the conditional distribution of β_k given $\beta_{k-1}, \ldots, \beta_{k-p}$ is vM(μ_k, κ), where

$$\mu_k = \mu + g\left(\varphi_1 g^{-1}(\beta_{k-1} - \mu) + \cdots + \varphi_p g^{-1}(\beta_{k-p} - \mu)\right),$$

for some $\varphi_1, \ldots, \varphi_p \in \mathbb{R}$, $\mu \in (-\pi, \pi]$, $\kappa > 0$ and for a link function g of (4.2.10). Similar circular time series could be obtained by replacing the vM distribution by other circular distributions with mean direction μ_k and

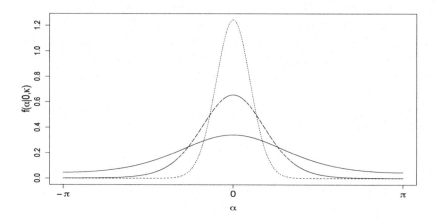

Fig. 4.5:   vM$(0, \kappa)$ density with $\kappa = 1$ (solid line), $\kappa = 3$ (dashed line) and $\kappa = 10$ (dotted line).

concentration $\kappa > 0$. Note that the CAR(1) time series with $\mu = 0$ and $\varphi_1 = 1$ has $\mu_k = \beta_{k-1}$ and thus β_k given β_{k-1} is vM(β_{k-1}, κ), $\forall k \in \mathbb{Z}$.

4.2.6   *vM time series*

Another circular time series $\{\beta_k\}_{k \in \mathbb{Z}}$ with AR structure and vM conditional distributions is obtained as follows. Let $k \in \mathbb{Z}$, $p \in \mathbb{N}^*$ and consider the parameters $\varphi_0, \varphi_1, \ldots, \varphi_p \geq 0$ and $\mu \in (-\pi, \pi]$. Define the complex-valued random variable

$$V_k = \varphi_0 e^{i\mu} + \varphi_1 e^{i\beta_{k-1}} + \cdots + \varphi_p e^{i\beta_{k-p}}.$$

The vM(p) time series is obtained when the conditional distribution of β_k given $\beta_{k-1}, \ldots, \beta_{k-p}$ is vM(μ_k, κ_k), where

$$\mu_k = \arg V_k \quad \text{and} \quad \kappa_k = |V_k|.$$

Similar circular time series could be obtained by replacing the vM distribution by other circular distributions with mean direction μ_k and concentration κ_k.

If μ_0 and $\beta_{k-1}, \ldots, \beta_{k-p}$ are pointing approximately towards the same main direction, which is is given by μ_k, then their concentration around it, viz. κ_k, is large. Thus, β_k is likely to point approximately towards μ_k.

Otherwise, if μ and $\beta_{k-1}, \ldots, \beta_{k-p}$ are well spread around the circle, then their main direction μ_k is fuzzy and their concentration around it, namely κ_k, is small. Thus, if wind directions would be modelled with the vM(p), then periods with a prevailing wind direction would be followed by periods with rapid variations of directions and conversely.

4.2.7 *Model selection*

The availability of the five categories of circular time series presented in the previous sections renders the problem of model selection more ample than in the linear case, when restricting to ARMA models.

One difficulty arises from the fact that these five categories are not necessarily exclusive, in the sense that a particular time series can belong to more than one category. For instance, the vM(1) time series with $\varphi_0 = 0$ corresponds to the CAR(1) time series with $\mu = 0$ and $\varphi_1 = 1$. Moreover, two time series can be different by their formulae but very close by their numerical evaluation. This is the case with vM and WAR time series. Indeed, one can show that vM(p) time series can be accurately approximated by the WAR(p) based on a Gaussian linear (or unwrapped) time series, provided that the coefficients of the WAR(p) time series are nonnegative. The proof this approximation is based on the fact that the vM density (4.2.12) and the wrapped normal density (4.2.7) become very close when $\kappa = \sigma^{-2}$; refer e.g. to p. 45 of Jammalamadaka and SenGupta (2001) for details.

The major source of information for model selection may be the c.a.c.r.f. Regarding the Gaussian radially projected time series, the particularity of the c.a.c.r.f. (4.2.4) is that it is always nonnegative. It is a consequence of isotropy. As seen in Figure 4.2, the c.a.c.r.f. of radially projected Gaussian AR and MA have same shapes of a.c.r.f. of their unprojected AR and MA. Regarding the WAR, it is seen in Figure 4.4 that the c.a.c.r.f. has same shape of the unwrapped AR a.c.r.f. We can deduce from the LARMA c.a.c.r.f. (4.2.11) that various features of the a.c.r.f. of the ARMA time series should be retrieved in the c.a.c.r.f. of the LARMA. In particular, the LAR(1) c.a.c.r.f. should resemble to the AR(1) a.c.r.f. and therefore display some regular decay towards zero, which has alternating nature when $\varphi_1 < 0$; cf. e.g. Figure 2.10. Also, the LMA(q) c.a.c.r.f. inherits the central property of the MA(q) a.c.r.f., namely that it vanishes at all lags of order $q + 1$ and larger; cf. Proposition 2.2.20 and Figure 2.9.

4.3 Long range dependence

A stationary stochastic process with long range dependence or, alternatively, with long memory process, possesses relatively few high frequencies. This short section provides two alternative definitions of this notion and some more comments.

As mentioned at the end of Section 3.3, the spectral distribution is typically light-tailed and, as a consequence, spectral moments of high order do exist. Thus the process admits high order $\mathcal{L}_2$-derivatives and it is smooth in this sense. It is difficult to distinguish a nonstationary process with a trend from a stationary process with long range dependence, given that any portion of a slow cycle can be confused with a trend. A statistical analysis would require a sample of very large size. Just like trend, long range dependence can be removed by differentiation of the process. For differentiation of time series we refer to 2.1.2.

The definition of long range dependence refers to the following class of functions of slow variation, presented in Definition 4.3.1.

Definition 4.3.1 (Function of slow variation). *Consider the function l :* $\mathbb{R}_+ \to \mathbb{R}$ *and* $a \in [0, \infty]$. *Assume* $l(x) \geq 0$ *whenever* $x \geq 0$ *satisfies*

$$\begin{cases} |x - a| < \delta, & \text{for some } \delta > 0, \quad \text{if } a \in [0, \infty), \\ x > x_0, & \text{for some } x_0 > 0, \quad \text{if } a = \infty. \end{cases}$$

Then the function l possesses slow variation at a if

$$\frac{l(xt)}{l(x)} \overset{x \to a}{\longrightarrow} 1, \quad \forall t > 0.$$

Examples 4.3.2 (Functions of slow variation). *Some functions of slow variation at $a = \infty$ are the following.*

(1) $l(x) = \log x$: $\dfrac{\log(tx)}{\log x} = \dfrac{\log x + \log t}{\log x} \overset{x \to \infty}{\longrightarrow} 1.$

(2) $l(x) = \log \log x$: $\lim\limits_{x \to \infty} \dfrac{\log \log(tx)}{\log \log x} = \lim\limits_{x \to \infty} \dfrac{\frac{1}{\log(tx)} \frac{1}{x}}{\frac{1}{\log x} \frac{1}{x}} = 1.$

(3) $l(x) = \log^{(k)} x = \log \log^{(k-1)} x$, *for* $k = 2, 3, \dots,$ *where* $\log^{(1)} x = \log x$ *and* $\log^{(0)} x = x$: $\dfrac{\mathrm{d}}{\mathrm{d}x} \log^{(k)} x = \dfrac{1}{\log^{(k-1)} x} \dfrac{1}{\log^{(k-2)} x} \cdots \dfrac{1}{\log x} \dfrac{1}{x}$

and so $\lim\limits_{x \to \infty} \dfrac{\log^{(k)}(tx)}{\log^{(k)} x} = \lim\limits_{x \to \infty} \dfrac{\frac{1}{\log^{(k-1)}(tx)} \cdots \frac{1}{\log(tx)} \frac{1}{x}}{\frac{1}{\log^{(k-1)} x} \cdots \frac{1}{\log x} \frac{1}{x}} = 1.$

There are in fact two possible definitions of the concept of long range dependence. One definition is based on the spectral density, whose existence is assumed, and the other one on the a.c.v.f. These two definitions are equivalent under some minor regularity conditions. In the spectral domain, long range dependence means that the spectral density tends to infinity, as the frequency tends to zero, and is nevertheless integrable. In the time domain, long range dependence means that the a.c.v.f. decays in a regular manner as the time lag increases and that it is not integrable, for a continuous time process, or not summable, for a time series.

Definition 4.3.3 (Long range dependence via spectral density). *A stationary stochastic process with spectral density f has long range dependence or long memory, if there exists a constant $h \in (1/2, 1)$ and a function l of slow variation at 0, such that the spectral density admits the following re-expression,*

$$f(\alpha) = l(|\alpha|)|\alpha|^{1-2h}, \quad \forall \alpha \in \mathbb{R}.$$

Definition 4.3.4 (Long range dependence via a.c.v.f.). *A stationary stochastic process with a.c.v.f γ has long range dependence or long memory, if there exists a constant $h \in (1/2, 1)$ and a function l with slow variation at ∞, such that the a.c.v.f. admits the following re-expression,*

$$\gamma(s) = l(|s|)|s|^{2(h-1)}, \quad \forall s \in \mathbb{R}.$$

Definitions 4.3.3 and 4.3.4 are equivalent and moreover with common constant h, under some regularity conditions.

Near to the boundaries of $(1/2, 1)$ we note the following behavior. Consider $\varepsilon > 0$ small. If $h = \begin{cases} 1/2 + \varepsilon \\ 1 - \varepsilon \end{cases}$, then the spectral density increases, as α approaches 0, at the $\begin{cases} \text{slow rate } |\alpha|^{-2\varepsilon} \\ \text{fast rate } |\alpha|^{-1+2\varepsilon} \end{cases}$ and the a.c.v.f. decreases, as s approaches $\pm\infty$, at the $\begin{cases} \text{fast rate } |s|^{-1+2\varepsilon} \\ \text{slow rate } |s|^{-2\varepsilon} \end{cases}$.

Thus, the longest range dependence is obtained as $h \in (1/2, 1)$ approaches one, exactly how it happens with the h-selfsimilar fractional Brownian motion, cf. Proposition 3.2.27. The index h is a measure of long range dependence and it is called Hurst exponent. It originates from Hurst (1951), which studies ancient data over a long period of time on the hydrology of the Nile river.

Consider the random variables $X_1, \ldots, X_n$ sampled from the stationary process $\{X_t\}_{t \in T}$, with mean μ, a.c.v.f. γ. Consider the sample mean $M_n = \sum_{j=1}^{n} X_j/n$, which is an unbiased estimator of μ. Then a formula for $n \operatorname{var}(M_n)$ is given in (2.5.12) and it converges to $\gamma(0) + 2 \sum_{j=1}^{\infty} \gamma(j)$, whenever the a.c.v.f. is summable, as given by Theorem 2.5.4.2. Therefore the variance has the same asymptotic rate as in the i.i.d. case, viz. $\operatorname{var}(M_n) = O(n^{-1})$, as $n \to \infty$, and this rate is practical for obtaining methods of statistical inference. As explained just above, under long range dependence the a.c.v.f. decreases too slow for summability. In this case, one can show the surrogate limit for the variance of the mean,

$$\lim_{n \to \infty} \frac{\operatorname{var}(M_n)}{\gamma(n)} = \frac{1}{h(2h-2)}.$$

In most practical studies, h represents an unknown parameter that must be estimated. Statistical inference becomes more complicated than with summable a.c.v.f. A reference on this topic is Beran (1994).

4.4   Nonintegrable spectral density and intrinsic stationarity

A stationary stochastic process with long range dependence can be obtained by a spectral density that is unbounded at the origin. However this density must be integrable, in order to obtain a square-integrable stochastic process, as desired. It is precisely Bochner's theorem 3.5.1 that tells that the mass of the spectral distribution must be finite. This section explains how a nonintegrable spectral density can be used for constructing a stationary process with finite variance.

The proposed method makes use of a time invariant filter with null gain, as defined in Example 4.4.1 below.

Example 4.4.1 (Nonintegrable power type spectral density). *Consider the power type spectral density*

$$f(\alpha) = |\alpha|^{-p}, \quad \forall \alpha \in \mathbb{R}, \tag{4.4.1}$$

for some $p \in (1,3)$, which is unbounded at the origin and nonintegrable. Consider the time invariant linear filter with impulse response function

$$\psi(v) = \sum_{j=1}^{n} \psi_j \delta(v - t_j), \quad \forall v \in \mathbb{R}, \tag{4.4.2}$$

for some $-\infty < t_1 < \cdots < t_n < \infty$, $\psi_1, \ldots, \psi_n \in \mathbb{R}$ *and* $n \geq 1$. *We assume the standardization*

$$\xi(0) = \sum_{j=1}^{n} \psi_j = 0,$$

where ξ *denotes the transfer function (3.9.4). We remind that, in view of (3.9.10),* $\xi(0)$ *is called the gain of the filter. Thus the standardization yields a linear filter with null gain. Null gain yields asymptotic linearity of the transfer function near the origin, precisely*

$$\xi(\alpha) = \sum_{j=1}^{n} \psi_j e^{-it_j \alpha} \sim -i \sum_{j=1}^{n} \psi_j t_j \alpha, \quad \text{as } \alpha \to 0.$$

This asymptotic linearity makes the spectral density

$$f_Y(\alpha) = |\xi(\alpha)|^2 f(\alpha) = \left| \sum_{j=1}^{n} \psi_j e^{-it_j \alpha} \right|^2 |\alpha|^{-p}, \quad \forall \alpha \in \mathbb{R},$$

integrable and so this spectral density can be considered as the one of a stationary stochastic process; cf. (3.9.8). According to (3.9.7), the resulting stationary process at time $t \in \mathbb{R}$ *has a.c.v.f. given by*

$$\gamma_Y(h) = \int_{\mathbb{R}} e^{ih\alpha} |\xi(\alpha)|^2 |\alpha|^{-p} d\alpha, \quad \forall h \in \mathbb{R}.$$

In particular, its variance is given by

$$\gamma_Y(0) = \int_{-\infty}^{\infty} |\xi(\alpha)|^2 |\alpha|^{-p} d\alpha < \infty. \tag{4.4.3}$$

   Let us recapitulate the results. Formal application of formulae for linear filters that hold for stationary input processes have allowed to identify a new stationary process, denoted $\{Y_t\}_{t \in \mathbb{R}}$, *through its spectral density and its a.c.v.f. It follows from (4.4.3) that* $\text{var}(Y_t) = \gamma_Y(0) < \infty$, $\forall t \in \mathbb{R}$. *Assume that the input process with nonintegrable spectral density (4.4.1) is available and denote it by* $\{X_t\}_{t \in \mathbb{R}}$. *Then the process* $\{Y_t\}_{t \in \mathbb{R}}$ *can be obtained by applying the linear filter to the input* $\{X_t\}_{t \in \mathbb{R}}$, *yielding the equation*

$$Y_t = \sum_{j=1}^{n} \psi_j X_{t-t_j}, \quad \forall t \in \mathbb{R}.$$

Moreover, by using (3.9.4) we can evaluate the variance by the following steps,

$$\text{var}(Y_t) = \int_{-\infty}^{\infty} |\xi(\alpha)|^2 |\alpha|^{-p} \mathrm{d}\alpha$$

$$= \int_{-\infty}^{\infty} \left| \sum_{j=1}^{n} \psi_j e^{-it_j \alpha} \right|^2 |\alpha|^{-p} \mathrm{d}\alpha$$

$$= \sum_{j=1}^{n} \sum_{k=1}^{n} \psi_j \psi_k \int_{-\infty}^{\infty} \{\cos(t_j - t_k)\alpha - 1\} |\alpha|^{-p} \mathrm{d}\alpha$$

$$= -4 \sum_{j=1}^{n} \sum_{k=1}^{n} \psi_j \psi_k \int_{0}^{\infty} \sin^2 \frac{(t_j - t_k)\alpha}{2} \alpha^{-p} \mathrm{d}\alpha$$

$$= \sum_{j=1}^{n} \sum_{k=1}^{n} \psi_j \psi_k \eta(t_j - t_k), \quad \forall t \in \mathbb{R}, \qquad (4.4.4)$$

where

$$\eta(s) = -4 \int_{0}^{\infty} \sin^2 \frac{s\alpha}{2} \alpha^{-p} \mathrm{d}\alpha = -\frac{\pi}{\Gamma(p) \sin \frac{\pi(p-1)}{2}} |s|^{p-1}, \quad \forall s \in \mathbb{R}.$$

The last equality can be obtained with the formulae 3.823 and 8.334-3 of Gradshteyn et al. (2007). We can give to (4.4.4) the interpretation of (3.9.9).

The main idea obtained from this presentation is the following. The functions η and f possess precisely the relation between a.c.v.f. and spectral density, eventhough η is not a proper a.c.v.f. and f is not a proper spectral density. We explained that this relation between η and f holds under the restriction that the linear filter has null gain. The function η is called the generalized a.c.v.f. of the nonintegrable spectral density f. The generalized a.c.v.f. η is generally not n.n.d., because the validity of (4.4.4) and thus its nonnegativity can be guaranteed only when $\sum_{j=1}^{n} \psi_j = 0$.

The considerations of Example 4.4.1 motivate the following alternative notion of stationarity.

Definition 4.4.2 (Intrinsic stationarity). *The stochastic process $\{X_t\}_{t \in T}$, not necessarily square-integrable or stationary, is intrinsically stationary if there exists a time invariant linear filter with null gain Ψ such that the filtered process $\{Y_t\}_{t \in T} = \Psi(\{X_t\}_{t \in T})$ is (weakly) stationary. In this situation, $\{X_t\}_{t \in T}$ is intrinsically stationary w.r.t. the filter Ψ.*

Examples 4.4.3 (Intrinsically stationary processes).

(1) Deseasonalization

For the stochastic process $\{X_t\}_{t \in \mathbb{R}}$ and for $d > 0$, define the lag d difference operator ∇_d such that

$$\nabla_d X_t = X_t - X_{t-d}, \quad \forall t \in \mathbb{R}.$$

This operator is given in Definition 2.1.15 for time series (i.e. for $T = \mathbb{Z}$) and thus for positive integer lag d only. The lag d operator removes the harmonic of period d, which usually has the interpretation of seasonality. Thus if $Y_t = \nabla_d X_t$, $\forall t \in \mathbb{R}$, is a stationary process, then $\{X_t\}_{t \in \mathbb{R}}$ is intrinsically stationary. Indeed, the linear filter (4.4.2) with $n = 2$, $\psi_1 = 1$, $\psi_2 = -1$, $t_1 = 0$ and $t_2 = d$ has null gain and its application to $\{X_t\}_{t \in \mathbb{R}}$ yields $\{Y_t\}_{t \in \mathbb{R}}$.

(2) Finite differentiation and ARIMA(p, d, q)

The backward shift operator B, the finite difference operator $\nabla = 1 - B$ and their d-th order composition B^d and ∇^d, for $d \in \mathbb{N}^*$, are given in Definition 2.1.14 for time series (i.e. with $T = \mathbb{Z}$). But they also apply to continuous time stochastic processes (i.e. with $T = \mathbb{R}$). The operator ∇^d is a time invariant linear filter with null gain. Indeed, we have

$$(1 - z)^d = \sum_{j=0}^{d} c_j z^j, \quad \text{where} \quad c_j = (-1)^j \binom{d}{j}, \quad \text{for} \quad j = 0, \ldots, d.$$

Therefore, for $n = d + 1$, one sets $\psi_j = c_{j-1}$, for $j = 1, \ldots, n$. In this case $\sum_{j=1}^{n} \psi_j = 0$ holds. Also, $t_j = j - 1$, for $j = 1, \ldots, n$. The operator ∇^d is thus a linear filter with $n = d + 1$. It removes any polynomial of degree d from the trend of a nonstationary time series. It may also remove periodicities of orders d or smaller. We remind that $\{X_k\}_{k \in \mathbb{Z}}$ is an ARIMA(p, d, q) time series if $Y_k = \nabla^d X_k$, $\forall k \in \mathbb{Z}$, is a causal ARMA(p, q) time series. With continuous time (i.e. $T = \mathbb{R}$), a polynomial of degree d is removed with the derivative of order d.

4.5 Unstable system

In this section we present the concept of unstable system and unstable stochastic process.

An unstable process is defined through a linear differential equation with a driving process, precisely through (3.9.23). The term unstable refers to the stability of the differential equation, according to Definition 3.9.8.

An unstable processes is subject to the fluctuations of the driving process without attenuation over time. We introduce this concept with the class of Whittle-Matérn stochastic processes and we then provide a generalization.

Definition 4.5.1 (Whittle-Matérn processes). *Any stationary process* $\{Y_t\}_{t \in \mathbb{R}}$ *that solves the linear differential equation of order two given by*

$$-Y_t'' + Y_t = W_t', \quad \forall t \in \mathbb{R}, \tag{4.5.1}$$

where $\{W_t'\}_{t \in \mathbb{R}}$ *is Gaussian WN with variance* σ^2, *for some* $\sigma > 0$, *cf. Section 3.4.2, is a Whittle-Matérn process.*

The roots of the characteristic equation of (4.5.1), generally defined in (3.9.24), are -1 and 1. Because these two roots are not both negative, the corresponding differential equation or system is unstable, according to Definition 3.9.8. The stability assumption guarantees that the fluctuations of the driving WN process $\{W_t'\}_{t \in \mathbb{R}}$ are attenuated as time passes. With an unstable process, there is no certitude for this attenuation.

Any Whittle-Matérn process $\{Y_t\}_{t \in \mathbb{R}}$ is stationary and we assume that it possesses a spectral density that we denote by f_Y. According to (3.9.25), we obtain that the spectral density of the process on the left side of the stochastic differential equation (4.5.1) is given by $(1 + \alpha^2)^2 f_Y(\alpha)$, $\forall \alpha \in \mathbb{R}$. Thus we have $(1 + \alpha^2)^2 f_Y(\alpha) = f_{W'}(\alpha)$, where $f_{W'}(\alpha) = \sigma^2/(2\pi)$, $\forall \alpha \in \mathbb{R}$, is the spectral density of $\{W_t'\}_{t \in \mathbb{R}}$; cf. (3.9.12). So we obtain the spectral density of $\{Y_t\}_{t \in \mathbb{R}}$ in the following form,

$$f_Y(\alpha) = \frac{\sigma^2}{2\pi} \frac{1}{(1 + \alpha^2)^2}, \quad \forall \alpha \in \mathbb{R}. \tag{4.5.2}$$

We note that the Whittle-Matérn process is not uniquely defined. Indeed, if $\{Z_t\}_{t \in \mathbb{R}}$ solves the homogeneous correspondent of the differential equation (4.5.1), then $Z_t + C_1 e^{-t} + C_2 e^t$, $\forall t \in \mathbb{R}$, for any two random variables C_1 and C_2, is also solution; cf. Theorem A.9.1. We have precisely the class of Whittle-Matérn processes: they all solve the differential equation equation (4.5.1) and, consequently, possess spectral density (4.5.2).

Denote by F_Y d.f. of f_Y. With the integral formula $\int 2/(1 + x^2)^2 dx = x/(1+x^2)+\arctan x$, we obtain the spectral d.f. of a Whittle-Matérn process $\{Y_t\}_{t \in \mathbb{R}}$ as follows,

$$F_Y(\alpha) = \frac{\sigma^2}{2} \left(\frac{\pi}{2} + \frac{\alpha}{1 + \alpha^2} + \arctan \alpha \right), \quad \forall \alpha \in \mathbb{R}.$$

The spectral density (4.5.2) can be generalized to

$$f(\alpha) = \frac{\sigma^2}{2\pi} \frac{1}{(a^2 + \alpha^2)^{\nu+\frac{1}{2}}}, \quad \forall \alpha \in \mathbb{R}.$$

for some constants $a, \sigma, \nu > 0$. We note that all spectral moments of order smaller than 2ν of f do exist. Referring to Remark 3.3.11.2, we deduce that all $\mathcal{L}_2$-derivatives of the stochastic process of order smaller than ν do exist. Thus, the parameter ν regulates the smoothness of the stochastic process.

The a.c.v.f. of f is given by

$$\gamma(s) = \frac{\sigma^2}{2^\nu \sqrt{\pi} a^{2\nu} \Gamma(\nu + \frac{1}{2})} (a|s|)^\nu K_\nu(a|s|), \quad \forall s \in \mathbb{R}^*,$$

where

$$K_\nu(z) = \int_0^\infty e^{-z \cosh t} \cosh(\nu t) dt, \quad \forall z, \ \nu \in \mathbb{C} \text{ such that } |\arg z| < \frac{\pi}{2},$$

is the modified Bessel function K, cf. e.g. 9.6.24 at p. 376 of Abramowitz and Stegum (1972). The following useful properties can be found in Abramowitz and Stegum (1972): $K_\nu(z) > 0$, for $z > 0$ and $\nu > -1$, cf. p. 374; $K_\nu(z) \to 0$, as $z \to \pm\infty$, for $|\arg z| < \pi/2$, cf. p. 374; and $K_\nu(z) \sim (1/2)\Gamma(\nu)(2/z)^\nu$, as $z \to 0$, for $\operatorname{Re} \nu > 0$, cf. 9.6.9, p. 375. This last property provides the variance formula

$$\gamma(0) = \frac{\sigma^2}{2\sqrt{\pi} a^{2\nu}} \frac{\Gamma(\nu)}{\Gamma\left(\nu + \frac{1}{2}\right)}.$$

4.6 Hilbert transform and envelope

In this section we introduce the Hilbert transform and the envelope of a complex-valued stationary process. The Hilbert transform is obtained by the application of a particular time invariant linear filter with null gain to the stationary process. This operation leads to another stationary process that has similar spectral distribution as the original one. The Hilbert transform is useful for the determination of the so-called envelope, of the original stationary process. The envelope of some possibly highly oscillating stationary process is a stochastic process that bounds the modulus of the original process. The envelope is smoother and simultaneously very close to the original process.

Definition 4.6.1 (Hilbert transform). *Let $\{X_t\}_{t \in \mathbb{R}}$ be a complex-valued stationary process with spectral process $\{Z_\alpha\}_{\alpha \in \mathbb{R}}$. Consider the transfer* *function*

$$\xi(\alpha) = -\mathrm{i}\,\mathrm{sgn}\,\alpha = \begin{cases} \mathrm{i}, & \text{if } \alpha < 0, \\ 0, & \text{if } \alpha = 0, \\ -\mathrm{i}, & \text{if } \alpha > 0. \end{cases} \qquad (4.6.1)$$

Then the Hilbert transform of $\{X_t\}_{t\in\mathbb{R}}$ *is given by*

$$\hat{X}_t = \int_{\mathbb{R}} \xi(\alpha)\mathrm{e}^{\mathrm{i}\alpha t}\mathrm{d}Z_\alpha$$

$$= \mathrm{i}\left(\int_{(-\infty,0)} \mathrm{e}^{\mathrm{i}\alpha t}\mathrm{d}Z_\alpha - \int_{(0,\infty)} \mathrm{e}^{\mathrm{i}\alpha t}\mathrm{d}Z_\alpha\right), \quad \forall t \in \mathbb{R}.$$

To the transfer function (4.6.1) corresponds the impulse response function

$$\psi(v) = \frac{1}{\pi v}, \quad \forall v \in \mathbb{R}^*,$$

in the sense that this transfer function is the principal value of the integral given by the Fourier transform (3.9.4).[2] Because ψ is not integrable, the Fourier transform of the transfer function (3.9.4) does not exist as improper integral but it does exist as principal value. Thus, the Hilbert transform is defined as principal value of the convolution

$$\hat{X}_t = \frac{1}{\pi} \int_{-\infty}^{\infty} \frac{X_s}{t-s}\mathrm{d}s, \quad \forall s \in \mathbb{R},$$

whenever the stochastic integral exists. But the existence of a pole at $s = t$ implies that this integral may indeed not exist as an improper integral, for many stationary stochastic processes $\{X_t\}_{t\in\mathbb{R}}$.

Another important associated stochastic process is given by

$$X_t^* = X_t + \mathrm{i}\hat{X}_t = \Delta Z_0 + 2\int_{(0,\infty)} \mathrm{e}^{\mathrm{i}\alpha t}\mathrm{d}Z_\alpha, \quad \forall t \in \mathbb{R}. \qquad (4.6.2)$$

Assume for instance that the stationary process $\{X_t\}_{t\in\mathbb{R}}$ is real-valued. In this situation we can use the specific representation provided by (3.7.1) and (3.7.2), namely

$$X_t = \Delta U_0 + \int_{(0,\infty)} \cos \alpha t\, \mathrm{d}U_\alpha + \int_{(0,\infty)} \sin \alpha t\, \mathrm{d}V_\alpha, \quad \forall t \in \mathbb{R},$$

---

[2]The integration formula $\int_0^\infty x^{-1}\sin ax\,\mathrm{d}x = \pi/2\,\mathrm{sgn}\,a$, $\forall a \in \mathbb{R}$, is useful in the evaluation of that principal value.

where $\Delta U_0 = \Delta Z_0$ is a real-valued random variable. With any real-valued process the Hermitian properties given in (3.7.3) and (3.7.4) hold. We therefore obtain

$$
\begin{aligned}
\hat{X}_t &= \mathrm{i} \left(\int_{(\infty,0)} \mathrm{e}^{-\mathrm{i}\alpha t} \mathrm{d} Z_{-\alpha} - \int_{(0,\infty)} \mathrm{e}^{\mathrm{i}\alpha t} \mathrm{d} Z_\alpha \right) \\
&= \mathrm{i} \left(\overline{\int_{(\infty,0)} \mathrm{e}^{\mathrm{i}\alpha t} \mathrm{d} Z_\alpha} - \int_{(0,\infty)} \mathrm{e}^{\mathrm{i}\alpha t} \mathrm{d} Z_\alpha \right) \\
&= 2 \operatorname{Im} \int_{(0,\infty)} \mathrm{e}^{\mathrm{i}\alpha t} \mathrm{d} Z_\alpha \\
&= \operatorname{Im} \int_{(0,\infty)} (\cos \alpha t + \mathrm{i} \sin \alpha t)(\mathrm{d} U_\alpha - \mathrm{i} \mathrm{d} V_\alpha) \\
&= \int_{(0,\infty)} \sin \alpha t \, \mathrm{d} U_\alpha - \int_{(0,\infty)} \cos \alpha t \, \mathrm{d} V_\alpha, \quad \forall t \in \mathbb{R}.
\end{aligned}
$$

Thus the associated process $\{X_t^*\}_{t\in\mathbb{R}}$ is generally complex-valued, with $\operatorname{Re} X_t^* = X_t$ and $\operatorname{Im} X_t^* = \hat{X}_t$, $\forall t \in \mathbb{R}$.

Because the Hilbert transform $\{\hat{X}_t\}_{t\in\mathbb{R}}$ is merely an application of a time invariant linear filter, it has the following properties.

Proposition 4.6.2 (Properties of Hilbert transform). *Consider the complex-valued stochastic process $\{X_t\}_{t\in\mathbb{R}}$ as stationary, with mean zero, with a.c.v.f. γ, with spectral d.f. F, with one-sided spectral d.f. F_1 (cf. Definition 3.5.7), with spectral process with jump at zero given by ΔZ_0 and with Hilbert transform $\{\hat{X}_t\}_{t\in\mathbb{R}}$. Then $\{\hat{X}_t\}_{t\in\mathbb{R}}$ has the following properties.*

(1) The process $\{\hat{X}_t\}_{t\in\mathbb{R}}$ has same mean, i.e. zero.
(2) The process $\{\hat{X}_t\}_{t\in\mathbb{R}}$ has same spectral d.f. $F_{\hat{X}}(\alpha) = F(\alpha)$, $\forall \alpha \neq 0$, however without jump at the null frequency, i.e. $\Delta F_{\hat{X}}(0) = 0$.
(3) The process $\{\hat{X}_t\}_{t\in\mathbb{R}}$ has the a.c.v.f.

$$
\gamma_{\hat{X}}(h) = \gamma(h) - \Delta F(0) = \int_{\mathbb{R}} \mathrm{e}^{\mathrm{i}h\alpha} \mathrm{d} F(\alpha) - \Delta F(0), \quad \forall h \in \mathbb{R}.
$$

If $\{\hat{X}_t\}_{t\in\mathbb{R}}$ is real-valued, then this formula simplifies to

$$
\gamma_{\hat{X}}(h) = \int_{(0,\infty)} \cos h\alpha \, \mathrm{d} F_1(\alpha), \quad \forall h \in \mathbb{R},
$$

where F_1 is the one-sided spectral d.f. of Definition 3.5.7.
(4) The process $\{\hat{X}_t\}_{t\in\mathbb{R}}$ has the Hilbert transform

$$
\hat{\hat{X}}_t = -X_t + \Delta Z_0, \quad \forall t \in \mathbb{R}.
$$

Proof. 1. This part follows from (3.9.10).

2. This part follows from (3.9.8).

3. This part follows from (3.9.7) and from Definition 3.5.7 of the one-sided spectral d.f.

4. If the spectral process of $\{X_t\}_{t\in\mathbb{R}}$ is $\{Z_\alpha\}_{\alpha\in\mathbb{R}}$, then the spectral process of $\{\hat{X}_t\}_{t\in\mathbb{R}}$ is given by $\hat{Z}_\alpha = -\mathrm{i}\operatorname{sgn}\alpha\, Z_\alpha$, $\forall \alpha \in \mathbb{R}$. Thus the Hilbert transform of $\{\hat{X}_t\}_{t\in\mathbb{R}}$ is given by

$$\hat{\hat{X}}_t = \int_{\mathbb{R}} (-\mathrm{i}\operatorname{sgn}\alpha)\mathrm{e}^{\mathrm{i}\alpha t}\mathrm{d}\hat{Z}_\alpha$$

$$= \int_{\mathbb{R}} (-\mathrm{i}\operatorname{sgn}\alpha)\mathrm{e}^{\mathrm{i}\alpha t}\mathrm{d}\{(-\mathrm{i}\operatorname{sgn}\alpha)Z_\alpha\}$$

$$= -\int_{\mathbb{R}^*} \mathrm{e}^{\mathrm{i}\alpha t}\mathrm{d}Z_\alpha$$

$$= -X_t + \Delta Z_0, \quad \forall t \in \mathbb{R}.$$

$\square$

A consequence of Proposition 4.6.2.4 is that the Hilbert transform applied twice to a stationary process with $\Delta Z_0 = 0$ returns the original process with opposed sign. By denoting by H the operator of Hilbert transform and by I the identity operator, we have $H^2 = -I$. The Hilbert transform applied four times on the same process returns the original process, $H^2 H^2 = H^4 = I$. Further, $H^3 H = I$ implying that $H^{-1} = H^3$, meaning that repeated application yields the inverse Hilbert transform.

From here we assume that the stochastic process $\{X_t\}_{t\in\mathbb{R}}$ is real-valued. In general, the Hilbert transform is correlated with the original process. However, the random variables X_t and $\hat{X}_t$ are uncorrelated, $\forall t \in \mathbb{R}$. This is a consequence of the following proposition. We remind the Definition 3.4.3 of the crosscovariance function.

Proposition 4.6.3 (Crosscovariance between real-valued process and Hilbert transform). *Let $\{X_t\}_{t\in\mathbb{R}}$ be real-valued, stationary, with mean zero, with one-sided spectral d.f. F_1 (cf. Definition 3.5.7) and with Hilbert transform $\{\hat{X}_t\}_{t\in\mathbb{R}}$. Then the crosscovariance function between the process and its Hilbert transform is given by*

$$\gamma_{X\hat{X}}(s,t) = \operatorname{cov}(X_s, \hat{X}_t) = -\int_{(0,\infty)} \sin(s-t)\alpha\, \mathrm{d}F_1(\alpha), \quad \forall s,t \in \mathbb{R}.$$

Proof. Let $s, t \in \mathbb{R}$ and denote by $\{Z_\alpha\}_{\alpha\in\mathbb{R}}$ the spectral process of $\{X_t\}_{t\in\mathbb{R}}$. Given $\mathsf{E}[X_s] = \mathsf{E}[\hat{X}_t] = 0$, given (3.6.13), given the symmetry of F and by considering the remarks of Appendix A.6 regarding the change of integration variable, we have

$$\gamma_{X\hat{X}}(s,t) = \mathsf{E}\left[X_s\overline{\hat{X}_t}\right]$$

$$= \mathsf{E}\left[\int_{\mathbb{R}} e^{is\alpha}\mathrm{d}Z_\alpha\overline{\int_{\mathbb{R}^*} -i\,\mathrm{sgn}\alpha\, e^{it\alpha}\mathrm{d}Z_\alpha}\right]$$

$$= i\int_{\mathbb{R}}\int_{\mathbb{R}^*} \mathrm{sgn}\beta\, e^{i(s\alpha-t\beta)}\mathsf{E}\left[\mathrm{d}Z_\alpha\overline{\mathrm{d}Z_\beta}\right]$$

$$= i\int_{\mathbb{R}^*} \mathrm{sgn}\alpha\, e^{i(s-t)\alpha}\mathrm{d}F(\alpha)$$

$$= i\left(\int_{(0,\infty)} e^{i(s-t)\alpha}\mathrm{d}F(\alpha) - \int_{(-\infty,0)} e^{i(s-t)\alpha}\mathrm{d}F(\alpha)\right)$$

$$= i\left(\int_{(0,\infty)} e^{i(s-t)\alpha}\mathrm{d}F(\alpha) - \int_{(0,\infty)} e^{-i(s-t)\alpha}\mathrm{d}F(-\alpha)\right)$$

$$= i\left(\int_{(0,\infty)} e^{i(s-t)\alpha}\mathrm{d}F(\alpha) - \int_{(0,\infty)} e^{-i(s-t)\alpha}\mathrm{d}F(\alpha)\right)$$

$$= i\left(\int_{(0,\infty)} \frac{e^{i(s-t)\alpha} - e^{-i(s-t)\alpha}}{2}\mathrm{d}F_1(\alpha)\right).$$

$\square$

A practical application of the Hilbert transform concerns the determination of the envelope of a stationary process, which is defined as follows.

Definition 4.6.4 (Envelope of real-valued stationary process). *Let $\{X_t\}_{t\in\mathbb{R}}$ be a real-valued stationary process with spectral process with Hilbert transform $\{\hat{X}_t\}_{t\in\mathbb{R}}$ and associated process $\{X_t^*\}_{t\in\mathbb{R}}$ as given in (4.6.2). Then the envelope of $\{X_t\}_{t\in\mathbb{R}}$ is the stochastic process given by*

$$R_t = |X_t^*| = \sqrt{X_t^2 + \hat{X}_t^2}, \quad \forall t \in \mathbb{R}.$$

The terminology is due to the trivial and practical consequence that, $\forall t \in \mathbb{R}$,

$$|X_t| \leq R_t,$$

with equality iff $\hat{X}_t = 0$. The envelope provides a confinement of the original stationary process that is fine enough to reveal the important structure of the process. An elementary illustration of these concepts is provided by the following example.

Example 4.6.5 (Envelopes of cosine and sine). *Consider the deterministic process* $X_t = \cos \alpha_0 t$, $\forall t \in \mathbb{R}$, *for some* $\alpha_0 \in \mathbb{R}$. *Then the spectral process is given by*

$$Z_\alpha = \frac{1}{2}(\mathsf{I}\{-\alpha_0 \le \alpha\} + \mathsf{I}\{\alpha_0 \le \alpha\}), \quad \forall \alpha \in \mathbb{R}.$$

Thus when $\alpha_0 > 0$ *we have*

$$\hat{X}_t = \mathrm{i}\left(\frac{\mathrm{e}^{-\mathrm{i}\alpha_0 t}}{2} - \frac{\mathrm{e}^{\mathrm{i}\alpha_0 t}}{2}\right) = \sin \alpha_0 t, \quad \forall t \in \mathbb{R}.$$

If $\alpha_0 < 0$, *then we obtain*

$$\hat{X}_t = -\sin \alpha_0 t, \quad \forall t \in \mathbb{R}.$$

Thus when $\alpha_0 \in \mathbb{R}$ *we have*

$$\hat{X}_t = \sin |\alpha_0| t,$$
$$X_t^* = \cos \alpha_0 t + \mathrm{i} \sin |\alpha_0| t = \mathrm{e}^{\mathrm{i}|\alpha_0|t},$$
$$R_t = \left|\mathrm{e}^{\mathrm{i}|\alpha_0|t}\right| = 1$$

and the envelope yields the trivial inequality $|X_t| \le 1$, $\forall t \in \mathbb{R}$.

   Consider $X_t = \sin \alpha_0 t$, $\forall t \in \mathbb{R}$, *for some* $\alpha_0 \in \mathbb{R}$. *Then the spectral process is given by*

$$Z_\alpha = \frac{1}{2\mathrm{i}}(-\mathsf{I}\{-\alpha_0 \le \alpha\} + \mathsf{I}\{\alpha_0 \le \alpha\}), \quad \forall \alpha \in \mathbb{R}.$$

Thus when $\alpha_0 > 0$ *we have*

$$\hat{X}_t = \mathrm{i}\left(\frac{\mathrm{e}^{-\mathrm{i}\alpha_0 t}}{-2\mathrm{i}} - \frac{\mathrm{e}^{\mathrm{i}\alpha_0 t}}{2\mathrm{i}}\right) = -\cos \alpha_0 t, \quad \forall t \in \mathbb{R}.$$

If $\alpha_0 < 0$, *then we obtain*

$$\hat{X}_t = \cos \alpha_0 t, \quad \forall t \in \mathbb{R}.$$

Thus we have, for $\alpha_0 \in \mathbb{R}$,

$$\hat{X}_t = -\mathrm{sgn}\alpha_0 \cos \alpha_0 t,$$
$$X_t^* = \sin \alpha_0 t + \mathrm{i}(-\mathrm{sgn}\alpha_0 \cos \alpha_0 t) = -\mathrm{i}\,\mathrm{sgn}\alpha_0\, \mathrm{e}^{\mathrm{i}|\alpha_0|t},$$
$$R_t = |-\mathrm{i}\,\mathrm{sgn}\alpha_0\, \mathrm{e}^{\mathrm{i}|\alpha_0|t}| = 1$$

and the envelope yields $|X_t| \le 1$, $\forall t \in \mathbb{R}$.
   Consequently, for $X_t = \mathrm{e}^{\mathrm{i}\alpha_0 t}$, $\forall t \in \mathbb{R}$, *with* $\alpha_0 \in \mathbb{R}$, *we obtain*

$$\hat{X}_t = \sin |\alpha_0| t - \mathrm{i}\,\mathrm{sgn}\alpha_0 \cos \alpha_0 t = -\mathrm{i}\,\mathrm{sgn}\alpha_0\, \mathrm{e}^{\mathrm{i}\alpha_0 t}, \quad \forall t \in \mathbb{R}.$$

We do not compute the envelope of this complex exponential process, because it is defined for real-valued processes only.

Example 4.6.6 (Envelope of sum of cosines narrow banded process). *A simple generalization of the cosine function of Example 4.6.5 would be the real-valued stationary process given by*

$$X_t = B(\alpha_1)\cos(\alpha_1 t + \theta(\alpha_1)) + B(\alpha_2)\cos(\alpha_2 t + \theta(\alpha_2)), \quad \forall t \in \mathbb{R},$$

for some $0 < \alpha_1 < \alpha_2 < \infty$, for some real-valued and square-integrable random variables $B(\alpha_1), B(\alpha_2)$ and for $\theta(\alpha_1), \theta(\alpha_2)$ uniformly distributed over $[0, 2\pi)$, where these four random variables are independent. This process is a particular case of the process with countably many frequencies (3.6.1), in which we remove the factor 2 for simplicity. From the previous computations and from the linearity of the Hilbert transform, we obtain, $\forall t \in \mathbb{R}$,

$$
\begin{aligned}
\hat{X}_t &= B(\alpha_1)\sin\{\alpha_1 t + \theta(\alpha_1)\} + B(\alpha_2)\sin\{\alpha_2 t + \theta(\alpha_2)\} \\
&= B(\alpha_1)\cos\left\{\alpha_1 t + \theta(\alpha_1) - \frac{\pi}{2}\right\} + B(\alpha_2)\cos\left\{\alpha_2 t + \theta(\alpha_2) - \frac{\pi}{2}\right\}, \\
X_t^* &= B(\alpha_1)e^{i\{\alpha_1 t + \theta(\alpha_1)\}} + B(\alpha_2)e^{i\{\alpha_2 t + \theta(\alpha_2)\}} \text{ and} \\
R_t &= |B(\alpha_1)e^{i\{\alpha_1 t + \theta(\alpha_1)\}} + B(\alpha_2)e^{i\{\alpha_2 t + \theta(\alpha_2)\}}| \\
&= \sqrt{B^2(\alpha_1) + B^2(\alpha_2) + 2B(\alpha_1)B(\alpha_2)\cos\{(\alpha_1 - \alpha_2)t + \theta(\alpha_1) - \theta(\alpha_2)\}}.
\end{aligned}
$$

These results lead to two important remarks. The first one is that the Hilbert transform $\{\hat{X}_t\}_{t\in\mathbb{R}}$ is the original process $\{X_t\}_{t\in\mathbb{R}}$ with phases reduced by $\pi/2$.

The second one is that the envelope $\{R_t\}_{t\in\mathbb{R}}$ has slower fluctuation than $\{X_t\}_{t\in\mathbb{R}}$, as its frequency is given by $\alpha_1 - \alpha_2$. The bandwidth is the difference between the largest and the smallest frequencies in stationary process. In this example it is $\alpha_2 - \alpha_1$ and when the bandwidth is small, then the process is called narrow-banded. The envelope indicates the slowly changing amplitude of the process. The original process has frequency typically higher than the one of the envelope. Another way of illustrating this is by defining α_c and δ such that $\alpha_1 = \alpha_c - \delta$ and $\alpha_2 = \alpha_c + \delta$. Thus

$$X_t^* = \underbrace{e^{i\alpha_c t}}_{part\ 1}\underbrace{\left(B(\alpha_1)e^{i\{\frac{\delta}{2}t+\theta(\alpha_1)\}} + B(\alpha_2)e^{i\{-\frac{\delta}{2}t+\theta(\alpha_2)\}}\right)}_{part\ 2}, \quad \forall t \in \mathbb{R}.$$

We distinguish two parts in the above expression: part 1 is oscillating fast, but having unit modulus, it has ho influence on the envelope; part 2 is oscillating slow, it has the amplitudes of the original process and it determines the amplitudes of the envelope.

Example 4.6.7 (Envelope of Gaussian process). *Let $\{X_t\}_{t\in\mathbb{R}}$ be a real-valued, stationary and Gaussian process with mean zero, with a.c.v.f. γ and whose spectral d.f. F possesses no jump at the origin, i.e. such that $\Delta F(0) = 0$. The Hilbert transform $\{\hat{X}_t\}_{t\in\mathbb{R}}$ has same mean, same spectral d.f. and same a.c.v.f.; cf. Proposition 4.6.2. Thus both $\{X_t\}_{t\in\mathbb{R}}$ and $\{\hat{X}_t\}_{t\in\mathbb{R}}$ have the same Gaussian f.d.d. Denote $\sigma^2 = \gamma(0)$ and let $t \in \mathbb{R}$, then*

$$X_t \sim \hat{X}_t \sim \mathcal{N}(0, \sigma^2),$$

and R_t, the envelope at time t, has density $x/\sigma^2 \, e^{-x^2/(2\sigma^2)}$, $\forall x \geq 0$, which is the rescaled Rayleigh density given in (3.2.7).

4.7 Simulation of stationary Gaussian processes

It is shown that for any n.n.d. function $\kappa : \mathbb{R} \to \mathbb{C}$, there exists a Gaussian stationary process $\{X_t\}_{t\in\mathbb{R}}$ with mean null and with a.c.v.f. κ. In other terms, when the a.c.v.f. of a stationary process is given, we can always associate it to a Gaussian process. The Gaussian assumption is adequate in many applied fields. For instance, Section 3.2.3 mentions that the fractional Brownian motion has an important role in various scientific fields, because of its selfsimilarity and because of its long range dependence. Thus, the stochastic simulation of Gaussian stationary processes such the fractional Brownian motion is an important numerical problem.

A simulation algorithm usually returns a vector representing the sampled stochastic process over a finite time horizon. Precisely, when we want to generate the stationary stochastic process $\{X_t\}_{t\in\mathbb{R}}$ over the time interval $[0, t^\dagger)$, the algorithm generates n values of the process over the lattice of span $\delta > 0$ given by $\{k\delta\}_{k=0,\dots,n-1}$, such that the time horizon is $t^\dagger = n\delta$. In this context, we remind Remark 3.5.13 between sampling a stochastic process at times of a lattice and the Nyquist frequency. Sampling a stationary stochastic process over $\{k\delta\}_{k=0,\dots,n-1}$ yields a discrete time stochastic process whose spectral d.f. is obtained by wrapping the spectral d.f. of the continuous time process. The wrapped spectral d.f. is given by (3.5.8) and, when the spectral density of the original process exists, the wrapped spectral density is given by (3.5.9), both these wrapped functions being defined over $(-\pi/\delta, \pi/\delta]$. Wrapping is meant around the circle of circumference $2\pi/\delta$. Should the original spectral distribution have bounded domain, so to vanish outside $(-\pi/\delta, \pi/\delta]$, then this wrapping operation would have no effect. This means that none of the harmonics of the original process would be lost by sampling. In the practice, by choosing δ sufficiently small, the sampled process $\left\{X_{k\delta}^{(n)}\right\}_{k=0,\ldots,n-1}$ can be obtained arbitrarily close to the original process $\{X_t\}_{t\in[0,t^\dagger)}$, in the sense that sufficiently many high frequency harmonics are retrieved in the sampled process. All frequencies appearing in the spectral decomposition of $\left\{X_{k\delta}^{(n)}\right\}_{k=0,\ldots,n-1}$ do belong to the interval $(-\pi/\delta, \pi/\delta]$. The details of this explanation are given in Section 4.7.3.

This section presents three principal algorithms for simulating Gaussian stationary processes. The first one is based on the Choleski factorization of the covariance matrix of a sample of the process and it is given in Section 4.7.1. The second algorithm is based on embedding the covariance matrix into a circulant matrix and it is given in Section 4.7.2. The third algorithm of this section is based on the spectral decomposition of a stationary process and on its spectral distribution and it is presented in Section 4.7.3.

4.7.1 *Simulation by Choleski factorization*

The computer generation of a Gaussian process, stationary or not, corresponds to the generation of a high-dimensional Gaussian random vector. Thus, suppose that we want to generate the real-valued random vector $X \sim \mathcal{N}(\mathbf{0}, \Sigma)$, where the covariance matrix $\Sigma \in \mathbb{R}^{d\times d}$ is assumed positive definite (p.d.) and where dimension index d is a large integer. The method presented in this section is an application of Choleski's factorization. This presentation mainly follows pp. 49–50 and pp. 311–313 of Asmussen and Glynn (2007).

It may be instructive to start with the presentation of the simple case with dimension $d = 2$. Let $\sigma_1, \sigma_2 > 0$, $\rho \in (-1, 1)$ and define the covariance matrix

$$\Sigma = \begin{pmatrix} \sigma_1^2 & \rho\sigma_1\sigma_2 \\ \rho\sigma_1\sigma_2 & \sigma_2^2 \end{pmatrix}.$$

Thus Σ is p.d. and we assume that the random variables Z_1, Z_2, Z_3 are independent and standard normal. One can then easily control that, for

$$X_1 = \sigma_1(\sqrt{1-|\rho|}\, Z_1 + \sqrt{|\rho|}\, Z_2),$$

and

$$X_2 = \sigma_2(\sqrt{1-|\rho|}\, Z_3 + \operatorname{sgn}\rho\, \sqrt{|\rho|}\, Z_2),$$

the random vector

$$X = (X_1, X_2)^\top \sim \mathcal{N}(\mathbf{0}, \Sigma).$$

Thus, generating three standard normal random variables leads to any desired normal random vector of dimension two. Standard normal random variables can be generated with the Box-Müller algorithm, which is depicted at p. 182.

We now turn to the general and relevant case with dimension $d > 2$. If a matrix $C \in \mathbb{R}^{d \times d}$ such that $\Sigma = CC^\top$ would be available, then the generation of $X \sim \mathcal{N}(0, \Sigma)$ would be simple. The major problem is indeed the determination of a matrix C with the desired property. An obvious choice is $C = \Sigma^{1/2}$. However, this choice should not be retained for large values of d because it is numerically inefficient. A computationally efficient choice is provided by the Choleski factorization, which is presented in Theorem 4.7.1.

Theorem 4.7.1 (Choleski factorization). *Let $\Sigma \in \mathbb{R}^{d \times d}$ be p.d., then there exists a matrix $C \in \mathbb{R}^{d \times d}$ that is lower triangular, i.e. $c_{ij} = 0$, for $1 \leq i < j \leq d$, where $C = (c_{ij})_{i,j=1,\ldots,d}$, and that satisfies*

$$\Sigma = CC^\top.$$

The proof of the Choleski factorization provides simultaneously the generation of algorithm of a Gaussian random vector and we so present this proof directly under Algorithm 4.7.2.

Algorithm 4.7.2 (Generation of Gaussian random vector by Choleski factorization). *Suppose that the first $k - 1$ rows of the lower triangular matrix $C = (c_{ij})_{i,j=1,\ldots,d}$ have been computed, for $k \in \{2, \ldots, d\}$.*

- *Then*

$$\sigma_{kj} = \sum_{i=1}^{j} c_{ki} c_{ji}, \quad \text{for } j = 1, \ldots, k,$$

  where $\Sigma = (\sigma_{ij})_{i,j=1,\ldots,d}$. We have k equations with k unknowns $c_{k1}, \ldots, c_{k,k-1}, c_{kk}^2$.
- *The solution can be obtained recursively by*

$$c_{k1} = \frac{\sigma_{k1}}{c_{11}},$$

$$c_{kj} = \frac{1}{c_{jj}}(\sigma_{kj} - c_{k1}c_{j1} - \cdots - c_{k,j-1}c_{j,j-1}), \quad \text{for } j = 2, \ldots, k-1, \text{ and}$$

$$c_{kk}^2 = \sigma_{kk} - c_{k1}^2 - \cdots - c_{k,k-1}^2.$$

Then we can proceed as follows.

(1) Generate $Z_1, \ldots, Z_d$ independent standard normal.
(2) Define $\mathbf{Z} = (Z_1, \ldots, Z_d)^\top$ and compute $\mathbf{X} = \mathbf{C}\mathbf{Z}$.

It then follows $\mathbf{X} \sim \mathcal{N}(\mathbf{0}, \mathbf{\Sigma})$. When the covariance matrix $\mathbf{\Sigma}$ has the Toeplitz form, then we can use this algorithm for generating a Gaussian stationary process.

An important application of Choleski's factorization is for the prediction or forecast problem, where the following Lemma 4.7.3 is required. This Lemma is a simple multidimensional generalization of Proposition A.8.4.4.

Lemma 4.7.3 (Gaussian conditional subvector). *Consider the Gaussian or normal random vector $\mathbf{X}$ and its partition*

$$\mathbf{X} = \begin{pmatrix} \mathbf{X}_1 \\ \mathbf{X}_2 \end{pmatrix} \sim \mathcal{N}\left(\begin{pmatrix} \boldsymbol{\mu}_1 \\ \boldsymbol{\mu}_2 \end{pmatrix}, \begin{pmatrix} \mathbf{\Sigma}_{11} \ \mathbf{\Sigma}_{12} \\ \mathbf{\Sigma}_{21} \ \mathbf{\Sigma}_{22} \end{pmatrix} \right),$$

where the covariance matrix of $\mathbf{X}$ is assumed p.d. Then we have the following conditional distributions

$$\mathbf{X}_1 | \mathbf{X}_2 \sim \mathcal{N}\left(\boldsymbol{\mu}_1 + \mathbf{\Sigma}_{12}\mathbf{\Sigma}_{22}^{-1}(\mathbf{X}_2 - \boldsymbol{\mu}_2), \mathbf{\Sigma}_{11} - \mathbf{\Sigma}_{12}\mathbf{\Sigma}_{22}^{-1}\mathbf{\Sigma}_{21} \right) \quad \text{and}$$
$$\mathbf{X}_2 | \mathbf{X}_1 \sim \mathcal{N}\left(\boldsymbol{\mu}_2 + \mathbf{\Sigma}_{21}\mathbf{\Sigma}_{11}^{-1}(\mathbf{X}_1 - \boldsymbol{\mu}_1), \mathbf{\Sigma}_{22} - \mathbf{\Sigma}_{21}\mathbf{\Sigma}_{11}^{-1}\mathbf{\Sigma}_{12} \right).$$

Obviously, $\mathbf{\Sigma}_{11}$ and $\mathbf{\Sigma}_{22}$ are p.d.

In the prediction problem we want to generate the random variable X_{n+1} given the random variables $X_0, \ldots, X_n$ that are all normal and, for simplicity with mean null. We define

$$\mathbf{\Sigma}(n) = \mathsf{var}((X_0, \ldots, X_n)) \quad \text{and}$$
$$\mathbf{\Sigma}(n+1) = \begin{pmatrix} \mathbf{\Sigma}(n) & \boldsymbol{\tau}(n) \\ \boldsymbol{\tau}^\top(n) \ \tau(n+1, n+1) \end{pmatrix} = \mathsf{var}((X_0, \ldots, X_n, X_{n+1})).$$

We obtain from Lemma 4.7.3 that

$$X_{n+1} | X_0, \ldots, X_n \sim \mathcal{N}\left(\hat{X}_{n+1}, \sigma_{n+1}^2 \right), \tag{4.7.1}$$

where

$$\hat{X}_{n+1} = \mathsf{E}[X_{n+1} | X_0, \ldots, X_n] = \boldsymbol{\tau}^\top(n) \Sigma^{-1}(n) \begin{pmatrix} X_0 \\ \vdots \\ X_n \end{pmatrix} \quad \text{and}$$

$$\sigma_{n+1}^2 = \mathsf{var}(X_{n+1} | X_0, \ldots, X_n) = \tau(n+1, n+1) - \boldsymbol{\tau}^\top(n)\Sigma^{-1}(n)\boldsymbol{\tau}(n).$$

The major numerical hurdle for generating X_{n+1} from the conditional normal distribution (4.7.1) is the evaluation of $\mathbf{\Sigma}^{-1}(n)$. This problem can be addressed by means of the Choleski factorization of $\boldsymbol{\Sigma}(n)$. For $\boldsymbol{\Sigma}(n) = \boldsymbol{C}(n)\boldsymbol{C}^\top(n)$, where $\boldsymbol{C}(n)$ is a $(n+1) \times (n+1)$ lower triangular matrix,

$$\boldsymbol{\Sigma}^{-1}(n) = \{\boldsymbol{C}^\top(n)\}^{-1}\boldsymbol{C}^{-1}(n),$$

is simple to compute. (The eigenvalues of $\boldsymbol{C}(n)$ are the elements of the main diagonal.) Moreover, this algorithm is convenient for computing iterated predictions. Indeed, the very convenient structure

$$\boldsymbol{C}(n+1) = \left(\begin{array}{ccc|c} & \boldsymbol{C}(n) & & \mathbf{0} \\ \hline c_{n+2,1} & \cdots & c_{n+2,n+1} & c_{n+2,n+2} \end{array}\right),$$

is a direct consequence of the recursive form of Choleski's factorization.

4.7.2   Simulation by circulant embedding

Another efficient algorithm for generating Gaussian stationary processes is given by the method of circulant embedding. In some cases, this method can be a more efficient method than the method based on Choleski's decomposition. Some references on this techniques are pp. 314–316 of Asmussen and Glynn (2007), Dietrich and Newsam (1997) and Wood and Chan (1994). The central object is the circulant, which is a particular Toeplitz matrix.

Definition 4.7.4 (Circulant). *A circulant of dimension $n \geq 1$ is any matrix in $\mathbb{C}^{n \times n}$ of the form*

$$C = \begin{pmatrix} c_0 & c_{n-1} & \cdots & c_2 & c_1 \\ c_1 & c_0 & \cdots & c_3 & c_2 \\ \vdots & \vdots & \ddots & \vdots & \vdots \\ c_{n-2} & c_{n-3} & \cdots & c_0 & c_{n-1} \\ c_{n-1} & c_{n-2} & \cdots & c_1 & c_0 \end{pmatrix}. \tag{4.7.2}$$

By denoting $\boldsymbol{C} = (c_{ij})_{i,j=1,\ldots,n}$, we have

$$c_{ij} = c_{(i-j)\bmod n}, \quad \text{for } i,j = 0, \ldots, n-1.$$

The columns of $\boldsymbol{C}$ are thus the cyclic permutations of the first one, which is $\boldsymbol{c} = (c_0, \ldots, c_{n-1})^\top$.

Proposition 4.7.5 (Eigenvalues of a circulant). *The eigenvalues of the circulant of order n given by (4.7.2) and denoted $\boldsymbol{C}$, are*

$$\lambda_j = \sum_{k=0}^{n-1} c_k e^{-ik\omega_j}, \quad \text{for } j = 0, \ldots, n-1,$$

where $\omega_j = 2\pi j/n$, for $j = 0, \ldots, n-1$, are the Fourier frequencies. The eigenvector associated with the eigenvalue λ_j is

$$\left(1, e^{i\omega_j}, \ldots, e^{i(n-1)\omega_j}\right),$$

which is the $(j+1)$-th column of the matrix $\boldsymbol{F} = \left(e^{ik\omega_j}\right)_{j,k=0,\ldots,n-1}$, for $j = 0, \ldots, n-1$. In matrix notation we have

$$\boldsymbol{C} = \frac{1}{n}\boldsymbol{F}\boldsymbol{\Lambda}\overline{\boldsymbol{F}},$$

where

$$\boldsymbol{\Lambda} = \begin{pmatrix} \lambda_0 & & \boldsymbol{0} \\ & \ddots & \\ \boldsymbol{0} & & \lambda_{n-1} \end{pmatrix}.$$

Note that for $\boldsymbol{\lambda} = (\lambda_0, \ldots, \lambda_{n-1})^{\top}$, we have

$$\boldsymbol{\lambda} = \overline{\boldsymbol{F}}\boldsymbol{c}. \qquad (4.7.3)$$

Proof. For $\boldsymbol{F} = (f_{jk})_{j,k=1,\ldots,n}$ and for $j, k = 1, \ldots, n$, we have

$$(\boldsymbol{C}\boldsymbol{F})_{jk} = \sum_{l=1}^{n} c_{jl} f_{lk} = \sum_{l=1}^{n} c_{(j-l) \bmod n} e^{i(l-1)\omega_{k-1}}$$

$$= \sum_{l=1}^{n} c_{(j-1-[l-1]) \bmod n} e^{i(l-1)\omega_{k-1}} = \sum_{l=0}^{n-1} c_{(j-1-l) \bmod n} e^{il\omega_{k-1}}$$

$$= \sum_{l=0}^{n-1} c_l e^{i(j-1-l)\omega_{k-1}} = e^{i(j-1)\omega_{k-1}} \sum_{l=0}^{n-1} c_l e^{-il\omega_{k-1}}$$

$$= e^{i(j-1)\omega_{k-1}} \lambda_{k-1} = (\boldsymbol{F}\boldsymbol{\Lambda})_{jk},$$

from the periodicity of order n of the summands. Note that $\boldsymbol{F}\overline{\boldsymbol{F}} = n\boldsymbol{I}$, i.e. $\boldsymbol{F}^{-1} = n^{-1}\overline{\boldsymbol{F}}$; refer to (A.10.2) for the details. $\qquad \square$

We note that, for $n \geq 2$, the Toeplitz covariance matrix

$$\mathrm{var}((X_0, \ldots, X_n)) = \begin{pmatrix} r_0 & r_1 & \cdots & r_n \\ r_1 & r_0 & \cdots & r_{n-1} \\ \vdots & \vdots & \ddots & \vdots \\ r_n & r_{n-1} & \cdots & r_0 \end{pmatrix}, \qquad (4.7.4)$$

where $r_0, \ldots, r_n \in \mathbb{C}$, is generally not a circulant. But it can be embedded into a circulant of dimension $2n$. We can indeed construct the $2n \times 2n$ symmetric circulant

$$
C = \left(\begin{array}{cccc|cccc}
r_0 & r_1 & \cdots & r_n & r_{n-1} & \cdots & & r_1 \\
r_1 & r_0 & \cdots & r_{n-1} & r_n & \cdots & & r_2 \\
\vdots & \vdots & \ddots & \vdots & \vdots & & & \vdots \\
r_n & r_{n-1} & \cdots & r_0 & r_1 & \cdots & & r_{n-1} \\
\hline
r_{n-1} & r_n & \cdots & r_1 & r_0 & \cdots & & r_{n-2} \\
\vdots & \vdots & & \vdots & \vdots & \ddots & & \vdots \\
r_1 & r_2 & \cdots & r_{n-1} & r_{n-2} & \cdots & & r_0
\end{array} \right).
$$

The columns of C are indeed the cyclic permutations of the first one, which is $c = (r_0 \, r_1 \cdots r_n \, r_{n-1} \cdots r_1)^{\top}$. It may be helpful for the intuition to copy the first column of C around a circle and to rotate this labeled circle in order to obtain the other columns. It follows from Proposition 4.7.5 that the vector of the $2n$ eigenvalues of C is given by (4.7.3). By assuming these eigenvalues nonnegative, we obtain the matrix

$$
D = n^{-\frac{1}{2}} F \Lambda^{\frac{1}{2}},
$$

and thus

$$
D\overline{D}^{\top} = \frac{1}{n} F \Lambda^{\frac{1}{2}} \overline{\Lambda^{\frac{1}{2}}} \, \overline{F}^{\top} = \frac{1}{n} F \Lambda \overline{F} = C. \qquad (4.7.5)
$$

We remind that the eigenvalues and vectors of a real symmetric matrix are always real. Thus, the eigenvalues of a symmetric circulant are real but not necessarily nonnegative. Because the vector of eigenvalues (4.7.3) can be efficiently computed by the FFT, cf. Section A.10, Algorithm 4.7.6 below inherits the high efficiency of the FFT. This algorithm can be used for the generation of real-valued Gaussian stationary processes.

Algorithm 4.7.6 (Generation of Gaussian random vectors by circulant embedding). *The real-valued Gaussian random vector $(X_0, \ldots, X_n)$ with mean zero and Toeplitz covariance matrix (4.7.4) can be generated by the following steps.*

(1) *Compute $\lambda = \overline{F}c$ and $D = n^{-\frac{1}{2}} F \Lambda^{\frac{1}{2}}$ with the FFT, cf. Section A.10. Set $D_R = \operatorname{Re} D$ and $D_I = \operatorname{Im} D$.*

(2) *Generate Z_R and Z_I independent $\mathcal{N}(0, I_{2n})$.*

(3) *Set $Z = Z_R + \mathrm{i} Z_I$ and $X = DZ = D_R Z_R - D_I Z_I + \mathrm{i}(D_R Z_I + D_I Z_R)$.*

(4) Thus $\boldsymbol{X}_R = \operatorname{Re} \boldsymbol{X} = \boldsymbol{D}_R \boldsymbol{Z}_R - \boldsymbol{D}_I \boldsymbol{Z}_I$ *and* $\boldsymbol{X}_I = \operatorname{Im} \boldsymbol{X} = \boldsymbol{D}_R \boldsymbol{Z}_I + \boldsymbol{D}_I \boldsymbol{Z}_R.$

The first $n + 1$ *elements of* $\boldsymbol{X}_R$ *and* $\boldsymbol{X}_I$ *provide two realizations of* $(X_0, \ldots, X_n)$ *that are however dependent.*

We show the last paragraph of the algorithm by computing

$$\operatorname{var}(\boldsymbol{X}_R) = \operatorname{var}(\boldsymbol{X}_I) = \boldsymbol{D}_R \boldsymbol{D}_R^\top + \boldsymbol{D}_I \boldsymbol{D}_I^\top,$$

and

$$\boldsymbol{C} = \boldsymbol{D}\overline{\boldsymbol{D}}^\top = \boldsymbol{D}_R \boldsymbol{D}_R^\top + \boldsymbol{D}_I \boldsymbol{D}_I^\top + \mathrm{i}(\boldsymbol{D}_I \boldsymbol{D}_R^\top - \boldsymbol{D}_R \boldsymbol{D}_I^\top),$$

from (4.7.5). These results and the fact that $\boldsymbol{C}$ is real lead to $\boldsymbol{D}_I \boldsymbol{D}_R^\top - \boldsymbol{D}_R \boldsymbol{D}_I^\top = 0$ and thus to $\operatorname{var}(\boldsymbol{X}_R) = \operatorname{var}(\boldsymbol{X}_I) = \boldsymbol{C}$.

Compared with the method of Choleski's decomposition, given in Section 4.7.1, the method of circulant embedding has superior numerical efficiency, whenever the FFT is applied. On the side of the disadvantages we can mention the limitation to the case where $\lambda_0, \ldots, \lambda_{2n-1} \geq 0$ and the lack of recursivity, w.r.t. increasing order n. We can provide the following sufficient condition for $\lambda_0, \ldots, \lambda_{2n-1} \geq 0$: the covariances $r_0, \ldots, r_n \geq 0$ are decreasing and also convex.

4.7.3 *Simulation by spectral decomposition*

This section presents an algorithm for generating a real-valued, Gaussian and stationary process with continuous time through its spectral decomposition. The main idea of this algorithm is already presented in Remark 3.7.1. The first step is to approximate the spectral decomposition of the process, which is either an infinite sum or a mean square integral, by a finite sum and precisely by a sum of a finite number of frequencies. Then one generates summands of this finite sum, precisely the complex coefficients of the harmonics, as given in (4.7.8). This yields an approximate sample path of the desired stationary process. The approximation can made be as close as desired to the true stationary process, by choosing a sufficiently large number of summands (denoted n). This amounts to choose a sufficiently small discretization unit of the time domain (denoted δ). Although an approximation of the process with fine discretization increases the computing time, the use of the FFT, presented in Section A.10, improves substantially the numerical efficiency.

Let us first re-express the spectral decomposition of the process as a discrete Fourier transform. The starting point can be the real-valued and stationary process with countable number of frequencies $\{X_t\}_{t\in\mathbb{R}}$ given by (3.6.1) that we rewrite here for convenience. We thus assume

$$X_t = B(\alpha_0) + 2\sum_{j=1}^{\infty} B(\alpha_j)\cos(\alpha_j t + \theta(\alpha_j)), \quad \forall t \in \mathbb{R}, \qquad (4.7.6)$$

where $0 = \alpha_0 < \alpha_1 < \cdots < \infty$ are the selected frequencies, $\theta(\alpha_j)$, for $j = 1, 2, \ldots$, are uniformly distributed over $[0, 2\pi)$, $B(\alpha_j)$, for $j = 0, 1, \ldots$, are real-valued and square-integrable random variables. All these random variables are independent. Further details, also from Section 3.6.1, are the following. We have $\mathsf{E}[X_t] = \mathsf{E}[B(0)]$, $\forall t \in \mathbb{R}$, and we assume $\mathsf{E}[B(0)] = 0$. By defining, as in (3.6.5), $\sigma_j^2 = \mathsf{E}\left[B^2(\alpha_j)\right]$, with $\sigma_j > 0$, $\sigma_{-j} = \sigma_j$, and $\alpha_{-j} = -\alpha_j$, for $j = 0, 1, \ldots$, we obtain the spectral d.f. $F(\alpha) = \sum_{j=-\infty}^{\infty} \sigma_j^2 \, \mathsf{I}\{\alpha_j \leq \alpha\}$, $\forall \alpha \in \mathbb{R}$, cf. (3.6.8).

This stationary Gaussian process originally appears in (3.2.5) and, by following the arguments given just after (3.2.5), we find that (4.7.6) is indeed a Gaussian stationary process under two conditions. The first one is

$$B(\alpha_j) \sim \frac{\sigma_j}{\sqrt{2}} R, \quad \text{for } j = 1, 2, \ldots, \qquad (4.7.7)$$

where R is a Rayleigh distributed random variable with density (3.2.7).[3] The second condition is $B(0) \sim \mathcal{N}(0, \sigma_0^2)$.

By defining $C_0 = B(\alpha_0)$ and $C_j = 2B(\alpha_j)e^{i\theta(\alpha_j)}$, for $j = 1, 2, \ldots$, we obtain

$$X_t^* = X_t + \mathrm{i}\hat{X}_t = \sum_{j=0}^{\infty} C_j e^{\mathrm{i}\alpha_j t}, \quad \forall t \in \mathbb{R},$$

where

$$\hat{X}_t = B(\alpha_0) + 2\sum_{j=1}^{\infty} B(\alpha_j)\sin(\alpha_j t + \theta(\alpha_j)), \quad \forall t \in \mathbb{R},$$

is a real-valued ancillary process.

---

[3]In order to see this, rewrite the process as

$$X_t = B(\alpha_0) + \sum_{j=1}^{\infty} 2B(\alpha_j)\cos-\theta(\alpha_j)\cos\alpha_j t + 2B(\alpha_j)\sin-\theta(\alpha_j)\sin\alpha_j t, \quad \forall t \in \mathbb{R},$$

and set

$$2B(\alpha_j)\cos-\theta(\alpha_j) \sim 2B(\alpha_j)\sin-\theta(\alpha_j) \sim \mathcal{N}(0, \zeta_j^2), \quad \text{for } j = 1, 2, \ldots,$$

for some $\zeta_1, \zeta_2, \ldots > 0$. Let R be the Rayleigh random variable with density (3.2.7) and let $j \in \mathbb{N}^*$. The above normality holds if $2B_j \sim \zeta_j R$. Because $\mathsf{E}[B_j^2] = \sigma_j^2$, we have $\zeta_j = \sqrt{2}\sigma_j$, which gives (4.7.7).

As shown in Section 4.6, if we have a complex-valued stationary process $\{X_t^*\}_{t\in\mathbb{R}}$ whose real part $\{X_t\}_{t\in\mathbb{R}}$ is the stationary process of interest, then the complex part $\{\hat{X}_t\}_{t\in\mathbb{R}}$ is the Hilbert transform of $\{X_t\}_{t\in\mathbb{R}}$; cf. Definition 4.6.1. Our ancillary process is thus the Hilbert transform of the process of interest. It allows us to determine the envelope of $\{X_t\}_{t\in\mathbb{R}}$, in the sense of Definition 4.6.4.

For the time horizon of interest $t^\dagger \in (0,\infty)$: we want to generate $\{X_t\}_{t\in\mathbb{R}}$ restricted to the time interval $[0, t^\dagger)$. We restrict the interval to the lattice of span $\delta > 0$, such that, for some integer n, $t^\dagger = n\delta$. Consider

$$\mathcal{F}_n = \left\{ -\left\lfloor \frac{n-1}{2} \right\rfloor, \ldots, \left\lfloor \frac{n}{2} \right\rfloor \right\},$$

and the Fourier frequencies $\omega_j = 2\pi j/n$, $\forall j \in \mathcal{F}_n$. Note that card $\mathcal{F}_n = n$ and that $\omega_j \in (-\pi, \pi]$, $\forall j \in \mathcal{F}_n$. Define the discrete time process

$$X_{k\delta}^{(n)} = \operatorname{Re} \sum_{j\in\mathcal{F}_n} C_j \exp\left\{ \mathrm{i}\frac{\omega_j}{\delta}(k\delta) \right\} = \operatorname{Re} \sum_{j\in\mathcal{F}_n} C_j \exp\left\{ \mathrm{i}\frac{2jk\pi}{n} \right\}, \quad (4.7.8)$$

for $k = 0, \ldots, n-1$. Then $\left\{ X_{k\delta}^{(n)} \right\}_{k=0,\ldots,n-1}$ approximates $\{X_t\}_{t\in[0,t^\dagger)}$ arbitrarily well for large values of n.

Note that we could have also started this explanation from the decomposition (3.7.1), where the possible frequencies are the continuum.

In the presentation of Section 4.7, we mention the importance of Remark 3.5.13 on sampling and on the Nyquist frequency. Sampling a stationary process over a time lattice with span $\delta > 0$ produces a process with wrapped spectral d.f. F° given by (3.5.8) and with wrapped spectral density f° given by (3.5.9), provided that the spectral density of the original process exists. This wrapping is around the circle of circumference $2\pi/\delta$ and so both wrapped d.f. and density are defined over $(-\pi/\delta, \pi/\delta]$. Should the original spectral distribution be very small outside $(-\pi/\delta, \pi/\delta]$, then none of the important harmonics of high frequency would be lost and the generated process $\left\{ X_{k\delta}^{(n)} \right\}_{k=0,\ldots,n-1}$ would be very close to the original process $\{X_t\}_{t\in\mathbb{R}}$. The frequencies that appear in the spectral decomposition (4.7.8) of $\left\{ X_{k\delta}^{(n)} \right\}_{k=0,\ldots,n-1}$ are ω_j/δ, $\forall j \in \mathcal{F}_n$, and belong to interval $(-\pi/\delta, \pi/\delta]$.

The spectral d.f. of $\left\{ X_{k\delta}^{(n)} \right\}_{k=0,\ldots,n-1}$ is a step function with jumps at the frequencies ω_j/δ, for $j \in \mathcal{F}_n$. It can be written as

$$F_n(\alpha) = \sum_{j\in\mathcal{F}_n} \sigma_{j,n}^2 \mathrm{I}\left\{ \frac{\omega_j}{\delta} \le \alpha \right\}, \quad \forall \alpha \in \left(-\frac{\pi}{\delta}, \frac{\pi}{\delta} \right],$$

where $\sigma_{j,n} \in \mathbb{R}_+$, $\forall j \in \mathcal{F}_n$. These unknown coefficients are required in the algorithm for generating $\left\{X_{k\delta}^{(n)}\right\}_{k=0,\dots,n-1}$ and so one needs to compute them. Denote by F the spectral d.f. of the true process $\{X_t\}_{t \in \mathbb{R}}$. Given that $\left\{X_{k\delta}^{(n)}\right\}_{k=0,\dots,n-1}$ approximates $\{X_t\}_{[0,t^\dagger)}$, the spectral d.f. F_n must be close to the wrapped transform of F, i.e. F° given by (3.5.8). In other terms, the jumps of F_n are expected close to the increments of F° taken over nearby intervals of length $2\pi/(n\delta)$. We thus approximate $\sigma_{j,n}^2$, $\forall j \in \mathcal{F}_n$, by the values

$$\tilde{\sigma}_{j,n}^2 = F^\circ \left(\frac{\omega_j}{\delta}\right) - F^\circ \left(\frac{\omega_{j-1}}{\delta}\right), \quad \forall j \in \mathcal{F}_n. \qquad (4.7.9)$$

We can now present simulation algorithm for the stationary process (4.7.6), when it is Gaussian, together with its envelope.

Algorithm 4.7.7 (Generation of Gaussian stationary process and envelope by spectral distribution). *Let $[0, t^\dagger)$ be the time interval of interest.*

(1) Select a positive integer l, compute $n = 2^l$ and $\delta = t^\dagger/n$.

(2) Compute F° by wrapping the spectral d.f. of the true process F, according to (3.5.8).

(3) Compute $\hat{\sigma}_{j,n} \geq 0$, $\forall j \in \mathcal{F}_n$, by using F° as in (4.7.9).

(4) Generate $C_0 \sim \mathcal{N}(0, \tilde{\sigma}_{0,n}^2)$.

(5) Repeat for $j \in \mathcal{F}_n$.

  Generate U_j uniform over $[0,1)$ and R_j Rayleigh distributed, with density (3.2.7).

  Set $\theta_j = 2\pi U_j$ and $B_j = \tilde{\sigma}_{j,n} R_j/\sqrt{2}$.

  Set $C_j = 2B_j e^{i\theta_j}$.

(6) Apply the FFT to C_j, for $j \in \mathcal{F}_n$.

  Take the real part of the FFT, in order to obtain $\left\{X_{k\delta}^{(n)}\right\}_{k=0,\dots,n-1}$, given in (4.7.8).

  Take the complex part of the FFT, in order to obtain $\left\{\hat{X}_{k\delta}^{(n)}\right\}_{k=0,\dots,n-1}$, which corresponds to the Hilbert transform.

  Compute the envelope $R_{k\delta}^{(n)} = \sqrt{X_{k\delta}^{(n)2} + \hat{X}_{k\delta}^{(n)2}}$, for $k = 0,\dots,n-1$.

The FTT is explained in Section A.10. It is precisely the restriction to $n \in 2^{\mathbb{N}}$ that allows for the use of the FTT. This algorithm can be rewritten for $j \in \{0,\dots,n-1\}$ instead of $j \in \mathcal{F}_n$; cf. Remark A.10.1.

4.7.4 *Simulation by ARMA approximation*

Theorem 2.4.30 states that, under weak conditions, the ARMA(p, q) time series $\{X_k\}_{k \in \mathbb{Z}}$ has spectral density

$$f(\alpha) = \frac{\sigma^2}{2\pi} \frac{|\theta(e^{-i\alpha})|^2}{|\varphi(e^{-i\alpha})|^2}, \quad \forall \alpha \in [-\pi, \pi], \tag{4.7.10}$$

where φ and θ are the autoregressive and moving average polynomials, respectively, and where σ^2 is the variance of the WN. Thus, if the spectral density with the precise form (4.7.10) would be given available, then one could determine the corresponding ARMA(p, q) time series. The distribution of the WN could be chosen Gaussian. In this ideal situation, by simulating the Gaussian WN of the time series, one generates the time series $\{X_k\}_{k \in \mathbb{Z}}$ by using the ARMA(p, q) equations. We have thus obtained the desired Gaussian time series, in the sense that it possesses the desired spectral density (4.7.10).

Generally, the spectral density for the simulation is not given with the precise form (4.7.10). In order to simulate a Gaussian stationary time series, Krenk and Clausen (1987) suggest to proceed in two steps: first to approximate the available spectral density with the form (4.7.10) and then to generate the ARMA(p, q) time series, from the polynomials φ and θ that are obtained from the approximation of the spectral density.

It appears quite intuitive that, by choosing the orders p and q sufficiently large, the desired density would be accurately approximated. This problem resembles to the approximation of a function by a rational one, which is a standard topic of numerical analysis. It is known that for a fixed computational effort, one can usually construct a rational approximation that is more accurate than a polynomial approximation. The method of Padé is usually applied in this context.

We can note that Theorem 2.4.37 gives us a theoretical basis for this method: for any continuous spectral density f, there exists a causal AR(p) time series with spectral density arbitrarily close to f. The order p that would be required in order to obtain a good approximation to f could be large. It could be substantially larger then the two degrees of an approximation by rational function. The same remark holds for the MA(q) approximation, which is provided by Theorem 2.4.38.

4.8 Large deviations theory for time series

Some asymptotic normal theory for the mean and for the a.c.v.f. of stationary time series is presented in Section 2.5.2. Section 4.8.1 provides a brief

introduction to large deviations approximations and the application to the
AR(1) time series is given in Section 4.8.2.

4.8.1 *Some notions of large deviations theory*

Large deviations theory analyzes the exponential decay of probabilities of
certain kinds of rare events. More precisely, it provides limiting values or
bounds to $n^{-1} \log p_n$, when $n \to \infty$, where $\{p_n\}_{n \geq 1}$ is an asymptotically
vanishing sequence of probabilities that satisfies some regularity conditions.
The resulting approximations are called of logarithmic asymptotics. The
origin of large deviations theory goes back to Laplace, Cramér, Chernov
and Varadhan; refer e.g. to the monograph Varadhan (1984) for a complete presentation. We only mention that the first theoretical results were
presented in Cramér (1944) in the context of actuarial risk modelling. Further general references on the theory of large deviations are e.g. Shwartz
and Weiss (1995), Dembo and Zeitouni (2009) and in particular Bucklew
(2004), which is partially followed in this section. Applications can be found
in many fields, chief among them are statistical inference, information theory, statistical mechanics and actuarial risk theory. This section begins
by providing motivations for large deviations approximations and then it
presents the Theorem of Gärtner and Ellis, which is one of the major result
of large deviations theory.

First steps towards large deviations

Let us explain the objective of large deviations approximations and thus
explain the limited validity of the Central limit theorem. Let $X_1, X_2, \ldots$
be i.d.d. random variables with cumulant generating function $K(v) =
\log \mathsf{E}\left[e^{vX_1}\right] \in \mathbb{R} \cup \{\infty\}$, $\forall v \in \mathbb{R}$. Let $M_n = n^{-1} \sum_{j=1}^{n} X_j$, for some $n \geq 1$,
and $y > \mathsf{E}[X_1]$. We are interested in the asymptotic decay of $\mathsf{P}[M_n > y]$ as
$n \to \infty$.

Let $k \in \mathbb{N}^*$ such that n/k is an integer. Then
$$X_{jk+1} + \cdots + X_{jk+k} > ky, \quad \text{for } j = 0, \ldots, \frac{n}{k} - 1 \Longrightarrow X_1 + \cdots + X_n > ny.$$
Consequently,

$$\mathsf{P}[M_n > y] \geq \mathsf{P}\left[\bigcap_{j=0}^{\frac{n}{k}-1} \{X_{jk+1} + \cdots + X_{jk+k} > ky\}\right]$$

$$= \left(\mathsf{P}\left[X_1 + \cdots + X_k > ky\right]\right)^{\frac{n}{k}}$$

$$= \exp\{-na_k(y)\},$$

with $a_k(y) = -k^{-1} \log \mathsf{P}\left[X_1 + \cdots + X_k > ky\right] > 0$, which holds whenever $\mathsf{P}\left[X_1 + \cdots + X_k > ky\right] > 0$. Thus, the decay of the tail probability of M_n is at most exponential. Let $\xi > 0$ such that $K(\xi)$ is finite, then Markov's inequality A.2.5 gives

$$\mathsf{P}[M_n > y] \leq \frac{\mathsf{E}\left[e^{n\xi M_n}\right]}{e^{n\xi y}} = \exp\left\{-n\left[\xi y - K(\xi)\right]\right\}. \tag{4.8.1}$$

If ξ is sufficiently small, then $\xi y - K(\xi) > 0$. This is guaranteed when K is finite over a neighborhood of the origin, which can be obtained when the distribution of X_1 decays sufficiently fast in the tails. Such probability distributions are referred to as light-tailed. Thus, the decay of the upper tail probability of M_n is at least exponential. We have shown that it is at most and at least exponential, so exactly exponential.

The upper bound (4.8.1) is similar in form to the large deviations approximation of Theorem 4.8.2 that follows. In this context we need the convex conjugate of a function.

Definition 4.8.1 (Convex conjugate or Legendre-Fenchel transform). *The convex conjugate or Legendre-Fenchel transform of the function $f : \mathbb{R} \to \mathbb{R} \cup \{-\infty, \infty\}$ is the function*

$$\check{f} : \mathbb{R} \to \mathbb{R} \cup \{-\infty, \infty\},$$
$$x \mapsto \sup_{v \in \mathbb{R}} vx - f(v).$$

A case of interest is obtained with the cumulant generating function K: the Legendre-Fenchel transform of K is given by

$$I(x) = \check{K}(x) = \sup_{v \in \mathbb{R}} vx - K(v) \in \mathbb{R}_+ \cup \{\infty\}, \quad \forall x \in \mathbb{R}. \tag{4.8.2}$$

Indeed $vx - K(v)|_{v=0} = 0$ implies $I(x) \geq 0$, $\forall x \geq 0$. The Legendre-Fenchel transform is also called rate function, in the context of large deviations theory, because it gives the rate of exponential decay of sequences of probabilities of rare events, according to Chernov's theorem 4.8.2 below. The cumulant generating function K is strictly convex over the interior of $\mathrm{dom}K$, unless it is the cumulant generating function of the Dirac distribution. One can indeed justify that, $\forall v \in (\mathrm{dom}K)^\circ$, $K''(v) > 0$, because $K''(v)$ has the interpretation of a variance, under the so-called exponentially tilted distribution. Assume that the supremum in the Legendre-Fenchel transform (4.8.2) is attained at point $v = \xi_x \in (\mathrm{dom}K)^\circ$. Strict convexity of K over $(\mathrm{dom}K)^\circ$ tells that ξ_x is unique. These considerations lead to

$$I(x) = \breve{K}(x) = \sup_{v \in \mathbb{R}} vx - K(v) = \xi_x x - K(\xi_x),$$

where ξ_x is the unique solution w.r.t. v of $K'(v) = x$. We can now give the first large deviations result.

Theorem 4.8.2 (Chernov). *Consider the previous definitions and assumptions. Then we have*

$$\mathsf{P}[M_n > y] = \mathrm{e}^{-n\{I(y)+\mathrm{o}(1)\}} = \mathrm{e}^{-nI(y)}\{1 + \mathrm{o}(1)\}, \quad \text{as} \quad n \to \infty. \quad (4.8.3)$$

If the supremum of (4.8.2) is attained at point $v = \xi_y \in (\mathrm{dom}K)^\circ$, then

$$I(y) = \xi_y y - K(\xi_y),$$

where ξ_y is he unique solution w.r.t. v of

$$K'(v) = y.$$

The point ξ_y of Theorem 4.8.2 is sometimes called saddlepoint, because it represents a saddlepoint of a surface defined over the complex plane, in the context of the saddlepoint approximation which is commented below.

Remarks 4.8.3 (Large and normal deviations approximations).

(1) *We see immediately that the Central limit theorem A.3.12 does not provide a valid alternative to Chernov's theorem 4.8.2. Denote $\mu = \mathsf{E}[X_1]$ and $\sigma = \sqrt{\mathsf{var}(X_1)}$ and let $x > \mu$, then we have*

$$\mathsf{P}\left[M_n \geq \mu + n^{-\frac{1}{2}}\sigma x\right] \overset{n \to \infty}{\longrightarrow} \Phi(x).$$

This approximation to the upper tail probabilities of M_n holds exclusively at points $y_n = \mu + n^{-1/2}\sigma x$, that approach μ at rate $n^{-1/2}$. All these points determine the normal deviations region of the mean. All points y of Chernov's theorem 4.8.2 do not depend on n and in this sense they lie in the large deviations region, which goes beyond the normal deviations region.

(2) *We also note that it is the relative error in Chernov's theorem 4.8.2 that vanishes asymptotically, whereas in the Central limit theorem it is only the absolute error that vanishes. This distinction is very important, because it makes the large deviations approximation substantially more reliable for approximating small and even very small tail probabilities, for example of order 10^{-6}.*

The theory of large deviations possesses other approximations than Chernov's theorem 4.8.2, that address different problems under different assumptions. An important one is the theorem of Gärtner and Ellis, which is presented below. Further important approximations are Varadhan's theorem for the approximation of integrals of exponential type, Sanov's theorem for the approximation of probabilities of empirical distributions and the contraction principle. A particular one is the saddlepoint approximation. Over the large deviations region, the saddlepoint approximation yields the relative error $O(n^{-1})$, which is smaller than relative error $o(1)$ of Chernov's theorem 4.8.2. It was introduced in statistics and probability by Daniels (1954) and since then, it has become a proper domain of research. Other important contributions are Lugannani and Rice (1980) and, in the context of stationary models, Daniels (1956), which gives the saddlepoint approximation to the distribution of the estimator of the AR(1) coefficient. The distribution of that estimator is also investigated by Phillips (1978). We refer to Gatto (2015) for a short introduction to the saddlepoint approximation, whereas two complete references are Chapter 4 of Barndorff-Nielsen and Cox (1989) and Field and Ronchetti (1990), for example.

The theorem of Gärtner and Ellis

An important large deviations approximation that generalizes Chernov's theorem is provided by Gärtner-Ellis theorem 4.8.8. It is now presented in detail and then applied to the mean of a sample from the AR(1). Consider the random variables $Y_1, Y_2, \ldots$ and their standardized cumulant generating function

$$L_n(v) = \frac{1}{n} \log \mathsf{E} \left[e^{v Y_n} \right] \in \mathbb{R} \cup \{\infty\}, \quad \forall v \in \mathbb{R},$$

for $n = 1, 2, \ldots$. Thus $n L_n$ is a cumulant generating function.

For example, consider for $n = 1, 2, \ldots$, the i.i.d. random variables $X_{n,1}, \ldots, X_{n,n}$ and $Y_n = \sum_{j=1}^{n} X_{n,j}$, then

$$L_n(v) = \log \mathsf{E} \left[e^{v X_{n,1}} \right] \in \mathbb{R} \cup \{\infty\}, \quad \forall v \in \mathbb{R},$$

for $n = 1, 2, \ldots$.

Define

$$L(v) = \lim_{n \to \infty} L_n(v) \in \mathbb{R} \cup \{\infty\}, \quad \forall v \in \mathbb{R}.$$

Let $u, v \in \mathrm{dom}L_n$, for $n = 1, 2, \ldots$, and also in $\mathrm{dom}L$. Let $\lambda \in [0, 1]$. Then strict convexity of L_n, for $n = 1, 2, \ldots$, leads to

$$L(\lambda u + (1 - \lambda)v) \leq \lambda L(v) + (1 - \lambda)L(u) < \infty.$$

Thus $\lambda u + (1 - \lambda)v \in \mathrm{dom}L$, meaning that $\mathrm{dom}L$ is a convex set. This and the above weak inequality means that L a convex function. Consequently, L is continuous over $(\mathrm{dom}L)^\circ$. Moreover, $L(0) = 0$ implies $0 \in \mathrm{dom}L$.

We need two new definitions.

Definition 4.8.4 (Steep function). *Let the real function f be differentiable over $(\mathrm{dom}f)^\circ$. Then f is steep if, for any sequence $\{x_n\}_{n \geq 1}$ with values in $(\mathrm{dom}f)^\circ$ that converges to a boundary point of $\mathrm{dom}f$, we have*

$$\lim_{n \to \infty} \left| \frac{\mathrm{d}}{\mathrm{d}x} f(x) \right| \Big|_{x = x_n} = \infty.$$

Examples 4.8.5. *(1) The cumulant generating function of the exponential distribution with density $\eta e^{-\eta x}$, $\forall x > 0$, where $\eta > 0$, is given by*

$$K(v) = \log \eta - \log(\eta - v), \quad \forall v < \eta.$$

The function K is steep at $v = \eta$.

(2) The inverse Gaussian density is given by

$$f(x) = \sqrt{\frac{\theta}{2\pi x^3}} \exp \left\{ -\frac{\theta}{2x} \left(\frac{x - \mu}{\mu} \right)^2 \right\}, \quad \forall x > 0,$$

where $\mu > 0$ is the expectation and $\theta > 0$. One can obtain the cumulant generating function

$$K(v) = \frac{\theta}{\mu} \left(1 - \sqrt{1 - 2\frac{\mu^2}{\theta}v} \right), \quad \forall v \leq \frac{1}{2}\frac{\theta}{\mu^2}.$$

The function K is steep at $v = \theta/(2\mu^2)$ because

$$K' \left(\frac{1}{2}\frac{\theta}{\mu^2} - \right) = \lim_{v \to \frac{1}{2}\frac{\theta}{\mu^2}, v < 0} \mu \left(1 - 2\frac{\mu^2}{\theta}v \right)^{-\frac{1}{2}} = \infty.$$

Definition 4.8.6 (Lower semicontinuity). *The real function f is lower semicontinuous at $x \in \mathrm{dom}f$ if, for any sequence $\{x_n\}_{n \geq 1}$ with values in $\mathrm{dom}f$ that converges to x, we have*

$$f(x) \leq \liminf_{n \to \infty} f(x_n).$$

Examples 4.8.7 (Lower semicontinuous functions).

(1) Consider the real and continuous function g and a ∈ domg. Then, the function

$$f(x) = \begin{cases} g(x), & \text{if } x \neq a, \\ g(x) - 1, & \text{if } x = a, \end{cases}$$

which has a removable discontinuity at a (see p. 166), is lower semi-continuous.

(2) The indicator function $f = \mathsf{I}_{(a,b)}$, for some $a < b \in \mathbb{R}$, which has two discontinuities of the first kind at a and b (see p. 166), is lower semicontinuous.

(3) The function

$$f(x) = \begin{cases} \sin \frac{1}{x}, & \text{if } x \neq 0, \\ -1, & \text{if } x = 0, \end{cases}$$

which has a discontinuity of the second kind at 0 (see p. 166), is lower semicontinuous.

Thus, in presence of a discontinuity of the first kind, the value at the point of jump can only be given to the lower part of the graph.

Theorem 4.8.8 (Gärtner-Ellis). *Consider the previous definitions and the rate function*

$$I(x) = \check{L}(x) = \sup_{v \in \mathbb{R}} vx - L(v) \in \mathbb{R} \cup \{\infty\}, \quad \forall x \in \mathbb{R}. \tag{4.8.4}$$

(1) If $C \subset \mathbb{R}$ is compact,[4] then

$$\limsup_{n \to \infty} \frac{1}{n} \log \mathsf{P}\left[\frac{Y_n}{n} \in C\right] \leq - \inf_{x \in C} I(x). \tag{4.8.5}$$

(2) Assume L lower semicontinuous. If $C \subset \mathbb{R}$ is closed, then (4.8.5) holds.

(3) Assume $(\mathrm{dom}L)^\circ \neq \emptyset$, L differentiable over $(\mathrm{dom}L)^\circ$ and steep. If $O \subset \mathbb{R}$ is open, then

$$\liminf_{n \to \infty} \frac{1}{n} \log \mathsf{P}\left[\frac{Y_n}{n} \in O\right] \geq - \inf_{x \in O} I(x). \tag{4.8.6}$$

---

[4]Compact is equivalent to closed and bounded, in $\mathbb{R}$.

Remarks 4.8.9.

(1) *The point x at which the infimum in (4.8.5) or in (4.8.6) is attained is called the dominating point of the set, because it determines the asymptotic probability of the entire set, namely of C or of O respectively.*

(2) *If the function L is strictly convex and satisfies the conditions of Theorem 4.8.8.3, then the rate function (4.8.4) at any $x \in \mathbb{R}$ can be re-expressed as*

$$I(x) = \xi_x x - L(\xi_x),$$

where ξ_x is the unique solution w.r.t. v of

$$L'(v) = x.$$

(3) *Let $B \in \mathcal{B}(\mathbb{R})$ such that*

$$\inf_{x \in B^\circ} I(x) = \inf_{x \in \bar{B}} I(x). \tag{4.8.7}$$

Then Theorem 4.8.8.2 and 4.8.8.3 give us

$$\lim_{n \to \infty} \frac{1}{n} \log \mathsf{P}\left[\frac{Y_n}{n} \in B\right] = \inf_{x \in B} I(x).$$

Any $B \in \mathcal{B}(\mathbb{R})$ such that such that (4.8.7) holds is called an I-continuity set.

(4) *Theorem 4.8.8.2 and 4.8.8.3 with $C = O = \mathbb{R}$ yield*

$$0 = \limsup_{n \to \infty} \frac{1}{n} \log 1 \leq -\inf_{x \in \mathbb{R}} I(x) \leq \liminf_{n \to \infty} \frac{1}{n} \log 1 = 0.$$

Thus,

$$\inf_{x \in \mathbb{R}} I(x) = 0.$$

4.8.2 Large deviations approximation for AR(1) time series

Let us remind the AR(1) equation $X_k = \varphi_1 X_{k-1} + Z_k, \forall k \in \mathbb{Z}$. We restrict to $|\varphi_1| < 1$ so that, according to Proposition 2.2.3, $\{X_k\}_{k \in \mathbb{Z}}$ becomes stationary. We restrict the time domain to $\mathbb{N}$ and take the fixed starting point $X_0 = 0$. Then $X_1 = Z_1$, $X_2 = \varphi_1 Z_1 + Z_2$ and in general

$$X_k = \varphi_1^{k-1} Z_1 + \cdots + \varphi_1 Z_{k-1} + Z_k = \sum_{j=1}^{k} \varphi_1^{k-j} Z_j, \quad \text{for } k = 1, 2, \ldots.$$

Let $n \in \mathbb{N}^*$ and consider

$$Y_n = \sum_{k=1}^{n} X_k = \sum_{k=1}^{n} \sum_{j=1}^{k} \varphi_1^{k-j} Z_j = \sum_{j=1}^{n} Z_j \sum_{l=0}^{n-j} \varphi_1^l = \sum_{j=1}^{n} Z_j \frac{\varphi_1^{n+1-j} - 1}{\varphi_1 - 1}.$$

Thus we have

$$L_n(v) = \frac{1}{n} \log \mathsf{E}\left[e^{vY_n}\right] = \frac{1}{n} \log \prod_{k=1}^{n} M_Z\left(\frac{1 - \varphi_1^k}{1 - \varphi_1} v\right)$$

$$= \frac{1}{n} \sum_{k=1}^{n} K_Z\left(\frac{1 - \varphi_1^k}{1 - \varphi_1} v\right) \xrightarrow{n\to\infty} K_Z\left(\frac{v}{1 - \varphi_1}\right),$$

$\forall v \in \mathbb{R}$ where the above functions exist, because convergence implies Cèsàro summability (cf. p. 8). So we define $L(v) = K_Z(v/(1 - \varphi_1))$, for $v \in \mathbb{R}$. Consequently, for the sample mean $M_n = Y_n/n$ and for given $y \in \mathbb{R}$, we have obtained that

$$\mathsf{P}\left[M_n > y\right] \quad \text{and} \quad \mathsf{P}\left[\frac{1}{1 - \varphi_1} \frac{1}{n} \sum_{k=1}^{n} Z_k > y\right],$$

are equivalent in logarithmic asymptotics, for large values of n. The two following examples consider the AR(1) with Gaussian and with double exponential WN.

Example 4.8.10 (AR(1) with Gaussian WN). *Assume that* $\{Z_k\}_{k\in\mathbb{Z}}$ *is Gaussian WN(1). Then*

$$K_Z(v) = \frac{v^2}{2} \quad \text{and} \quad L(v) = \frac{v^2}{2(1 - \varphi_1)^2}, \quad \forall v \in \mathbb{R}.$$

We also find $\xi(x) = (1 - \varphi_1)^2 x$ *and*

$$I(x) = \xi(x)x - L(\xi(x)) = \frac{1}{2}(1 - \varphi_1)^2 x^2, \quad \forall x \in \mathbb{R}.$$

This rate function I is continuous over $\mathbb{R}$. Thus any Borel set B is an I-continuity set and Gärtner-Ellis theorem 4.8.8 yields

$$\lim_{n\to\infty} \frac{1}{n} \log \mathsf{P}\left[\frac{Y_n}{n} \in B\right] = -\inf_{x \in B} I(x) = -\frac{1}{2}(1 - \varphi_1)^2 \inf_{x \in B} x^2.$$

Therefore, the large deviations approximation to the upper tail probability is obtained with $B = (y, \infty)$, for some $y > 0$, and it is given by

$$\mathsf{P}\left[M_n > y\right] = \exp\left\{-n\left[(1 - \varphi_1)^2 \frac{y^2}{2}\right]\right\} \{1 + o(1)\}, \quad \text{as } n \to \infty.$$

Example 4.8.11 (AR(1) with double exponential WN). *Assume that* $\{Z_k\}_{k\in\mathbb{Z}}$ *is double exponential or Laplace WN, with density*

$$\frac{1}{2}e^{-|x|}, \quad \forall x \in \mathbb{R}.$$

This alternative light-tailed distribution has slower decay than the normal. Thus

$$K_Z(v) = -\log(1 - v^2), \quad \forall v \in (-1, 1),$$

and

$$L(v) = K_Z\left(\frac{v}{1 - \varphi_1}\right) = -\log\left\{1 - \left(\frac{v}{1 - \varphi_1}\right)^2\right\}, \quad \forall v \in (-(1-\varphi_1), 1-\varphi_1).$$

Let $x \in \mathbb{R}^*$, *then by solving* $K_Z'(v) = 2v/(1 - v^2) = x$ *w.r.t.* v *one finds*

$$v = -\frac{1}{x} \pm \sqrt{1 + \frac{1}{x^2}}.$$

The restriction to $(-1, 1)\backslash\{0\}$ *gives*

$$v = -\frac{1}{x} + \sqrt{1 + \frac{1}{x^2}}, \quad \text{if } x > 0, \quad \text{and } v = -\frac{1}{x} - \sqrt{1 + \frac{1}{x^2}}, \quad \text{if } x < 0.$$

Therefore, by solving $L'(v) = x$ *w.r.t.* v, *restricted to* $(-(1-\varphi_1), 1-\varphi_1)\backslash\{0\}$, *one finds*

$$\xi(x) = -\frac{1}{x} + (1 - \varphi_1)\sqrt{1 + \frac{1}{(1 - \varphi_1)^2 x^2}}, \quad \forall x > 0,$$

and

$$\xi(x) = -\frac{1}{x} - (1 - \varphi_1)\sqrt{1 + \frac{1}{(1 - \varphi_1)^2 x^2}}, \quad \forall x < 0.$$

Thus we obtain the rate function

$$\begin{aligned}
I(x) &= \xi(x)x - L(\xi(x)) \\
&= \text{sgn}x\sqrt{1 + (1 - \varphi_1)^2 x^2} - 1 \\
&\quad + \log\left\{1 - \left(-\frac{1}{(1 - \varphi_1)x} + \text{sgn}x\sqrt{1 + \frac{1}{(1 - \varphi_1)^2 x^2}}\right)^2\right\} \\
&= \text{sgn}x\sqrt{1 + (1 - \varphi_1)^2 x^2} - 1 \\
&\quad + \log\left\{\frac{2}{(1 - \varphi_1)^2 x^2}\left(\text{sgn}x\sqrt{1 + (1 - \varphi_1)^2 x^2} - 1\right)\right\}, \quad \forall x \in \mathbb{R}^*.
\end{aligned}$$

This rate function I is continuous over $\mathbb{R}^*$ and therefore any set $B \in \mathcal{B}(\mathbb{R}^*)$ is an I-continuity set. Thus Gärtner-Ellis theorem 4.8.8 yields

$$\lim_{n \to \infty} \frac{1}{n} \log \mathsf{P}\left[\frac{Y_n}{n} \in B\right] = -\inf_{x \in B} I(x).$$

Let us give the large deviations approximation to the upper tail probability obtained by $B = (y, \infty)$, for some $y > 0$. Because the rate function I is convex with minimum at zero, cf. Remark 4.8.9, we obtain that

$$\inf_{x \in B} I(x) = I(y).$$

Thus, the large deviations approximation to the upper tail probability of $M_n = Y_n/n$ is given by

$$\mathsf{P}[M_n > y] = \exp\left\{-n\left[I(y) + \mathrm{o}(1)\right]\right\}$$

$$= \exp\left\{-n\left[\sqrt{1 + (1 - \varphi_1)^2 y^2} - 1 + \log\right.\right.$$

$$\left.\left.\left\{\frac{2}{(1 - \varphi_1)^2 y^2}\left(\sqrt{1 + (1 - \varphi_1)^2 y^2} - 1\right)\right\} + \mathrm{o}(1)\right]\right\}$$

$$= \exp\left\{-n\left[\sqrt{1 + (1 - \varphi_1)^2 y^2} - 1\right]\right\}$$

$$\cdot \left(\frac{(1 - \varphi_1)^2 y^2}{2\left(\sqrt{1 + (1 - \varphi_1)^2 y^2} - 1\right)}\right)^n \{1 + \mathrm{o}(1)\}, \quad \text{as } n \to \infty.$$

$$(4.8.8)$$

Approximation (4.8.8) is evaluated with various values of φ_1 and n in Figure 4.6. The upper graph shows these approximate upper tail probabilities with the coefficient $\varphi_1 = 0.5$ and with the sample sizes $n = 5$, 10, 20, represented respectively by dotted, dashed and solid lines. The lower graph shows the approximations with the sample size $n = 10$ and with the coefficients $\varphi_1 = 0.4$, 0.5, 0.6, drawn respectively by solid, dashed and dotted lines.

4.9   Information theoretic results for time series

This section presents some results on information theory for stationary time series. It is based on Gatto (2022), where proofs and more details can be obtained. Section 4.9.1 introduces an important circular distribution in this context together with quantities of information theory, adapted to circular distributions. Section 4.9.2 presents the important theorems that lead to the optimal time series, according to information theoretic principles.

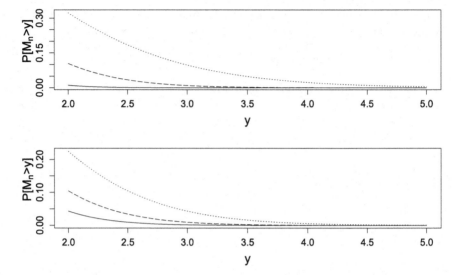

Fig. 4.6: Large deviations approximation to upper tail probability of the mean of AR(1) time series under double exponential WN, $P\,[M_n > y]$, with $\varphi_1 = 0.5$ and $n = 5$ (dotted line), $n = 10$ (dashed line) and $n = 20$ (solid line), in the upper graph, and with $n = 10$ and $\varphi_1 = 0.4$ (solid line), $\varphi_1 = 0.5$ (dashed line) and $\varphi_1 = 0.6$ (dotted line), in the lower graph.

4.9.1   *Circular distributions and information theory*

The spectral distribution that corresponds to the most uncertain or unpredictable time series with some values of the a.c.v.f. fixed, is the generalized von Mises (GvM) spectral distribution given in Definition 4.9.1. It is thus a maximum entropy spectral distribution and the corresponding stationary time series is called the generalized von Mises time series. The Gaussian-GvM time series is the stationary time series that maximizes entropies in frequency and time domains, respectively referred to as spectral and temporal entropies.

Let $\{X_j\}_{j\in\mathbb{Z}}$ be a complex-valued time series whose elements belong to $\mathcal{L}_2$ and which is (weakly) stationary with a.c.v.f. γ. We denote $\mu = E[X_j]$, $\sigma^2 = \gamma(0)$, for some $\sigma \in (0, \infty)$, F_σ the spectral d.f. and f_σ the spectral density, whose existence is assumed. The spectral distribution of a complex-valued time series can be viewed as rescaled circular probability distribution, namely a probability distribution over the circle. With a real-valued time series, the spectral distribution is a rescaled axially symmetric circular distribution. A class of circular distributions that possesses various theoretical properties has densities is the following.

Definition 4.9.1 (GvM distribution). *For $k = 1, 2, \ldots$, we define the generalized von Mises density of order k (GvM$_k$) by*

$$f_1^{(k)} (\alpha \mid \mu_1, \ldots, \mu_k, \kappa_1, \ldots, \kappa_k)$$

$$= \frac{1}{2\pi G_0^{(k)} (\delta_1, \ldots, \delta_{k-1}, \kappa_1, \ldots, \kappa_k)} \exp \left\{ \sum_{j=1}^{k} \kappa_j \cos j(\alpha - \mu_j) \right\}, \quad (4.9.1)$$

$\forall \alpha \in (-\pi, \pi]$ *(or any other interval of length 2π), where $\mu_j \in (-\pi/j, \pi/j]$, $\kappa_j \geq 0$, for $j = 1, \ldots, k$. The normalizing constant is*

$$G_0^{(k)}(\delta_1, \ldots, \delta_{k-1}, \kappa_1, \ldots, \kappa_k)$$

$$= \frac{1}{2\pi} \int_0^{2\pi} \exp\{\kappa_1 \cos \alpha + \kappa_2 \cos 2(\alpha + \delta_1) + \cdots + \kappa_k \cos k(\alpha + \delta_{k-1})\} d\alpha,$$

where $\delta_j = (\mu_1 - \mu_{j+1}) \mathrm{mod}(2\pi/(j+1))$, for $j = 1, \ldots, k-1$, whenever $k \geq 2$, and for $k = 1$ it is $G_0^{(1)}(\kappa_1) = I_0(\kappa_1)$, where $I_n(z) = (2\pi)^{-1} \int_0^{2\pi} \cos n\alpha \exp\{z \cos \alpha\} d\alpha$, $\forall z \in \mathbb{C}$, is the modified Bessel function I of integer order n; see e.g. Abramowitz and Stegum (1972), p. 376.

Let us denote a circular random variable θ following the density (4.9.1) as

$$\theta \sim \mathrm{GvM}_k(\mu_1, \ldots, \mu_k, \kappa_1, \ldots, \kappa_k).$$

The circular density (4.9.1) for $k \geq 2$ is thoroughly analyzed by Gatto and Jammalamadaka (2007) and by Gatto (2009). The GvM$_1$ density is the vM density that appears also in the context of circular time series in Section 4.2.5. It is given by

$$f_1^{(1)}(\alpha \mid \mu_1, \kappa_1) = \frac{1}{2\pi I_0(\kappa_1)} \exp\{\kappa_1 \cos(\alpha - \mu_1)\}, \quad \forall \alpha \in (-\pi, \pi],$$

where $\mu_1 \in (-\pi, \pi]$ and $\kappa_1 \geq 0$. Figure 4.5 shows three vM densities centered at $\mu_1 = 0$ and with concentration $\kappa_1 = 1$ by solid line, $\kappa_1 = 3$ by dashed line and $\kappa_1 = 10$ by dotted line. Compared to the vM, which is axially symmetric and unimodal whenever $\kappa_1 > 0$, the GvM$_2$ distribution allows for higher adjustability, in particular in terms of asymmetry and bimodality. We refer to Figure 4.7 for the graphs of three particular GvM$_2$ densities. The parameters $\mu_1 = 0$ and $\mu_2 = \pi/4$ are common to the three

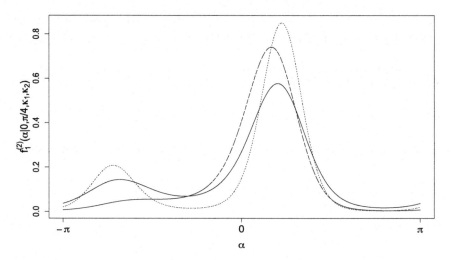

Fig. 4.7: $\mathrm{GvM}_2(0, \pi/4, \kappa_1, \kappa_2)$ density with $\kappa_1 = \kappa_2 = 1$ (solid line), $\kappa_1 = 2$, $\kappa_2 = 1$ (dashed line) and $\kappa_1 = 1$, $\kappa_2 = 2$ (dotted line).

densities. Then, the density with solid line has $\kappa_1 = \kappa_2 = 1$, the density with dashed line has $\kappa_1 = 2$, $\kappa_2 = 1$ and the density with dotted line has $\kappa_1 = 1$, $\kappa_2 = 2$. Thus, the GvM_2 distribution is a practical circular distribution that has found various applications: in meteorology, cf. Zhang et al. (2018), in oceanography, cf. Lin and Dong (2019), in offshore engineering, cf. Astfalck et al. (2018), in signal processing, cf. Christmas (2014), and in machine learning, cf. Christmas (2021).

The GvM_k spectral density is given by $f_\sigma^{(k)} = \sigma^2 f_1^{(k)}$, for some $\sigma \in (0, \infty)$: it is the GvM_k circular density $f_1^{(k)}$ given in (4.9.1) that is rescaled to have any desired total mass σ^2. When the GvM_k spectral density is axially symmetric around the null axis, then the corresponding time series $\{X_j\}_{j \in \mathbb{Z}}$ is real-valued. The GvM_2 density with $\kappa_1, \kappa_2 > 0$ is axially symmetric iff $\delta_1 = 0$ or $\delta_1 = \pi/2$, cf. Salvador and Gatto (2021). In both cases, the axis of symmetry has angle μ_1 w.r.t. the null direction. The GvM_2 spectral density has a practical role time series because of its uni- and bimodal shape. It is worth mentioning that the GvM spectral distribution has similarities with the exponential model of Bloomfield (1973), which is a truncated Fourier series of the logarithm of some spectral distribution. Bloomfield motivates the low truncation of the Fourier series by the fact that "the logarithm of an estimated spectral density function is often found to be a fairly well-behaved function". A closely related reference is Bogert et al. (1963). However, Bloomfield's model is given for real-valued time series only.

The estimation of the spectral distribution is an important problem in the analysis of stationary time series. Information theoretic quantities like Kullback-Leibler's information, cf. Kullback and Leibler (1951), or Shannon's entropy, cf. Shannon (1948), are very useful in this context. These quantities are defined for probability distributions but they can be considered for distributions with finite mass. These are spectral distributions and we assume them absolutely continuous. Thus, let f_σ and g_σ be two spectral densities whose integrals over $(-\pi, \pi]$ are both equal to σ^2.

Definition 4.9.2 (Spectral Kullback-Leibler information). *The spectral Kullback-Leibler information of f_σ w.r.t. g_σ is given by*

$$I(f_\sigma|g_\sigma) = \int_{-\pi}^{\pi} \log \frac{f_\sigma(\alpha)}{g_\sigma(\alpha)} f_\sigma(\alpha) d\alpha = \sigma^2 I(f_1|g_1), \qquad (4.9.2)$$

where $0 \log 0 = 0$ is assumed and where the support of f_σ is included in the support of g_σ, otherwise $I(f_\sigma|g_\sigma) = \infty$.

The quantity $I(f_1|g_1)$ is the mean logarithmic likelihood ratio or mean information of a realization from f_1 for discriminating in favor of f_1 against g_1. Gibbs inequality tells that

$$I(f_\sigma|g_\sigma) \geq 0,$$

for all spectral densities f_σ and g_σ, with equality iff $f_1 = g_1$ a.e. The Kullback-Leibler information is also called relative entropy, Kullback-Leibler divergence or distance, eventhough it is not a metric: it violates the symmetry and the triangle rules. Thus (4.9.2) is a measure of divergence for distributions with same total mass σ^2.

Shannon's entropy can be also extended to spectral distributions.

Definition 4.9.3 (Spectral Shannon's entropy). *Shannon's entropy can be defined for the spectral density f_σ by*

$$S(f_\sigma) = -\int_{-\pi}^{\pi} \log \frac{f_\sigma(\alpha)}{(2\pi)^{-1}\sigma^2} f_\sigma(\alpha) d\alpha = -I(f_\sigma|u_\sigma) = -\sigma^2 I(f_1|u_1),$$

$$(4.9.3)$$

where u_σ is the uniform density with total mass σ^2 over $(-\pi, \pi]$, i.e. $u_\sigma = \sigma^2/(2\pi)I_{(-\pi,\pi]}$, for some $\sigma > 0$, I_A denoting the indicator of set A.

Shannon's entropy of the circular density f_1 over $(-\pi, \pi]$ is originally defined as $-\int_{-\pi}^{\pi} \log f_1(\alpha) f_1(\alpha) d\alpha = -(2\pi)^{-1} - I(f_1 | u_1)$. It measures the uncertainty inherent in the probability distribution with density f_1. Equivalently, $S(f_1)$ measures the expected amount of information gained on obtaining an observation from f_1, based on the principle that the rarer an event, the more informative its occurrence. The spectral entropy defined in (4.9.3) slightly differs the original formula of Shannon's entropy for probability distributions: inside the logarithm, f_σ is divided by the uniform density with total mass σ^2. With this modification the spectral entropy becomes scale invariant w.r.t. σ^2, just like the spectral Kullback-Leibler information (4.9.2). The spectral entropy satisfies $S(f_\sigma) \leq 0$, with equality iff $f_\sigma = u_\sigma$ a.e. This follows from Gibbs inequality.

Section 4.9.2 provides information theoretic results for spectral distributions and introduces the related GvM and the Gaussian-GvM time series. After reminding some general concepts, the optimal spectral distributions under constraints on the a.c.v.f. are provided. The GvM spectral distribution maximizes Shannon's entropy under constraints on the first few values of the a.c.v.f. Then, the Gaussian-GvM time series is motivated from the fact that it follows the maximal entropy principle in both time and frequency domains. The final part of this section concerns computational aspects. Some series expansions for integral functions appearing in the context of the GvM$_2$ time series are given and an estimator for the parameters of the GvM spectral distribution is presented.

4.9.2   *The GvM and the Gaussian-GvM time series*

We can deduce from Herglotz theorem 2.4.10 and from the characterization of the a.c.v.f. in terms of n.n.d. function given by Theorem 2.4.6 that if we consider the spectral d.f. $F_\sigma^{(k)} = \sigma^2 F_1^{(k)}$, where $F_1^{(k)}$ is the GvM$_k$ d.f. with density $f_1^{(k)}$ given by (4.9.1), then there exists a stationary time series $\{X_j\}_{j \in \mathbb{Z}}$ with spectral d.f. $F_\sigma^{(k)}$ and density $f_\sigma^{(k)} = \sigma f_1^{(k)}$ that we call GvM or, more precisely, GvM$_k$ time series. Thus the GvM$_k$ time series is stationary by definition, it has variance $\sigma^2 = F_\sigma^{(k)}(\pi)$ and it is generally complex-valued, unless the GvM$_k$ spectral distribution is axially symmetric around the null direction.

The complex-valued GvM$_k$ stationary time series $\{X_j\}_{j \in \mathbb{Z}}$ can be chosen with mean zero, variance σ^2 and Gaussian, meaning that the double f.d.d.

$$(U_{j_1}, \ldots, U_{j_n}, V_{j_1}, \ldots, V_{j_n}), \tag{4.9.4}$$

$\forall j_1 < \cdots < j_n \in \mathbb{Z}$, $n \geq 1$, where $U_j = \operatorname{Re} X_j$ and $V_j = \operatorname{Im} X_j$, $\forall j \in \mathbb{Z}$, are Gaussian. In this case, the distribution of $\{X_j\}_{j \in \mathbb{Z}}$ is however not entirely determined by its a.c.v.f. $\gamma^{(k)}$ or, alternatively, by its spectral d.f. $F_\sigma^{(k)}$. (The formula for the a.c.v.f. is given later in Corollary 4.9.5.4.) In order to entirely determine this distribution, one also needs the pseudo-covariance $\mathsf{E}[X_{j+r}X_j]$, $\forall j, r \in \mathbb{Z}$. So an arbitrary Gaussian, with mean zero and (weakly) stationary time series $\{X_j\}_{j \in \mathbb{Z}}$ is not necessarily strictly stationary: $\{X_j\}_{j \in \mathbb{Z}}$ is strictly stationary iff the covariance $\mathsf{E}\left[X_{j+r}\overline{X_j}\right]$ and the pseudo-covariance $\mathsf{E}[X_{j+r}X_j]$ do not depend on $j \in \mathbb{Z}$, $\forall r \in \mathbb{Z}$. Precise justifications are given in Theorem 3.2.5 and its proof. This is indeed equivalent to the independence on $j \in \mathbb{Z}$ of

$$\psi_{UU}(r) = \mathsf{E}[U_{j+r}U_j], \ \psi_{VV}(r) = \mathsf{E}[V_{j+r}V_j],$$

$$\psi_{UV}(r) = \mathsf{E}[U_{j+r}V_j] \text{ and } \psi_{VU}(r) = \mathsf{E}[V_{j+r}U_j], \quad \forall r \in \mathbb{Z}, \qquad (4.9.5)$$

where $U_j = \operatorname{Re} X_j$ and $V_j = \operatorname{Im} X_j$, $\forall j \in \mathbb{Z}$. Note that under this independence on $j \in \mathbb{Z}$, we have $\psi_{VU}(r) = \psi_{UV}(-r)$, $\forall r \in \mathbb{Z}$. However, according Herglotz theorem 2.4.10, if the a.c.v.f. $\gamma^{(k)}$ is obtained by Fourier inversion of the GvM$_k$ spectral density, then it is n.n.d. By the above characterization of the a.c.v.f., a strictly stationary GvM$_k$ time series always exists. The existence of a particular (precisely radially symmetric) strictly stationary Gaussian-GvM$_k$ time series that satisfies some constraints on the a.c.v.f. is shown later.

Next, for any given Gaussian-GvM$_k$ time series with spectral d.f. $F_\sigma^{(k)}$, there exists a spectral process $\{Z_\alpha\}_{\alpha \in [-\pi, \pi]}$ that is complex-valued and Gaussian. We remind from Section 2.4.1 that the process that modulates the harmonics, $\{Z_\alpha\}_{\alpha \in [-\pi, \pi]}$, is defined through the mean square stochastic integral

$$X_j = \int_{(-\pi, \pi]} e^{i\alpha j} dZ_\alpha, \quad \text{a.s.}, \ \forall j \in \mathbb{Z}, \qquad (4.9.6)$$

and by the following conditions: $\mathsf{E}[Z_\alpha] = 0$, $\forall \alpha \in [-\pi, \pi]$, $\mathsf{E}\left[(Z_{\alpha_2} - Z_{\alpha_1})\overline{(Z_{\alpha_4} - Z_{\alpha_3})}\right] = 0$, $\forall -\pi \leq \alpha_1 < \alpha_2 < \alpha_3 < \alpha_4 \leq \pi$, i.e. it has orthogonal increments, and

$$\mathsf{E}\left[|Z_{\alpha_2} - Z_{\alpha_1}|^2\right] = F_\sigma^{(k)}(\alpha_2) - F_\sigma^{(k)}(\alpha_1), \ \forall -\pi \leq \alpha_1 < \alpha_2 \leq \pi. \quad (4.9.7)$$

There are several reasons for considering the Gaussian-GvM time series. A practical one is that their simulation can be done with the algorithms presented in Section 4.7. In particular, the algorithm of Section 4.7.3 makes use of the decomposition (4.9.6). A theoretical reason for considering normality is that it leads to a second maximal entropy principle, this one no longer in frequency domain but in time domain. This topic is presented later.

Let g_σ be the spectral density of some stationary time series with variance σ^2, for some $\sigma \in (0, \infty)$. For a chosen $k \in \{1, 2, \ldots\}$, consider the a.c.v.f. conditions or constraints

$$\mathcal{C}_k : \int_{-\pi}^{\pi} e^{ir\alpha} g_\sigma(\alpha) d\alpha = \psi_r, \text{ for } r = 1, \ldots, k, \quad (4.9.8)$$

for some $\psi_1, \ldots, \psi_k \in \mathbb{C}$ satisfying $|\psi_r| \leq \sigma^2$, for $r = 1, \ldots, k$, and such that the $(k+1) \times (k+1)$ matrix

$$\begin{pmatrix} \sigma^2 & \psi_1 & \cdots & \psi_k \\ \overline{\psi_1} & \sigma^2 & \cdots & \psi_{k-1} \\ \vdots & \vdots & \ddots & \vdots \\ \overline{\psi_k} & \overline{\psi_{k-1}} & \cdots & \sigma^2 \end{pmatrix}, \quad (4.9.9)$$

is n.n.d. One can re-express these conditions as

$$\mathcal{C}_k : \int_{-\pi}^{\pi} \cos r\alpha\, g_\sigma(\alpha) d\alpha = \nu_r \text{ and } \int_{-\pi}^{\pi} \sin r\alpha\, g_\sigma(\alpha) d\alpha = \xi_r,$$

$$\text{for } r = 1, \ldots, k, \quad (4.9.10)$$

where $\nu_r = \operatorname{Re} \psi_r$ and $\xi_r = \operatorname{Im} \psi_r$, giving thus $\sqrt{\nu_r^2 + \xi_r^2} \leq \sigma^2$, for $r = 1, \ldots, k$, and with n.n.d. matrix (4.9.9).

One encounters the two following practical problems. In an applied field where a specific spectral density h_σ is traditionally used, one may search for the spectral density g_σ that satisfies $\mathcal{C}_k$ and that is the closest to the traditional density h_σ. Alternatively, the spectral density g_σ is unknown but the values of $\psi_1, \ldots, \psi_k$ are available, either because they constitute a priori knowledge about the time series or because they are obtained from a sample of the stationary time series. In this second case, the values of $\psi_1, \ldots, \psi_k$ can be obtained by taking them equal to the corresponding values of the empirical or sample a.c.v.f. For the sample $X_1, \ldots, X_n$ of the time series, the sample a.c.v.f. is given by

$$\hat{\gamma}_n(r) = \frac{1}{n} \sum_{j=1}^{n-r} (X_{j+r} - M_n)\overline{(X_j - M_n)} \text{ and } \hat{\gamma}_n(-r) = \overline{\hat{\gamma}_n(r)},$$

$$\text{for } r = 0, \ldots, n-1, \quad (4.9.11)$$

where $M_n = n^{-1} \sum_{j=1}^{n} X_j$. Thus we set $\psi_r = \hat{\gamma}_n(r)$, for $r = 1, \ldots, k$ and for $k \leq n - 1$. Note that the matrix (4.9.9) is n.n.d. in this case.

Note also that the sample a.c.v.f. is a biased estimator of the true a.c.v.f. (but asymptotically unbiased).

Theorem 4.9.4 below addresses the first of these two problems and it is the central part of this article. The second problem is addressed by Corollary 4.9.5. The following definitions are required. For $k = 1, 2, \ldots$ and for an arbitrary circular density g_1, define the following integral functions:

$$G_r^{(k)}(\delta_1, \ldots, \delta_{k-1}, \kappa_1, \ldots, \kappa_k; g_1)$$
$$= \int_0^{2\pi} \cos r\alpha \cdot \exp\{\kappa_1 \cos\alpha + \kappa_2 \cos 2(\alpha + \delta_1)$$
$$+ \cdots + \kappa_k \cos k(\alpha + \delta_{k-1})\} g_1(\alpha) d\alpha,$$

$$H_r^{(k)}(\delta_1, \ldots, \delta_{k-1}, \kappa_1, \ldots, \kappa_k; g_1)$$
$$= \int_0^{2\pi} \sin r\alpha \cdot \exp\{\kappa_1 \cos\alpha + \kappa_2 \cos 2(\alpha + \delta_1)$$
$$+ \cdots + \kappa_k \cos k(\alpha + \delta_{k-1})\} g_1(\alpha) d\alpha,$$

$$A_r^{(k)}(\delta_1, \ldots, \delta_{k-1}, \kappa_1, \ldots, \kappa_k; g_1) = \frac{G_r^{(k)}(\delta_1, \ldots, \delta_{k-1}, \kappa_1, \ldots, \kappa_k; g_1)}{G_0^{(k)}(\delta_1, \ldots, \delta_{k-1}, \kappa_1, \ldots, \kappa_k; g_1)},$$

and

$$B_r^{(k)}(\delta_1, \ldots, \delta_{k-1}, \kappa_1, \ldots, \kappa_k; g_1) = \frac{H_r^{(k)}(\delta_1, \ldots, \delta_{k-1}, \kappa_1, \ldots, \kappa_k; g_1)}{G_0^{(k)}(\delta_1, \ldots, \delta_{k-1}, \kappa_1, \ldots, \kappa_k; g_1)},$$

for $r = 1, \ldots, k$, where $\delta_j = (\mu_1 - \mu_{j+1}) \mod (2\pi/(j+1))$, for $j = 1, \ldots, k-1$ and $\kappa_1, \ldots, \kappa_k \geq 0$. For these constants we make the conventions that the arguments $\delta_1, \ldots, \delta_{k-1}$ vanish when $k = 1$ and that the argument g_1 is omitted when equal to the circular uniform density u_1. For example, $G_0^{(1)}(\kappa_1) = (2\pi)^{-1} \int_0^{2\pi} e^{\kappa_1 \cos\alpha} d\alpha = I_0(\kappa_1)$. Define the matrix of counterclockwise rotation of angle α as

$$\boldsymbol{R}(\alpha) = \begin{pmatrix} \cos\alpha & -\sin\alpha \\ \sin\alpha & \cos\alpha \end{pmatrix}. \tag{4.9.12}$$

Theorem 4.9.4 (Kullback-Leibler closest spectral distribution). *Let $\sigma \in (0, \infty)$ and let g_σ and h_σ be two spectral densities with total mass σ^2.*

(1) The spectral density g_σ that satisfies $\mathcal{C}_k$, given in and that is the closest to another spectral density h_σ, in the sense of minimizing the Kullback-Leibler information $I(g_\sigma | h_\sigma)$, is the exponential tilt of h_σ that takes the form

$$g_\sigma(\alpha)$$
$$= \frac{1}{G_0^{(k)}(\delta_1, \ldots, \delta_{k-1}, \kappa_1, \ldots, \kappa_k; h_1)} \exp\left\{\sum_{j=1}^k \kappa_j \cos j(\alpha - \mu_j)\right\} h_\sigma(\alpha),$$

$$\tag{4.9.13}$$

$\forall \alpha \in (-\pi, \pi]$, where $\delta_j = (\mu_1 - \mu_{j+1}) \mathrm{mod}(2\pi/(j+1))$, for $j = 1, \ldots, k-1$, $\mu_j \in (-\pi/j, \pi/j]$ and $\kappa_j \geq 0$, for $j = 1, \ldots, k$. The values of these parameters are the solutions of

$$\begin{pmatrix} \nu_r \\ \xi_r \end{pmatrix} = \sigma^2 \boldsymbol{R}(r\mu_1) \begin{pmatrix} A_r^{(k)}(\delta_1, \ldots, \delta_{k-1}, \kappa_1, \ldots, \kappa_k; h_1) \\ B_r^{(k)}(\delta_1, \ldots, \delta_{k-1}, \kappa_1, \ldots, \kappa_k; h_1) \end{pmatrix}, \quad (4.9.14)$$

where $\boldsymbol{R}(r\mu_1)$ denotes the rotation matrix (4.9.12) at $\alpha = r\mu_1$, for $r = 1, \ldots, k$, and where $\nu_1, \ldots, \nu_k$ and $\xi_1, \ldots, \xi_k$ are given by (2.4.14).
(2) For any spectral density g_σ that satisfies $\mathcal{C}_k$, the minimal Kullback-Leibler information of g_σ w.r.t. h_σ is given by

$$- \sigma^2 \log G_0^{(k)}(\delta_1, \ldots, \delta_{k-1}, \kappa_1, \ldots, \kappa_k; h_1)$$

$$+ \sum_{r=1}^k \kappa_r (\nu_r \cos r\mu_r + \xi_r \sin r\mu_r) \leq I(g_\sigma | h_\sigma), \quad (4.9.15)$$

with equality iff g_σ is a.e. given by (4.9.13), where the values of the parameters $\mu_j \in (-\pi/j, \pi/j]$ and $\kappa_j \geq 0$, for $j = 1, \ldots, k$, are solutions of (4.9.14).

Theorem 4.9.4 is a rather direct consequence or generalization of Theorem 2.1 of Gatto (2009), in which the trigonometric moments are replaced by the a.c.v.f. and the circular distribution is replaced by the spectral distribution. Indeed, along with the generalization of the circular distribution to the spectral distribution, the a.c.v.f. of a stationary time series generalizes the trigonometric moment. Precisely, the r-th trigonometric moment of the circular random variable θ with density g_1 is given by

$$\varphi_r = \gamma_r + \mathrm{i}\sigma_r = \mathsf{E}\left[\mathrm{e}^{\mathrm{i}r\theta} \right] = \int_{-\pi}^{\pi} \mathrm{e}^{\mathrm{i}r\alpha} g_1(\alpha) \mathrm{d}\alpha, \quad (4.9.16)$$

for some $\gamma_r, \sigma_r \in \mathbb{R}$ and $\forall r \in \mathbb{Z}$, whereas the a.c.v.f. of the stationary time series with the spectral density $g_\sigma = \sigma^2 g_1$ is given by

$$\gamma(r) = \sigma^2 \varphi_r = \sigma^2 (\gamma_r + \mathrm{i}\sigma_r), \quad \forall r \in \mathbb{Z}. \quad (4.9.17)$$

Clearly, $\gamma(0) = \sigma^2$ and $|\gamma(r)| \leq \gamma(0)$, $\forall r \in \mathbb{Z}$. The claim that (4.9.17) is indeed the a.c.v.f. of a stationary time series is rigorously justified by the above mentioned Herglotz theorem and characterization of the a.c.v.f.

The existence and the unicity of the solution to (4.9.14), i.e. of the parameter values satisfying $\mathcal{C}_k$, can be justified by the fact (4.9.14) can be reparametrized in terms of the saddlepoint equation (or exponential tilting equation) given by (14) of Gatto (2009). This is the saddlepoint equation of a distribution with bounded domain. In this case, the solution, called saddlepoint, exists and it is unique. These facts are well know in the theory of large deviations.

In the context of the justification of Theorem 4.9.4.1, we can note that an equivalent expression to (4.9.14) is given by

$$
\psi_r = \sigma^2 e^{ir\mu_1} \cdot \Big\{ A_r^{(k)}(\delta_1, \ldots, \delta_{k-1}, \kappa_1, \ldots, \kappa_k; h_1)
$$
$$
+ i B_r^{(k)}(\delta_1, \ldots, \delta_{k-1}, \kappa_1, \ldots, \kappa_k; h_1) \Big\},
$$

for $r = 1, \ldots, k$, which can be seen equivalent to $\mathcal{C}_k$.

When analyzing a time series with periodic components, leading for example to certain monthly or weekly constraints, then the set of k constraints $\mathcal{C}_k$ may no longer be appropriate. Instead of it, one may still need the constraints in the form given in (4.9.8) but exclusively for r limited to some subset of $\{1, \ldots, k\}$, which is possibly different than $\{1, \ldots, j\}$, $\forall j \in \{1, \ldots, k\}$. Theorem 4.9.4 can be easily generalized to this situation. For simplicity, assume that only the l-th constraint must be removed from $\mathcal{C}_k$, for some $l \in \{1, \ldots, k-1\}$, and thus assume $k \geq 2$. Then Theorem 4.9.4 has to be adapted by setting $\kappa_l = 0$ in (4.9.13) and by removing the equation (4.9.14) whenever $r = l$. In addition, if $l = 1$, then μ_1 appearing in δ_j given just after (4.9.13) and appearing in (4.9.14) must be replaced by μ_m, with m arbitrary selected in $\{2, \ldots, k\}$. Similar adaptations could be considered for the next results of this chapter, essentially to Corollary 4.9.5 and to Theorem 4.9.9.

A major consequence of Theorem 4.9.4 is that the GvM$_k$ spectral distribution is a maximum entropy distribution. This fact and related results are given in Corollary 4.9.5.

Corollary 4.9.5 (Maximal Shannon's spectral entropy distribution and GvM a.c.v.f.). *Let $\sigma \in (0, \infty)$ and g_σ a spectral density with total mass σ^2.*

(1) *The spectral density g_σ that maximizes Shannon's entropy $S(g_\sigma)$ under $\mathcal{C}_k$, given in (4.9.8), is the GvM$_{\sigma,k}(\mu_1, \ldots, \mu_k, \kappa_1, \ldots, \kappa_k)$ density, i.e. $f_\sigma^{(k)}(\cdot \mid \mu_1, \ldots, \mu_k, \kappa_1, \ldots, \kappa_k)$, where $\mu_j \in (-\pi/j, \pi/j]$ and $\kappa_j \geq 0$, for $j = 1, \ldots, k$. The values of these parameters are determined by (4.9.14).*

(2) *If g_σ is a spectral density satisfying $\mathcal{C}_k$, then its entropy is bounded from above as follows,*

$$S(g_\sigma) \leq$$

$$\sigma^2 \log G_0^{(k)}(\delta_1,\ldots,\delta_{k-1},\kappa_1,\ldots,\kappa_k) - \sum_{r=1}^{k} \kappa_r (\nu_r \cos r\mu_r + \xi_r \sin r\mu_r),$$

with equality iff $g_\sigma = f_\sigma^{(k)}(\cdot|\mu_1,\ldots,\mu_k,\kappa_1,\ldots,\kappa_k)$ *a.e. The values of the parameters are determined by (4.9.14) with* $h_1 = u_1$, *i.e. the circular uniform density, where* $\nu_1,\ldots,\nu_k$ *and* $\xi_1,\ldots,\xi_k$ *are given by (2.4.14).*

(3) *The entropy of the* $\text{GvM}_{\sigma,k}(\mu_1,\ldots,\mu_k,\kappa_1,\ldots,\kappa_k)$ *spectral density is given by*

$$\begin{aligned}
S\left(f_\sigma^{(k)}\right) = \sigma^2 \Bigg\{ &\log G_0^{(k)}(\delta_1,\ldots,\delta_{k-1},\kappa_1,\ldots,\kappa_k) \\
&- \kappa_1 A_1^{(k)}(\delta_1,\ldots,\delta_{k-1},\kappa_1,\ldots,\kappa_k) \\
&- \sum_{r=2}^{k} \kappa_r \left[A_r^{(k)}(\delta_1,\ldots,\delta_{k-1},\kappa_1,\ldots,\kappa_k) \cos r\delta_{r-1} \right. \\
&\left. - B_r^{(k)}(\delta_1,\ldots,\delta_{k-1},\kappa_1,\ldots,\kappa_k) \sin r\delta_{r-1} \right] \Bigg\},
\end{aligned}$$

where $\sum_{r=2}^{k}$ *vanishes whenever* $k < 2$.

(4) *The a.c.v.f.* $\gamma_\sigma^{(k)}$ *of the* $\text{GvM}_{\sigma,k}(\mu_1,\ldots,\mu_k,\kappa_1,\ldots,\kappa_k)$ *spectral distribution can be obtained by*

$$\begin{pmatrix} \text{Re}\,\gamma_\sigma^{(k)}(r) \\ \text{Im}\,\gamma_\sigma^{(k)}(r) \end{pmatrix} = \sigma^2 \boldsymbol{R}(r\mu_1) \begin{pmatrix} A_r^{(k)}(\delta_1,\ldots,\delta_{k-1},\kappa_1,\ldots,\kappa_k) \\ B_r^{(k)}(\delta_1,\ldots,\delta_{k-1},\kappa_1,\ldots,\kappa_k) \end{pmatrix},$$

and $\gamma_\sigma^{(k)}(-r) = \overline{\gamma_\sigma^{(k)}(r)}$, *for* $r = 1, 2, \ldots$.

Corollary 4.9.5 can be obtained from Theorem 4.9.4 as follows. Theorem 4.9.4.1 and the relation between Kullback-Leibler information and entropy (4.9.3) tell that the GvM_k spectral distribution maximizes the entropy, under the given constraints on the a.c.v.f. The upper bound for the entropy of a spectral distribution satisfying the given constraints is provided by Theorem 4.9.4.2. Thus, by considering $h_1 = u_1$ in Theorem 4.9.4, we obtain Corollary 4.9.5.1 and part 4.9.5.2. Corollary 4.9.5.3 is a consequence of part 4.9.5.2. It is obtained by replacing ν_r and ξ_r, for $r = 1,\ldots,k$, that appear in the upper bound of the entropy, by expressions depending on the parameters of the GvM_k distribution, through the identity (4.9.14).

In the practice, when partial prior information in the form of $\mathcal{C}_k$ is available and it is desired to determine the most noninformative spectral distribution that satisfies the known prior information, then the GvM$_k$ spectral distribution is the optimal choice. It is in fact the most credible distribution or the one that nature would have generated, when some prior information and only that information would be available. Maximal entropy distributions are important in many contexts: in statistical mechanics, the choice of a maximum entropy distribution subject to constraints is a classical approach referred to as the maximum entropy principle. One can find various studies on spectral distributions with maximal entropy. It is later explained that the AR(k) model maximizes an alternative entropy among all stationary time series satisfying $\mathcal{C}_k$. Franke (1985) showed that ARMA time series maximizes that entropy among all stationary time series satisfying these same constraints and additional constraints on the impulse responses. Other references on spectral distributions with maximal entropy are given in Gatto (2022).

The simplest example is the following.

Example 4.9.6 (vM spectral distribution). *Corollary 4.9.5.3 with $k = 1$ yields the entropy of the vM spectral distribution,*

$$S\left(f_\sigma^{(1)}\right) = \sigma^2 \left\{ \log G_0^{(1)}(\kappa_1) - \kappa_1 A_1^{(1)}(\kappa_1) \right\} = \sigma^2 \left\{ \log I_0(\kappa_1) - \kappa_1 \frac{I_1(\kappa_1)}{I_0(\kappa_1)} \right\},$$

for $\kappa_1 \geq 0$. By noting that $B_r^{(1)}(\kappa_1) = 0$, for $r = 1, 2, \ldots$, Corollary 4.9.5.4 with $k = 1$ gives the a.c.v.f. of the vM spectral distribution as

$$\begin{pmatrix} \operatorname{Re} \gamma_\sigma^{(1)}(r) \\ \operatorname{Im} \gamma_\sigma^{(1)}(r) \end{pmatrix} = \sigma^2 A_r^{(1)}(\kappa_1) \begin{pmatrix} \cos r\mu_1 \\ \sin r\mu_1 \end{pmatrix} = \sigma^2 \frac{I_r(\kappa_1)}{I_0(\kappa_1)} \begin{pmatrix} \cos r\mu_1 \\ \sin r\mu_1 \end{pmatrix}, \quad (4.9.18)$$

and $\gamma_\sigma^{(1)}(-r) = \overline{\gamma_\sigma^{(1)}(r)}$, for $r = 1, 2, \ldots$. When $\kappa_1 > 0$, the vM spectral distribution is axially symmetric about the origin iff $\mu_1 = 0$.

We now provide a strictly stationary Gaussian-GvM time series that follows the maximal entropy principle in the time domain, in addition to the maximal entropy principle in frequency domain, under common constraints on the a.c.v.f. Consider the complex-valued Gaussian time series $\{X_j\}_{j \in \mathbb{Z}}$ in $\mathcal{L}_2$ that is strictly stationary with mean zero. This time series is introduced at the end of Section 2.1. Define $U_j = \operatorname{Re} X_j$ and $V_j = \operatorname{Im} X_j$, $\forall j \in \mathbb{Z}$. Let $n \geq 1$ and $j_1 < \cdots < j_n \in \mathbb{Z}$. Consider the random vector $(U_{j_1}, \ldots, U_{j_n}, V_{j_1}, \ldots, V_{j_n})$ and denote by $p_{j_1, \ldots, j_n}$ its joint density.

Thus $p_{j_1,\ldots,j_n}$ is the $2n$-dimensional normal density with mean zero and $2n \times 2n$ covariance matrix

$$\boldsymbol{\Sigma}_{j_1,\ldots,j_n} = \mathsf{var}\left((U_{j_1},\ldots,U_{j_n},V_{j_1},\ldots,V_{j_n})\right) = \mathsf{E}\left[\begin{pmatrix} \boldsymbol{U}\boldsymbol{U}^\top & \boldsymbol{U}\boldsymbol{V}^\top \\ \boldsymbol{V}\boldsymbol{U}^\top & \boldsymbol{V}\boldsymbol{V}^\top \end{pmatrix}\right],$$

(4.9.19)

where $\boldsymbol{U} = (U_{j_1},\ldots,U_{j_n})^\top$ and $\boldsymbol{V} = (V_{j_1},\ldots,V_{j_n})^\top$. According to (4.9.5), the elements of $\boldsymbol{\Sigma}_{j_1,\ldots,j_n}$ are given by

$$\mathsf{E}[U_{j_l}U_{j_m}] = \psi_{UU}(j_l - j_m),\ \ \mathsf{E}[V_{j_l}V_{j_m}] = \psi_{VV}(j_l - j_m),$$

$$\mathsf{E}[U_{j_l}V_{j_m}] = \psi_{UV}(j_l - j_m)\ \ \text{and}\ \ \mathsf{E}[V_{j_l}U_{j_m}] = \psi_{VU}(j_l - j_m),$$

with $\psi_{VU}(j_l - j_m) = \psi_{UV}(j_m - j_l)$, for $l, m = 1,\ldots,n$. Because $\boldsymbol{\Sigma}_{j_1,\ldots,j_n}$ depends on $j_1,\ldots,j_n$ only through $l_1 = j_2 - j_1,\ldots,l_{n-1} = j_n - j_{n-1}$, we consider the alternative notation $\boldsymbol{\Sigma}^{l_1,\ldots,l_{n-1}} = \boldsymbol{\Sigma}_{j_1,\ldots,j_n}$.

An important subclass of complex-valued normal or Gaussian random vectors is given by the radially symmetric ones, which is obtained by setting the mean and the pseudo-covariance matrix equal to zero. That is, the complex Gaussian random vector $\boldsymbol{X} = (X_{j_1},\ldots,X_{j_n})^\top$, where $X_{j_l} = U_{j_l} + iV_{j_l}$, for $l = 1,\ldots,n$, is radially symmetric iff $\mathsf{E}[\boldsymbol{X}] = \boldsymbol{0}$ and $\mathsf{E}\left[\boldsymbol{X}\boldsymbol{X}^\top\right] = \boldsymbol{0}$. A radially symmetric complex-valued random vector $\boldsymbol{X}$ is characterized by the fact that,

$$\forall \alpha \in (-\pi, \pi],\ \mathrm{e}^{i\alpha}\boldsymbol{X} \sim \boldsymbol{X}.$$

Because these vectors and the related processes are often used in signal processing, we consider them in this section.

Generally, without assuming neither stationarity nor normality, one defines the temporal entropy of a complex-valued time series as follows.

Definition 4.9.7 (Shannon's temporal entropy). *Let $\{X_j\}_{j\in\mathbb{Z}}$ be a complex-valued time series. Then its Shannon's temporal entropy at times $j_1 < \cdots < j_n \in \mathbb{Z}$ is given by Shannon's entropy of $(U_{j_1},\ldots,U_{j_n}, V_{j_1},\ldots,V_{j_n})$, precisely by*

$$T_{j_1,\ldots,j_n} = -\int_{-\infty}^{\infty} \cdots \int_{-\infty}^{\infty} \log p_{j_1,\ldots,j_n}(u_1,\ldots,u_n,v_1,\ldots,v_n)$$

$$p_{j_1,\ldots,j_n}(u_1,\ldots,u_n,v_1,\ldots,v_n)\mathrm{d}u_1\cdots\mathrm{d}u_n\mathrm{d}v_1\cdots\mathrm{d}v_n, \quad (4.9.20)$$

where $U_j = \mathrm{Re}\, X_j$, $V_j = \mathrm{Im}\, X_j$, $\forall j \in \mathbb{Z}$, and whenever the joint density $p_{j_1,\ldots,j_n}$ of $(U_{j_1},\ldots,U_{j_n}, V_{j_1},\ldots,V_{j_n})$ exists.

Under strict stationarity, the temporal entropy (4.9.20) becomes invariant under time shift and we can thus define the alternative notation

$$T^{l_1,\ldots,l_{n-1}} = T_{j_1,\ldots,j_n}.$$

Theorem 4.9.8 mentions two known and important information theoretic results for the Gaussian distribution.

Theorem 4.9.8 (Maximum Shannon's entropy of Gaussian distribution). *1. If $p_{j_1,\ldots,j_n}$ denotes the $2n$-dimensional Gaussian density with arbitrary mean and covariance matrix $\Sigma_{j_1,\ldots,j_n}$, then Shannon's temporal entropy (4.9.20) is given by*

$$T_{j_1,\ldots,j_n} = \{1 + \log(2\pi)\}\, n + \frac{1}{2} \log \det \Sigma_{j_1,\ldots,j_n}. \qquad (4.9.21)$$

2. Among all random vectors $(U_{j_1}, \ldots, U_{j_n}, V_{j_1}, \ldots, V_{j_n})$ having arbitrary density with fixed covariance matrix $\Sigma_{j_1,\ldots,j_n}$, the one that is normally distributed maximizes Shannon's entropy (4.9.20). The maximum of Shannon's entropy is given by (4.9.21).

We now consider the constraints on the a.c.v.f. (4.9.8) and search for the (strictly) stationary time series, with mean and pseudo-covariances null, that maximizes the temporal entropy.

Theorem 4.9.9 (Maximal Shannon's temporal entropy distribution). *Consider the class of complex-valued and stationary time series $\{X_j\}_{j\in\mathbb{Z}}$ with mean null, variance σ^2, for some $\sigma \in (0,\infty)$, and pseudo-covariances null. Denote by γ the a.c.v.f. of $\{X_j\}_{j\in\mathbb{Z}}$, $\nu = \mathrm{Re}\,\gamma$ and $\xi = \mathrm{Im}\,\gamma$.*

(1) If the a.c.v.f. γ satisfies $\mathcal{C}_k$ given in (4.9.8) or in (2.4.14), thus $\gamma(1) = \psi_1 = \nu_1 + \mathrm{i}\,\xi_1, \ldots, \gamma(k) = \psi_k = \nu_k + \mathrm{i}\,\xi_k$, then the time series $\{X_j\}_{j\in\mathbb{Z}}$ in the above class that maximizes Shannon's temporal entropy (4.9.20) with $n = k+1$ and $j_1 = 1, \ldots, j_{k+1} = k+1$ is the one for which the corresponding double f.d.d. (3.9.19) with $j_1 = 1, \ldots, j_{k+1} = k+1$ is Gaussian, with mean zero and with $2(k+1) \times 2(k+1)$ covariance matrix $\Sigma(k) = \Sigma^{1,\ldots,1}$ given by (4.9.19) with

$$\psi_{UU}(r) = \psi_{VV}(r) = \frac{\nu_r}{2} \quad \text{and} \quad \psi_{UV}(r) = \psi_{VU}(-r) = -\frac{\xi_r}{2},$$

for $r = 1, \ldots, k$.

(2) The corresponding value of the temporal entropy is given by

$$T(k) = \{1 + \log(2\pi)\}\,(1+k) + \frac{1}{2} \log \det \Sigma(k).$$

The proof of Theorem 4.9.9 is given in Gatto (2022). So when $\{X_j\}_{j\in\mathbb{Z}}$ is the strictly stationary Gaussian-GvM time series, both spectral and temporal Shannon's entropies are maximized under the constraints $\mathcal{C}_k$.

Regarding computational aspects, we can give the following comments. First, some series expansions for some of the constants $G_r^{(k)}$, for $r = 0,\ldots,k$, $H_r^{(k)}$, for $r = 1,\ldots,k$, and also for the GvM_k spectral d.f. are given in Gatto (2022). All these cases are integrals over bounded domain of smooth integrands and therefore numerical integration should be a simple alternative to these expansions.

We now study the estimation problem: we review some classical results of spectral estimation and then present an estimator to the parameters of the GvM spectral distribution.

A classical estimator of the spectral density is the periodogram, introduced in Section 2.4.5. It is given in Definition 2.4.39. Proposition 2.4.40 provides its alternative representation as discrete Fourier transform of the sample a.c.v.f., namely

$$\Lambda_n(j) = \sum_{r=-(n-1)}^{n-1} \hat{\gamma}_n(r)\mathrm{e}^{-\mathrm{i}\frac{2\pi jr}{n}},$$

for $j = \lfloor(n-1)/2\rfloor,\ldots,-1,1,\ldots,\lfloor n/2\rfloor$, $\hat{\gamma}_n$ being the sample a.c.v.f. (4.9.11), n the sample size and $\lfloor\cdot\rfloor$ the floor function. Because of its non-parametric nature, the periodogram is well-suited for detecting particular features, such as periodicities, that may not be identified by a parametric estimator. However, its irregular nature may not be desirable in some contexts and it does not result from an important optimality criterion.

One of the earliest studies on maximum entropy spectral distributions is due to Burg (1967), who considered $B(f_\sigma) = \int_{-\pi}^{\pi} \log f_\sigma(\alpha)\mathrm{d}\alpha$ as measure of entropy of the spectral density f_σ. This entropy is different than the present adaptation of Shannon's entropy, i.e. $S(f_\sigma)$ given in (4.9.3), but we can relate Burg's entropy to Kullback-Leibler information by

$$B(f_\sigma) = 2\pi\left\{\log\frac{\sigma^2}{2\pi} - \frac{1}{\sigma^2}I(u_\sigma|f_\sigma)\right\} = 2\pi\left\{\log\frac{\sigma^2}{2\pi} - I(u_1|f_1)\right\}. \quad (4.9.22)$$

Thus, maximizing Burg's entropy amounts to minimize the re-directed Kullback-Leibler information, instead of the usual Shannon's entropy. For real-valued time series, it turns out that the spectral density estimator that maximizes the entropy (4.9.22) subject to the constraints $\mathcal{C}_k$ in (4.9.8), with $\psi_r = \hat{\gamma}_n(r)$, for $r = 1,\ldots,n-1$ and $k = n-1$, is equal to the autoregressive estimator of order k. This autoregressive estimator is given by the formula of the spectral density of the $\mathrm{AR}(k)$ model that has been fitted to the sample of n consecutive values of the time series. For more details refer e.g. to pp. 365–366 of Brockwell and Davis (1991).

Estimators of the parameters of the GvM_k spectral distribution can be obtained from the generalization of the trigonometric method of moments estimator for the GvM_k circular distribution, which is introduced by Gatto (2008). This estimator is the circular version of the method of moments estimators. Consider the GvM_k spectral distribution with unknown parameters $\mu_1, \ldots, \mu_k$ and $\kappa_1, \ldots, \kappa_k$, for some $k \leq n - 1$. Consider the a.c.v.f. conditions $\mathcal{C}_k$ in which the spectral density g_σ is taken equal to the GvM_k spectral density with variance σ^2, viz. σ^2 times the circular density (4.9.1), and in which the quantity $\psi_r \in \mathbb{C}$ is replaced by the sample a.c.v.f. at r, namely by $\hat{\gamma}_n(r)$, for $r = 1, \ldots, k$, cf. (4.9.11). The resulting r-th equation can be re-expressed in a similar way to (4.9.14), precisely as

$$\begin{pmatrix} \operatorname{Re} \hat{\gamma}_n(r) \\ \operatorname{Im} \hat{\gamma}_n(r) \end{pmatrix} = \sigma^2 \boldsymbol{R}(r\mu_1) \begin{pmatrix} A_r^{(k)}(\delta_1, \ldots, \delta_{k-1}, \kappa_1, \ldots, \kappa_k) \\ B_r^{(k)}(\delta_1, \ldots, \delta_{k-1}, \kappa_1, \ldots, \kappa_k) \end{pmatrix}, \quad (4.9.23)$$

for $r = 1, \ldots, k$. This gives a system of $2k$ real equations and $2k$ unknown real parameter. The values of $\mu_1, \delta_1, \ldots, \delta_{k-1}, \kappa_1, \ldots, \kappa_k$ that solve this system of equations are the resulting estimators and they can be denoted $\hat{\mu}_1, \hat{\delta}_1, \ldots, \hat{\delta}_{k-1}, \hat{\kappa}_1, \ldots, \hat{\kappa}_k$. We now give two examples.

Example 4.9.10 (vM spectral distribution). *When $k = 1$ we have the vM spectral distribution. Because $B_1^{(1)}(\kappa_1) = 0$, we obtain the two estimating equations*

$$\begin{pmatrix} \operatorname{Re} \hat{\gamma}_n(1) \\ \operatorname{Im} \hat{\gamma}_n(1) \end{pmatrix} = \sigma^2 A_1^{(1)}(\kappa_1) \begin{pmatrix} \cos \mu_1 \\ \sin \mu_1 \end{pmatrix} = \sigma^2 \frac{I_1(\kappa_1)}{I_0(\kappa_1)} \begin{pmatrix} \cos \mu_1 \\ \sin \mu_1 \end{pmatrix},$$

with the two unknown parameters μ_1 and κ_1. The solutions are the estimators $\hat{\mu}_1$ and $\hat{\kappa}_1$. For $\kappa_1 > 0$, if $\mu_1 = 0$ is given, then we have axial symmetry about the null axis, thus real-valued time-series, and the two estimating equations reduce to the single equation

$$\frac{\hat{\gamma}_n(1)}{\sigma^2} = A_1^{(1)}(\kappa_1), \quad (4.9.24)$$

whose solution is the estimator $\hat{\kappa}_1$. It is shown in Amos (1974) that $A_1^{(1)}$ has positive derivative over $(0, \infty)$. It follows that $A_1^{(1)}$ is a strictly increasing and differentiable probability d.f. over $[0, \infty)$. So its inverse function is easily computed numerically.

Example 4.9.11 (GvM$_2$ spectral distribution). *When $k = 2$ we have the GvM$_2$ spectral distribution. The estimating equations (4.9.23) can be solved with the expansions of the constants given in Gatto (2022). As previously mentioned, with $\kappa_1, \kappa_2 > 0$, the GvM$_2$ distribution is axially symmetric around the axis μ_1 iff $\delta_1 = \delta^{(1)} = 0$ or $\delta_1 = \delta^{(2)} = \pi/2$. With these values of δ_1 and with $\mu_1 = 0$, the axial symmetry is about the null axis and the time series becomes real-valued. We note that $B_r^{(2)}\left(\delta^{(j)}, \kappa_1, \kappa_2\right) = 0$, for $r = 1, 2$ and for the cases $j = 1, 2$. These equalities and $\operatorname{Im}\hat{\gamma}_n(r) = 0$, for $r = 1, 2$, allow to simplify (4.9.23) to*

$$\frac{\hat{\gamma}_n(r)}{\sigma^2} = A_r^{(2)}\left(\delta^{(j)}, \kappa_1, \kappa_2\right),$$

for $r = 1, 2$ and for the two cases $j = 1, 2$. These estimating equations generalize the estimation equation (4.9.24) of the real-valued vM time series. For each one of these two cases, we have two equations and two unknown values, namely κ_1 and κ_2, giving the estimators $\hat{\kappa}_1$ and $\hat{\kappa}_2$.

We conclude this section on estimation by mentioning the test that the spectral density is a GvM one, precisely the test of the null hypothesis

$$H_0 : f_\sigma = f_\sigma^{(k)}\left(\cdot \mid \mu_1, \ldots, \mu_k, \kappa_1, \ldots, \kappa_k\right),$$

where all parameters are specified. In Anderson (1993), this problem is addressed with the Cramér-von Mises and with the Kolmogorov-Smirnov criteria. Both criteria are based on

$$\sqrt{n}\{\hat{F}_n - F_\sigma^{(k)}(\cdot \mid \mu_1, \ldots, \mu_k, \kappa_1, \ldots, \kappa_k)\}, \tag{4.9.25}$$

where n is the sample size, $\hat{F}_n$ is the estimator of the spectral d.f. obtained by integration of the periodogram and where $F_\sigma^{(k)}(\cdot \mid \mu_1, \ldots, \mu_k, \kappa_1, \ldots, \kappa_k)$ is the d.f. of $f_\sigma^{(k)}(\cdot \mid \mu_1, \ldots, \mu_k, \kappa_1, \ldots, \kappa_k)$. Under the main assumption that the time series admits the AR(∞) representation, the asymptotic distribution of (4.9.25) is obtained not only for the GvM but for any specified spectral density. Note that Anderson (1993) considers real-valued time series only, but the results can be directly adapted to complex-valued time series.

Appendix A

Mathematical complements

A.1 Hilbert space, $\mathcal{L}_2$ and $\mathcal{L}_p$

This section briefly introduces the Hilbert space, the important case $\mathcal{L}_2$ and the $\mathcal{L}_p$ space, with $p > 0$. Related results on orthogonal projection and conditional expectation are also summarized.

A.1.1 *Hilbert space*

A Hilbert space allows for infinite dimensional analogues of $\mathbb{R}^n$ or $\mathbb{C}^n$, in the sense that many properties of $\mathbb{R}^n$ or $\mathbb{C}^n$ are retrieved in the Hilbert space, in particular those of Euclidean geometry. Generalized versions of Pythagora's theorem and parallelogram law hold in the Hilbert space. Thus the spaces $\mathbb{R}^n$ or $\mathbb{C}^n$ are two simple instances, but important Hilbert spaces are spaces of functions, e.g. of random variables. The orthogonal projection onto a sub-Hilbert space has a central role in the study of stationary models.

Definition A.1.1 (Banach and Hilbert spaces). *A Banach space is a normed vector space $\mathcal{B}$ such that any Cauchy sequence in $\mathcal{B}$ converges to an element of $\mathcal{B}$, i.e. that is complete.*

A Hilbert space $\mathcal{H}$ is a vector space equipped with a scalar product $\langle \cdot, \cdot \rangle$ that makes $\mathcal{H}$ a Banach space w.r.t. the norm or the seminorm[1] $\| \cdot \| = \langle \cdot, \cdot \rangle^{1/2}$.

---

[1] A seminorm satisfies $\|x\| \geq 0$, $\|cx\| = |c| \|x\|$ and $\|x + y\| \leq \|x\| + \|y\|$. Here are x and y elements of a vector space and c is in $\mathbb{R}$ or $\mathbb{C}$. A norm can be obtained by adding $\|x\| = 0 \Rightarrow x = 0$. (The converse is always true: $\|0\| = \|0 \cdot 0\| = 0\|0\| = 0$.) Thus $d(x, y) = \|x - y\|$ is a pseudometric, viz. a metric for which $d(x, y) = 0$ is possible for some $x \neq y$. This pseudometric defines the equivalence classes that are obtained by the equivalence relation $x \sim y \Leftrightarrow d(x, y) = 0$.

The scalar product is continuous w.r.t. both of its arguments: this is shown for the Hilbert space $\mathcal{L}_2$ in Lemma A.1.5 and the general proof would be similar.

Let $\{x_n\}_{n \in \mathbb{N}^*}$ be a sequence in the Hilbert space $\mathcal{H}$ and $x \in \mathcal{H}$. Then

$$\|x_n - x\| \overset{n \to \infty}{\longrightarrow} 0 \iff \|x_n - x_m\| \overset{m,n \to \infty}{\longrightarrow} 0.$$

The part $(\Rightarrow)$ follows from $\|x_n - x_m\| \leq \|x_n - x\| + \|x_m - x\|$. The part $(\Leftarrow)$ is the completeness of $\mathcal{H}$.

Let $\mathcal{H}$ be an Hilbert space and let X be any subset of $\mathcal{H}$. The set $X^\perp$ is the orthogonal part of X and it is defined as follows,

$$X^\perp = \{y \in \mathcal{H} \mid \langle x, y \rangle = 0, \ \forall x \in X\}.$$

Let X be a nonempty and complete subspace of $\mathcal{H}$. It is a subspace in the sense that it is closed under addition of its elements and under multiplication of its elements by any complex number. Note that X complete is equivalent to X closed, whenever X is a subset of complete metric space, such as the Hilbert space $\mathcal{H}$.

The operator pro_X is the orthogonal projection from $\mathcal{H}$ onto its subspace X. It is defined by

$$\mathsf{pro}_X(y) \in X, \ \forall y \in \mathcal{H}, \ \text{and} \ \langle y - \mathsf{pro}_X(y), x \rangle = 0, \ \forall x \in X, \ y \in \mathcal{H}.$$

The difference $y - \mathsf{pro}_X y$ is called the residual of the projection of y onto X.

One can show that:

- the orthogonal projection always exists;
- the orthogonal projection is unique or unique up to elements of its equivalence class, in the case of the seminorm.

Three important properties of the orthogonal projection are the following.

Proposition A.1.2 (Properties of orthogonal projection). *Let $X \subset Y$ be any subspaces of the Hilbert space $\mathcal{H}$, let $x, x_1, x_2, \ldots \in \mathcal{H}$ and let $c_1, c_2 \in \mathbb{C}$. The following properties hold:*

(1) $\mathsf{pro}_X(c_1 x_1 + c_2 x_2) = c_1 \mathsf{pro}_X(x_1) + c_2 \mathsf{pro}_X(x_2)$, *i.e. the linearity;*
(2) $\mathsf{pro}_X(\mathsf{pro}_X(x)) = \mathsf{pro}_X(x)$, *i.e. the idempotence;*
(3) $\mathsf{pro}_X(\mathsf{pro}_Y(x)) = \mathsf{pro}_X(x)$, *i.e. the iterativity in subspaces;*
(4) $\|x_n - x\| \overset{n \to \infty}{\longrightarrow} 0 \implies \mathsf{pro}_X(x_n) \overset{n \to \infty}{\longrightarrow} \mathsf{pro}_X(x)$, *i.e. the continuity.*

Let X and Y be two orthogonal subspaces of the Hilbert space $\mathcal{H}$, in the sense that $Y \subset X^{\perp}$. The Cartesian product $X \times Y$ becomes a new space, by defining the operations componentwise: $(x_1, y_1) + (x_2, y_2) = (x_1 + x_2, x_2 + y_2)$ and $c(x, y) = (cx, cy)$, $\forall x, x_1, x_2 \in X, y, y_1, y_2 \in Y$ and $c \in \mathbb{C}$. This space is called direct sum of X and Y and it is denoted $X \oplus Y$.

Definition A.1.3 (Isometric operator and isometry). *Let $\mathcal{H}_1$ and $\mathcal{H}_2$ be two Hilbert spaces with respective scalar products $\langle \cdot, \cdot \rangle_1$ and $\langle \cdot, \cdot \rangle_2$. The operator $A : \mathcal{H}_1 \to \mathcal{H}_2$ is called isometric, i.e. is an isometry, if*

$$\langle Au, Av \rangle_2 = \langle u, v \rangle_1, \quad \forall u, v \in \mathcal{H}_1,$$

i.e. if the scalar product is preserved.

Two Hilbert spaces $\mathcal{H}_1$ and $\mathcal{H}_2$ are called isometric if there exists an isometry from $\mathcal{H}_1$ to $\mathcal{H}_2$.

Note that the isometric operator A is necessarily a linear operator. It is thus continuous and bijective. To see this, let $u', u'', v \in \mathcal{H}_1$ and $z', z'' \in \mathbb{C}$. Then we find

$$\begin{aligned}
\langle A(z'u' + z''u''), Av \rangle_2 &= \langle z'u' + z''u'', v \rangle_1 \\
&= z'\langle u', v \rangle_1 + z''\langle u'', v \rangle_1 \\
&= z'\langle Au', Av \rangle_2 + z''\langle Au'', Av \rangle_2 \\
&= \langle z'Au' + z''Au'', av \rangle_2.
\end{aligned}$$

A.1.2 Space $\mathcal{L}_2$

The theory of stationary processes takes place over the Hilbert space $\mathcal{L}_2$ of random variables, namely the space of square-integrable random variables.

Definition A.1.4 (Hilbert space $\mathcal{L}_2$). *Let $(\Omega, \mathcal{F}, \mathsf{P})$ be a probability space and define the space of random variables $X : \Omega \to \mathbb{C}$ such that $\mathsf{E}[|X|^2] < \infty$, denoted $\mathcal{L}_2(\Omega, \mathcal{F}, \mathsf{P})$ or shortly $\mathcal{L}_2$. The following properties hold.*

(1) $\mathcal{L}_2$ is a vector space w.r.t. the seminorm $\|X\| = \mathsf{E}^{1/2}[|X|^2]$.
(2) $\forall X, Y \in \mathcal{L}_2$, $\langle X, Y \rangle = \mathsf{E}\left[X\overline{Y}\right]$ is a scalar product.
(3) Each Cauchy sequence of $\mathcal{L}_2$ converges towards an element of $\mathcal{L}_2$, w.r.t. the seminorm $\|X\| = \langle X, X \rangle^{1/2}$ or w.r.t. the pseudometric $d(X, Y) = \|X - Y\|$, $\forall X, Y \in \mathcal{L}_2$, making $\mathcal{L}_2$ a complete space.

The three above properties make $\mathcal{L}_2$ an Hilbert space.

Let $X, Y \in \mathcal{L}_2$, then $\|X\| = 0$ iff $X = 0$ a.s. Also, $\|X - Y\| = 0$ iff $X = Y$ a.s.

Lemma A.1.5 (Continuity of scalar product in $\mathcal{L}_2$). *Let* $X, X_1,$ $X_2, \ldots, Y, Y_1, Y_2, \ldots$ *be complex-valued random variables in* $\mathcal{L}_2$. *If* $X_n \overset{\mathcal{L}_2}{\to} X$ *and* $Y_n \overset{\mathcal{L}_2}{\to} Y$, *then*

$$\mathsf{E}[X_n] \overset{n \to \infty}{\longrightarrow} \mathsf{E}[X], \tag{A.1.1}$$

$$\|X_n\| \overset{n \to \infty}{\longrightarrow} \|X\| \quad \text{and} \tag{A.1.2}$$

$$\langle X_n, Y_n \rangle \overset{n \to \infty}{\longrightarrow} \langle X, Y \rangle. \tag{A.1.3}$$

Proof. Let $n \geq 1$, then it follows from Minkowski inequality A.2.2 that

$$\big| \|X_n\| - \|X\| \big| \leq \|X_n - X\|,$$

which yields (A.1.2), as $n \to \infty$.

In order to show (A.1.3), consider Cauchy-Schwarz inequality A.2.1 and (A.1.2). Thus

$$
\begin{aligned}
|\langle X_n, Y_n \rangle - \langle X, Y \rangle| &= |\langle X_n, Y_n - Y \rangle + \langle X_n - X, Y \rangle| \\
&\leq |\langle X_n, Y_n - Y \rangle| + |\langle X_n - X, Y \rangle| \\
&\leq \underbrace{\|X_n\|}_{\to \|X\| < \infty} \underbrace{\|Y_n - Y\|}_{\to 0} + \underbrace{\|X_n - X\|}_{\to 0} \underbrace{\|Y\|}_{< \infty} \\
&\longrightarrow 0, \quad \text{as } n \to \infty.
\end{aligned}
$$

By inserting $Y = Y_1 = Y_2 = \cdots = 1$ in (A.1.3) we obtain (A.1.1). $\square$

Lemma A.1.6 (Loève criterion). *Let* $X_1, X_2, \ldots$ *be complex-valued random variables in* $\mathcal{L}_2$. *Then the following equivalence holds,*

$$X_n \overset{\mathcal{L}_2}{\longrightarrow} X, \text{ for some } X \in \mathcal{L}_2 \Longleftrightarrow \lim_{m,n \to \infty} \langle X_m, X_n \rangle = c, \text{ for some } c \in \mathbb{C}.$$

Proof. ($\Leftarrow$) $\|X_m - X_n\|^2 = \langle X_m - X_n, X_m - X_n \rangle = \|X_m\|^2 + \|X_n\|^2 - \langle X_n, X_m \rangle - \langle X_m, X_n \rangle \overset{m,n \to \infty}{\longrightarrow} c + c - c - c = 0.$
($\Rightarrow$) By using the continuity of the scalar product, we obtain $\langle X_m, X_n \rangle \overset{m,n \to \infty}{\longrightarrow} \langle X, X \rangle \in \mathbb{C}.$ $\square$

Definition A.1.7 (Closed span in $\mathcal{L}_2$). *Given a set of random variables* $\{X_t\}_{t \in T}$ *of* $\mathcal{L}_2$, *for some* $T \subset \mathbb{R}$, *we denote by*

$$\overline{\mathrm{sp}}\{X_t\}_{t \in T},$$

the smallest closed subspace of $\mathcal{L}_2$ *that contains* $\{X_t\}_{t \in T}$, *which we call closed span of* $\{X_t\}_{t \in T}$. *It is a closed subspace in the sense that* $\mathcal{L}_2$-*limits of its elements lie in the subspace.*

Conditional expectation in $\mathcal{L}_2$

For simplicity, consider real-valued random variables.

Definition A.1.8 (Conditional expectation in $\mathcal{L}_2$). *Let $X_1, \ldots, X_n, Y \in \mathcal{L}_2$ be real-valued and let $M(X_1, \ldots, X_n)$ be the (closed) subspace of $\mathcal{L}_2$ of all measurable functions $\mathbb{R}^n \to \mathbb{R}$ of $X_1, \ldots, X_n$. The conditional expectation of Y given $X_1, \ldots, X_n$ is obtained by*

$$\mathsf{E}[Y|X_1, \ldots, X_n] = \mathsf{pro}_{M(X_1, \ldots, X_n)}(Y).$$

An alternative notation for $\mathsf{E}[Y|X_1, \ldots, X_n]$ is $\mathsf{E}[Y|\sigma(X_1, \ldots, X_n)]$, where $\sigma(X_1, \ldots, X_n)$ is the σ-algebra generated by $X_1, \ldots, X_n$. In particular, by denoting by 1 any random variable a.s. equal to 1, the expectation of $Y \in \mathcal{L}_2$ is given by

$$\mathsf{E}[Y] = \mathsf{pro}_{M(1)}(Y).$$

The conditional expectation of jointly Gaussian random variables has a simple form.

Lemma A.1.9 (Conditional expectation of Gaussian random variables). *Let $X_1, \ldots, X_n$ and Y be jointly Gaussian random variables, then*

$$\mathsf{E}[Y \,|\, X_1, \ldots, X_n] = \mathsf{pro}_{\overline{\mathrm{sp}}\{1, X_1, \ldots, X_n\}}(Y),$$

where 1 denotes any a.s. unitary random variable.

Proof. Let $x_1, \ldots, x_n \in \mathbb{R}$, then the conditional distribution of the Gaussian random variable Y given the values of the Gaussian random variables $X_1 = x_1, \ldots, X_n = x_n$ is Gaussian with expectation $a_0 + a_1 x_1 + \cdots + a_n x_n$, for some $a_0, \ldots, a_n \in \mathbb{R}$. Thus

$$\mathsf{E}[Y \,|\, X_1, \ldots, X_n] \in \overline{\mathrm{sp}}\{1, X_1, \ldots, X_n\}.$$

It then follows from $\mathsf{E}[Y \,|\, X_1, \ldots, X_n] = \mathsf{pro}_{M(X_1, \ldots, X_n)}Y$ that $\mathsf{pro}_{M(X_1, \ldots, X_n)}Y \in \overline{\mathrm{sp}}\{1, X_1, \ldots, X_n\}$. Thus, the idempotence property given in Proposition A.1.2.2 implies that

$$\mathsf{pro}_{M(X_1, \ldots, X_n)}Y = \mathsf{pro}_{\overline{\mathrm{sp}}\{1, X_1, \ldots, X_n\}}Y.$$

$\square$

A.1.3 Space $\mathcal{L}_p$

The space $\mathcal{L}_p$, with $p > 0$, is a space of functions that is defined by using a natural generalization of the p-norm of finite dimensional vector spaces. In what follows, the functions are the random variables and the space $\mathcal{L}_p(\Omega, \mathcal{F}, \mathsf{P})$, or shortly $\mathcal{L}_p$, consists of all complex-valued random variables X for which the moment of order p, exists, viz. $\mathsf{E}[|X|^p] < \infty$, for a $p > 0$. It follows from Minkowski's inequality, i.e. Theorem A.2.2, that the space $\mathcal{L}_p$ with $p \geq 1$ is a vector space w.r.t. the seminorm

$$\|X\|_p = \mathsf{E}^{\frac{1}{p}}[|X|^p].$$

Over the space $\mathcal{L}_p$ we consider the equivalence classes of random variables that are a.s. equal. This space is complete w.r.t. the pseudometric $d_p(X, Y) = \|X - Y\|_p$. As already mentioned, the space $\mathcal{L}_2(\Omega, \mathcal{F}, \mathsf{P})$ or $\mathcal{L}_2$ is a Hilbert space with scalar or inner product $\langle X, Y \rangle = \mathsf{E}[X\overline{Y}]$.

Definition A.1.10 (Convergence in $\mathcal{L}_p$). *Consider the random variables X and $X_1, X_2, \ldots \in \mathcal{L}_p(\Omega, \mathcal{F}, \mathsf{P})$ and $p > 0$. The sequence $X_1, X_2, \ldots$ converges in $\mathcal{L}_p$ or in p-th mean to X, if*

$$\lim_{n \to \infty} \|X_n - X\|_p = 0.$$

The convergence in $\mathcal{L}_p$ is denoted $X_n \xrightarrow{\mathcal{L}_p} X$, where $n \to \infty$ is implicitly meant.

It follows directly from Lyapunov inequality, cf. Theorem A.2.4, that $\forall p \in [1, q]$,

$$\mathcal{L}_q \subset \mathcal{L}_p,$$

and that

$$X_n \xrightarrow{\mathcal{L}_q} X \Longrightarrow X_n \xrightarrow{\mathcal{L}_p} X.$$

Further properties of the convergence in $\mathcal{L}_p$ are presented in Section A.3.

Conditional expectation in $\mathcal{L}_1$

The conditional expectation in $\mathcal{L}_2$ is given in Definition A.1.8 in terms of the orthogonal projection. The definition in $\mathcal{L}_1$ avoids the orthogonal projection operator. We consider real-valued random variables.

Definition A.1.11 (Conditional expectation in $\mathcal{L}_1$). *Consider $Y \in \mathcal{L}_1(\Omega, \mathcal{F}, \mathsf{P})$ real-valued and the sub-σ-algebra $\mathcal{G} \subset \mathcal{F}$. The conditional expectation of Y given $\mathcal{G}$ is the a.s. unique random variable $\mathsf{E}[Y|\mathcal{G}]$ such that:*

(1) $E[Y|\mathcal{G}]$ *is $\mathcal{G}$-measurable;*
(2) $E[Y;A] = E[E[Y|\mathcal{G}];A], \forall A \in \mathcal{G}$.

Definition A.1.11.2 can be re-expressed as follows:

$E[YZ] = E[ZE[Y|\mathcal{G}]]$, for any bounded and $\mathcal{G}$-measurable random variable Z over $(\Omega, \mathcal{F}, \mathsf{P})$.

Let X be a real-valued random variable over $(\Omega, \mathcal{F}, \mathsf{P})$ and let $\mathcal{G} = \sigma(X)$. Definition A.1.11.1 can be re-expressed as:

$E[Y|\mathcal{G}]$ is a Borel function of X.

In this case we denote $E[Y|X] = E[Y|\sigma(X)]$.

Example A.1.12 (Expectation of simple random variable). *Let $X = \sum_{k=1}^{n} a_k \mathsf{I}_{A_k}$, where $A_1, \ldots, A_n$ form a partition of Ω and $a_1, \ldots, a_n \in \mathbb{R}^*$ are different. It then follow from Definition A.1.11.1, that*

$$\exists c_1, \ldots, c_n \in \mathbb{R} \text{ such that } E[Y|X] = \sum_{k=1}^{n} c_k \mathsf{I}\{X = a_k\} = \sum_{k=1}^{n} c_k \mathsf{I}_{A_k}.$$

It follows from Definition A.1.11.2 that, for $k = 1, \ldots, n$,

$$E[Y;A_k] = E\left[\sum_{j=1}^{n} c_j \mathsf{I}_{A_j}; A_k\right] = c_k \mathsf{P}[A_k],$$

and thus

$$c_k = \frac{E[Y;A_k]}{\mathsf{P}[A_k]}.$$

We thus obtain the intuitive result

$$E[Y|X] = \sum_{k=1}^{n} \frac{E[Y;A_k]}{\mathsf{P}[A_k]} \mathsf{I}_{A_k} = \sum_{k=1}^{n} E[Y|A_k] \mathsf{I}_{A_k}.$$

Some important properties are the following.

Proposition A.1.13 (Properties of conditional expectation). *Let $Y \in \mathcal{L}_1(\Omega, \mathcal{F}, \mathsf{P})$, let X be another random variable in $(\Omega, \mathcal{F}, \mathsf{P})$ and let $\mathcal{G} \subset \mathcal{F}$ be a σ-algebra. The conditional expectation $E[\cdot|\mathcal{G}]$ satisfies the following properties:*

(1) $E[c_1 X + c_2 Y|\mathcal{G}] = c_1 E[X|\mathcal{G}] + c_2 E[Y|\mathcal{G}], \forall c_1, c_2 \in \mathbb{R}$, *whenever X is integrable, i.e. the linearity;*
(2) $E[E[Y|\mathcal{G}]|\mathcal{G}] = E[Y|\mathcal{G}]$, *i.e. the idempotence property;*

(3) $\mathcal{H} \subset \mathcal{G}$ *is a σ-algebra* $\implies$ $\mathsf{E}[\mathsf{E}[Y|\mathcal{G}]|\mathcal{H}] = \mathsf{E}[Y|\mathcal{H}]$, *i.e. the iterativity property;*

(4) $\mathsf{E}[|XY|] < \infty$ *and* X *is* $\mathcal{G}$-*measurable* $\implies$ $\mathsf{E}[XY|\mathcal{G}] = X\mathsf{E}[Y|\mathcal{G}]$.

These properties can be compared with the one of the orthogonal projection of Proposition A.1.2.

A.2   Inequalities with random variables

The mostly used inequalities obtained with random variables are summarized in this paragraph.

Theorem A.2.1 (Cauchy-Schwarz inequality). *Let* X *and* Y *be complex-valued variables such that If* $\mathsf{E}[|X|^2] < \infty$ *and* $\mathsf{E}[|Y|^2] < \infty$, *then*

$$|\mathsf{E}[XY]| \leq \sqrt{\mathsf{E}\left[|X|^2\right]\mathsf{E}\left[|Y|^2\right]}.$$

Note that Cauchy-Schwarz inequality is also given with $\mathsf{E}[|XY|]$ on the left side of the inequality (which is obtained by replacing X and Y by $|X|$ and $|Y|$).

Theorem A.2.2 (Minkowski inequality). *Let* X *and* Y *be complex-valued random variables. If* $\mathsf{E}[|X|^p] < \infty$ *and* $\mathsf{E}[|Y|^p] < \infty$, *for* $p \geq 1$, *then it holds that* $\mathsf{E}[|X + Y|^p] < \infty$ *and*

$$\mathsf{E}^{\frac{1}{p}}[|X + Y|^p] \leq \mathsf{E}^{\frac{1}{p}}[|X|^p] + \mathsf{E}^{\frac{1}{p}}[|Y|^p].$$

Theorem A.2.3 (Jensen inequality). *Let* $g : I \to \mathbb{R}$ *be a convex function, where* $I \subset \mathbb{R}$ *is an interval containing the range of the real-valued random variable* X. *Then*

$$\mathsf{E}[g(X)] \geq g(\mathsf{E}[X]),$$

holds.

Proof. It follows from the convexity of g that $\exists\, a \in \mathbb{R}$ such that $\forall x, x_0 \in I$,

$$g(x) \geq g(x_0) + a(x - x_0).$$

In simple cases one has $a = g'(x_0)$, otherwise a can be a one-sided derivative. The claim follows by setting $x = X$ and $x_0 = \mathsf{E}[X]$. $\square$

Theorem A.2.4 (Lyapunov inequality). *Let* X *be a complex-valued random variable. Then* $\forall\, 0 < p \leq q$ *we have*

$$\mathsf{E}^{\frac{1}{p}}[|X|^p] \leq \mathsf{E}^{\frac{1}{q}}[|X|^q].$$

Proof. Let $0 < p \leq q$. One can apply Jensen's inequality, viz. Theorem A.2.3, to the random variable $Y = |X|^p$ with $g(y) = |y|^{q/p}$. □

Theorem A.2.5 (Markov inequality). *Let $g : \mathbb{R} \to \mathbb{R}_+$ be measurable and nondecreasing and let X be a real-valued random variable. Then, $\forall \varepsilon > 0$,*

$$P[X \geq \varepsilon] \leq \frac{E[g(X)]}{g(\varepsilon)}.$$

Proof. Let $\varepsilon > 0$, then $E[g(X)] \geq E[g(X)I\{g(X) \geq \varepsilon\}] \geq g(\varepsilon)P[g(X) \geq \varepsilon]$. □

A.3 Sequences of events and of random variables

This section presents the Lemma of Borel-Cantelli for sequences of events and the different types of convergences of sequences of random variables.

The upper and lower limits of the sequence of events $\{A_n\}_{n \geq 1}$ are defined by

$$\limsup_{n \to \infty} A_n = \{\omega \in \Omega | \omega \in A_n \text{ for infinitely many } n\} = \bigcap_{n \geq 1} \bigcup_{m \geq n} A_m,$$

$$\liminf_{n \to \infty} A_n = \{\omega \in \Omega | \omega \in A_n \text{ for large enough } n\} = \bigcup_{n \geq 1} \bigcap_{m \geq n} A_m.$$

Alternatively, we denote $\limsup_{n \to \infty} A_n$ as A_n, $n \geq 1$, infinitely often (i.o.). The following result holds in this context.

Lemma A.3.1 (Borel-Cantelli). *Let A_n, for $n = 1, 2, \ldots$, be a sequence of events, then the following statements hold.*

(1) If $\sum_{n=1}^{\infty} P[A_n] < \infty$, then $P[A_n, n \geq 1, i.o.] = 0$.
(2) If $\sum_{n=1}^{\infty} P[A_n] = \infty$ and A_n, for $n = 1, 2, \ldots$, are independent, then $P[A_n, n \geq 1, i.o.] = 1$.

Proof. 1. We note that

$$P[A_n, n \geq 1, \text{i.o.}] = P\left[\bigcap_{n=1}^{\infty} \bigcup_{m=n}^{\infty} A_m\right] = \lim_{n \to \infty} P\left[\bigcup_{m=n}^{\infty} A_m\right]$$

$$\leq \lim_{n \to \infty} \sum_{m=n}^{\infty} P[A_m] = 0.$$

2. The complements $A_1^c, A_2^c, \ldots$ are also independent. Thus for $n = 1, 2, \ldots,$ we have

$$P\left[\bigcap_{m=n}^{\infty} A_m^c\right] = \prod_{m=n}^{\infty} (1 - P[A_m]).$$

Further, because $\log(1 - p) \leq -p$, $\forall p \in (0, 1)$, we have

$$\log \prod_{m=n}^{\infty} (1 - P[A_m]) = \sum_{m=n}^{\infty} \log(1 - P[A_m]) \leq - \sum_{m=n}^{\infty} P[A_m] = -\infty.$$

Thus $P[\cap_{m=n}^{\infty} A_m^c] = 0$. It follows from this that $P[\cup_{m=n}^{\infty} A_m] = 1$ and $P[A_n, n \geq 1, \text{i.o.}] = 1$.   □

Almost sure convergence

Definition A.3.2 (Almost sure convergence). *Consider the random variables X and $X_1, X_2, \ldots$, all defined over $(\Omega, \mathcal{F}, P)$. The sequence $\{X_n\}_{n \geq 1}$ converges a.s. to X if*

$$P\left[\left\{\omega \in \Omega \,\middle|\, \lim_{n \to \infty} X_n(\omega) = X(\omega)\right\}\right] = 1.$$

The a.s. convergence is denoted $X_n \xrightarrow{\text{as}} X$, with $n \to \infty$ implicitly meant. A synonym of a.s. convergence is convergence with probability one.

One can easily show that $X_n \xrightarrow{\text{as}} X$ can be also expressed as

$$P\left[\{|X_n - X| > \varepsilon\}, n \geq 1, \text{ i.o.}\right] = 0, \ \forall \varepsilon > 0.$$

A basic result in this context is the following theorem.

Theorem A.3.3 (Strong law of large numbers). *Let $\{X_n\}_{n \geq 1}$ be a sequence of i.i.d. random variables such that $E[|X_1|] < \infty$. Let $S_n = \sum_{i=1}^{n} X_n$, for $n = 1, 2, \ldots$. Then*

$$\frac{S_n}{n} \xrightarrow{\text{as}} E[X_1].$$

Convergence in probability

Definition A.3.4 (Convergence in probability). *Let X and $X_1, X_2, \ldots$ be random variables over $(\Omega, \mathcal{F}, P)$. The sequence $X_1, X_2, \ldots$ converges in probability to X if*

$$\forall \varepsilon > 0, \ \lim_{n \to \infty} P[|X_n - X| > \varepsilon] = 0.$$

The convergence in probability is denoted $X_n \xrightarrow{P} X$, where $n \to \infty$ is implicitly meant. Theorem A.3.5 is given in this context.

Theorem A.3.5 (Weak law of large numbers). *Let $\{X_n\}_{n\geq 1}$ be a sequence of i.i.d. random variables, where $\mathsf{E}[X_1]$ exists. Let $S_n = \sum_{i=1}^{n} X_n$, for $n = 1, 2, \ldots$. Then*

$$\frac{S_n}{n} \xrightarrow{\text{P}} \mathsf{E}[X_1].$$

Convergence in probability and almost sure convergence are rather different concepts. Almost sure convergence is the existence of $\Omega_0 \in \mathcal{F}$ such that $\mathsf{P}[\Omega_0] = 1$ and $X_n(\omega) \to X(\omega)$, $\forall \omega \in \Omega_0$. Convergence in probability is that the sequence of probabilities of $\{|X_n - X| > \varepsilon\}$, for $n = 1, 2, \ldots$, vanishes, $\forall \varepsilon > 0$. The next example illustrates that convergence in probability does not imply almost sure convergence.

Example A.3.6. *Consider $\Omega = [0, 1]$ with the Lebesgue measure and*

$$X_n(\omega) = \begin{cases} 1, & \text{if } \omega \in [j2^{-k}, (j+1)2^{-k}], \\ 0, & \text{otherwise,} \end{cases}$$

where $n = 2^k + j$, with $k = 0, 1, \ldots$ and $j = 0, \ldots, 2^k - 1$. Then

$$\mathsf{P}[|X_n| > 0] = 2^{-k} \xrightarrow{n \to \infty} 0.$$

But at any $\omega \in \Omega$, $\{X_n(\omega)\}_{n\geq 1}$ contains infinitely many 0 and 1: for no $\omega \in \Omega$ does the sequence converge.

Theorem A.3.7. *Almost sure convergence implies convergence in probability.*

Convergence in $\mathcal{L}_p$

The space $\mathcal{L}_p$ is introduced in Section A.1.3 as the space defined using a natural generalization of the p-norm of finite dimensional vector spaces. This section provides further properties of the convergence in $\mathcal{L}_p$, which is given in Definition A.1.10.

Theorem A.3.8. *Convergence in $\mathcal{L}_p$ implies convergence in probability.*

Proof. It follows from the inequality of Markov that

$$\mathsf{P}[|X_n - X| > \varepsilon] = \mathsf{P}[|X_n - X|^p > \varepsilon^p] \leq \varepsilon^{-p}\mathsf{E}[|X_n - X|^p], \quad \forall \varepsilon > 0.$$

$\square$

The converse of Theorem A.3.8 for $p = 1$ is obtained under the following condition.

Definition A.3.9 (Uniform integrability). *The sequence of random variables* $\{X_n\}_{n\geq 0}$ *is uniformly integrable if each of its random variables is integrable, i.e. in* $\mathcal{L}_1$, *and if*

$$\lim_{x\to\infty} \sup_{n\geq 1} \mathsf{E}[|X_n|; |X_n| > x] = 0.$$

Uniform integrability is weaker then the condition of Dominated convergence theorem A.5.3. Indeed, if $|X_n| \leq Y$ a.s., for $n = 1, 2, \ldots$ and $\mathsf{E}[Y] < \infty$, then

$$\lim_{x\to\infty} \sup_{n\geq 1} \mathsf{E}[|X_n|; |X_n| > x] \leq \lim_{x\to\infty} \mathsf{E}[Y; Y > x] = 0.$$

Lemma A.3.10 (Characterization of $\mathcal{L}_1$ convergence). *Let* $X, X_1, X_2, \ldots \in \mathcal{L}_1(\Omega, \mathcal{F}, \mathsf{P})$, *then*

$$X_n \xrightarrow{\mathsf{P}} X \text{ and } \{X_n\}_{n\geq 0} \text{ is uniformly integrable} \iff X_n \xrightarrow{\mathcal{L}_1} X.$$

Convergence in $\mathcal{L}_p$ is neither sufficient nor necessary for a.s. convergence. Nevertheless, it follows from Dominated convergence theorem A.5.3 that $X_n \xrightarrow{\text{as}} X$ and $|X_n| \leq Y$, for $n = 1, 2, \ldots$, with $\mathsf{E}[Y^p] < \infty$, imply $X_n \xrightarrow{\mathcal{L}_p} X$.

Lemma A.3.11 (Equality of $\mathcal{L}_p$ and a.s. limits). *Consider the random variables* $X, X', X_1, X_2, \ldots \in \mathcal{L}_p(\Omega, \mathcal{F}, \mathsf{P})$, *for some* $p > 0$. *If* $X_n \xrightarrow{\mathcal{L}_p} X$ *and* $X_n \xrightarrow{\text{as}} X'$, *then* $X = X'$ *a.s.*

Proof. It follows from Fatou's lemma A.5.5, that

$$\|X - X'\|_p^p = \mathsf{E}[|X - X'|^p] = \mathsf{E}\left[\liminf_{n\to\infty} |X - X_n|^p\right]$$

$$\leq \liminf_{n\to\infty} \mathsf{E}\left[|X - X_n|^p\right] = 0.$$

Thus we have $X = X'$ a.s. $\square$

Weak convergence

The famous illustration of weak convergence or convergence in distribution is given by the following theorem.

Theorem A.3.12 (Central limit theorem). *Let the random variables* $X_1, X_2, \ldots \in \mathcal{L}_2(\Omega, \mathcal{F}, \mathsf{P})$ *be i.i.d. and denote* $\mu = \mathsf{E}[X_1]$ *and* $\sigma^2 = \mathsf{var}(X_1)$. *Then it holds that*

$$\lim_{n\to\infty} \mathsf{P}\left[\frac{S_n - n\mu}{\sqrt{n}\sigma} \leq x\right] = \Phi(x), \quad \forall x \in \mathbb{R}.$$

Definition A.3.13 (Weak convergence and convergence in distribution).

(1) The sequence $\{F_n\}_{n\geq 1}$ of d.f. over $\mathbb{R}$ converges weakly to the d.f. F over $\mathbb{R}$, if

$$F_n(x) \overset{n\to\infty}{\Longrightarrow} F(x),$$

$\forall x \in \mathbb{R}$ *continuity point of F.*

(2) The sequence $\{\mathsf{P}_n\}_{n\geq 1}$ of probability measures over $\mathbb{R}$ converges weakly to the probability measure P over $\mathbb{R}$, if the corresponding sequence of d.f. $\{F_n\}_{n\geq 1}$ converges weakly to the d.f. F of P.

(3) The sequence $\{X_n\}_{n\geq 1}$ of random variables converges in distribution to the random variable X $\{\mathsf{P}_{X_n}\}n \geq 1$ converges weakly to P_X, where P_Y denotes the induced probability of the random variable Y.

The cases 1, 2 and 3 above are denoted $F_n \overset{w}{\longrightarrow} F$, $\mathsf{P}_n \overset{w}{\longrightarrow} \mathsf{P}$ and $X_n \overset{d}{\longrightarrow} X$ (as $n \to \infty$), respectively.

Theorem A.3.14. *If $X_n \overset{P}{\longrightarrow} X$, then $X_n \overset{d}{\longrightarrow} X$. If X is a.s. equal to a constant, then the converse holds as well.*

Lemma A.3.15 (Characterization of weak convergence). $\mathsf{P}_n \overset{w}{\longrightarrow} \mathsf{P}$ *iff*

$$\lim_{n\to\infty} \int_{\mathbb{R}} g(\omega)\mathrm{d}\mathsf{P}_n(\omega) = \int_{\mathbb{R}} g(\omega)\mathrm{d}\mathsf{P}(\omega), \ \forall\, g : \mathbb{R} \longrightarrow \mathbb{R} \ \textit{bounded and continuous.}$$

Corollary A.3.16 (Lévy convergence). $X_n \overset{d}{\longrightarrow} X$ *iff*

$$\lim_{n\to\infty} \mathsf{E}\left[e^{ivX_n}\right] = \mathsf{E}\left[e^{ivX}\right], \ \forall v \in \mathbb{R}.$$

Theorem A.3.17 (Lévy continuity). *If $\varphi(v) = \lim_{n\to\infty} \mathsf{E}[e^{ivX_n}]$ exists, $\forall v \in \mathbb{R}$, and if $\varphi(v)$ is continuous at $v = 0$, then there exists a random variable X such that $\varphi(v) = \mathsf{E}[e^{ivX}]$, $\forall v \in \mathbb{R}$.*

In view of Corollary A.3.16, Theorem A.3.17 yields the relation $X_n \overset{d}{\longrightarrow} X$, between the random variables $X_1, X_2, \ldots$ and X.

Proof. This proof is made by three main parts. We show only two of these three parts.

The first part refers to Helly's theorem A.3.19. If F_n is the d.f. of X_n, for $n = 1, 2, \ldots$, then there exists a subsequence of d.f. $\{F_{n_k}\}_{k\geq 1}$ and a d.f. F, perhaps defective, in the sense that $F(\infty) < 1$, such that $\lim_{k\to\infty} F_{n_k}(x) = F(x)$, $\forall x \in \mathbb{R}$ continuity point of F.

The second part consists in showing that the d.f. F is in fact a proper one, i.e. a probability d.f., i.e. $F(\infty) = 1$. The assumption of continuity at $v = 0$ of $\varphi(v)$ is used in this part of the proof.

For the third and last part, we have from Corollary A.3.16 that $F_{n_k} \xrightarrow{w} F \Leftrightarrow \forall v \in \mathbb{R}, \ \mathsf{E}[e^{ivX_{n_k}}] \to \mathsf{E}[e^{ivX}]$, as $k \to \infty$. This and the first assumption of the theorem tell that $\varphi(v) = \mathsf{E}[e^{ivX}], \ \forall v \in \mathbb{R}$. $\qquad\square$

Theorem A.3.18 (Slutski). *If $X_n \xrightarrow{d} X$ and $Y_n \xrightarrow{P} y$, where y is a constant, then*

$$X_n + Y_n \xrightarrow{d} X + y \quad \text{and} \quad X_n Y_n \xrightarrow{d} Xy.$$

A theorem that closely relates to weak convergence is Theorem A.3.19 of Helly. This theorem is the analogue version of Bolzano-Weierstrass theorem for sequences of numbers: from any bounded sequence in $\mathbb{R}^\infty$ one can extract a convergent subsequence. In this theorem, d.f. are not necessarily probability d.f.

Theorem A.3.19 (Helly). *Let $F_1, F_2, \ldots$ be d.f. over $\mathbb{R}$, such that $F_n(-\infty) = 0$ and $F_n(\infty) \le y$, for some $y < \infty$ and for $n = 1, 2, \ldots$. Then there exist a d.f. F over $\mathbb{R}$ and a subsequence $\{F_{n_k}\}_{k \ge 1}$ such that*

$$F_{n_k}(x) \xrightarrow{k \to \infty} F(x),$$

$\forall x \in \mathbb{R}$ *continuity point of F.*

Remark A.3.20 (Limit at infinity). *Note that $F_{n_k}(\infty) \xrightarrow{k \to \infty} F(\infty)$ may not hold. Should it hold, then $F_{n_k} \xrightarrow{w} F$ would hold whenever $F(\infty) = 1$. For example, let $F_n(x) = \Delta_n(x) = \mathbb{I}\{x \ge n\}, \ \forall x \in \mathbb{R}$ and $n \in \mathbb{N}^*$. If $F_{n_k}(x) \xrightarrow{k \to \infty} F(x), \ \forall x \in \mathbb{R}$ continuity point of F, then $F(x) = 0, \ \forall x \in \mathbb{R}$. However, $F_{n_k}(\infty) = 1, \ \forall k \in \mathbb{N}^*$. But this problem happens only when the mass of goes to infinity. For example, it does not happen with d.f. with bounded support, like in the proof of Herglotz's theorem 2.4.10. It does not happen either whenever the d.f. is tight.*

Summary of relations between convergences

Let X and $X_1, X_2, \ldots$ be random variables over $(\Omega, \mathcal{F}, \mathsf{P})$ and $p > 0$. Figure A.1 provides a summary of the implications between stochastic convergences: the convergences appearing in the upper boxes imply the convergences appearing in the lower boxes.

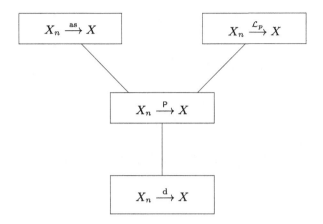

Fig. A.1: Hierarchy of stochastic convergences; convergences in upper boxes imply convergences in lower boxes.

A.4 Characteristic function

The characteristic function is an essential function in probability theory. The pair characteristic function and probability distribution is similar to the pair a.c.v.f. and spectral distribution.

Definition A.4.1 (Characteristic function). *The characteristic function of the random variable X is given by*

$$\varphi(v) = \mathsf{E}\left[e^{ivX}\right], \quad \forall v \in \mathbb{R}.$$

Proposition A.4.2 (Properties of characteristic function). *Any characteristic function $\varphi : \mathbb{R} \to \mathbb{C}$ possesses the following properties:*

(1) φ is Hermitian, viz. $\varphi(-v) = \overline{\varphi(v)}$, $\forall v \in \mathbb{R}$;
(2) φ is uniformly continuous;
(3) φ is n.n.d., cf. Definition 2.1.20.

Proof. 1. The proof of this property is rather trivial.
2. $\forall v, h \in \mathbb{R}$,

$$\left|\varphi(v+h) - \varphi(v)\right| = \left|\mathsf{E}\left[e^{ivX}(e^{ihX} - 1)\right]\right| \leq \mathsf{E}\left[\left|e^{ivX}(e^{ihX} - 1)\right|\right]$$
$$\leq \mathsf{E}\left[\left|e^{ihX} - 1\right|\right] \xrightarrow{h \to 0} 0,$$

by Dominated convergence.

3. $\forall c_1, \ldots, c_n \in \mathbb{C}$, $v_1, \ldots, v_n \in \mathbb{R}$ and $n \in \mathbb{N}^*$, we have

$$\sum_{j=1}^{n}\sum_{k=1}^{n} c_j \bar{c}_k \varphi(v_j - v_k) = \mathsf{E}\left[\left|\sum_{j=1}^{n} c_j e^{iv_j X}\right|^2\right] \in \left[0, \left(\sum_{k=1}^{n} |c_j|\right)^2\right].$$

□

Note that Lévy continuity theorem A.3.17 can be re-expressed as follows.

Let $\varphi_1, \varphi_2, \ldots$ be characteristic functions such that $\varphi(v) = \lim_{n \to \infty} \varphi_n(v)$ exists, $\forall v \in \mathbb{R}$, and $\varphi(v)$ is continuous at $v = 0$. Then φ is a characteristic function.

Theorem A.4.3 (Inversion formula for probability d.f.). *Let F be a d.f. with characteristic function φ. Let $F^*(x) = \{F(x) + F(x-)\}/2$, $\forall x \in \mathbb{R}$. Then, $\forall x < y \in \mathbb{R}$,*

$$F^*(y) - F^*(x) = \frac{1}{2\pi} \lim_{u \to \infty} \int_{-u}^{u} \frac{\exp\{-iyv\} - \exp\{-ixv\}}{-iv} \varphi(v) dv.$$

In Theorem A.4.3, $\forall x \in \mathbb{R}$ continuity point of F, we have $F^*(x) = F(x)$. In this context we can mention Proposition 3.1.22 regarding the number of discontinuity points.

Corollary A.4.4 (Inversion formula for probability density). *Let F be a d.f. with characteristic function φ. If $\int_{-\infty}^{\infty} |\varphi(v)| dv < \infty$, then F is absolutely continuous with bounded density f and it holds that, $\forall x \in \mathbb{R}$,*

$$f(x) = \frac{1}{2\pi} \int_{-\infty}^{\infty} \exp\{-ixv\}\varphi(v) dv. \tag{A.4.1}$$

Proof. Let $x < y \in \mathbb{R}$ be two continuity points of F. Then Theorem A.4.3 and Theorem A.5.8 of Fubini-Tonelli give

$$\begin{aligned}
F(y) - F(x) &= \frac{1}{2\pi} \lim_{u \to \infty} \int_{-u}^{u} \frac{\exp\{-iyv\} - \exp\{-ixv\}}{-iv} \varphi(v) dv \\
&= \frac{1}{2\pi} \lim_{u \to \infty} \int_{-u}^{u} \int_{x}^{y} \exp\{-itv\} dt \, \varphi(v) dv \\
&= \int_{x}^{y} \frac{1}{2\pi} \int_{-\infty}^{\infty} \exp\{-itv\}\varphi(v) \, dv \, dt.
\end{aligned}$$

□

The integrability condition of the characteristic function that appears in Corollary A.4.4 is a sufficient but not a necessary condition. In fact, (A.4.1) holds at any point $x \in \mathbb{R}$ where f is continuous, left- and right-differentiable, without further condition.

A consequence of Theorem A.4.3 is the following.

Corollary A.4.5 (Unicity of the characteristic function). *Two random variables X and Y possess the same characteristic function iff their d.f. are identical.*

Consequently, a distribution is completely determined by its characteristic function.

Proof. Let X have d.f. F and let Y have d.f. G, both with characteristic function φ. Theorem A.4.3 gives $F(y) - F(x) = G(y) - G(x)$, $\forall x < y \in \mathbb{R}$ continuity points of F and G. By letting $x \to -\infty$, we obtain $F = G$ at all common continuity points. Given that there are at most countably many such discontinuity points, cf. Proposition 3.1.22, we have $F = G$ everywhere. $\qquad\square$

Theorem A.4.6 (Expansion of the characteristic function). *Let X be a random variable with characteristic function φ. If $\mathsf{E}[|X|^k] < \infty$, for some $k \in \mathbb{N}^*$, then*

$$\varphi(v) = \sum_{j=0}^{k} \frac{\mathsf{E}[X^j]}{j!}(iv)^j + \mathrm{o}(v^k), \quad \forall v \in \mathbb{R}.$$

In connection with stochastic processes, we have the following lemma.

Lemma A.4.7. *If the stochastic process $\{X_t\}_{t \in \mathbb{R}}$ is stochastically continuous at all $t \in \mathbb{R}$, then, $\forall v \in \mathbb{R}$, $t \mapsto \mathsf{E}[e^{ivX_t}]$ is continuous at all $t \in \mathbb{R}$.*

A.5 Theorems of integration

This paragraph gives some standard results for the computation of expectations, namely of Lebesgue integrals.

Theorem A.5.1 (Monotone convergence). *Let X, Y and $X_1, X_2, \ldots$ be random variables over $(\Omega, \mathcal{F}, \mathsf{P})$.*

(1) If $X_n \geq Y$ a.s., for $n = 1, 2, \ldots$, $\mathsf{E}[Y] > -\infty$ and $X_n \uparrow X$ a.s., as $n \to \infty$, then

$$\mathsf{E}[X_n] \uparrow \mathsf{E}[X], \quad \text{as } n \to \infty.$$

(2) If $X_n \leq Y$ a.s., for $n = 1, 2, \ldots$, $\mathsf{E}[Y] < \infty$ and $X_n \downarrow X$ a.s., for $n \to \infty$, then

$$\mathsf{E}[X_n] \downarrow \mathsf{E}[X], \quad \text{as } n \to \infty.$$

Example A.5.2 (Expectation of monotonic series). *Let $X_1, X_2, \ldots$ be nonnegative random variables over $(\Omega, \mathcal{F}, \mathsf{P})$. Then*

$$\mathsf{E}\left[\sum_{n=1}^{\infty} X_n\right] = \sum_{n=1}^{\infty} \mathsf{E}[X_n].$$

Theorem A.5.3 (Dominated convergence). *Let X, Y and $X_1, X_2, \ldots$ be random variables over $(\Omega, \mathcal{F}, \mathsf{P})$. If $|X_n| \leq Y$ a.s., for $n = 1, 2, \ldots$, $\mathsf{E}[|X|], \mathsf{E}[Y] < \infty$ and $X_n \xrightarrow{as} X$, then*

$$\mathsf{E}[X] = \lim_{n \to \infty} \mathsf{E}[X_n].$$

A direct consequence of Theorem A.5.3 is the following result on interchange between differentiation and integration.

Corollary A.5.4 (Change of differentiation and integration order). *Let $\{X_t\}_{t \in \mathbb{R}}$ be integrable, in the sense that $X_t \in \mathcal{L}_1$, $\forall t \in \mathbb{R}$, and differentiable, in the sense that $X'_t = \mathrm{d}/\mathrm{d}t \, X_t$ exists, $\forall t \in \mathbb{R}$. Assume further that the process of derivatives $\{X'_t\}_{t \in \mathbb{R}}$ is continuous w.r.t. t and satisfies $|X'_t| \leq Y$, $\forall t \in \mathbb{R}$, for some integrable random variable Y. The a.s. validity of these assumptions is sufficient. Then*

$$\frac{\mathrm{d}}{\mathrm{d}t}\mathsf{E}[X_t] = \mathsf{E}[X'_t], \quad \forall t \in \mathbb{R}.$$

Proof. Let $\{h_n\}_{n \geq 1}$ be a vanishing real sequence, $n \in \mathbb{N}^*$ and $t \in \mathbb{R}$. From the Mean value theorem we obtain $(X_{t+h_n} - X_t)/h_n = X'_{t_n}$, a.s., for some t_n between t and $t + h_n$. From Dominated convergence theorem A.5.3 and from a.s. continuity of $\{X'_t\}_{t \in \mathbb{R}}$, we obtain

$$\frac{\mathrm{d}}{\mathrm{d}t}\mathsf{E}[X_t] = \lim_{n \to \infty} \frac{\mathsf{E}[X_{t+h_n}] - \mathsf{E}[X_t]}{h_n} = \lim_{n \to \infty} \mathsf{E}\left[\frac{X_{t+h_n} - X_t}{h_n}\right]$$

$$= \lim_{n \to \infty} \mathsf{E}[X'_{t_n}] = \mathsf{E}\left[\lim_{n \to \infty} X'_{t_n}\right] = \mathsf{E}[X'_t].$$

$\square$

Lemma A.5.5 (Fatou). *Let $X_1, X_2, \ldots$ be nonnegative random variables over $(\Omega, \mathcal{F}, \mathsf{P})$, then*

$$\mathsf{E}\left[\liminf_{n \to \infty} X_n\right] \leq \liminf_{n \to \infty} \mathsf{E}[X_n].$$

Consider the probability spaces $(\Omega_1, \mathcal{F}_1, \mathsf{P}_1)$ and $(\Omega_2, \mathcal{F}_2, \mathsf{P}_2)$. The product σ-algebra $\mathcal{F}_1 \times \mathcal{F}_2$ is the smallest σ-algebra that contains $\{A_1 \times A_2 | A_1 \in \mathcal{F}_1 \wedge A_2 \in \mathcal{F}_2\}$. If $\Omega_1 = \Omega_2 = \mathbb{R}$, then $\mathcal{F}_1 \times \mathcal{F}_2 = \mathcal{B}(\mathbb{R}^2)$. The sections of $A \in \mathcal{F}_1 \times \mathcal{F}_2$ are $A_{\omega_2} = \{\omega_1 \in \Omega_1 | (\omega_1, \omega_2) \in A\}$, $\forall \omega_2 \in \Omega_2$, and $A_{\omega_1} = \{\omega_2 \in \Omega_2 | (\omega_1, \omega_2) \in A\}$, $\forall \omega_1 \in \Omega_1$. The product measure over $(\Omega_1 \times \Omega_2, \mathcal{F}_1 \times \mathcal{F}_2)$ is

$$\mathsf{P}_1 \times \mathsf{P}_2[A] = \int_{\Omega_2} \mathsf{P}_1[A_{\omega_2}] \mathrm{d}\mathsf{P}_2(\omega_2) = \int_{\Omega_1} \mathsf{P}_2[A_{\omega_1}] \mathrm{d}\mathsf{P}_1(\omega_1), \ \forall A \in \mathcal{F}_1 \times \mathcal{F}_2.$$

Consequently $\mathsf{P}_1 \times \mathsf{P}_2[A_1 \times A_2] = \mathsf{P}_1[A_1]\mathsf{P}_2[A_2]$, $\forall A_1 \in \mathcal{F}_1, A_2 \in \mathcal{F}_2$. The following theorems by Fubini and Tonelli allow for the computation of an integral over $(\Omega_1 \times \Omega_2, \mathcal{F}_1 \times \mathcal{F}_2, \mathsf{P}_1 \times \mathsf{P}_2)$ by means of integrals over the marginal spaces.

Theorem A.5.6 (Fubini). *Consider the random variable $X \in \mathcal{L}_1(\Omega_1 \times \Omega_2)$ and the functions*

$$X_1 : \Omega_1 \to \mathbb{R}, \qquad X_2 : \Omega_2 \to \mathbb{R},$$
$$\omega_1 \mapsto \int_{\Omega_2} X(\omega_1, \omega_2) \mathrm{d}\mathsf{P}_2(\omega_2) \quad and \quad \omega_2 \mapsto \int_{\Omega_1} X(\omega_1, \omega_2) \mathrm{d}\mathsf{P}_1(\omega_1).$$

Then $X_1 \in \mathcal{L}_1(\Omega_1)$, $X_2 \in \mathcal{L}_1(\Omega_2)$ and

$$\int_{\Omega_1 \times \Omega_2} X \mathrm{d}\mathsf{P}_1 \times \mathsf{P}_2 = \int_{\Omega_1} \left(\int_{\Omega_2} X \mathrm{d}\mathsf{P}_2 \right) \mathrm{d}\mathsf{P}_1 = \int_{\Omega_2} \left(\int_{\Omega_1} X \mathrm{d}\mathsf{P}_1 \right) \mathrm{d}\mathsf{P}_2.$$
$$(A.5.1)$$

When X is a nonnegative random variable, the integrability condition becomes superfluous and the next theorem can be applied.

Theorem A.5.7 (Tonelli). *When the random variable X is nonnegative over $\Omega_1 \times \Omega_2$, then (A.5.1) holds.*

Though, each of the three integrals in (A.5.1) can take the value ∞. The composition of Fubini's and Tonelli's theorems leads directly to the following theorem.

Theorem A.5.8 (Fubini-Tonelli). *If $X \in \mathcal{L}_1(\Omega_1 \times \Omega_2)$ or $X_1 \in \mathcal{L}_1(\Omega_1)$ or $X_2 \in \mathcal{L}_1(\Omega_2)$, then (A.5.1) holds.*

In Theorems A.5.6, A.5.7 and A.5.8, P_1 and P_2 do not need to be probability measures: P_1 or P_2 can be replaced by σ-finite measures, defined as follows.

Definition A.5.9 (σ-finite measure). *A measure μ over $(\Omega, \mathcal{F})$ is σ-finite, if the set Ω can be covered with at most countably many measurable subsets, each one having finite measure μ. In other terms, there exists sets $A_1, A_2, \ldots \in \mathcal{F}$ with $\mu(A_1) < \infty$, $\mu(A_2) < \infty, \ldots$, such that*

$$\bigcup_{n=1}^{\infty} A_n = \Omega.$$

A.6   Remark on Riemann-Stieltjes integration

This section provides a minor but useful remark on the change of variable with Riemann-Stieltjes integration.

The Riemann-Stieltjes integral is a practical generalization of the Riemann integral. One can refer e.g. to Chapter 6 of Rudin (1964). Let $-\infty < a < b < \infty$. The Riemann-Stieltjes integral over the interval $[a, b]$ of $f : \mathbb{R} \to \mathbb{R}$ w.r.t. $g : \mathbb{R} \to \mathbb{R}$, assumed monotone over the interval $[a, b]$, is denoted

$$\int_{[a,b]} f(x) \, dg(x),$$

and it is given by the limit, as the mesh of the partition $a = x_0 < \cdots < x_n = b$ vanishes, of

$$\sum_{j=1}^{n} f(\xi_j) \{g(x_j) - g(x_{j-1})\},$$

where $\xi_j \in [x_{j-1}, x_j]$, for $j = 1, \ldots, n$. Note that g may have jumps.

Consider now $0 < a < b < \infty$, then

$$\int_{[-b,-a]} f(x) \, dg(x) = \lim \sum_{j=1}^{n} f(\psi_j) \{g(y_j) - g(y_{j-1})\}$$

$$= \lim \sum_{j=1}^{n} f(-\xi_j) \{g(-x_{j-1}) - g(-x_j)\}$$

$$= \int_{[a,b]} f(-x) \, dg(-x), \qquad (A.6.1)$$

where $-b = y_0 < \cdots < y_n = -a$, with $\psi_j \in [y_{j-1}, y_j]$, for $j = 1, \ldots, n$, and where the limits are for vanishing meshes of partitions. In the last equality of (A.6.1), the symbol $dg(-x)$ indicates that the partition of $[a, b]$ is reflected around the origin, but any finite increment of g in the sum still follows the principle: g at right point $(-x_{j-1})$ minus g at left point $(-x_j)$.

When g is differentiable, the symbol $dg(-x)$ represents the differential of g evaluated at $-x$. This is not the differential of g compounded with sign inversion, at point x. If g' denotes the derivative, then the differential of g compounded with sign inversion, at point x, is $dg(-x) = -g'(-x)dx$. In this case, the symbol $dg(-x)$ of the integral is $g'(-x)dx$, so the opposite of the differential. Consider the example where g is a symmetric d.f. The Riemann-Stieltjes increment in (A.6.1) is $dg(-x) = dg(x)$, whereas $dg(-x) = -dg(x)$ is the differential.

We have thus the following definition.

Definition A.6.1. *Consider the functions $f : \mathbb{R} \to \mathbb{R}$ and $g : \mathbb{R} \to \mathbb{R}$, the latter monotone over the interval $[a, b]$, for some $a < b \in \mathbb{R}$. Then*

$$\int_{[a,b]} f(x)\,dg(-x) = \int_{[-b,-a]} f(-x)\,dg(x).$$

A.7 Taylor and Laurent series

The Laurent expansion or series appears in the analysis of time series in the time domain. This section summarizes some useful results in this context, that can be found in several books of complex analysis, such as Rudin (1987). It then provides some specific results for time series.

We start with the Taylor expansion or series of a complex function. The function $f : \mathbb{C} \to \mathbb{C}$ is called analytic at $z_0 \in \mathbb{C}$ if it is differentiable over a neighborhood of z_0 and it is called analytic over $U \subset \mathbb{C}$ open if it is differentiable at any $z \in U$. Theorem A.7.1 tells that an analytic function possesses a Taylor series.

Theorem A.7.1 (Taylor series). *Let $U \subset \mathbb{C}$ be open, $z_0 \in \mathbb{C}$ and assume $f : U \to \mathbb{C}$ analytic. Then $\exists\, a_0, a_1, \ldots \in \mathbb{C}$ such that*

$$f(z) = \sum_{n=0}^{\infty} a_n (z - z_0)^n,$$

called Taylor series, converges at any z in the disk $\{z \in \mathbb{C} | |z - z_0| \leq r\}$ that is contained in U, for some $r > 0$.

The coefficients of the Taylor series can be expressed as

$$a_n = \frac{f^{(n)}(z_0)}{n!} = \frac{1}{2\pi i} \int_{C_{z_0,r}} \frac{f(z)}{(z - z_0)^{n+1}}\,dz, \quad \forall n \in \mathbb{N},$$

where the path of integration is the circle $C_{z_0,r} = \{z \in U \mid |z - z_0| = r\}$
(that is traversed in the counterclockwise sense).

Corollary A.7.2 (Unicity of Taylor series). *The power series representing
an analytic function around any point z_0, in the interior of the domain of
definition, is unique.*

If the function f of Taylor's theorem A.7.1 possesses a pole at $z_0 \in U$
and is analytic over $U \backslash \{z_0\}$, then one can find the Laurent series of f.

Theorem A.7.3 (Laurent series). *Let $U \subset \mathbb{C}$, $z_0 \in U$ and assume $f : U \to
\mathbb{C}$ analytic in the annulus $\mathcal{A}_{z_0,r_1,r_2} = \{z \in \mathbb{C} \mid r_1 < |z - z_0| < r_2\}$ that is
contained in U, for some $0 < r_1 < r_2$. Then $\exists \ldots, a_{-1}, a_0, a_1, \ldots \in \mathbb{C}$ such
that*

$$f(z) = \sum_{n=-\infty}^{\infty} a_n (z - z_0)^n,$$

called Laurent series, converges at any $z \in \mathcal{A}_{z_0,r_1,r_2}$.
   The coefficients of the series admit the integral representation

$$a_n = \frac{1}{2\pi i} \int_{C_{z_0,r}} \frac{f(z)}{(z - z_0)^{n+1}} dz, \quad \forall n \in \mathbb{Z},$$

where the circle $C_{z_0,r}$ lies inside the annulus $\mathcal{A}_{z_0,r_1,r_2}$, i.e. for $r_1 < r < r_2$.
   *The series $\sum_{n=0}^{\infty} a_n (z - z_0)^n$, called analytic part, converges to an ana-
lytic function for $|z - z_0| < r_2$.*
   *The series $\sum_{n=-\infty}^{-1} a_n (z - z_0)^n$, called principal part, converges to an
analytic function for $|z - z_0| > r_1$.*

Corollary A.7.4 (Unicity of Laurent series). *The Laurent series representing an analytic function over an annulus around any point z_0, in the
interior of the domain of definition, is unique.*

Theorem A.7.5. *Let $U \subset \mathbb{C}$ be a connected open set and let $f : U \to \mathbb{C}$ be
an analytic function. Let $\{z_n\}_{n\in\mathbb{N}} \in U^{\infty}$ be a sequence of pairwise different
elements converging to some point of U. Assume that $f(z_n) = 0$, $\forall n \in \mathbb{N}$.
Then $f(z) = 0$, $\forall z \in U$.*

Proposition A.7.6. *Let $p_m(z) = \sum_{i=0}^{m} \alpha_i z^i$ and $q_n(z) = \sum_{j=0}^{n} \beta_j z^j$ be two
polynomials in $\mathbb{C}$ with real coefficients, $\alpha_1, \ldots, \alpha_m, \beta_1, \ldots, \beta_n \in \mathbb{R}$. Assume
that $q_n(z) \neq 0$, $\forall z \in \mathcal{A}_r^\circ$, the interior of the annulus (2.2.2), for some
$r > 1$. Then the rational function p_m/q_n is analytic in $\mathcal{A}_r^\circ$ and there exists*

a sequence $\{\psi_j\}_{j \in \mathbb{Z}} \in \mathbb{R}^\infty$ *such that*

$$\frac{p_m(z)}{q_n(z)} = \sum_{k=-\infty}^{\infty} \psi_k z^k, \quad \forall z \in \mathcal{A}_r^\circ.$$

Thus the coefficients of this Laurent expansion are real.

Proof. Theorem A.7.3 implies that there exists a sequence $\{\lambda_j\}_{j \in \mathbb{Z}} \in \mathbb{C}^\infty$ such that

$$\frac{p_m(z)}{q_n(z)} = \sum_{k=-\infty}^{\infty} \lambda_k z^k, \quad \forall z \in \mathcal{A}_r^\circ.$$

Let $\psi_j = \text{Re}\,\lambda_j$, $\forall j \in \mathbb{Z}$. Since $p_m(z)/q_n(z) \in \mathbb{R}$, $\forall z \in \mathbb{R} \cap \mathcal{A}_r^\circ$, then

$$\sum_{k=-\infty}^{\infty} \lambda_k z^k = \text{Re}\left(\sum_{k=-\infty}^{\infty} \lambda_k z^k\right) = \sum_{k=-\infty}^{\infty} \psi_k z^k, \quad \forall z \in \mathbb{R} \cap \mathcal{A}_r^\circ.$$

Thus,

$$\frac{p_m(z)}{q_n(z)} - \sum_{k=-\infty}^{\infty} \psi_k z^k = 0, \quad \forall z \in \mathbb{R} \cap \mathcal{A}_r^\circ.$$

Therefore Theorem A.7.5, with $U = \mathcal{A}_r^\circ$ and $\{z_n\}_{n \in \mathbb{N}}$ any convergent sequence in $\mathbb{R} \cap U$, yields

$$\frac{p_m(z)}{q_n(z)} = \sum_{k=-\infty}^{\infty} \psi_k z^k, \quad \forall z \in \mathcal{A}_r^\circ.$$

$\square$

Proposition A.7.7. *Let* $\{\alpha_j\}_{j \in \mathbb{N}} \in \mathbb{R}^\infty$ *and assume that the series* $\alpha(z) = \sum_{j=0}^{\infty} \alpha_j z^j$ *is convergent at any* z *in* $\mathcal{D}_\varepsilon^\circ$, *the interior of the disk (2.2.1), for some* $\varepsilon > 0$. *Assume further that* $\alpha(z) \neq 0$, $\forall z \in \mathcal{D}_\varepsilon^\circ$. *Then,* $1/\alpha$ *is analytic over* $\mathcal{D}_\varepsilon^\circ$ *and it admits the Taylor series*

$$\frac{1}{\alpha(z)} = \sum_{j=0}^{\infty} \beta_j z^j, \quad \forall z \in \mathcal{D}_\varepsilon^\circ,$$

where $\beta_0, \beta_1, \ldots \in \mathbb{R}$.

Proof. The rational function $1/\alpha$ is indeed analytic and Taylor's theorem A.7.1 with $U = \mathcal{D}_\varepsilon^\circ$ implies that there exists a sequence $\{\lambda_j\}_{j \in \mathbb{N}} \in \mathbb{C}^\infty$, such that

$$\frac{1}{\alpha(z)} = \sum_{j=0}^{\infty} \lambda_j z^j, \quad \forall z \in \mathcal{D}_\varepsilon^\circ.$$

Let $\beta_j = \text{Re}\,\lambda_j$, $\forall j \in \mathbb{N}$. Then $\beta_j = \lambda_j$, $\forall j \in \mathbb{N}$, follows from arguments similar to those given in the proof of Proposition A.7.6. $\square$

A.8 Special functions and distributions

This section presents some special functions and special distributions that appear with stationary models.

A.8.1 *Function sinc*

Definition A.8.1. *The sinc function is given by*

$$\operatorname{sinc} x = \begin{cases} \frac{\sin x}{x}, & \text{if } x \neq 0, \\ \lim_{x \to 0} \frac{\sin x}{x} = 1, & \text{if } x = 0. \end{cases}$$

It is symmetric around the point $x = 0$, which is its maximum. It tends to zero as $|x|$ increases, by alternating around the null level. Its root are $x = k\pi$, $\forall k \in \mathbb{Z}^*$. We refer to Figure A.2.

The sinc function is not Lebesgue integrable, i.e. not in $\mathcal{L}_1$. Indeed,

$$\int_{-\infty}^{\infty} |\operatorname{sinc} x| dx = \sum_{k=-\infty}^{\infty} \int_{k\pi}^{(k+1)\pi} \left| \frac{\sin x}{x} \right| dx$$

$$> \sum_{k=-\infty}^{\infty} \frac{1}{|k+1|\pi} \int_{k\pi}^{(k+1)\pi} |\sin x| \, dx$$

$$= \frac{2}{\pi} \sum_{k=-\infty}^{\infty} \frac{1}{|k+1|} = \infty. \tag{A.8.1}$$

However, the above integral without absolute value does converge. Indeed, an important integral is

$$\int_0^{\infty} \operatorname{sinc} x dx = \frac{\pi}{2}.$$

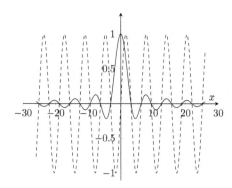

Fig. A.2: Functions sin (dashed line) and sinc (solid line).

In order to show this equation, let us denote by $\hat{f}(v) = \int_0^\infty e^{-vx} f(x) dx$ the Laplace transform of the function f, at any $v \in \mathbb{R}$ where the integral exists. Then

$$\left(\widehat{\frac{f(x)}{x}} \right)(v) = \int_v^\infty \hat{f}(u) du,$$

at any $v \geq 0$ where the integral exists. One easily computes $\widehat{\sin}(v) = 1/(1 + v^2)$, $\forall v > 0$. Thus, we find

$$\int_0^\infty e^{-vx} \operatorname{sinc} x dx = \int_v^\infty \widehat{\operatorname{sin}} u du = [\arctan u]_v^\infty = \frac{\pi}{2} - \arctan v \xrightarrow{v \to 0} \frac{\pi}{2}.$$

Note however that sinc $\in \mathcal{L}_2$. This can be easily verified by using the inequality $\operatorname{sinc} x \leq x \wedge 1$, $\forall x \geq 0$.

Let $a > 0$. It is direct to see that $\operatorname{sinc} av$ is the characteristic function at $v \in \mathbb{R}$ of the uniform density over $(-a, a)$. In other terms, the a.c.v.f.

$$\gamma(h) = \operatorname{sinc} ah, \quad \forall h \in \mathbb{R},$$

has the spectral density

$$f(\alpha) = \frac{1}{2a} \mathbb{1}\{|\alpha| \leq a\}, \quad \forall \alpha \in \mathbb{R}. \tag{A.8.2}$$

A.8.2 *Dirac distribution and Dirac function*

This section provides the precise definitions of the Dirac distribution and the Dirac function.

Definition A.8.2 (Dirac distribution). *Let $(\Omega, \mathcal{F})$ be a measure space and let $\omega \in \Omega$. The Dirac distribution at point ω is given by*

$$\Delta_\omega : \mathcal{F} \to \{0, 1\}$$

$$A \mapsto \begin{cases} 1, & \text{if } \omega \in A, \\ 0, & \text{if } \omega \notin A. \end{cases}$$

The Dirac distribution Δ_ω is indeed a probability distribution:

- $\Delta_\omega[A] \geq 0$, $\forall A \in \mathcal{F}$.
- $\Delta_\omega[\Omega] = 1$.
- Let $\{A_n\}_{n \geq 1}$ be a sequence of disjoint events of $\mathcal{F}$, then it holds either that $\omega \notin \bigcup_{n=1}^\infty A_n$, therefore $\Delta_\omega[\bigcup_{n=1}^\infty A_n] = 0$ and $\Delta_\omega[A_n] = 0$, $\forall n \geq 1$, or that $\omega \in \bigcup_{n=1}^\infty A_n$, therefore $\omega \in A_m$ for exactly one $m \geq 1$ and so $\Delta_\omega[\cup_{n=1}^\infty A_n] = 1$, $\Delta_\omega[A_m] = 1$ and $\Delta_\omega[A_n] = 0$, $\forall n \neq m$. It therefore holds that

$$\Delta_\omega \left[\bigcup_{n=1}^\infty A_n \right] = \sum_{n=1}^\infty \Delta_\omega[A_n].$$

Thus $(\Omega, \mathcal{F}, \Delta_\omega)$ is a probability space.

Consider the measurable function $f \colon \Omega \to \mathbb{R}$ and $A \in \mathcal{F}$, then it holds that

$$\int_A f \, \mathrm{d}\Delta_\omega = f(\omega).$$

Definition A.8.3 (Dirac function). *Let* $(\Omega, \mathcal{F}) = (\mathbb{R}, \mathcal{B}(\mathbb{R}))$, λ *be the Lebesgue measure and* f *be a Borel function. The Dirac function* δ *satisfies*

$$\int_{\mathbb{R}} \delta(t - x) f(t) \mathrm{d}\lambda(t) = f(x), \quad \forall x \in \mathbb{R}.$$

For $B \in \mathcal{B}(\mathbb{R})$ it holds that

$$\int_B \delta(t - x) f(t) \, \mathrm{d}\lambda(t) = \int_B f \, \mathrm{d}\Delta_x = f(x)\Delta_x[B].$$

If $f(t) = 1$, $\forall t \in B$, then

$$\int_B \delta(t - x) \, \mathrm{d}\lambda(t) = \int_B \mathrm{d}\Delta_x = \Delta_x[B].$$

Consequently,

$$\delta_x(t) = \delta(t - x),$$

satisfies the same integral equation of the Radon-Nikodym derivative $\mathrm{d}\Delta_x/\mathrm{d}\lambda$. For this reason it is called Dirac density, although the Dirac distribution does not posses a density: Δ_x is not absolutely continuous w.r.t. the Lebesgue measure λ. (The Dirac function δ either is not function in the precise sense.)

The d.f. of the Dirac distribution over $(\mathbb{R}, \mathcal{B}(\mathbb{R}))$ and with mass 1 one at point 0 Δ: $\Delta(x) = 1\{x \geq 0\}$, $\forall x \in \mathbb{R}$.

A.8.3 *Gaussian or normal distribution*

This section presents the Gaussian or normal distribution and provides some properties and results. The complex Gaussian distribution is also presented.

Let $\boldsymbol{\mu} \in \mathbb{R}^d$, $\boldsymbol{\Sigma} \in \mathbb{R}^{d \times d}$ a n.n.d. matrix and let $\boldsymbol{X}$ be a random vector of $\mathbb{R}^d$. We denote by $\boldsymbol{X} \sim \mathcal{N}(\boldsymbol{\mu}, \boldsymbol{\Sigma})$ the fact the random vector $\boldsymbol{X}$ is Gaussian or, equivalently, normal with mean vector $\boldsymbol{\mu}$ and covariance matrix $\boldsymbol{\Sigma}$. The characteristic function of $\boldsymbol{X} \sim \mathcal{N}(\boldsymbol{\mu}, \boldsymbol{\Sigma})$ is given by

$$\mathsf{E}[\exp\{\mathrm{i}\langle \boldsymbol{v}, \boldsymbol{X} \rangle\}] = \exp\left\{\mathrm{i}\langle \boldsymbol{\mu}, \boldsymbol{v} \rangle - \frac{1}{2}\boldsymbol{v}^\top \boldsymbol{\Sigma} \boldsymbol{v}\right\}, \quad \forall \boldsymbol{v} \in \mathbb{R}^d.$$

When $\boldsymbol{\Sigma}$ is p.d., the density of $\boldsymbol{X}$ is given by

$$f(\boldsymbol{x}) = (2\pi)^{-\frac{d}{2}} (\det \boldsymbol{\Sigma})^{-\frac{1}{2}} \exp\left\{ -\frac{1}{2}\boldsymbol{x}^\top \boldsymbol{\Sigma}^{-1}\boldsymbol{x} \right\}, \quad \forall \boldsymbol{x} \in \mathbb{R}^d. \qquad (A.8.3)$$

When the n.n.d. covariance matrix $\boldsymbol{\Sigma}$ is not p.d., the distribution of $\boldsymbol{X}$ is called singular normal. The singular normal or Gaussian distribution assigns mass one to a subspace of $\mathbb{R}^d$. Its density takes the form of (A.8.3), however with $\boldsymbol{\Sigma}^{-1}$ replaced by a generalized inverse of $\boldsymbol{\Sigma}^2$ and with $\det \boldsymbol{\Sigma}$ replaced by the product of the nonnull eigenvalues of $\boldsymbol{\Sigma}$.

An important characterization of a Gaussian random vector is that every linear combination of its elements has a univariate normal distribution. A similar closure property holds in terms of conditional distributions. It is given in Proposition A.8.4.4, for two-dimensional Gaussian vectors, and in Lemma 4.7.3, for arbitrary dimension.

Proposition A.8.4. *Let* $\boldsymbol{X} = (X_1, X_2)^\top$ *be a normal random vector with expectation* $\boldsymbol{\mu} = (\mu_1, \mu_2)^\top$ *and p.d. covariance matrix*

$$\boldsymbol{\Sigma} = \begin{pmatrix} \sigma_{11} & \sigma_{12} \\ \sigma_{12} & \sigma_{22} \end{pmatrix}.$$

Define

$$X_{2.1} = X_2 - \sigma_{12}\sigma_{11}^{-1}X_1, \ \mu_{2.1} = \mu_2 - \sigma_{12}\sigma_{11}^{-1}\mu_1 \ \text{and} \ \sigma_{22.1} = \sigma_{22} - \sigma_{12}^2\sigma_{11}^{-1}.$$

(Note that $\boldsymbol{\Sigma}$ *p.d. implies* $\sigma_{11} > 0$*). Then the following properties hold.*

(1) $X_{2.1} \sim \mathcal{N}(\mu_{2.1}, \sigma_{22.1})$.
(2) $\forall \boldsymbol{a}, \boldsymbol{b} \in \mathbb{R}^2$, $\langle \boldsymbol{a}, \boldsymbol{X} \rangle$ *and* $\langle \boldsymbol{b}, \boldsymbol{X} \rangle$ *are independent iff* $\boldsymbol{a}^\top \boldsymbol{\Sigma} \boldsymbol{b} = 0$.
(3) X_1 *and* $X_{2.1}$ *are independent.*
(4) The conditional distribution of X_2 *given* $X_1 = x_1$ *is given by*

$$\mathcal{N}\left(\mu_2 + \sigma_{12}\sigma_{11}^{-1}(x_1 - \mu_1), \sigma_{22.1}\right).$$

In Proposition A.8.4.4 we can note that the conditional expectation does depend on the value of x_1 whereas the conditional variance does not.

Proof. 1. Any linear combination of the elements of $\boldsymbol{X}$ and in particular $X_{2.1}$ is normally distributed. Then, $\mathsf{E}[X_{2.1}] = \mu_{2.1}$ and $\mathsf{var}(X_{2.1}) = \sigma_{22.1}$.

---

[2] The matrix $\boldsymbol{\Sigma}^-$ is a generalized inverse of $\boldsymbol{\Sigma}$ if it satisfies $\boldsymbol{\Sigma}\boldsymbol{\Sigma}^-\boldsymbol{\Sigma} = \boldsymbol{\Sigma}$.

2. It follows from
$$\mathsf{cov}(\langle a, X \rangle, \langle b, X \rangle) = a^\top \Sigma b,$$
that $\langle a, X \rangle$ and $\langle b, X \rangle$ are uncorrelated iff the right side is zero. Note that $\langle a, X \rangle$ and $\langle b, X \rangle$ are jointly normal, so they are independent iff they are uncorrelated.

3. Apply Proposition A.8.4.2 to $a = (1, 0)^\top$ and $b = (-\sigma_{12}\sigma_{11}^{-1}, 1)^\top$.

4. It follows from Proposition A.8.4.3 that, $\forall B \in \mathcal{B}(\mathbb{R})$ and $x_1 \in \mathbb{R}$,
$$\mathsf{P}\left[X_2 \in B \mid X_1 = x_1\right] = \mathsf{P}\left[X_{2.1} + \sigma_{12}\sigma_{11}^{-1}x_1 \in B\right].$$

This with Proposition A.8.4.1 implies that the conditional distribution of X_2 given $X_1 = x_1$ is
$$\mathcal{N}\left(\mu_{2.1} + \sigma_{12}\sigma_{11}^{-1}x_1, \sigma_{22.1}\right) = \mathcal{N}\left(\mu_2 + \sigma_{12}\sigma_{11}^{-1}(x_1 - \mu_1), \sigma_{22.1}\right).$$
$\square$

A d-dimensional complex-valued random vector $X = (X_1, \dots, X_d)^\top$ is called complex Gaussian or complex normal if the $2d$-dimensional vector
$$\left(\mathrm{Re}\, X^\top, \mathrm{Im}\, X^\top\right) = \left(\mathrm{Re}\, X_1, \dots, \mathrm{Re}\, X_d, \mathrm{Im}\, X_1, \dots, \mathrm{Im}\, X_d\right),$$
follows a $2d$-dimensional Gaussian distribution. It was introduced by Wooding (1956). The distribution of the complex Gaussian random vector is entirely determined by its expectation $\mu = \mathsf{E}[X]$, its covariance matrix
$$\Sigma = \mathsf{var}(X) = \mathsf{E}\left[(X - \mu)\overline{(X - \mu)}^\top\right],$$
and its pseudo-covariance matrix
$$\Sigma^* = \mathsf{var}^*(X) = \mathsf{E}\left[(X - \mu)(X - \mu)^\top\right].$$

A.9   Linear differential equations

This section summarizes the methods for solving linear differential equations that are useful in our context.

Consider the first order linear differential equation
$$y'(x) + b(x)y(x) = z(x). \tag{A.9.1}$$
The general solution is the sum of the general solution of the homogeneous equation, viz. of (A.9.1) with $z(x) = 0$, and of a particular solution. The general solution of the homogeneous equation is $ae^{-\int b(x)\mathrm{d}x}$.

The method of variation of constants gives a particular solution. If $y_0(x) = a(x)e^{-\int b(x)\mathrm{d}x}$ solves (A.9.1), then by inserting it in (A.9.1), one obtains $a'(x) = z(x)e^{\int b(x)\mathrm{d}x}$.

Alternatively, the general solution can be obtained by the method of integrating factors. It consists in multiplying the equation by the integrating factor $e^{\int b(x)\mathrm{d}x}$ and then in integrating.

Theorem A.9.1 (Second order linear differential equation). *The general solution of the second order linear and homogeneous differential equation*

$$y''(x) + by'(x) + cy(x) = 0, \quad \forall x \in \mathbb{R}, \tag{A.9.2}$$

where $b, c \in \mathbb{R}$, is given by

$$y(x) = a_1 e^{r_1 x} + a_2 e^{r_2 x}, \quad \forall x \in \mathbb{R},$$

where $a_1, a_2 \in \mathbb{C}$ and $r_1, r_2 \in \mathbb{C}$ are the solutions of the characteristic equation

$$r^2 + br + c = 0,$$

whenever $r_1 \neq r_2$. If $r_1 = r_2$, then the general solution is

$$y(x) = (a_1 x + a_2) e^{r_1 x}, \quad \forall x \in \mathbb{R}.$$

Proof. Let $y(x) = e^{rx} m(x)$. By replacing $y'(x)$ and $y''(x)$ into (A.9.2) we find

$$m''(x) + (b + 2r)m'(x) + \underbrace{(r^2 + br + c)}_{\substack{=0, \\ \text{for } r = r_1 \text{ or } r = r_2}} m(x) = 0.$$

We then have

$$r^2 + br + c = (r - r_1)(r - r_2) = r^2 \underbrace{- (r_1 + r_2)}_{=b} r + \underbrace{r_1 r_2}_{=c}.$$

For $r = r_2$ we obtain

$$m''(x) + (r_1 - r_2)m'(x) = 0 \iff m'(x) = c_1 e^{(r_2 - r_1)x}$$

$$\iff m(x) = \underbrace{\frac{c_1}{r_2 - r_1} e^{(r_2 - r_1)x} + a_1}_{=a_2}.$$

It follows that $y(x) = a_2 e^{r_1 x} e^{(r_2 - r_1)x} + a_1 e^{r_1 x} = a_1 e^{r_1 x} + a_2 e^{r_2 x}$. $\qquad \square$

Any solution of the inhomogeneous equation

$$y''(x) + by'(x) + cy(x) = z(x), \tag{A.9.3}$$

is the sum of the solution of the homogeneous equation and of a particular solution.

If we can guess the form of a particular solution to (A.9.3), then the method of undetermined coefficients produces a particular solution. The particular guess is indeed given up to two unknown coefficients that can be identified upon inserting the guess into (A.9.3).

If no guess can be found, then the method of variation of constants produces a particular solution. Given the two solutions $y_1(x)$ and $y_2(x)$ of the homogeneous equation, let us guess that (A.9.3) is solved by

$$y_0(x) = a_1(x)y_1(x) + a_2(x)y_2(x),$$

Then, this is true if

$$a_1'(x)y_1(x) + a_2'(x)y_2(x) = 0 \quad \text{and} \quad a_1'(x)y_1'(x) + a_2'(x)y_2'(x) = z(x).$$

This is seen by computing $y_0'(x)$ and $y_0''(x)$ under these two constraints and by inserting the two results obtained, together with $y_0(x)$, into (A.9.3).

With the initial conditions $y(x_0) = k_0$ and $y'(x_0) = k_1$ we obtain an unique solution called solution of the initial value problem.

We now give a result for difference equations.

Theorem A.9.2 (Second order linear difference equation). *The general solution of the second order linear and homogeneous difference equation*

$$y_k + by_{k-1} + cy_{k-2} = 0, \quad \forall k \in \mathbb{Z}, \tag{A.9.4}$$

where $b \in \mathbb{R}$ and $c \in \mathbb{R}^*$, is given by

$$y_k = a_1 r_1^k + a_2 r_2^k, \quad \forall k \in \mathbb{Z},$$

where $a_1, a_2 \in \mathbb{C}$ and $r_1, r_2 \in \mathbb{C}^*$ are the solutions of the characteristic equation

$$r^2 + br + c = 0,$$

whenever $r_1 \neq r_2$. If $r_1 = r_2$, then the general solution is

$$y_k = (a_1 + a_2 k)r_1^k, \quad \forall k \in \mathbb{Z}.$$

Proof. Consider the solution $y_k = r^k$, for some $r \in \mathbb{C}^*$. By inserting it into (A.9.4), we obtain $r^{k-2}(r^2 + br + c) = 0$. The roots of the quadratic are $r_1, r_2 \neq 0$, because $c \neq 0$. Given $r \neq 0$, we have two possible cases. If $r_1 \neq r_2$, then $\{r_1^k\}_{k \in \mathbb{Z}}$ and $\{r_2^k\}_{k \in \mathbb{Z}}$ are linearly independent and span the space of solutions, which has dimension two. Thus the general solution takes the form $a_1 r_1^k + a_2 r_2^k$. If $r_1 = r_2$ (and are thus real), then $\{r_1^k\}_{k \in \mathbb{Z}}$ and $\{kr_1^k\}_{k \in \mathbb{Z}}$ are linearly independent and span the space of solutions. Thus the general solution takes the form $a_1 r_1^k + a_2 kr_1^k$. $\qquad\square$

A.10 Fast Fourier transform

This section briefly presents the main ideas of the FFT. The FFT of Cooley and Tukey (1965) is a classical numerical technique that allows for efficient evaluation of the discrete Fourier transform or of the inverse discrete Fourier transform. In the latter case it is often called inverse FFT.

Let $n \in \mathbb{N}^*$. The discrete Fourier transform $\varphi_{(n)}$ of the data $p_0, \ldots, p_{n-1} \in \mathbb{C}$ is defined by

$$\varphi_{(n)}(\alpha) = \sum_{k=0}^{n-1} e^{i\alpha k} p_k, \quad \forall \alpha \in \mathbb{R}. \tag{A.10.1}$$

The knowledge of (A.10.1) at the Fourier frequencies $\omega_j = 2\pi j/n$, for $j = 0, \ldots, n-1$, is sufficient for the reconstruction of $p_0, \ldots, p_{n-1}$. Note that we arbitrarily choose the Fourier frequencies in $[0, 2\pi)$. Any other interval of length 2π, like $(-\pi, \pi]$, could also be considered, as explained in Remark A.10.1. Let $\boldsymbol{p} = (p_0, \ldots, p_{n-1})^\top$, $\boldsymbol{\varphi} = (\varphi_0, \ldots, \varphi_{n-1})^\top = (\varphi_{(n)}(\omega_0), \varphi_{(n)}(\omega_1), \ldots, \varphi_{(n)}(\omega_{n-1}))^\top$ and $\boldsymbol{F} = \left(e^{ik\omega_j}\right)_{j,k=0,\ldots,n-1}$, which is a symmetric matrix of dimension $n \times n$. Then we have $\boldsymbol{\varphi} = \boldsymbol{F}\boldsymbol{p}$. It holds for $j, k = 0, \ldots, n-1$,

$$\sum_{l=0}^{n-1} \overline{e^{il\omega_j}} \, e^{ik\omega_l} = \sum_{l=0}^{n-1} e^{i(\omega_k - \omega_j)l} = \begin{cases} \dfrac{e^{in(\omega_k - \omega_j)} - 1}{e^{i(\omega_k - \omega_j)} - 1} = 0, & \text{if } j \neq k, \\ n, & \text{if } j = k. \end{cases} \tag{A.10.2}$$

Thus $\overline{\boldsymbol{F}}\boldsymbol{F} = n\boldsymbol{I}$, where $\overline{\boldsymbol{F}}$ denotes the matrix of complex conjugates of $\boldsymbol{F}$. Therefore we can reconstruct $\boldsymbol{p}$ with $\boldsymbol{p} = n^{-1}\overline{\boldsymbol{F}}\boldsymbol{\varphi}$. It follows that the inverse discrete Fourier transform of $\boldsymbol{\varphi}$ is given by

$$p_k = \frac{1}{n} \sum_{j=0}^{n-1} e^{-i\omega_j k} \varphi_{(n)}(\omega_j), \quad \text{for } k = 0, \ldots, n-1. \tag{A.10.3}$$

Furthermore, the computation of $\boldsymbol{p}$ with (A.10.3) requires $\mathrm{O}(n^2)$ operations. The number of operations can be substantially reduced and precisely to $\mathrm{O}(n \log n)$, by means of the following inverse FFT-algorithm, which is due to Cooley and Tukey (1965). Let us choose $n \in 2^{\mathbb{N}^*}$. It follows from

(A.10.3) that, for $l = 0, \ldots, n/2 - 1$, we have

$$
p_{2l} = \frac{1}{n} \sum_{j=0}^{n-1} e^{-i\omega_j 2l} \varphi_{(n)}(\omega_j)
$$

$$
= \frac{1}{n} \sum_{j=0}^{\frac{n}{2}-1} \left\{ e^{-i\omega_j 2l} \varphi_{(n)}(\omega_j) + e^{-i\omega_{\frac{n}{2}+j} 2l} \varphi_{(n)}(\omega_{\frac{n}{2}+j}) \right\}
$$

$$
= \frac{1}{n} \sum_{j=0}^{\frac{n}{2}-1} e^{-i\omega_j 2l} \left\{ \varphi_{(n)}(\omega_j) + \varphi_{(n)}(\omega_{\frac{n}{2}+j}) \right\}, \qquad (A.10.4)
$$

and, from similar considerations, we also have

$$
p_{2l+1} = \frac{1}{n} \sum_{j=0}^{\frac{n}{2}-1} e^{-i\omega_j (2l+1)} \left\{ \varphi_{(n)}(\omega_j) + \varphi_{(n)}(\omega_{\frac{n}{2}+j}) \right\}. \qquad (A.10.5)
$$

The transforms (A.10.4) and (A.10.5) have length $n/2$. The assumption $n = 2^m$ allows for m consecutive conversions of transforms into new transforms of half length. At the final stage there are m transforms of length one. The computation of $\boldsymbol{p}$ by these consecutive halvings requires only $O(n \log n)$ operations. Programs for computing the FFT are available available in most numerical software.

Remark A.10.1 (Choice of interval of frequencies). *The Fourier frequencies can be chosen in any half-open interval of $\mathbb{R}$ of length 2π. The choice is arbitrary. In order to control this, let*

$$
\omega_j^{(h)} = \frac{2\pi(j+h)}{n}, \qquad \text{for } j = 0, \ldots, n-1,
$$

let $\boldsymbol{\varphi}^{(h)} = (\varphi_0^{(h)}, \ldots, \varphi_{n-1}^{(h)})^\top = (\varphi_{(n)}(\omega_0^{(h)}), \varphi_{(n)}(\omega_1^{(h)}), \ldots, \varphi_{(n)}(\omega_{n-1}^{(h)}))^\top$ *and let* $\boldsymbol{F}^{(h)} = \left(e^{ik\omega_j^{(h)}} \right)_{j,k=0,\ldots,n-1}$. *Then we retrieve the same important equalities as with $h = 0$: $\boldsymbol{\varphi}^{(h)} = \boldsymbol{F}^{(h)} \boldsymbol{p}$, $\overline{\boldsymbol{F}^{(h)}} \boldsymbol{F}^{(h)} = n\boldsymbol{I}$, where $\overline{\boldsymbol{F}^{(h)}}$ is the matrix of complex conjugates of $\boldsymbol{F}^{(h)}$, and $\boldsymbol{p} = n^{-1} \overline{\boldsymbol{F}^{(h)}} \boldsymbol{\varphi}^{(h)}$. Consequently, (A.10.4) and (A.10.5) can be applied with the frequencies $\omega_0^{(0)}, \ldots, \omega_{n-1}^{(0)}$ as well. After $h = 0$, the second common choice is $h = -\lfloor (n-1)/2 \rfloor$, so to have*

$$
\omega_j^{(h)} \in [-\pi, \pi), \ \forall j \in \mathcal{F}_n = \left\{ -\left\lfloor \frac{n-1}{2} \right\rfloor, \ldots, \left\lfloor \frac{n}{2} \right\rfloor \right\}.
$$

Remark A.10.2 (Aliasing error). *Consider $p_0, p_1, \ldots \in \mathbb{C}$ such that $\sum_{k=0}^{\infty} |p_k| < \infty$ and define the Fourier transform*

$$\varphi(\alpha) = \sum_{k=0}^{\infty} e^{i\alpha k} p_k, \quad \forall \alpha \in \mathbb{R}.$$

If the discrete Fourier transform $\varphi_{(n)}$ given in (A.10.1) is not available but the Fourier transform φ is known, then we can nevertheless accurately approximate $p_0, \ldots, p_{n-1}$ for n large by

$$\tilde{p}_k = \frac{1}{n} \sum_{j=0}^{n-1} e^{-i\omega_j k} \varphi(\omega_j), \quad \text{for} \quad k = 0, \ldots, n-1.$$

Indeed,

$$\varphi(\omega_j) = \sum_{l=0}^{\infty} \sum_{k=0}^{n-1} e^{i\omega_j(k+nl)} p_{k+nl} = \sum_{k=0}^{n-1} e^{i\omega_j k} \hat{p}_k, \quad \text{for} \quad j = 0, \ldots, n-1,$$

where $\hat{p}_k = \sum_{l=0}^{\infty} p_{k+nl}$, for $k = 0, \ldots, n-1$. It follows that $\hat{p}_k = \tilde{p}_k$ and that, because of the summability assumption, the errors $\tilde{p}_k - p_k = \sum_{l=1}^{\infty} p_{k+nl}$, for $k = 0, \ldots, n-1$, become negligible as $n \to \infty$. This type of error is called aliasing error, which is to distinguish from the discretization error. It could be summarized as the wrapping effect of the truncation of the data $p_0, p_1, \ldots$ to $p_0, \ldots, p_{n-1}$.

Appendix B

Abbreviations, mathematical notation and data

B.1 Abbreviations

- a.c.r.f: autocorrelation function
- a.c.v.f: autocovariance function
- a.c.v.g.f.: autocovariance generating function
- p.a.c.r.f: partial autocorrelation function
- c.a.c.r.f: circular autocorrelation function
- d.f.: distribution function
- f.d.d.: finite dimensional distributions
- f.d.d.f.: finite dimensional d.f.
- i.d.d.: independent and identically distributed
- iff: if and only if
- i.o.: infinitely often
- n.n.d.: nonnegative definite
- p.d.: positive definite
- AR: autoregressive
- MA: moving average
- ARMA: autoregressive and moving average
- ARIMA: integrated autoregressive and moving average
- WAR: wrapped autoregressive
- LAR: linked autoregressive
- LMA: linked moving average
- LARMA: linked autoregressive and moving average
- CAR: circular autoregressive
- WN: white noise
- f.p.e.: final prediction error
- P.r.m.: Poisson random measure

- FFT: fast Fourier transform
- vM: von Mises
- GvM: generalized von Mises

B.2 Mathematical notation

- i: imaginary unit
- $\mathbb{N} = \{0, 1, \ldots\}$
- $\mathbb{Z} = \{\ldots, -1, 0, 1, \ldots\}$
- $\mathbb{R} = (-\infty, \infty)$
- $\bar{\mathbb{R}} = \{-\infty\} \cup (-\infty, \infty) \cup \{\infty\}$
- $\mathbb{R}_+ = [0, \infty)$
- $\mathbb{R}_- = (-\infty, 0]$
- $\mathbb{C} = \{x + \mathrm{i}y \mid x, y \in \mathbb{R}\}$
- $\mathbb{A}^* = \mathbb{A}\backslash\{0\}$, where $\mathbb{A}$ denotes any of the above sets.
- S°: interior of the set S
- S^c: complement of the set S
- $\mathcal{B}(S)$: Borelian σ-algebra of the set S
- Re z: real part of $z \in \mathbb{C}$
- sgn x: sign of $x \in \mathbb{R}$
- $\lfloor x \rfloor = \max\{k \in \mathbb{Z} \mid k \leq x\}$: rounding of $x \in \mathbb{R}$ to the next smaller or equal integer
- $\lceil x \rceil = \min\{k \in \mathbb{Z} \mid k \geq x\}$: rounding of $x \in \mathbb{R}$ to the next larger or equal integer
- $\mathrm{id}_\mathbb{R}$: identity function over $\mathbb{R}$
- $\langle x, y \rangle$: scalar product of the elements x and y of some Hilbert space
- card S: cardinal number of the set S
- $\boldsymbol{I}$: identity matrix
- $\boldsymbol{A}^\top$: transpose of the matrix $\boldsymbol{A}$
- rank $\boldsymbol{A}$: rank of the matrix $\boldsymbol{A}$
- $\Gamma(z) = \displaystyle\int_0^\infty \mathrm{e}^{-x} x^{z-1} \mathrm{d}z$, $\forall z \in \mathbb{C}$ such that $\mathrm{Re}\, z > 0$: gamma function
- $(z)_n = \dfrac{\Gamma(z+n)}{\Gamma(z)} = \begin{cases} z(z+1)\cdots(z+n-1), & \text{if } n = 1, 2, \ldots, \\ 1, & \text{if } n = 0, \end{cases}$ $\forall z \in \mathbb{C}$:
Pochhammer symbol or ascending factorial

- $[z]_n = \dfrac{\Gamma(z+1)}{\Gamma(z-n+1)} = \begin{cases} z(z-1)\cdots(z-n+1), & \text{if } n = 1, 2, \ldots, \\ 1, & \text{if } n = 0, \end{cases} \forall z \in \mathbb{C}:$
  descending factorial

- $\displaystyle \binom{x}{k} = \begin{cases} \frac{[x]_k}{k!}, & \text{if } k = 1, 2, \ldots, \\ 1, & \text{if } k = 0, \\ 0, & \text{if } k = -1, -2, \ldots, \end{cases} \forall x \in \mathbb{R}:$ binomial coefficient

- ${}_2F_1(a, b; c; z) = \displaystyle \sum_{n=0}^{\infty} \frac{(a)_n (b)_n}{(c)_n} \frac{z^n}{n!}, \; \forall z \in \mathcal{D}_0^{\circ}$, where $a, b \in \mathbb{C}$ and $c \in \mathbb{C} \setminus \{0, -1, \ldots\}$: Gauss hypergeometric function; cf. e.g. 15.1.1 at p. 556 of Abramowitz and Stegum (1972)

- $J_\nu(z) = \dfrac{1}{\pi} \displaystyle\int_0^\pi \cos(z \sin \alpha - \nu \alpha) \mathrm{d}\alpha - \dfrac{\sin(\nu\pi)}{\pi} \int_0^\infty e^{-z \sinh t - \nu t} \mathrm{d}t, \; \forall z \in \mathbb{C}$ such that $|\mathrm{arg}\, z| < \frac{\pi}{2}$: Bessel function J of order $\nu \in \mathbb{C}$; cf. e.g. 9.1.22 at p. 360 of Abramowitz and Stegum (1972).

- $I_n(z) = \dfrac{1}{2\pi} \displaystyle\int_0^{2\pi} e^{z \cos \alpha} \cos n\alpha \; \mathrm{d}\alpha, \; \forall z \in \mathbb{C}$: modified Bessel function I of order $n \in \mathbb{N}$; cf. e.g. 9.6.19 at p. 376 of Abramowitz and Stegum (1972).

- $K_\nu(z) = \displaystyle\int_0^\infty e^{-z \cosh t} \cosh \nu t \; \mathrm{d}t, \; \forall z \in \mathbb{C}$ such that $|\mathrm{arg}\, z| < \frac{\pi}{2}$: modified Bessel function K of order $\nu \in \mathbb{C}$; cf. e.g. 9.6.24 at p. 376 of Abramowitz and Stegum (1972)

- $g'(x) = \dfrac{\mathrm{d}}{\mathrm{d}x} g(x), \; g''(x) = \left(\dfrac{\mathrm{d}}{\mathrm{d}x}\right)^2 g(x), \; g'''(x) = \left(\dfrac{\mathrm{d}}{\mathrm{d}x}\right)^3 g(x)$
- g^{-1}: inverse of function g
- $g(x-) = \displaystyle\lim_{h \to 0, h > 0} g(x - h)$
- $g(x+) = \displaystyle\lim_{h \to 0, h > 0} g(x + h)$
- $\Delta g(t) = g(t+) - g(t-)$: jump of the function g at t
- $\mathrm{dom}\, g = \{x \in \mathbb{R} \mid |g(x)| < \infty\}$; domain of definition of $g \colon \mathbb{R} \to \mathbb{R}$
- $f(x) \sim g(x)$, as $x \to a$: f is asymptotically equivalent to g, as the argument tends to a, i.e. $\displaystyle\lim_{x \to a} \frac{f(x)}{g(x)} = 1$, where $a \in \{-\infty\} \cup \mathbb{R} \cup \{\infty\}$
- $f(x) = o(g(x))$, as $x \to a$: f is asymptotically smaller than g, as the argument tends to a, i.e. $\displaystyle\lim_{x \to a} \frac{f(x)}{g(x)} = 0$, where $a \in \{-\infty\} \cup \mathbb{R} \cup \{\infty\}$

- $f(x) = O(g(x))$, as $x \to a$: f is asymptotically bounded by g, as the argument tends to a, i.e.

$$\begin{cases} \exists\, c, d > 0, \text{ such that } |x - a| < d \Longrightarrow |f(x)| \le c|g(x)|, & \text{if } a \in \mathbb{R}, \\ \exists\, c, x_0 > 0, \text{ such that } x > x_0 \Longrightarrow |f(x)| \le c|g(x)|, & \text{if } a = \infty, \\ \exists\, c > 0, x_0 < 0, \text{ such that } x < x_0 \Longrightarrow |f(x)| \le c|g(x)|, & \text{if } a = -\infty \end{cases}$$

- $f(x) \approx \sum_{k=0}^{\infty} c_k g_k(x)$, as $x \to a$: f admits the asymptotic expansion $\sum_{k=0}^{\infty} c_k g_k(x)$, if the argument tends to a, i.e. $g_{n+1}(x) = o(g_n(x))$, as $x \to a$, for $n = 0, 1, \ldots$, where $a \in \{-\infty\} \cup \mathbb{R} \cup \{\infty\}$

- $\mathsf{I}\{S\} = \begin{cases} 1, \text{ if } S \text{ is true}, \\ 0, \text{ if } S \text{ is false}: \end{cases}$ indicator of the statement S

- $\mathsf{I}_S(\omega) = \mathsf{I}\{\omega \in S\}$: indicator function of the set S

- E, var, cov, corr and cov*: expectation, variance, covariance, correlation and pseudo-covariance functionals

- ccorr and ccorr': circular correlation functionals

- P: probability measure

- $\mathsf{E}[X; S] = \mathsf{E}[X\mathsf{I}_S]$: expectation of the random variable X over the set S

- $\mathsf{E}[X|\Psi]$: conditional expectation of X given the set or the random variable or the σ-algebra Ψ

- $F * G(x) = \int_{\mathbb{R}} F(x - y)\mathrm{d}G(y)$: convolution of the distribution functions F and G

- $F^{*n} = F * \cdots * F$: nth convolution power of the distribution function F

- $r * s(x) = \int_{-\infty}^{\infty} r(x - y)s(y)\mathrm{d}y$: convolution of the functions r and s, not representing distribution functions

- $r^{*n} = r * \cdots * r$: n-th convolution power of the function r

- $\phi(x) = \dfrac{1}{\sqrt{2\pi}}e^{-\frac{x^2}{2}}$: standard Gaussian or normal density

- $\Phi(x) = \dfrac{1}{\sqrt{2\pi}} \int_{-\infty}^{x} e^{-\frac{y^2}{2}}\mathrm{d}y$: standard Gaussian or normal distribution function

- $\delta(x)$: Dirac function or density centered at 0

- $\Delta(x) = \mathsf{I}\{x \ge 0\}$: Dirac distribution function centered at 0

- $\mathrm{sinc}\, z = \dfrac{\sin z}{z}$

- A_n, $n \geq 1$, i.o. $= \bigcap\limits_{n=1}^{\infty} \bigcup\limits_{m=n}^{\infty} A_m$: occurrence of infinitely many of the events A_n, for $n = 1, 2, \dots$
- $\mathcal{L}_p(\Omega)$: space of random variables $X : \Omega \to \mathbb{R}$, such that $\mathsf{E}[|X|^p] < \infty$, where $p > 0$
- $\sigma(\{X_t\}_{t \in T})$: σ-algebra generated by the random variables X_t, $\forall t \in T$
- $\mathcal{N}(\mu, \sigma^2)$: Gaussian or normal random variable with expectation μ and variance σ^2
- χ_k^2: chi-square random variable with k degrees of freedom
- $\mathcal{S}_\alpha(\tau, \beta, \gamma)$: α-stable random variable with stability index α, skewness parameter β, scale parameter τ and location parameter μ
- $\mathrm{AR}(p)$: autoregressive time series of order p
- $\mathrm{MA}(q)$: moving average time series of order q
- $\mathrm{ARMA}(p, q)$: autoregressive and moving average time series of respective orders p and q
- $\mathrm{ARIMA}(p, d, q)$: autoregressive, moving average and integrated time series of respective orders p, q and d
- B: backward shift operator of time series, $BX_k = X_{k-1}$
- ∇: difference operator of time series, $\nabla X_k = X_k - X_{k-1}$
- ∇: lag d difference operator of time series, $\nabla_d X_k = X_k - X_{k-d}$
- $\mathrm{WN}(\sigma^2)$: white noise time series or continuous time process with variance σ^2
- $\mathrm{WAR}(p)$: wrapped autoregressive time series of order p
- $\mathrm{LAR}(p)$: linked autoregressive time series of order p
- $\mathrm{LMA}(q)$: linked moving average time series of order q
- $\mathrm{LARMA}(p, q)$: linked autoregressive and moving average time series of respective orders p and q
- $\mathrm{CAR}(p)$: circular autoregressive time series of order p
- $\mathrm{vM}(p)$: von Mises time series of order p
- $\mathrm{PRM}(\mu)$: Poisson random measure with mean measure μ
- $\mathrm{vM}(\mu, \kappa)$: von Mises circular random variable with mean direction μ and concentration κ
- $\mathrm{GvM}_k(\mu_1, \dots, \mu_k, \kappa_1, \dots, \kappa_k)$: generalized von Mises circular random variable of order k with parameters $\mu_1, \dots, \mu_k$ and $\kappa_1, \dots, \kappa_k$
- $X \sim Y$: the random variables X and Y are identically distributed
- $X_n \xrightarrow{\text{as}} X$: the sequence of random variables $\{X_n\}_{n \geq 1}$ converges almost surely, i.e. with probability one, to the random variable X

- $X_n \xrightarrow{d} X$: the sequence of random variables $\{X_n\}_{n \geq 1}$ converges in distribution to the random variable X
- $X_n \xrightarrow{\mathcal{L}_p} X$: the sequence of random variables $\{X_n\}_{n \geq 1}$ converges in $\mathcal{L}_p$ to the random variable X, where $p > 0$
- $X_n \xrightarrow{P} X$: the sequence of random variables $\{X_n\}_{n \geq 1}$ converges in probability to the random variable X under the probability P
- $\overline{\mathrm{sp}}\{X_t\}_{t \in T}$: closed span generated by the random variables X_t, $\forall t \in T$
- $S^{\perp}$: orthogonal part of the subset S of an Hilbert space
- pro_S: orthogonal projection from an Hilbert onto the subset S
- $S_1 \oplus S_2$: direct sum of the orthogonal subspaces S_1 and S_2 of an Hilbert space
- $x_{(1)} \leq \cdots \leq x_{(n)}$: from smallest to largest ordered values from $x_1, \ldots, x_n$
- $\mathcal{D}_\varepsilon = \{z \in \mathbb{C} \, | \, |z| \leq 1 + \varepsilon\}$: centered closed disk with radius $1 + \varepsilon$, for $\varepsilon \geq 0$
- $\mathcal{A}_r = \left\{z \in \mathbb{C} \, \middle| \, \dfrac{1}{r} \leq |z| \leq r \right\}$: centered closed annulus delimited by the radii $1/r$ and r, for $r > 1$

B.3   Data

The sample in Table B.1 is provided by the Swiss Office of Public Health.

Table B.1: Daily numbers of reported coronavirus (SARS-CoV-2) infections in Switzerland between March 29 and June 17 2021, in chronological order from left to right and then from top to bottom.

2531	2093	2132	1985	1205	1317	1022	1289	2689
2389	2490	2414	1682	1178	2744	2375	2214	2147
2121	1430	1048	2744	2391	2309	2266	2165	1456
1188	2441	2037	2009	1711	1678	1158	832	2017
1735	1651	1483	1572	1062	779	1718	1381	1442
672	1169	802	653	1671	1343	1212	1041	983
624	469	581	1146	925	850	750	499	357
855	651	549	470	490	311	187	583	418
375	304	318	194	120	328	233	191	155

Bibliography

Abramowitz, M., Stegun, I. E. (1972), *Handbook of Mathematical Functions with Formulas, Graphs, and Mathematical Tables*, Dover Publications (reprint).

Amos, D. E. (1974), "Computation of modified Bessel functions and their ratios", *Mathematics of Computation*, 28, 239–251.

Anderson, T. W. (1993), "Goodness of fit tests for spectral distributions", *The Annals of Statistics*, 21, 830–847.

Anderson, T. W. (1994), *The Statistical Analysis of Time Series*, Wiley and Sons.

Applebaum, D. (2004), *Lévy Processes and Stochastic Calculus*, Cambridge University Press.

Asmussen, S., Glynn P. W. (2007), *Stochastic Simulation. Algorithms and Analysis*, Springer.

Astfalck, L., Cripps, E., Gosling, J., Hodkiewicz, M., Milne, I. (2018), "Expert elicitation of directional metocean parameters", *Ocean Engineering*, 161, 268–276.

Barndorff-Nielsen, O. E., Cox, D. (1989), *Asymptotic Techniques for Use in Statistics*, Chapman and Hall.

Barndorff-Nielsen, O. E., Shephard, N. (2001), "Non-Gaussian Ornstein-Uhlenbeck-based models and some of their uses in financial economics", *Journal of the Royal Statistical Society*, Series B, 63, 167–241.

Beran, J. (1994), *Statistics for Long-Memory Processes*, Chapman and Hall.

Blatter, C. (1998), *Wavelets. A Primer*, A K Peters.

Bloomfield, P. (1973), "An exponential model for the spectrum of a scalar time series", *Biometrika*, 60, 217–226.

Bogert, B. P., Healy, M. J. R., Tukey, J. W., Rosenblatt, M. (1963), "The quefrency analysis of time series for echoes: Cepstrum, pseudoauto-covariance, cross-cepstrum and saphe cracking", in: *Proceedings of the Symposium on Time Series Analysis*, editor Rosenblatt, Wiley and Sons, pp. 209–243.

Bollerslev, T. (1986), "Generalized autoregressive conditional heteroskedasticity", *Journal of Econometrics*, 31, 307–327.

Box, G. E. P, Jenkins, G. M. (1970), *Time Series Analysis: Forecasting and Control*, Holden-Day.

Breckling, J. (1989), *The Analysis of Directional Time Series: Applications to Wind Speed and Direction, Lecture Notes in Statistics*, Springer, p. 61.

Brémaud, P. (2014), *Fourier Analysis of Stochastic Processes*, Springer, p. 3.

Brillinger, D. R. (1993), "The digital rainbow: some history and applications of numerical spectrum analysis", *Canadian Journal of Statistics*, 21, 1–19.

Brockwell, P. J., Davis, R. A. (1991), *Time Series: Theory and Methods*, second edition, Springer.

Brockwell, Davis (2002), *Introduction to Time Series and Forecasting*, second edition, Springer.

Broemeling, L. D. (2019), *Bayesian Analysis of Time Series*, CRC Press.

Bucklew, J. A. (2004), *Introduction to Rare Event Simulation*, Springer.

Burg, J. P. (1967), "Maximum entropy spectral analysis", unpublished presentation, 37-th meeting of the Society of Exploration Geophysicists, Oklahoma City, Oklahoma.

Carothers, N. L. (2000), *Real Analysis*, Cambridge University Press.

Christmas, J. (2014), "Bayesian spectral analysis with Student-t noise", *IEEE Transactions on Signal Processing*, 62, 2871–2878.

Christmas, J. (2022), "Non-stationary, online variational Bayesian learning with circular variables", *Pattern Recognition*, 122, 108340.

Cochrane, D., Orcutt, G. H. (1949), "Application of least squares regression to relationships containing auto-correlated error terms", *Journal of the American Statistical Association*, 245, 32–61.

Cooley, J. W., Tukey, J. W. (1965), "An algorithm for the machine calculation of complex Fourier series", in *Mathematics of Computation*, Vol. 19, pp. 297–301.

Cox, D. R. (1955), "Some statistical methods connected with series of events", *Journal of the Royal Statistical Society*, Series B, 129–164.

Cramér, H. (1940), "On the theory of stationary random processes", *Annals of Mathematics*, 41, 215–230.

Cramér, H. (1942), "On harmonic analysis in certain functional spaces", *Arkiv för Matematik, Astronomi och Fysik*, 28 B, 12.

Cramér, H. (1944), "On a new limit theorem of the theory of probability", *Uspekhi Matematicheskikh Nauk*, 10, 166–178.

Cramér, H., Leadbetter, M. R. (1967), *Stationary and Related Stochastic Processes*, Wiley and Sons.

Daniels, H. E. (1954), "Saddlepoint approximations in statistics", *Annals of Mathematical Statistics*, 25, 631–650.

Daniels, H. E. (1956), "The approximate distribution of serial correlation coefficients", *Biometrika*, 43, 169–185.

Dembo, A., Zeitouni, O. (2009), *Large Deviations Techniques and Applications*, second edition, Springer.

Dietrich, C. R., Newsam, G. N. (1997), "Fast and exact simulation of stationary Gaussian processes through circulant embedding of the covariance matrix", *SIAM Journal on Scientific Computing*, 18, 1088–1107.

Durbin, J., Watson, G. S. (1950), "Testing for serial correlation in least squares regression, I", *Biometrika*, 37, 409–428.

Durbin, J., Watson, G. S. (1951), "Testing for serial correlation in least squares regression, II", *Biometrika*, 38, 159–179.

Einstein, A. (1956), "Investigations on the theory of Brownian movement", Dover Publications (reprint).

Embrechts, P., Klüppelberg, C., Mikosch, T. (2013), *Modelling Extremal Events: for Insurance and Finance*, Springer.

Engle, R. F. (1982), "Autoregressive conditional heteroscedasticity with estimates of the variance of United Kingdom inflation", *Econometrica*, 50, 987–1007.

Feller, W. (1971), *An Introduction to Probability Theory and its Applications*, Volume II, second edition, Wiley and Sons.

Field, C. A., Ronchetti, E. (1990), *Small Sample Asymptotics*, Institute of Mathematical Statistics Lecture Notes-Monograph Series, Vol. 13.

Fisher, N. I. (1993), *Statistical Analysis of Circular Data*, Cambridge University Press.

Fisher, N. I., Lee, A. J. (1983), "A correlation coefficient for circular data", *Biometrika*, 70, 327–332.

Fisher, N. I., Lee, A. (1994), "Time series analysis of circular data", *Journal of the Royal Statistical Society*, Series B, 56, 327–339.

Fisher, R. A. (1929), "Tests of significance in harmonic analysis", *Proceedings of the Royal Society of London*, Series A, 125(796), 54–59.

Franke, J. (1985), "ARMA processes have maximal entropy among time series with prescribed autocovariances and impulse responses", *Advances in Applied Probability*, 17, 910–840.

Gatto, R. (2008), "Some computational aspects of the generalized von Mises distribution", *Statistics and Computing*, 18, 321–331.

Gatto, R. (2009), "Information theoretic results for circular distributions", *Statistics*, 43, 409–421.

Gatto, R. (2010), "A saddlepoint approximation to the distribution of inhomogeneous discounted compound Poisson processes", *Methodology and Computing in Applied Probability*, 12, 533–551.

Gatto, R. (2012), "Saddlepoint approximations to tail probabilities and quantiles of inhomogeneous discounted compound Poisson processes with periodic intensity functions", *Methodology and Computing in Applied Probability*, 14, 1053–1074.

Gatto, R. (2015), "Saddlepoint approximations", *StatsRef: Statistics Reference Online*, editors Balakrishnan et al., Wiley and Sons, pp. 1–7.

Gatto, R. (2020), *Stochastische Modelle der Aktuariellen Risikotheorie. Eine Mathematische Einführung*, second edition, Springer Spektrum.

Gatto, R. (2022), "Information theoretic results for stationary time series and the Gaussian-generalized von Mises time series", *Directional Statistics for Innovative Applications. A Bicentennial Tribute to Florence Nightingale*, editors Sen Gupta and Arnold, Springer.

Gatto, R., Jammalamadaka, S. R. (2003), "Inference for wrapped symmetric α-stable circular models", *Sankhyā*, 65, 333–355.

Gatto, R., Jammalamadaka, S. R. (2007), "The generalized von Mises distribution", *Statistical Methodology*, 4, 341–353.

Gatto, R., Jammalamadaka, S. R. (2015), "Directional statistics: introduction", *StatsRef: Statistics Reference Online*, editors Balakrishnan et al., Wiley and Sons, pp. 1–8.

Gonella, J. (1972), "A rotary-component method for analysing meteorological and oceanographic vector time series", *Deep-Sea Research*, 19, 833–846.

Gradshteyn, I. S., Ryzhik, I. M. (2007), *Table of Integrals, Series, and Products*, seventh edition, Academic Press.

Grandell, J. (1997), *Mixed Poisson Processes*, Chapman & Hall.

Grimmett, G. R., Stirzaker, D. R. (1991), *Probability and Random Processes*, second edition, Wiley and Sons.

Hurst, H. E. (1951), "Long-term storage capacity of reservoir", *Transactions of the American Society of Civil Engineers*, 116, 770–808.

James, J. F. (1995), *A Student's Guide to Fourier Transforms with Applications in Physics and Engineering*, Cambridge University Press.

Jammalamadaka, S. R., Sarma, Y. R. (1988), "A correlation coefficient for angular variables", *Statistical Theory and Data Analysis*, II, 349–364.

Jammalamadaka, S. R., SenGupta, A. (2001), *Topics in Circular Statistics*, World Scientific Press.

Kalman, R. E. (1960), "A new approach to linear filtering and prediction problems", *Transactions of the ASME–Journal of Basic Engineering*, 82, 35–45.

Khintchine, Y. A. (1960), *Mathematical Methods in the Theory of Queueing*, Charles Griffin, London.

Koopmans, L. H. (1995), *The Spectral Analysis of Time Series*, second edition, Academic Press.

Krenk, S., Clausen, J. (1987), "On the calibration of ARMA processes for simulation", *Reliability and Optimization of Structural Systems*, Springer, pp. 243–257.

Kullback, S., Leibler, R. A. (1951), "On information and sufficiency", *The Annals of Mathematical Statistics*, 22, 79–86.

Lin, Y., Dong, S. (2019), "Wave energy assessment based on trivariate distribution of significant wave height, mean period and direction", *Applied Ocean Research*, 87, 47–63.

Lindgren, G. (2012), *Stationary Stochastic Processes: Theory and Applications*, Chapman and Hall/CRC Press.

Lindgren, G., Rootzń, H., Sandsten, M. (2014), *Stationary Stochastic Processes for Scientists and Engineers*, Chapman and Hall/CRC Press.

Loève, P. (1948), "Fonctions aléatoires du second ordre", a note in Lévy, P., *Processus Stochastiques et Mouvement Brownien*, Gauthier-Villars.

Lugannani, R., Rice, S. (1980), "Saddle point approximation for the distribution of the sum of independent random variables", *Advances in Applied probability*, 12, 475–490.

Lundberg, F. (1903), *Approximerad Framställning av Sannolikehetsfunktionen. A Återförsäkering av Kollektivrisker*, doctoral thesis, Almqvist & Wiksell, Uppsala.

Mandelbrot, B. B., van Ness, J. W. (1968), "Fractional Brownian motions, fractional noises and applications", *SIAM Review*, 10, 422–437.

Mardia, K. V., Jupp, P. E. (2000), *Directional Statistics*, Wiley and Sons.

McKenzie, E. (1985), "Some simple models for discrete variate time series", *Water Resource Bulletin*, 21, 645–650.

Montgomery, D. C., Jennings, C. L., Kulahci, M. (2015), *Introduction to Time Series Analysis and Forecasting*, Wiley and Sons.

Paley, R. E. A. C., Wiener, N. (1934), *Fourier Transforms in the Complex Domain*, American Mathematical Society.

Pewsey, A., Garcia-Portugués, E. (2021), "Recent advances in directional statistics", *Test*, 30, 1–58.

Phillips, P. C. B. (1978), "Edgeworth and saddlepoint approximations in the first-order noncircular autoregression", *Biometrika*, 65, 91–98.

Pinkus, A., Zafrany, S. (1997), *Fourier Series and Integral Transforms*, Cambridge University Press.

Priestley, M. B. (1981), *Spectral Analysis and Time Series*, Academic Press, London.

Rice, S. O. (1944), "Mathematical analysis of random noise", *Bell Systems Technical Journal*, 23, 282–332.

Rice, S. O. (1945), "Mathematical analysis of random noise", *Bell Systems Technical Journal*, 24, 46–156.

Rice, S. O. (1954), "Mathematical analysis of random noise", *Selected Papers on Noise and Stochastic Processes*, editor Wax, Dover Publications, pp. 133–295.

Rowe, D. B. (2005), "Modeling both the magnitude and phase of complex-valued fMRI data", *NeuroImage*, 25, 1310–1324.

Rudin, W. (1964), *Principles of Mathematical Analysis*, McGraw-Hill, p. 2.

Rudin, W. (1987), *Real and Complex Analysis*, McGraw-Hill.

Shiryayev, A. N. (1984), *Probability*, Springer.

Salvador, S., Gatto, R. (2021), "Bayesian tests of symmetry for the generalized von Mises distribution", *Computational Statistics*, 37, 947–974.

Shwartz, A., Weiss, A. (1995), *Large Deviations for Performance Analysis: Queues, Communication and Computing*, CRC Press.

Shannon, C. E. (1948), "A mathematical theory of communication", *Bell System Technical Journal*, 27, 379–423, 623–656.

Shumway, R. H., Stoffer, D. S. (2019), *Time Series: a Data Analysis Approach Using R*, CRC Press.

Slutsky, E. (1927), "The summation of random causes as the source of cyclic processes", 3, 1, Moscow Conjuncture Institute.

St. Denis, M., Pierson, W. J. (1953), "On the motion of ships in confused seas", *Transactions of the Society of Naval Architects and Marine Engineers*, 61, 280–357.

Steutel, F. W., van Harn, K. (1979), "Discrete analogues of self-decomposability and stability", *Annals of Probability*, 7, 893–899.

Straumann, D. (2005), *Estimation in Conditionally Heteroscedastic Time Series Models*, Lecture Notes in Statistics, Vol. 181, Springer.

Uhlenbeck, G. E., Ornstein, L. S. (1930), "On the theory of the Brownian motion", *Physical review*, 36, 823.

Uhlenbeck, G. E., Ornstein, L. S. (1954), "On the theory of the Brownian motion", *Selected Papers on Noise and Stochastic Processes*, editor Wax, Dover Publications, pp. 93–111.

Varadhan, S. R. S. (1984), *Large Deviations and Applications*, SIAM Monograph.

Walker, G. (1931), "On periodicity in series of related terms", *Proceedings of the Royal Society of London*, Series A, 131, 518–532.

Watson, G. S., Beran, R. J. (1967), "Testing a sequence of unit vectors for serial correlation", *Journal of Geophysical Research*, 72, 5655–5659.

Weiß, C. H. (2018), *An Introduction to Discrete-Valued Time Series*, Wiley and Sons.

Whittle, P. (1951), *Hypothesis Testing in Time Series Analysis*, Almqvist & Wiksells, Boktryckeri.

Wold, H. (1938), *A Study in the Analysis of Stationary Time Series*, doctoral thesis, Almqvist & Wiksell, Stockholm.

Wood, A. T., Chan, G. (1994), "Simulation of stationary Gaussian processes in $[0,1]^d$", *Journal of Computational and Graphical Statistics*, 3, 409–432.

Wooding, R. A. (1956), "The multivariate distribution of complex normal variables", *Biometrika*, 212–215.

Yaglom, A. M. (1962), *An Introduction to the Theory of Stationary Random Functions*, Prentice-Hall.

Yaglom, A. M. (1987a), *Correlation Theory of Stationary and Related Random Functions. Volume I: Basic Results*, Springer.

Yaglom, A. M. (1987b), *Correlation Theory of Stationary and Related Random Functions. Volume II: Supplementary Notes and References*, Springer.

Yule, G. U. (1921), "On the time-correlation problem, with especial reference to the variate-difference correlation method, *Journal of the Royal Statistical Society*, 84, 497–537.

Yule, G. U. (1927), "On a method of investigating periodicities in disturbed series, with special reference to Wolfer's sunspot numbers", *Philosophical Transactions of the Royal Society of London*, Series A, 226, 267–298.

Zhang, L., Li, Q., Guo, Y., Yang, Z., Zhang, L. (2018), "An investigation of wind direction and speed in a featured wind farm using joint probability distribution methods", *Sustainability*, 10, 4338.

Index

Printed in the United States
by Baker & Taylor Publisher Services